Greek Letters

α	alpha	ζ	zeta
β	beta	ξ	xi
γ	gamma	π	pi
δ	delta	ρ	rho
χ	chi	τ	tau
λ	lambda	μ	mu
ν	nu	Π	PI
ϵ	epsilon	Φ	PHI
σ	sigma	Ψ	PSI
θ	theta	Σ	SIGMA
ψ	psi	Γ	GAMMA
ϕ	phi	Δ	DELTA
ω	omega	Ω	OMEGA

ELEMENTS OF
REAL ANALYSIS

ELEMENTS OF

REAL ANALYSIS

HERBERT S. GASKILL   **P. P. NARAYANASWAMI**

Department of Mathematics
Memorial University of Newfoundland

PRENTICE-HALL
Upper Saddle River, New Jersey 07458

Library of Congress Cataloging-in-Publication Data

Gaskill, Herbert S.
    Elements of real analysis   /   Herbert S. Gaskill,
P. P. Narayanaswami.
        p. cm.
    Includes bibliographical references and index.
    ISBN 0-13-897067-X (alk. paper)
    1. Mathematical analysis.   I. Narayanaswami, P. P.   II. Title.
QA300.G288   1998
515' .822—dc21   97-25465
                                                                                       CIP

Acquisition Editor: George Lobell
Editorial Assistant: Gale Epps
Editorial Director: Tim Bozik
Editor-in-Chief: Jerome Grant
AVP, Production and Manufacturing: David W. Riccardi
Production Editor: Elaine W. Wetterau
Senior Managing Editor: Linda Mihatov Behrens
Executive Managing Editor: Kathleen Schiaparelli
Manufacturing Buyer: Alan Fischer
Manufacturing Manager: Trudy Pisciotti
Director of Marketing: John Tweeddale
Marketing Manager: Melody Marcus
Marketing Assistant: Jennifer Pan
Creative Director: Paula Maylahn
Art Director, Cover: Jayne Conte
Cover Designer: Pat Wosczyk
Cover: "Interior with Window," acrylic on canvas,
    painting and photograph by Gary Stephan

©1998 by Prentice-Hall, Inc.
Simon & Schuster / A Viacom Company
Upper Saddle River, NJ 07458

Printed in the United States of America
10   9   8   7   6   5   4   3   2   1

ISBN 0-13-897067-X

Prentice-Hall International (UK) Limited, *London*
Prentice-Hall of Australia Pty. Limited, *Sydney*
Prentice-Hall Canada Inc. *Toronto*
Prentice-Hall Hispanoamericana, S.A., *Mexico*
Prentice-Hall of India Private Limited, *New Delhi*
Prentice-Hall of Japan, Inc., *Tokyo*
Simon & Schuster Asia Pte. Ltd., *Singapore*
Editora Prentice-Hall do Brasil, Ltda., *Rio de Janeiro*

Contents

Preface

To the Instructor

Twenty-five years ago, several classic books were written aimed at introducing undergraduate mathematics majors to rigorous mathematics in the form of analysis. At that time, when most would agree that all students were mathematically better prepared, it was recognized that even prospective majors had tremendous difficulties achieving competence at rigorous mathematics. Over the past twenty-five years, the general level of preparedness in mathematics of students entering most universities has declined as a result of decreased emphasis in the school curriculum on skill at basic algebra and symbol manipulation. Indeed, many mathematics and science majors exhibit evidence of this decline. Since facility in these two skills are essential prerequisites for mathematicians, a lack thereof can reduce a student's chance of success at beginning analysis. There are two plausible ways to address this problem.

The first is to reduce the level of material being presented to the student. There were several books that have taken this approach in various areas of mathematics. In analysis, such an approach can be accomplished by carefully selecting the material presented to avoid the most (more) difficult material and/or by carefully selecting the kinds of questions put to students to avoid asking those questions that are known to cause difficulty. We have rejected this approach because we believe it to be self-defeating to the stated goal of a basic advanced calculus/first real analysis course, which is to produce a mathematically mature student with a thorough knowledge and understanding of limiting processes.

The second approach to the problem of a decline in skill at symbol manipulation and basic algebra is to confront the problem head-on. This is the approach taken in this text. Specifically, we consistently stress the role of algebra and algebraic manipulation in the construction of proofs. Throughout, but especially early in the book, calculations are performed in detail so that students will not be left with a feeling that mathematics is mysterious and understanding is beyond their abilities. In addition, each definition, proof, and example is followed by a **DISCUSSION** section that presents material on intuition, on what thought processes might lead to the proof just presented, on what ideas a definition is trying to capture, and so forth. The volume of this material, and the number of examples, increases the length of the book by about 50%. Mature mathematicians can generate discussion material for themselves. It was expected that beginning students of twenty-five years ago could generate this material for themselves. It is recognized that most of today's beginning students cannot, particularly, given the wider audiences expected to take an elementary analysis course.

This book is long because it includes a great deal of material of a pedagogical nature to aid students in their work outside the classroom, because this is where most learning at universities takes place. Is the inclusion of this material worth it? Our students say it is. But no student has to read this material to succeed because the pedagogical material is carefully separated from the mathematics. So if there are students for whom the *Discussion* sections lack utility, they can simply omit them.

The book treats in order, the principles of logic, the axioms for the real numbers, limits of sequences, limits of functions, the topology of **R**, differentiation, integration, infinite series, convergence and uniform convergence of sequences of real-valued functions, and concludes with a chapter on transcendental functions.

The approach emphasizes rigor and a clear exposition of the various concepts in mathematical analysis. Foremost among these concepts is the notion of a **LIMIT**, and its study, in the form of delta-epsilon (δ-ϵ) analysis, is the principal theme of the book.

Perhaps the most difficult material in the book to teach is that in the first chapter, which covers the development of the real line and the algebraic tools useful in analysis. All the details are there. But to cover this material in its entirety would take far more time than any of us want to spend. Our approach is to spend one lecture on each of Sections 0.1–0.3 and 0.6; and two lectures each on 0.4 and 0.5, which cover material related to the supremum principle. Our expectation is you will be selective. Our reason for including the amount of material presented is so students will be able to find all the foundational material required to do analysis in one place, for example, the rules for dealing with arithmetic, inequalities, absolute values, geometric series, and other topics that used to be part of the common knowledge of entering students, but no longer is. Much of the material is covered in the problems and we encourage our students to read through the problems to become aware of tools contained in the problems. The remainder of the book consists of the standard topics, except for a section on Riemann-Stieltjes integration (5.7) and the Weierstrass' Approximation Theorem (8.4). Both of these can be omitted.

As you teach from the book, we would be delighted to hear from you and have included our email addresses at the end of this preface for that purpose. Also, an instructor's Solutions Manual is available through your Prentice-Hall representative.

To the Student

This book is about the theory of limiting processes on the real line. This material forms the heart of mathematics, and in its present state of development, it brings to fruition more than 2000 years of mathematical work in geometry and algebra. As such, it is an essential part of the training of any student of mathematics, whether pure or applied.

It is commonly said that mathematics is the only true deductive science. As working mathematicians, we know this is utterly false. Mathematical ideas, like any other ideas, are generated by mulling over observations. The simplest way to collect observations is to perform experiments. The way in which mathematics differs from other sciences is that the mathematician's experiments are thought experiments conducted by working out the details of simple examples related to the particular system or theory of interest. The intuitions gained as a result of these experiments must then be provable in a rigorous mathematical sense before they will be accepted as truth. Thus, there are two aspects to the mathematical process: generating insight via experiment and establishing insight as truth by means of constructing a rigorous proof of the statement.

Proofs in mathematics are not merely a bunch of statements put together to draw up the desired conclusion; rather, they are cleverly and delicately woven fabrics of thought and, hence, must be understood as a single theme. Construction of proofs requires insight as to why the particular fact *should be true*. Thus, even though throughout this book you will be given problems of the form 'Show ...', whence the true statement is known, you will not be able to generate the required proof of the statement without a fundamental understanding of the basic insight that is the essential reason why the given statement is true. Thus the first question that you should ask when confronting the problem 'Show ...' is, 'Why should this be true?' Or alternatively, 'What is the insight that led to this statement?' To answer these questions, you will be forced into performing the type of experiments that are essential to the generation of insight, and without which no proof would have even been found.

In the text, each theorem is followed by a standard type of mathematical proof. Most proofs, examples, and definitions are followed by a section labeled **Discussion**. This section is offered to explain the essential piece of insight that is being captured by the theorem, example, or definition. It also may contain clues about which experiments are most relevant to the ideas. One effect of these sections should be to convince you that mathematics is not nearly so much the result of brilliance as it is of hard work!

Throughout the book, you will find the terms (**WHY?**) and (**HOW?**), which are invitations to the reader to supply a missing argument or a simple calculation. Also, phrases such as 'it is readily seen', 'it follows easily', and 'it is straight-forward' are to be taken as warning signals, where you must stop and provide the appropriate reasoning in support of the conclusion that follows. This is to be expected, and even an experienced mathematician would be unable to read and fully understand the proofs in this book at the same reading speed used for reading novels. Thus, read with pencil and paper in hand, and expect the going to be slow.

The essence of our previous remarks has been succinctly captured by J. L. Kelley when he said, 'Mathematics is not a spectator sport'. With this in mind, we have supplied a goodly number of exercises for your attention. Your success will depend on the effort with which you attack these problems. All answers must be justified with a rigorous argument. Some are very difficult, and many cannot be done in five minutes but require much thought and work to generate the insight discussed above.

Finally, to succeed in this endeavor, you will require the two most important attributes of working mathematicians, patience and persistence: patience so as not to hurry a solution whose time has not yet arrived, persistence to keep trying even when you believe that **you** have no hope of succeeding. It is these two qualities that finally see all research mathematicians through the day.

As you learn from the book, we would be delighted to hear from you to know what works and what doesn't, and have included our email addresses at the end of this preface for that purpose.

Acknowledgments

No book could be completed without the help of others. We have had a great deal of help and it is a pleasure to thank those who have helped us.

First, and foremost, are our wives, Cathy and Padma, who did so many things to make this book possible. Second, our department heads, John Burry, Bruce Shawyer, and Bruce Watson, who from time to time organized our teaching loads to aid the effort. Third, our colleagues Richard Charron, Michael Clase, Renzo Piccinini, Don Rideout, and Bruce Shawyer, who made comments on the manuscript during the course of teaching. Fourth, there are the many students who made comments while learning from a draft of the book. Most especially, there was Tara Stuckless whose careful proofreading eliminated so many errors.

At our publisher, there was Mike Ryan, who brought us into the fold. Then there was George Lobell, who gave us so much good advice. Finally, there was Elaine Wetterau, whose search for perfection led us to producing the best book we could get. Thanks to you all.

Every book has reviewers, some who want to be acknowledged, others who don't. Our experience is all reviews, even very critical ones, are valuable. Thanks to you all, in particular,

Frank DeMeyer   Colorado State University
William H. Ruckle   Clemson University.

The manuscript for this book was produced using the technical typesetting program LaTeX 2_ε. The authors received much invaluable help in using LaTeX 2_ε from our colleagues Edgar Goodaire and Bruce Shawyer. Our computer system is managed by Randy Bouzanne. Without his technical assistance, we couldn't have completed this project. More generally, we thank the creators of LaTeX 2_ε, GNUPLOT, and all the other UNIX utilities that have proved so useful to our project.

H. S. GASKILL[1]

P. P. NARAYANASWAMI

---

[1]Our email addresses are: herb@ math.mun.ca and swami@ math.mun.ca

Chapter 0

Basic Concepts

This book is about limits and limiting processes, and some five basic types of limiting processes and their interrelationships will be studied. But more than this, *Elements of Real Analysis* is about the heart of mathematics and the structure that serves as the foundation not only for real analysis but also for all of modern science. This structure is none other than the real numbers. It is a structure with which all readers of this text should be long since familiar. And it is in the theory of this structure that all the mainstream branches of modern mathematics have their roots.

This book is also about how to do mathematics and start thinking like a mathematician. It is about the kinds of questions that mathematicians ask and the methods mathematicians use to find answers to their questions. It is about the activities mathematicians undertake to develop their intuitive understanding of mathematical objects and conjectures related to such objects. Finally, it is about how mathematicians test their understanding by constructing rigorous proofs of their intuitively arrived at conjectures.

The purpose of this chapter is to lay a logical and axiomatic foundation on which we can build real analysis. This axiomatic foundation will precisely describe the real numbers. It will also provide a framework that we will use to capture the geometry of our intuitive ideas. Thus, we will consistently draw pictures and perform sample calculations to aid and encourage our intuition, since drawing pictures and performing sample calculations are well-established methods by which mathematicians build intuition. We will also reason by analogy, taking ideas developed in a simple context and pushing them to their limit in a more complicated context. This, too, will develop our intuition. But as mathematicians, we will always test our intuition by constructing proofs from the axioms that we will adopt. For, it is by virtue of these proofs that we will convert intuitive ideas into incontrovertible facts.

0.1 Logic

This section presents the basic principles of logic that are indispensable to a working mathematician. Logic provides the methodology by which we will wield our axioms, and a minimal understanding of this methodology is essential to any working mathematician.

Mathematicians are primarily interested in two things: discovering and proving theorems. (Of course, this process does not take place in a vacuum, and the mathematician hopes that the theorems that he (she[1]) discovers and proves will provide information about the objects that peak his curiosity.) The statement of a theorem is a mathematical assertion or a formula, and so one requirement for mathematical success is an ability to manipulate these formulae in various ways. Generally, theorems tend to be rather complex statements that, however, can be broken down into basic units.

If we think of complex statements as being constructed from simple ones, then the constructions turn out to be simple and few in number. Suppose then that we let Φ and Ψ stand for two mathematical assertions, say, $x = 2$ and $y < 7$, respectively. Now we can construct new statements from Φ and Ψ as follows:

$$\Phi \ or^2 \ \Psi,$$
$$\Phi \ and \ \Psi,$$
$$not \ \Phi,$$
$$if \ \Phi, \ then \ \Psi,$$
$$\Phi \ if \ and \ only \ if \ \Psi.$$

Using the examples given for Φ and Ψ, the constructions yield the following mathematical assertions:

$$x = 2 \ or \ y < 7,$$
$$x = 2 \ and \ y < 7,$$
$$not \ (x = 2),$$
$$if \ x = 2, \ then \ y < 7,$$
$$x = 2 \ if \ and \ only \ if \ y < 7.$$

(We usually write $x \neq 2$ or $y \not< 7$ instead of *not* $(x = 2)$ or *not* $(y < 7)$.) These constructions carry the names **disjunction**, **conjunction**, **negation**, **implication**, and **logical equivalence**, respectively. Of these constructions, the implication construction is considered so important that it exists in several equivalent forms:

$$\Phi \ implies \ \Psi,$$
$$\Psi \ is \ necessary \ for \ \Phi,$$
$$\Phi \ is \ sufficient \ for \ \Psi.$$

Again we stress that all of these statements are identical in meaning to *if* Φ, *then* Ψ. Another statement related to this same implication is the **converse**:

$$if \ \Psi, \ then \ \Phi.$$

Finally, a construction that is often used instead of logical equivalence, but has the same meaning, is

$$\Phi \ is \ necessary \ and \ sufficient \ for \ \Psi.$$

---

[1]We will use 'he' and 'she' interchangeably throughout this book.

[2]In this first section, *or, and, if, then,* and other similar mathematical terms will be written in *italics* to remind the reader that they are being used with precise mathematical meanings and not their usual English meanings.

The key question of importance to a mathematician is: *How does the truth of the complex whole depend on the truth of the constituent parts?* The answer to this question is determined by how we reason in the real world and also by the fact that the truth of a statement should depend *only on the truth values of the respective parts and not on the meaning of the respective parts.* The latter comment becomes clearer when we consider that a column of numbers whose sum we must find may represent bushels of wheat, gallons of gas, or dollars and cents, but the outcome of the computation is independent of any meaning we attach to the numbers. That is, the value of the sum depends only on the individual numbers entering the sum, not on what the numbers represent. For this to be true of our logical analysis of statements, the truth of an *'or'* statement should depend only on the truth value of the two disjuncts[3] and not on the meaning of the two disjuncts. With this principle in mind, if we think about reality, a disjunction is true exactly if one of the disjuncts is true, or if both are; a conjunction is true exactly if both of the conjuncts are true; a negation is true exactly if the negated statement is false. The implication is more subtle and we treat it in detail.

A main feature of mathematics is the universal acceptance of the following.

Law of the Excluded Middle: *A given mathematical statement* Φ *is either true or false; there is no third alternative.*

It is impossible to do much mathematics without accepting this principle, and so we adopt it and use it in an axiomatic fashion. With this in mind, consider the statement

If it rains tomorrow, then I will not go swimming,

in which a **hypothesis** about rain implies a **conclusion** about swimming. There are various factual outcomes that can happen tomorrow, and these are shown in a two-way table. Inside Table 0.1.1, we have put the **truth values** for the implication that are associated with these outcomes.

	Rain	No rain
Do not swim	True	True
Swim	False	True

Table 0.1.1

It is clear that if it really does rain and I don't swim, then the implication should be true; also, if it really does rain and I do swim, then the implication is false. The issue is: in any other case have I lied? In real life, we generally have almost no interest in the statement when the hypothesis is false. This is the tack we take in mathematics. By labeling as true those cases in which the hypothesis is false, we are agreeing that *the only time an implication can be false is when the hypothesis is true and the conclusion simultaneously false.* Thus, in proving the truth of an implication, we may take as an assumption the truth of the hypothesis. If under a true hypothesis, the conclusion is always true, then the implication is true. If under the same hypothesis, there is an instance in

---

[3]The constituent parts of a disjunction (conjunction) are called disjuncts (conjuncts).

which the conclusion is false, then the implication is false. Thus, we see that of the two choices for truth values for the other cases, choosing 'true' instead of 'false' is really agreeing to proceed as we do in reality, namely, ignore the cases when the hypothesis is false.

The logical equivalence construction is really an abbreviation. Thus, Φ *if and only if* Ψ stands for

$$\Phi \ \textit{implies} \ \Psi \ .\textit{and.} \ \Psi \ \textit{implies} \ \Phi.$$

(The two periods before and after the 'and' are used in the above instead of parentheses to indicate the conjunction of two implications; we shall continue the practice of using periods instead of parentheses to avoid the ambiguities that arise when a statement contains several logical connectives.) From this, we see that Φ is logically equivalent to Ψ if the truth of Φ implies the truth of Ψ and the truth of Ψ implies the truth of Φ. While proving that two statements are logically equivalent, we will always proceed by writing out the two implications and proving them separately.

A key feature of this analysis is that it provides us with a method for testing whether two statements always yield the same truth value and, hence, are logically equivalent. An example of this is the **contrapositive**, a statement form that is constructed from an implication in the following manner:

$$\text{original implication:} \quad \textit{if} \ \Phi, \textit{then} \ \Psi$$
$$\text{contrapositive:} \quad \textit{if} \ (\textit{not} \ \Psi), \ \textit{then} \ (\textit{not} \ \Phi).$$

To see that these statements are logically equivalent, we ask the question: under what conditions is each statement false? The original implication is false exactly when the hypothesis is true and the conclusion is false. This means that Φ is true and Ψ is false. But then *not* Ψ is true and *not* Φ is false, whence the contrapositive is also false. On the other hand, the only time the contrapositive is false is when its hypothesis is true and its conclusion is false. But this is exactly the situation just described. We conclude that an implication is false exactly when its contrapositive is false. Thus, an implication and its contrapositive are equivalent.

The contrapositive is a very useful form, since the contrapositive of a statement is often much easier to prove than the original statement.

Example 1. Find the contrapositive of the statement

$$\textit{if} \ x = 2, \ \textit{then} \ y < 7.$$

Solution. The contrapositive is

$$\textit{if} \ y \not< 7, \ \textit{then} \ x \neq 2. \qquad \square^4$$

One of the most important manipulational skills in mathematics is to be able to take a given mathematical statement and correctly write down its negation. The process of forming the negation is simplified if one has a working knowledge

---

[4]Throughout the book, the symbol $\square$ indicates the end of a proof, a solution, or a discussion.

of some simple rules. These rules are easily established by considering how we negate disjunctions, conjunctions, negations, and implications.

The simplest statement to negate is a negation, that is, a statement of the form *not* Φ.

$$
\begin{array}{rl}
\text{original statement:} & \textit{not } \Phi \\
\text{negation:} & \textit{not } (\textit{not } \Phi) \\
\text{equivalent form:} & \Phi
\end{array}
$$

The fact that the negation of the negation of a statement is equivalent to the original statement, namely, *not* (*not* Φ) is logically equivalent to Φ, is usually expressed as the **Principle of the Double Negative**.

To negate a disjunction, we proceed as follows:

$$
\begin{array}{rl}
\text{original statement:} & \Phi \textit{ or } \Psi \\
\text{negation:} & \textit{not } (\Phi \textit{ or } \Psi) \\
\text{equivalent form:} & (\textit{not } \Phi) \textit{ and } (\textit{not } \Psi)
\end{array}
$$

The general rule to remember here is that when you negate an *or* statement you get an *and* statement. You may be familiar with this principle for sets where the complement of a union is the intersection of the complements. This is one of **DeMorgan's Laws**.

The other DeMorgan's Law concerns the negating of a conjunction:

$$
\begin{array}{rl}
\text{original statement:} & \Phi \textit{ and } \Psi \\
\text{negation:} & \textit{not } (\Phi \textit{ and } \Psi) \\
\text{equivalent form:} & (\textit{not } \Phi) \textit{ or } (\textit{not } \Psi)
\end{array}
$$

The fact that the negation of an *and* statement is an *or* statement is not too surprising once one has accepted the first DeMorgan's Law.

The last construction to be negated is an implication:

$$
\begin{array}{rl}
\text{original statement:} & \textit{if } \Phi, \textit{ then } \Psi \\
\text{negation:} & \textit{not } (\textit{if } \Phi, \textit{ then } \Psi) \\
\text{equivalent form:} & \Phi \textit{ and } (\textit{not } \Psi)
\end{array}
$$

The equivalent form is perhaps worth a further bit of explanation. Consider the assertion: *if $x = 2$, then $y < 7$*. We have already noted that this assertion is false only when the antecedent (hypothesis) is true and the conclusion is false. Now our equivalent form of the negation is $x = 2$ *and* $y \not< 7$. Evidently, this latter is true exactly when the antecedent (of the previous statement) is true and its conclusion is false. Thus, this form must be equivalent to the negation.

This tells us something very important about testing the truth or falsity of implications. If we want to disprove an implication, we are required to set up a situation in which the *hypothesis is true and the conclusion is false*. Thus, in the example above, if we want to disprove the assertion that *if $x = 2$ then $y < 7$*, we would have to produce a y that was greater than or equal to 7 while at the same time keeping x fixed at the value 2. The specific x and y that disprove the assertion form a **counterexample**.

Example 2. Find the negation of

$$\textit{if } x^2 - y^2 \textit{ is even, then } (x \neq 2 \textit{ or } y = 3).$$

Solution. The form of the statement is an implication in which the conclusion is a disjunction. Thus, the basic form of the negation obtained by using the rules for implication is

$$x^2 - y^2 \textit{ is even .and. not } (x \neq 2 \textit{ or } y = 3).$$

This can be further simplified by using the rules for disjunctions, to obtain

$$x^2 - y^2 \textit{ is even .and. } x = 2 \textit{ and } y \neq 3.$$

At this point, we see that the original statement is false exactly if each of these conjuncts can be made true simultaneously. □

In the above example, we have made use of parentheses. Parentheses in mathematical statements play the same role as punctuation in English. To remove them will generally introduce ambiguities and, at worst, will completely change the meaning of statements. Consider

$$\textit{not } (x \neq 2 \textit{ and } y = 3)$$

and

$$\textit{not } x \neq 2 \textit{ and } y = 3.$$

The meaning of the first statement is completely clear. While the meaning of the second is ambiguous, it would generally be taken to be the same as

$$(\textit{not } x \neq 2) \textit{ and } y = 3,$$

which has a different meaning from the first statement. It is safe to say that unless one works in one of the specially developed parenthesis-free languages, such as reverse Polish notation, which is employed on Hewlett-Packard calculators, the correct use of parentheses, and/or periods, is essential and should be cultivated.

We have already used the symbol '=', which stands for equality. The equality relation plays a fundamental role in mathematics as a *logical* tool. To fully specify this role requires a deeper excursion into the principles of logic than we intend to take. However, there are certain minimal requirements that the equality relation must satisfy. We summarize these:

Equality Principles: *Let x, y, and z denote arbitrary mathematical quantities. Then,*

  (i) $x = x$ (reflexivity);

 (ii) *if $x = y$, then $y = x$* (symmetry);

(iii) *if $x = y$ and $y = z$, then $x = z$* (transitivity).

The notion of *mathematical quantity* is imprecise; however, it can be made precise with effort, but the details are again beyond the scope of this book. The reader should think of *mathematical quantities* as sets, numbers, functions, or anything else mathematicians ordinarily discuss. The basic point then becomes that any statement we would make about the number 5 is unchanged with respect to truth if we rephrase it in terms of $2 + 3$ or $4 + 1$ or any other convenient expression for 5. More generally,

> *the truth of any mathematical statement is unchanged by the replace-*
> *ment of any mathematical quantity occurring in the statement with*
> *any other equal quantity.*

A consequence of this general fact is that equations can be added, or otherwise combined, using available operations. For example, from $a = b$ and $x = y$ we can deduce

$$a + x = b + y, \; a \cdot x = b \cdot y, \; a^x = b^y,$$

and so forth, simply by starting with an identity $(a + x = a + x)$ and then applying the general principle.

A key feature of mathematical statements that we have not yet discussed is their use of quantifiers. Mathematical statements consistently contain expressions like: *for all x, for every y, there is a w,* or *there exists a q.* These expressions are of two distinct types, but they have the same general purpose: namely, to **quantify** a variable. Phrases of the form *for every x, for each x,* or *for all x,* have the same function: they *universally quantify* the variable x. Hence, they are called **universal quantifiers**. Their intent is to assert that no matter what value we substitute for x, the given statement will be true.

The other type of quantifier is the **existential quantifier**. True to its name, it acts to assert the *existence* of an individual that when substituted for x (the variable named) will make the statement true. Phrases used to denote existential quantification are: *for some x, there is an x,* or *there exists x.*

Example 3.

> *There is an x, $x = \sqrt{2}$.*
> *For every x, if $x < 0$, then $x^2 > 0$.*
> *For all x, there exists y, $x - y = 0$.*
> *For all x, there exists y, $xy = 1$.*   □

Discussion. Whenever a mathematician writes statements such as those above, he intends them to be interpreted in some context. The context usually is determined by the structures being studied. We may want, for example, to study a general class of mathematical structures such as groups or fields or some particular field such as the real numbers, **R**, or the rational numbers, **Q**. In each case, the context of the study determines a set of legal substitutions for the variables that appear in the statements. In the case of groups, we substitute elements from an arbitrary group; in the case of fields, an arbitrary element from an arbitrary field; and in the case of a particular field, an arbitrary element from the field in question. *The context will, in general, have a substantial effect on the truth of assertions.* The first assertion—that the square root of two exists—is certainly true if the field under study is the real numbers. But it is false if we are studying the rational numbers, since as we shall later show, there is no rational number whose square is two. □

Mathematicians have developed a set of standard symbols to denote quantification. We will not generally employ these symbols in this book, preferring the English expressions. However, since they are in common use, particularly by mathematics professors writing on blackboards, we introduce them. The English phrase 'for all' is represented by the symbol ∀, which, according to lore, was arrived at by turning an 'A' upside down. The English phrase 'there exists' is symbolically represented by ∃, which, by similar reasoning, was arrived at by reversing the letter 'E'. Example 3, when written using these symbols, yields,

Example 4.

$$\exists x,\ x = \sqrt{2}.$$
$$\forall x,\ if\ x < 0,\ then\ x^2 > 0.$$
$$\forall x\,\exists y,\ x - y = 0.$$
$$\forall x\,\exists y,\ xy = 1. \qquad\qquad \square$$

The examples above are of a completely mathematical nature, which may suggest that quantifiers are not a part of our everyday thought and reasoning processes. Nothing could be further from the truth. We often make statements involving quantifiers. For example, the statement

All mathematicians are good citizens.

involves a universal quantifier, whereas the statement

I know a mathematician who is a spy.

also involves a quantifier, but in an indirect way. The use of quantifiers in normal language is complicated by the fact that it usually involves hidden additional logical structure. For example, consider the initial statement about mathematicians. What it really asserts is that every member of the class of mathematicians belongs to the class of good citizens. Thus, it could be rephrased as

For all x, if x is a mathematician, then x is a good citizen.

Notice that this form of the statement contains an implication that was implied by the initial statement. Of course, we never actually speak this way, but this is what is meant. Similarly, the second example could be rephrased as

There exists x, (x is a mathematician and x is a spy).

Again, there is additional logical structure in the form of an '*and*' statement.

As with other mathematical statements, it is essential to be able to negate statements containing quantifiers correctly. One useful device in the process is to write down the statement preceded by: *it is not true that*. We then carefully consider the meaning of the prefixed statement. For the first statement in Example 3, this procedure yields

It is not true that (there is an x, x = $\sqrt{2}$).

We would like to translate this into a simple statement in which all quantifiers precede the negation phrase; that is, *it is not true that*. Consider the meaning of the negated statement. Evidently, the only way the statement can be true will be that no matter what substitution we make for *x*, it is not the case that the substituted individual is the square root of two. Thus, an equivalent statement is

For every x, x ≠ $\sqrt{2}$.

Notice that the negation symbol is now completely contained within the simplest basic assertion of the original statement; that is, $x = \sqrt{2}$ becomes, after negation, $x \neq \sqrt{2}$. Further, as the negation symbol is moved inside the existential quantifier, *the quantifier is changed to a universal quantifier.*

Consider now, the second assertion

For every x, if x < 0, then x^2 > 0.

This becomes, after negation,

It is not true that (for every x, if x < 0, then x^2 > 0),

which, in turn, becomes

There is an x, (x < 0 and $x^2 \ngtr 0$).

Again, in the last statement, the negation symbol is now completely contained in the basic assertion. Notice that as the negation symbol moves inside the universal quantifier, *the quantifier is changed to an existential quantifier.* It seems apparent that of the two forms of the negation, the latter is much clearer in meaning, and so it is easier to see how to check its truth: we simply look for a number less than 0 whose square is not greater than 0.

The third and fourth of the sample statements are similar, so we consider only the last.

For all x, there exists y, xy = 1,

becomes, after negation,

It is not true that (for all x, there exists y, xy = 1),

becomes

There exists x, for all y, xy ≠ 1.

Again, we see that as the negation symbol is moved in, the type of quantification is reversed; that is, '*for every*' becomes '*there exists*', and '*there exists*' becomes '*for every*'. The name of the variable that is quantified is unchanged and *the order in which the variables are quantified is unchanged.* As before, it is much easier to decipher the meaning of the latter form than the former. Moreover, it is clear that to check the truth of the latter, we must produce a substitution for x such that no matter what we substitute for y, the product is not equal to 1. If the structure being discussed is, for example, the real numbers, then it is clear that substituting 0 for x will result in the obviously true statement

For every y, 0y ≠ 1.

To conclude our discussion of this example, we point out that *changing the order in which variables are quantified in a statement will often change the meaning of the statement.* To see this, let us consider the third statement of Example 3, again in the context of real numbers, together with the statement obtained from it by reversing the order in which the variables are quantified.

For all x, there exists y, $x - y = 0$.

There exists y, for all x, $x - y = 0$.

In the context of the real number system **R**, the first statement asserts that if we are given any arbitrary real number, then we can find another real number such that the difference of the two is 0. Note that the choice of y *depends* upon which x we were given in the first place. The second statement asserts that there is a *fixed y* such that no matter what value we choose for x, the difference will be 0. A little thought shows that while the first statement is clearly true, the second is definitely false.

In our consideration above on how to negate statements containing quantifiers, we argued based upon the meaning of particular statements. The use of meaning was illustrative. In fact, the methodology developed for negating statements containing quantifiers is completely independent of the meaning of these statements; rather, the methodology depends only on the *form* of the particular statement to be negated. Let us complete this discussion of negating statements involving quantifiers by again considering

All mathematicians are good citizens.

As we have already remarked, this statement is equivalent to

For all x, if x is a mathematician, then x is a good citizen.

When we negate this statement, we must deal not only with the quantifier, but with the implication as well. Thus, the negation is

There exists x, (x is a mathematician and x is not a good citizen).

Before concluding our section on logic, we want to discuss definitions as they occur in mathematics. The first important fact about definitions is that *they must be committed to memory, grasped thoroughly, and completely understood.* The reason for this is simple: in order to prove theorems, one must be able to think about the concepts involved or being discussed in these theorems. Thinking is a process that is internal to the brain, and in mathematics, there is evidence to suggest that much of the creative work is done at a subconscious level. In any case, the brain must be supplied with the basic tools for the job. Definitions are these tools. Trying to do mathematics while picking all the relevant definitions from a book is about as efficient as a computer whose memory functions by printing out questions the answers to which are found and punched back in by a key puncher with a high school education. In short, you must acquire the tools if you expect success. The axioms and theorems also fall in this category.

The second aspect of definitions that must be understood is their purpose. Mathematical languages, as with any language, are filled with nouns, that is, words that denote or select out some class of objects. The word *sequence* identifies a collection of objects in mathematics exactly as the word *chair* does in English. The difference is that *sequence* is defined in such a way that it is always possible to decide whether or not any given object is a *sequence*, and the result is a completely unambiguous *yes* or *no*. A little thought together with a visit to a modern furniture store will convince one that it is impossible to set down a completely unambiguous test for membership in the class of *chairs*. Thus, mathematical definitions achieve a level of *precision* that cannot be attained by ordinary languages, and it is this high level of precision that is one of the really beautiful features of mathematics. Thus, the role of a definition is to precisely delineate a class of objects in such a way that it is possible to decide for any object whether it is, or is not, a member of the class. To do this, the definition will generally list the properties that define membership in the class. In consequence, to assert that an object satisfies a given definition is equivalent to asserting that an object has all the properties listed in the definition.

Exercises

1. Give a *precise* negation of the following statements:
   (a) all snakes are not poisonous;
   (b) some problems are not easy;
   (c) it is not the case that I am not hardworking;
   (d) there exists $x, x > 0$ and $f(x) = 4$;
   (e) for all x, $f(x) = 7$ implies $x > 0$;
   (f) there exists x, $(x = 0$ or $f(x) = x)$;
   (g) for all x, there exists y, $f(x, y) = 0$;
   (h) there exists x, for all y, there exists z, $(f(x) = 0$ and $g(x, y) = 1$ and $h(x, y, z) = 0)$.

2. Obtain the contrapositive equivalents of the following implications:
   (a) whenever the phone rings, I run to answer it;
   (b) if $x^2 + y^2$ is negative, the earth will not rotate;
   (c) it is necessary for you to eat in order to live;
   (d) if $x^2 \neq 3$ and $y^2 \geq 5$, then ω is not an irrational number;
   (e) if the sum of any two even integers is odd, then there is a rational number whose square root is 7;
   (f) a sufficient condition for getting good grades is to be a genius.

3. Construct a truth table (similar to Table 0.0.1) for the following statements:
   (a) I will take a vacation, if I have money, and I do not work;
   (b) a monotonic and bounded sequence will have a limit;
   (c) $(P$ implies $Q)$ .implies. R;
   (d) $(P$ and not $R)$ .implies. (not $Q)$.

4. Let $f(x, y)$ stand for $x + y - xy$. Consider the eight statements:
   (a) for all x, for all y, $f(x, y) = 0$;
   (b) for all y, for all x, $f(x, y) = 0$;
   (c) there exists y, there exists x, $f(x, y) = 0$;
   (d) there exists x, there exists y, $f(x, y) = 0$;
   (e) there exists y, for all x, $f(x, y) = 0$;
   (f) there exists x, for all y, $f(x, y) = 0$;
   (g) for all x, there exists y, $f(x, y) = 0$;
   (h) for all y, there exists x, $f(x, y) = 0$.
   Which of these are true statements? Which of these statements are logically equivalent, independent of the domains of x and y?

5. Repeat Exercise 4 with the function $f(x, y) = x^2 + y - xy$.

6. Use truth tables to show the following statements are logically equivalent:
   (a) P implies Q;   (b) not $(P$ and $($not $Q))$;   (c) (not P) or Q.

7. Specify the hypothesis and the conclusion for each of the statements in Exercise 2.

8. Obtain the converse statements of each of the implications in Exercise 2.

9. Supply counterexamples to show each of the following statements is false:

   (a) all animals are carnivorous;

   (b) all birds can fly;

   (c) $n^2 + n + 41$ is always a prime number;

   (d) $(a + b + c)^n = a^n + b^n + c^n$ for all natural numbers n and all real numbers a, b, and c;

   (e) $1^2 + 2^2 + \cdots + n^2 = (n + 1)^2$ for all n.

10. The following mathematical definitions can be found in various places in this book. Use the symbols $\forall$ and $\exists$, respectively, to denote the universal and the existential quantifier, that is,

    $$\forall x \ P(x) \text{ stands for 'for all } x, \text{ the property } P(x) \text{ holds'.}$$
    $$\exists x \ P(x) \text{ stands for 'there exists } x, \text{ for which } P(x) \text{ holds'.}$$

    Rewrite the following statements using the symbols $\forall$ and $\exists$, and then obtain a *precise* negation of each of them.

    (a) If $(x, y) \in F$, and $(x, z) \in F$, then $y = z$. (F is a function.)

    (b) If $(x_1, y) \in F$, and $(x_2, y) \in F$, then $x_1 = x_2$. (F is one-to-one.)

    (c) Given $y \in A$, there exists an $x \in$ Dmn F such that $y = F(x)$. (F is onto A.)

    (d) If $x, y \in$ Dmn f, then $x < y$ .implies. $f(x) \le f(y)$. (f is monotonic increasing.)

    (e) s is a function whose domain is the set $\mathbf{N}$ of natural numbers and the range is contained in the set $\mathbf{R}$ of all real numbers. (s is a sequence.)

    (f) Given $\epsilon > 0$, there exists a real number $N > 0$ such that $n > N$ implies $|a_n - L| < \epsilon$. (The sequence $\{a_n\}$ converges to the limit L.)

    (g) Given $\epsilon > 0$, there exists a real number $\delta > 0$ such that for all $x \in$ Dmn f, $|x - c| < \delta$ implies $|f(x) - f(c)| < \epsilon$. (The function f is continuous at the real number c.)

    (h) There exists a real number L such that given $\epsilon > 0$, we can find $\delta > 0$, satisfying $0 < |h| < \delta$ .implies. $\left| \dfrac{f(x + h) - f(x)}{h} \right| < \epsilon$. ($f$ is differentiable at the point x.)

    (i) Given $\epsilon > 0$, there exists $N \in \mathbf{N}$ such that for all $x \in A, n > N$ implies $|f_n(x) - f_0(x)| < \epsilon$. (The sequence $\{f_n\}$ of functions is uniformly convergent to the function f_0 on the set A.)

    (j) Given $\epsilon > 0$, there exists $\delta > 0$ such that for all $x, y \in A, |x - y| < \delta$ .implies. $|f(x) - f(y)| < \epsilon$. (The function f is uniformly continuous on the set A.)

0.2 Field Axioms

The real numbers, which we will denote by **R**, have three important aspects: algebraic properties, order properties, and completeness properties. We want to specify a set of axioms to describe the real number system. Our axioms, therefore, will be divided into three groups: those dealing with algebra, those dealing with order, and those dealing with completeness.

The general purpose of a set of axioms is to lay down the basic properties of a class of objects to be studied. These initial properties, stated in the form of axioms, should be self-evident, since they will serve as the starting point for all future discussions. The basic objects for study, in this case the real numbers, are not defined. Instead, **number** is a primitive. Our axioms do not tell us what numbers are, rather, they tell us how numbers behave. The initial description of their behavior is contained in the axioms, and this description is augmented by the logical process of proving theorems. Thus, our overall intent is to build a rigorous description of **R** in the same way Euclid's *Elements* builds a description of geometry.

One of the basic things we know we can do with numbers is to combine them in various ways to get new ones. Formally, this involves the notion of a binary operation.

Definition. A **binary operation** on a set A is a function from $A \times A$ into A.

In the appendix on set theory, there is a complete treatment of the essential set theory required to support our development. In particular, operations with sets are discussed and appropriate set-theoretic definitions for the fundamental concepts such as cartesian product, relation, function, among others, are given.

There are two principal binary operations used to combine numbers: **addition** and **multiplication**. The study of the properties associated with these operations, generally, is the realm of algebra. However, algebra is an essential tool in analysis and supplies a portion of the foundation on which to base our rigorous development.

Definition. A **field** is a nonempty set **F** together with a binary operation $+$, called **addition**, a binary operation $\cdot$, called **multiplication**, and two distinct constants 0 and 1 in **F**, called **zero** and **one**, respectively, that satisfy the following axioms for all members $x, y, z \in$ **F**:

A1 *For every x, and for every y, $x + y = y + x$ and $x \cdot y = y \cdot x$.*

A2 *For every x, for every y, and for every z, $(x + y) + z = x + (y + z)$ and $(x \cdot y) \cdot z = x \cdot (y \cdot z)$.*

A3 *For every x, $x + 0 = x$ and $x \cdot 1 = x$.*

A4 *For every x, there exists y, $x + y = 0$.*

A5 *For every x, there exists y, if $x \neq 0$, then $x \cdot y = 1$.*

A6 *For every x, for every y, for every z, $x \cdot (y + z) = (x \cdot y) + (x \cdot z)$.*

Discussion. These axioms are probably familiar to the reader. **A1** is the **commutative** law for + and ·; **A2** is the **associative law** for the operations; **A3** asserts 0 and 1 are the **additive** and **multiplicative identities**, respectively; **A4** asserts the existence of **additive inverses**; **A5** asserts the existence of **multiplicative inverses** for elements other than 0; and **A6** is the **distributive law**. The constants 0 and 1 are not permitted to be equal so that the set **F** must contain at least two distinct elements. We, of course, use the standard names for 0 and 1, namely, **zero** and **one**, respectively. These are not to be confused with the 'usual' zero and 'usual' one in the real number system, which we use in our day-to-day life. We emphasize 0 is the only member of the field that does not possess a multiplicative inverse. As an immediate consequence of **A2**, either of the two equal members $(x + y) + z, x + (y + z)$ is unambiguously referred to as $x + y + z$, without using parenthesis. A similar remark applies to $x \cdot y \cdot z$. As we progress, when the context is clear, we will just write xy instead of $x \cdot y$.

Note, even though we say **F** is a field, to be more precise, we must refer to the field as the quintuple, $\langle \mathbf{F}, +, \cdot, 0, 1 \rangle$, since each of these entities has a role to play in the definition of a field. □

In this book we will be concerned with four familiar structures:

R	the collection of real numbers;
Q	the collection of rational numbers;
Z	the collection of whole numbers or integers;
N	the collection of natural numbers or positive integers.

Of these structures, the first two are fields, whereas the integers fail to have multiplicative inverses and the positive integers fail to have additive inverses and a zero. In fact, **Q** is a subfield of the field **R** (a **subfield** of a field **F** is a subset **G** of **F**, which, with the field operations of **F**, satisfies the field axioms).

In the development being followed, **R** will simply be realized as a structure satisfying the required axioms and the natural numbers, integers, and rational numbers will be obtained as subsets of the real numbers. This process amounts to making a definition. What the student should understand is the process of axiomatization, which may seem artificial and unfamiliar, is leading to the four mathematical structures with which all of us are familiar. What axiomatization adds, and is the only reason why one takes this course, is a foundation from which all of the remainder of analysis can be obtained rigorously. In particular, the basic facts of arithmetic can be obtained (see, for instance, Exercise 0.2.1[5]), and it is the manipulation of these facts that is one of the primary tools of analysis.

Example 1. Show that the quintuple, $\langle \{0, 1, 2\}, \oplus, \otimes, 0, 1 \rangle$, forms a field, where $\oplus$ denotes addition '*mod* 3' and $\otimes$ denotes multiplication '*mod* 3', namely, $x \oplus y$ (respectively, $x \otimes y$) is the remainder obtained when $x + y$ (respectively, $x \cdot y$) is divided by 3.

Solution. The following two tables define the two binary operations on $\{0, 1, 2\}$.

---

[5]Exercise x.y.z refers to Exercise z in section x.y. Exercise z refers to Exercise z in the current section.

$\oplus$	0	1	2
0	0	1	2
1	1	2	0
2	2	0	1

$\otimes$	0	1	2
0	0	0	0
1	0	1	2
2	0	2	1

Because the underlying set is finite, the axioms may be checked by **inspection**, that is, verifying all possible cases. □

Discussion. Once one has obtained the integers and their properties, one simply observes the required properties follow as consequences of the arithmetic properties of integers. For example, suppose one wants to establish the axiom **A2** on associativity for $\otimes$. First, one observes for any three integers, i, j, k, $(i \cdot j) \cdot k = i \cdot (j \cdot k)$. One then proves: if m and n have the same remainder on division by 3, then $n \cdot i$ and $m \cdot i$ will also have the same remainder on division by 3. These two facts now yield the required result.

While examples of different algebraic structures will be discussed in the exercises, the thrust of our discussions will be concerned with **R**, **Q**, **N**, and **Z**. □

Our intent is to study analysis rather than algebra, so we do not want to delve into the consequences of the field axioms in great detail. However, the facts of arithmetic are essential to the program, and the methods of obtaining them will illustrate the standard of rigor we intend to maintain.

For the remainder of this section, as well as Section 0.3, unless otherwise stated, x, y, z will denote arbitrary members of a field **F**, and 0 and 1 will denote the fixed elements, zero and one, of the field **F**. We begin with the additive version of a theorem on cancellation. The multiplicative version is left to Exercise 5.

Theorem 0.2.1. *For every x, y, z, if $x + y = x + z$, then $y = z$.*

Note. We have used the phrase *for every* only once, but from the context, it is clearly intended that we mean *for every x, for every y, for every z.* We will abuse our notation in this way whenever the meaning remains clear.

Proof. Let x, y, and z be fixed but arbitrary elements of **F** satisfying the hypothesis. By **A4**, there is a w in **F** such that $x + w = 0$. Now

$$
\begin{array}{rcll}
w + (x + y) & = & w + (x + z) & \text{by the Equality Principles,} \\
(w + x) + y & = & (w + x) + z & \text{by } \mathbf{A2}, \\
(x + w) + y & = & (x + w) + z & \text{by } \mathbf{A1}, \\
0 + y & = & 0 + z & \text{by choice of } w, \\
y + 0 & = & z + 0 & \text{by } \mathbf{A1}, \\
y & = & z & \text{by } \mathbf{A3},
\end{array}
$$

as desired. □

Discussion. We begin with some general comments and then follow with specifics. The proof above is really the sequence of statements on the left, together with the initial choice of w. Notice each of the statements is an instance of an axiom, namely, the axiom listed or referred to on the right, or it is the result of an application of a principle of logic, as in the appeal to the Equality Principles, or an application of the hypothesis of the theorem, as in the first line of the proof. Thus, a **proof** is a sequence of statements, each of which is either an instance of an axiom or previously established theorem, and the last line of which is the conclusion of the result to be proved, in this case, the statement $y = z$. The remarks on the right tell us how the statements are obtained from the axioms, and as such, they are an important aid to understanding the proof. These statements will, in later proofs, be omitted where they are obvious, and minor gaps in the reasoning will result. However, it is important the reader be able to supply all the missing reasons, and fill in all the gaps, since the ability to do so is an important test of understanding. Further, supplying the reasons and filling in any missing steps will make the proofs completely *convincing*, and this, after all, is the point of a proof, *to convince the skeptic.*

The motivation for this proof is very simple. We want to conclude $y = z$ from the hypothesis that $x + y = x + z$. In other words, we must try to eliminate x. Axiom **A4** guarantees the existence of a w, which 'cancels' x in the sense that $w + x = 0$. Also, **A3** states that the addition of 0 has no effect on the element. The above proof only formalizes these simple facts.

A particular feature of the proof is the use of the Equality Principles. Since the hypothesis of the theorem asserts

$$x + y = x + z,$$

we are permitted to substitute the right-hand side for the left-hand side in any expression and deduce equality between the original and resulting expressions. In particular, we start with the quantity $w + (x + y)$ and assert it is equal to $w + (x + z)$, by replacing $x + y$ with the equal quantity $x + z$. $\square$

Remark. We want to digress for a moment to discuss the general style and format of this book. It is traditional in that theorems are stated and proofs of these theorems are presented. It is an unfortunate fact that proofs can be very misleading. Proofs exist to establish once and for all, according to very high standards, that certain mathematical statements are irrefutable facts. What is unfortunate about this is that a proof, in spite of the fact that it is perfectly correct, does not in any way have to be enlightening. Thus, mathematicians, and mathematics students, are faced with two problems: the generation of proofs, and the generation of internal enlightenment. To understand a theorem *requires enlightenment*. If one has enlightenment, one knows why a particular theorem must be true. Understanding why a theorem is true will almost always lead to a proof. Sometimes, one's own understanding of a theorem will be significantly different from that presented in someone else's proof of the same theorem. Thus, understanding may lead to an alternate proof of a given theorem.

One can be enlightened about proofs as well as theorems. Without enlightenment, one is merely reduced to memorizing proofs. With enlightenment about a proof, its flow becomes clear, and it can become an item of astonishing beauty. In addition, the need to memorize disappears because the proof has become part of you.

What do the preceding comments have to do with the style and format of this book? First, as authors, we must confront the fact that we want to present the reader with correct proofs. Second, we want to create a situation of maximum possible enlightenment.

The first aim is achieved by presenting blocks of material that begin with the word **Proof** and end with a '□'. They are correct mathematical proofs, and they contain 'the truth and nothing but the truth', so to speak. As such, there is no attempt to motivate within a proof, since motivational material is not generally recognized as appropriate content for proofs. (This may have something to do with why research papers in mathematics are difficult to read.) As well, so that enlightenment is as easy to come by as possible, we have, by design, sacrificed elegance in favor of what is sometimes rather mundane computation.

The second aim is approached by the inclusion of blocks of material that start with **Discussion** and end with '□'. In these sections the motivation behind proofs is discussed. They also contain some of the words that we would say to our own students about the proof as we presented it in our classrooms. These are the sections that show how proofs are generated and make clear the proofs do not spring full blown from the heads of rather strange fellows called mathematicians. In the final analysis, the discussion sections can only offer pointers toward the direction of enlightenment. They can only act as aids for starting the search. The real work must be done by you, the reader. What we promise is that you, the reader, will know when you have achieved enlightenment and, second, that when you have achieved it, you will know why abstract mathematics has been the ultimate intellectual activity of humankind since before the time of Pythagoras and Archimedes.

In conclusion, then, as you read a proof, know that some suggestions toward a fuller understanding exist in the following discussion section. Look there for help. □

We turn now to proving that for a given member x in a field, there is exactly one member y belonging to the field that satisfies axiom **A4**. Similarly, if $x \neq 0$, there is a unique member y that answers axiom **A5**.

Theorem 0.2.2. *The additive inverse of any element of* **F** *is unique. Likewise, the multiplicative inverse of a nonzero element of* **F** *is unique.*

Proof. Let x be fixed and y and z be any two elements of **F** that witness **A4** for x. Then

$$
\begin{aligned}
x + y &= x + z \quad \text{since both sides are zero, whence} \\
y &= z \quad \text{by Theorem 0.2.1.}
\end{aligned}
$$

Thus, the additive inverse is unique, as claimed. The proof for multiplicative inverses is left to the reader (Exercise 6). □

Discussion. The theorem is a *uniqueness* theorem. It asserts the uniqueness of a certain object, in this case the additive inverse. Proofs of uniqueness generally start by assuming the existence of two distinct objects having the defining property. They are completed by showing the two objects must, in fact, be the same. In this case the defining property for the two objects is that of being the additive inverse of x. □

Notation. In the future, we will use $-x$ to denote the unique additive inverse of x and x^{-1} or $\frac{1}{x}$ to denote the unique multiplicative inverse of a nonzero x. As well, we use $\frac{y}{x}$ for $y \cdot \left(\frac{1}{x}\right)$. These notations are justified by the above theorem. Thus, it is clear that for any element x, we have $x + (-x) = (-x) + x = 0$, and for any nonzero x, $x \cdot \frac{1}{x} = \frac{1}{x} \cdot x = 1$. We emphasize 0^{-1} does not exist in any field.

Theorem 0.2.3. *For every x, $x \cdot 0 = 0$.*

Proof. Let x be fixed, but arbitrary. Now

$$
\begin{aligned}
0 + 0 &= 0 & \text{by } \mathbf{A3}, \\
x \cdot (0 + 0) &= x \cdot 0 & \text{by the Equality Principles}, \\
x \cdot 0 + x \cdot 0 &= x \cdot 0 & \text{by } \mathbf{A6}.
\end{aligned}
$$

Also,

$$
\begin{aligned}
x \cdot 0 + 0 &= x \cdot 0 & \text{by } \mathbf{A3}, \\
x \cdot 0 &= x \cdot 0 + 0 & \text{by the Equality Principles}.
\end{aligned}
$$

It follows

$$
x \cdot 0 + x \cdot 0 = x \cdot 0 + 0 \qquad \text{by the Equality Principles},
$$

whence

$$
x \cdot 0 = 0 \qquad \text{by Theorem 0.2.1.} \qquad \square
$$

Discussion. The key to this proof is the fact that $0 + 0 = 0$. Once we focus on this fact as a starting point, it is clear we only have to multiply through on both sides by x, use the distributive law, and then cancel. This leaves the question of how one arrives at the above as an appropriate staring point. To get there, one must start by having a full internal grasp of the axioms. Without this, one cannot hope to succeed. $\square$

Theorem 0.2.4. *For every x and y, $(-x) \cdot (-y) = x \cdot y$.*

Proof. Let x and y be arbitrary, but fixed. We claim $x \cdot y$ and $(-x) \cdot (-y)$ are both additive inverses of $(-x) \cdot y$. If this claim is true, then we are done, by Theorem 0.2.2 on the uniqueness of additive inverses. To establish the claim, note

$$
\begin{aligned}
(-x) \cdot y + (-x) \cdot (-y) &= (-x) \cdot [y + (-y)] & \text{by } \mathbf{A6}, \\
&= (-x) \cdot 0 & \text{by the definition of } -y, \\
&= 0 & \text{by Theorem 0.2.3.}
\end{aligned}
$$

On the other hand,

$$
\begin{aligned}
(-x) \cdot y + x \cdot y &= [(-x) + x] \cdot y & \text{by } \mathbf{A1} \text{ and } \mathbf{A6}, \\
&= 0 \cdot y & \text{by } \mathbf{A1} \text{ and the definition of } -x, \\
&= y \cdot 0 = 0 & \text{by } \mathbf{A1}, \text{ and Theorem 0.2.3.}
\end{aligned}
$$

Thus, the claim is established. $\square$

Discussion. This proof seems rather 'slick'. It stems, however, from the fact that $-(-x) = x$. If we apply this idea to the above situation, we get

$$
x \cdot y = -(-(x \cdot y)) = \underline{-((-x) \cdot y)} = (-x) \cdot (-y).
$$

Notice the underlined quantity is exactly the additive inverse of the quantity, $(-x) \cdot y$, that was added to both sides in the proof. The underlined quantity also has a common factor with both $x \cdot y$ and $(-x) \cdot (-y)$, which is what makes the proof go. Finding the quantity $(-x) \cdot y$ is, in large part, a matter of experimentation. Experimental calculations are often useful in getting an intuitive 'feel' for the reasoning underlying a proof. $\square$

Exercises

1. Prove the following statements where the letters a, b, c, ... stand for arbitrary members of a fixed field:

   (a) $-0 = 0$;
   (b) $-(-a) = a$;
   (c) $-a = (-1) \cdot a$;
   (d) $(-a) \cdot b = -(a \cdot b) = a \cdot (-b)$;
   (e) $1^{-1} = 1$;
   (f) $a \neq 0$ implies $\frac{a}{a} = 1$;

   (g) $a \neq 0$ and $a \cdot b = a \cdot c$ .implies. $b = c$; (Can the restriction $a \neq 0$ be removed?)

   (h) $b \neq 0$ and $d \neq 0$ .implies. $\frac{a}{b} \cdot \frac{c}{d} = \frac{a \cdot c}{b \cdot d}$;

   (i) $b \neq 0$ and $c \neq 0$ .implies. $\frac{a}{b} = \frac{a}{b} \cdot \frac{c}{c}$;

   (j) $a \cdot b = 0$ .implies. $a = 0$ or $b = 0$;

   (k) $b \neq 0$ and $\frac{a}{b} = 0$ .implies. $a = 0$;

   (l) $c \neq 0$ implies $\frac{a+b}{c} = \frac{a}{c} + \frac{b}{c}$;

   (m) $b \neq 0$ and $d \neq 0$ .implies. $\frac{a}{b} + \frac{c}{d} = \frac{a \cdot d + b \cdot c}{b \cdot d}$;

   (n) $b \neq 0$ and $d \neq 0$ .implies. $\frac{a}{b} = \frac{c}{d}$ if and only if $a \cdot d = b \cdot c$.

2. If a, b, c are arbitrary members of a field $\mathbf{F}$, show all the six expressions

$$a + b + c, \ a + c + b, \ b + a + c,$$
$$b + c + a, \ c + a + b, \ c + b + a$$

   represent the same element.

3. If a, b, c, d are arbitrary elements of a field $\mathbf{F}$, prove

   (a) $(a + b) + (c + d) = (a + c) + (b + d) = a + [(b + c) + d]$;

   (b) $(a \cdot b)(c \cdot d) = (a \cdot d)(b \cdot c) = a \cdot [(b \cdot c \cdot d] = [(a \cdot b) \cdot c] \cdot d$, thereby showing the expressions $a + b + c + d$ and $a \cdot b \cdot c \cdot d$ have an unambiguous meaning, when parentheses are not used;

   (c) $(a + b) \cdot (c + d) = (a \cdot c) + (a \cdot d) + (b \cdot c) + (b \cdot d)$.[6]

4. Let a and b denote arbitrary elements of a field. Define

$$a - b = a + (-b).$$

   This definition of **subtraction** abuses our notation by using the '$-$' in two distinct ways, as a **unary operator** (the initial use) and now as a binary operation symbol. Prove subtraction is not associative in any field in which $1 + 1 \neq 0$, but a form of the distributive law holds. Further, show if a, b, c, d are arbitrary members of a field $\mathbf{F}$:

---

[6]The use of parentheses on the right-hand side of this equation can be avoided by invoking standard rules for order of precedence of operations. In the future, these rules (unary operations followed by multiplicative operations followed by additive operations) will be assumed. Note the use of parentheses on the left cannot be avoided because they are being used to force addition to occur before multiplication.

(a) $-(a - b) = -a + b = b - a$;

(b) $(a - b) + (b - c) = a - c$;

(c) $a - (b + c) = (a - b) - c$;

(d) $a - (b - c) = (a - b) + c$;

(e) $a - b = c - d$ if and only if $a + d = b + c$;

(f) $(a - b) \cdot (a + b) = a \cdot a - b \cdot b$.

5. State and prove the multiplicative version of Theorem 0.2.1; see Exercise 1(g).

6. Prove the remainder of Theorem 0.2.2.

7. Prove if an element a of a field satisfies $a + x = x$ for every other x, then $a = 0$. Prove an analogous statement for the element 1.

8. If a and b are elements of a field such that $a \cdot a = b \cdot b$, prove either $a = b$, or $a = -b$. Can you make a similar statement if the hypothesis was changed to $a \cdot a \cdot a = b \cdot b \cdot b$?

9. Show the set $\{0, 1\}$ with a suitable definition of addition and multiplication satisfies the definition of a field. Is subtraction associative in this field?

10. Show the set $\mathbf{F} = \{0, 1, 2, \dots, 18\}$ forms a field, if addition (denoted by $\oplus$) and multiplication (denoted by $\otimes$) are defined 'modulo 19' (as in Example 1). Find the additive and multiplicative inverses of the elements $3, 10, 14$, and 18. Are there elements in this field which are their own additive (multiplicative) inverses? Perform the following arithmetic in this field:

$$\frac{1}{4} \otimes \left(\frac{3}{5} \oplus \frac{10}{17} \right)$$

[This field is usually denoted by $\mathbf{Z}_{19}$.]

11. Define $\mathbf{Z}_{12}$ in the same way as $\mathbf{Z}_{19}$. Show $\mathbf{Z}_{12}$ is not a field, and the quadratic equation $x^2 - 5x + 6 = 0$ admits more than two solutions in $\mathbf{Z}_{12}$. Find all solutions.

12. Construct an example of a field $\mathbf{F} = \langle \mathbf{F}, +, \cdot, 0, 1 \rangle$ where

$$1 + (1 + (1 + (1 + (1 + (1 + 1))))) = 0.$$

13. Let $\mathbf{F} = \langle \mathbf{F}, +, \cdot, 0, 1 \rangle$ be a field, and let a and b denote two distinct, fixed elements of $\mathbf{F}$. If new operations of addition and multiplication are defined on $\mathbf{F}$ as follows:

$$x \oplus y = x + y - a, \quad x \otimes y = a + \frac{(x - a)(y - b)}{b - a},$$

prove one obtains a new field $\mathbf{F}_1$. What are the zero and one in the field $\mathbf{F}_1$?

14. Let $\mathbf{C}$ denote the set of all **complex numbers**, that is, elements of the form $a + bi$, where $a, b \in \mathbf{R}$ and the symbol i satisfies the identity $i \cdot i = -1$. Show $\mathbf{C}$ is a field under the following addition and multiplication:

$$(a + bi) + (c + di) = (a + c) + (b + d)i,$$
$$(a + bi) \cdot (c + di) = (ac - bd) + (ad + bc)i.$$

Identify the zero and one in $\mathbf{C}$.

15. Let $\mathbf{F}$ be a given field, and let $\mathbf{C_F}$ denote the product $\mathbf{F} \times \mathbf{F}$, namely, $\{(x,y) : x \in \mathbf{F} \text{ and } y \in \mathbf{F}\}$. Define addition and multiplication on $\mathbf{C_F}$ as follows:

$$\begin{aligned} (a,b) + (c,d) &= (a+c, b+d), \\ (a,b) \cdot (c,d) &= (a \cdot c - b \cdot d, \ a \cdot d + b \cdot c). \end{aligned}$$

Prove $\mathbf{C_F}$ is a field under the above field operations. What are the zero and one in $\mathbf{C_F}$? Show the element $(0,1)$ satisfies the equation $x \cdot x + 1 = 0$. (The field $\mathbf{C_F}$ is referred to as the **field of complex numbers** over $\mathbf{F}$. If $\mathbf{F} = \mathbf{R}$, the resulting field $\mathbf{C_F}$ is the field $\mathbf{C}$ of complex numbers.)

16. Let $\mathbf{F}_{\sqrt{3}} = \{a + \sqrt{3}b : a, b \in \mathbf{Q}\}$. Define suitable addition and multiplication, so $\mathbf{F}_{\sqrt{3}}$ is a field.

17. For subsets A, B of a field $\mathbf{F}$, define the operations $+$ and $\cdot$ as follows:

$$\begin{aligned} A + B &= \{a + b : a \in A \text{ and } b \in B\}, \\ A \cdot B &= \{a \cdot b : a \in A \text{ and } b \in B\}. \end{aligned}$$

We also write $\{k\} \cdot A = kA$, $-1A = -A$, $A + (-B) = A - B$.
For arbitrary subsets A, B, C of a field $\mathbf{F}$, prove the following:
(a) $\mathbf{F} + \mathbf{F} = \mathbf{F} - \mathbf{F} = \mathbf{F}$;
(b) If $A \neq \emptyset$, then $A - A$ is not empty;
(c) $A \cdot \emptyset = \emptyset$;
(d) If $A \subseteq B$, $A + C \subseteq B + C$;
(e) $A \cdot (B + C) \subseteq A \cdot B + A \cdot C$.
Give an example to show equality need not happen in (e).
If $A \cdot C = B \cdot C$, does it follow $A = B$?

0.3 Order Axioms

We turn now to the second aspect of the real numbers: **order**. Much of man's initial concept of number arises in geometry where they are used to record and compare lengths. Thus, larger numbers correspond to longer lengths. Of course, we could start with numbers corresponding to the number of elements in a set, but this gives only the whole numbers, while the idea of length gives rise to not only rational numbers, but also irrationals such as $\sqrt{2}$ and π. As these geometrical ideas were developed over the centuries, mathematicians came to associate the real numbers with the **real number line.**

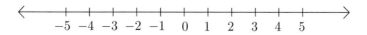

Figure 0.3.1 The real number line.

Thus, we think of a line stretching to infinity in both directions and the real numbers being assigned to points of the line. As we move along the line in one direction, we pass over the real numbers in a fixed order that is determined by the lengths they represent. Any set of axioms that we write down must make some attempt to capture these properties, and this is the thrust of the order axioms.

The basic feature of the picture is the fact that given a pair of real numbers, a and b, one lies '*to the left of*' the other. Thus, we could think of a binary relation (see appendix) which is characterized by the phrase 'to the left of'. Keeping this simple idea in mind, we proceed to write down statements that are obviously true for the geometric picture presented above.

Definition. An **ordered field** is a field $\mathbf{F}$, together with a binary relation, $<$, satisfying the following axioms (for x, y, z in $\mathbf{F}$):

O1 *For every x, and for every y, exactly one of the following is satisfied:*

$$x = y \ or \ x < y \ or \ y < x.$$

O2 *For every x, for every y, and for every z*

$$x < y \ and \ y < z \ .implies. \ x < z.$$

O3 *For every x, for every y, and for every z*

$$x < y \ implies \ x + z < y + z.$$

O4 *For every x, for every y, and for every z*

$$0 < z \ and \ x < y \ .implies. \ z \cdot x < z \cdot y.$$

Discussion. The principal concept underlying these axioms is that of a binary relation. Recall a binary relation is an arbitrary collection of ordered pairs. In this case, the ordered pairs are from $\mathbf{F} \times \mathbf{F}$. As a matter of convenience, we write $x < y$, instead of $(x, y) \in <$. These matters are discussed more fully in the appendix.

The first axiom is known as the **Law of Trichotomy.** The second states the **transitive** property. The third allows us to add across an inequality, while preserving the inequality. The last gives the restricted situations in which multiplication preserves the inequality.

The reader should interpret each of the axioms in terms of the 'to the left of' relation to see that they are intuitively true. □

While we write $a < b$, we say 'a is **less than** b'. We define the binary relation $>$ (**greater than**) by

$$a > b \text{ if and only if } b < a.$$

We define $\leq$ (**less than or equal**) by

$$a \leq b \ .if \ and \ only \ if. \ a < b \ or \ a = b.$$

Similarly, the reader can define $\geq$ (**greater than or equal**) in the obvious way. We often write

$$a < b < c$$

to mean

$$a < b \text{ and } b < c.$$

The other relations, $\leq$, $>$, and $\geq$, can also be used this way, and the expression can be extended to four or more terms. We say a is **positive** in case a is greater than 0, a is **negative** provided a is less than 0, and a is **nonnegative** if $a \geq 0$. It is a fact that

$$a < b \text{ if and only if } (b - a) \text{ is positive.}$$

The proof is left to the reader as Exercise 0.3.5(f).

The field of real numbers, $\mathbf{R}$, as well as the field of rational numbers, $\mathbf{Q}$, with the usual ordering relation $<$, furnish classical examples of an ordered field. There are other examples.

Let $\mathbf{F}$ (more precisely, $\langle \mathbf{F}, +, \cdot, <, 0, 1 \rangle$) denote an arbitrary ordered field. For the remainder of this section and the start of the next, our theorems and definitions relate to such an arbitrary ordered field. However, the reader should keep in mind that the motivation for proving these theorems is to establish the basic facts about the order properties of $\mathbf{R}$ and $\mathbf{Q}$, and these two structures should be thought of as the prime examples of ordered fields. Again, the facts established below and in the exercises will prove essential in the remainder of this text.

Theorem 0.3.1. *For every x, if x is positive, then $-x$ is negative. If x is negative, then $-x$ is positive.*

Proof. Fix x, which is negative. By definition, $x < 0$. Now,

$$
\begin{aligned}
x + (-x) &< 0 + (-x) &&\text{by } \mathbf{O3}, \text{ whence} \\
0 &< -x &&\text{by } \mathbf{A3} \text{ and } \mathbf{A4}.
\end{aligned}
$$

The fact that $-x$ is positive follows by definition. The rest of the theorem is left to Exercise 1. $\square$

Theorem 0.3.2. $\quad 0 < 1.$

Proof. By **O1**, exactly one of the following is true:

$$0 = 1; \; 0 < 1; \; 1 < 0.$$

The first possibility has already been discarded. Thus, it is enough to show the last can also be discarded. Hence, we assume for the sake of argument $1 < 0$. Now,

$$
\begin{aligned}
0 &< -1 &&\text{by Theorem 0.3.1, whence} \\
(-1) \cdot 0 &< (-1) \cdot (-1) &&\text{by } \mathbf{O4}.
\end{aligned}
$$

It follows

$$
\begin{aligned}
0 &< 1 \cdot 1 &&\text{by Theorems 0.2.3 and 0.2.4, whence} \\
0 &< 1 &&\text{by } \mathbf{A3}
\end{aligned}
$$

The last statement clearly contradicts our assumption $1 < 0$, and so this assumption must, in fact, be false (by **O1**). The only other possibility is $0 < 1$ as claimed. $\square$

Discussion. The proof above employs an **indirect argument** (proof by **contradiction**). We do not give a direct proof of the desired conclusion (in this case, $0 < 1$), but rather show any other alternative leads to nonsense (in this case, $1 < 0$ and simultaneously $0 < 1$, a possibility denied by **O1**). Such arguments are justified by our belief that our axioms are consistent, or simply put, our axioms will not permit the logical production of nonsense. We shall, on occasion, employ this method of reasoning and the reader should study the above simple example, so that he may use the technique as well. He should note that to use the technique correctly, care must be taken to state clearly the nature of the contradiction. $\square$

Theorem 0.3.3. *For every x, if $0 < x$, then $0 < \frac{1}{x}$.*

Theorem 0.3.4. *For every x, for every y, if $x < y$, then $x < \frac{x+y}{2} < y$, where $2 = 1 + 1$.*

Theorem 0.3.5. *For every x, for every y, if $0 < x < y$, then $0 < \frac{1}{y} < \frac{1}{x}$.*

Proofs of the above are left to the reader (Exercises 2, 3, and 4). It should be noted each of the above theorems is formalizing some fact that our intuition knows to be well established. It is also telling us our axioms are leading us in directions that we want to go, and in this sense, they are 'correct'.

Exercises

1. Complete the proof of Theorem 0.3.1.

2. Prove Theorem 0.3.3.

3. Prove Theorem 0.3.4.

4. Prove Theorem 0.3.5.

5. Prove the following where a, b, c, $\ldots$ denote arbitrary elements of a fixed ordered field:

   (a) $a \neq 0$ implies $0 < a \cdot a$;

   (b) $a < 0$ and $0 < b$ .implies. $0 > a \cdot b$;

   (c) $a < 0$ and $b < 0$ .implies. $0 < a \cdot b$;

   (d) $0 \leq a$ and $0 \leq b$ .implies. $0 \leq a + b$;

   (e) $a \leq b$ and $b \leq a$ .implies. $a = b$;

   (f) $a < b$ if and only if $b - a$ is positive;

   (g) $0 < a < 1$ .implies. $a \cdot a < a$;

   (h) $1 < a$ implies $a < a \cdot a$;

   (i) $a < a + 1$;

   (j) $0 < a < b$ .implies. $a \cdot a < b \cdot b$;

(k) $a \leq b$ and $c \leq d$ .implies. $a + c \leq b + d$;

(l) $0 < a$ and $0 < b$ .implies. $\frac{1}{a+b} < \frac{1}{a}$;

(m) $0 < b$ and $0 < c$ .implies. $\frac{a}{b} < \frac{a+c}{b}$;

(n) $0 < a, b, c, d$ .implies. $\frac{a}{b} < \frac{c}{d}$ if and only if $a \cdot d < b \cdot c$;

(o) $a \cdot a + 1 > 0$;

(p) $a \cdot b < 0$ .implies. $(a < 0$ and $b > 0)$ or $(a > 0$ and $b < 0)$;

(q) $a \cdot b > 0$ .implies. $(a > 0$ and $b > 0)$ or $(a < 0$ and $b < 0)$;

(r) $a < b$ and $0 < k < 1$ .implies. $a < k \cdot a + (1 - k) \cdot b < b$;

(s) $0 \leq a < b$ .implies. $\frac{a}{1+a} < \frac{b}{1+b}$;

(t) $0 < b, d$ and $\frac{a}{b} < \frac{c}{d}$ .implies. $\frac{a}{b} < \frac{a+c}{b+d} < \frac{c}{d}$.

6. Prove the following statements for an ordered field **F**:

   (a) $a + a = 0$ implies $a = 0$;

   (b) $a + a + a = 0$ implies $a = 0$;

   (c) $a \cdot a + b \cdot b \geq 0$ and $a \cdot a + b \cdot b > 0$, unless $a = b = 0$.

7. If a and b are arbitrary members of an ordered field **F**, prove either $\frac{a}{b} + \frac{b}{a} \geq 1 + 1$ or $\frac{a}{b} + \frac{b}{a} \leq (-1) + (-1)$. Use this to show $a \cdot a \cdot a = b \cdot b \cdot b$ implies $a = b$.

8. If $x < \epsilon$ for every $\epsilon > 0$ in an ordered field, prove $x \leq 0$.

9. If a and b are arbitrary elements of an ordered field **F** such that $a \leq b$, and for every $\epsilon > 0, a + \epsilon > b$, prove $a = b$.

10. Let **Q** denote what we ordinarily think of as the rational numbers. For $\frac{a}{b}$ and $\frac{x}{y}$ state the test used to decide their order relationship. Verify the validity of this test under the assumption **Q** is an ordered subfield of **R**, the ordered field of real numbers. (**Q** is formally defined in Exercise 0.4.32.)

11. Show if the equation $x \cdot x + 1 = 0$ has a solution in a field **F**, then **F** cannot be an ordered field.

12. Prove there are no finite ordered fields.

13. Show the field **C** of complex numbers defined in Exercise 0.2.15 cannot be an ordered field.

14. Let **F** be an ordered field. Show $\mathcal{C}_\mathbf{F} = \mathbf{F} \times \mathbf{F}$ becomes a field when addition and multiplication are defined as follows:

$$(a, b) + (c, d) = (a + c, b + d), \qquad (a, b) \cdot (c, d) = (a \cdot c - b \cdot d, a \cdot d + b \cdot c)$$

The field $\mathcal{C}_\mathbf{F}$ is called the the field of complex numbers over **F**.

The **dictionary ordering** on $\mathcal{C}_\mathbf{F}$ is defined by specifying $(a, b) > (c, d)$ when either $a > c$ or else $a = c$ and $b > d$. Show $\mathcal{C}_\mathbf{F}$ is not an ordered field under this ordering.

15. Some authors define a field **F** to be ordered if there exists a subset $P \subseteq \mathbf{F}$ (called the set of **positive elements**) such that

   (a) $P \cap (-P) = \emptyset$, where $-P = \{-x : x \in P\}$;

   (b) $P \cup \{0\} \cup (-P) = \mathbf{F}$;

   (c) $x, y \in P$ implies $x + y \in P$;

   (d) $x, y \in P$ implies $x \cdot y \in P$.

Define $x < y$ if and only if $y - x \in P$. Determine whether this set of axioms for an ordered field is equivalent to the one given in this section.

16. Let **F** denote the set of all **rational functions**, that is, expressions of the form

$$\frac{a_n x^n + a_{n-1} x^{n-1} + \cdots + a_0}{b_m x^m + b_{m-1} x^{m-1} + \cdots + b_0} \tag{$*$}$$

where the a's and the b's are real numbers and $b_m \neq 0$. The field operations are the 'usual' addition and 'usual' multiplication of such expressions. Define an order in **F** by prescribing $(*)$ to be positive if $a_n \cdot b_m > 0$. Verify **F** is an ordered field that contains the real numbers as a subfield.

Arrange the following elements of **F** in ascending order of magnitude:

$$\frac{x}{1}, \; \frac{1}{x}, \; \frac{x+2}{x-2}, \; \frac{2x-1}{x-3}, \; \frac{x}{1-3x}, \; \frac{x+1}{x^2-2}.$$

[Note: You may use the Principle of Induction (Section 0.4) to verify some of the field axioms.]

17. Let $\mathbf{Q}_{\sqrt{2}}$ denote the set $\{a + b\sqrt{2} : a, b \in \mathbf{Q}\}$, with the obvious field operations. Is it possible to prescribe an order in $\mathbf{Q}_{\sqrt{2}}$ so it is an ordered field?

18. In the ordered field **R**, of real numbers, obtain the solution set for the following inequalities:

(a) $(3x + 1)(2x - 3) > 0$;   (b) $7x(3x + 1) \geq 0$;

(c) $\frac{3x+11}{1-5x} > 0$;   (d) $\frac{4x}{8+7x} \leq 0$;

(e) $-\frac{1}{100} < 2x + 3 < \frac{1}{100}$;   (f) $-0.001 \leq \frac{5x+1}{2x} - \frac{11}{3} \leq 0.002$:

(g) $-\epsilon < \frac{3x+1}{2x} < \epsilon$, with $0 < \epsilon < \frac{1}{3}$;   (h) $\frac{3x-2}{2x} \geq \frac{1}{2}$ or $\frac{3x-2}{2x} \leq \frac{1}{2}$.

0.4   Completeness Axiom and Induction

So far, the axioms we have laid down are not enough to restrict our attention to the field **R** of all real numbers. The reason for this is there are many structures that satisfy the axioms for an ordered field that are provably distinct from the real numbers. As an example, think of the familiar field **Q**, the rational numbers. This field is precisely defined in Exercise 32, and it is well known (Exercise 32) the rational numbers form an ordered field. To see **Q** is distinct from **R**, we have only to write down the statement:

There is an x such that $x \cdot x = 1 + 1$.

It has been known since the days of Pythagoras that there is no rational number whose square is 2 (we prove this later). However, there is such a real number (as we will show) and the two fields cannot, therefore, be the same.

Let us return to our geometric example of the real line (see Figure 0.4.1). Suppose we draw a line at right angles to the real line that crosses it. The two

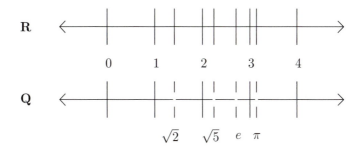

Figure 0.4.1 The rational 'subline' has holes.

lines will intersect in a point, and this point should correspond to a real number. If we think of the rational 'subline' as being all those points that correspond to rational numbers, then the assertion about a number corresponding to a point of intersection will not be true. For example, consider the vertical line that intersects the real line at the square root of 2; since there is no rational number whose square is 2, all we get where this vertical line crosses the rational subline is a hole. A similar statement holds for any other irrational number, such as the square root of 5. To 'fill in the holes' in the rational subline, we need a definition and our last axiom.

In this section, **F**, or more precisely $\mathbf{F} = \langle \mathbf{F}, +, \cdot, <, 0, 1 \rangle$, will always refer to an ordered field.

Definition. Let $S \subseteq \mathbf{F}$. An element $b \in \mathbf{F}$ is called an **upper bound** for S, if for every $x \in S, x \le b$. Further, b is called the **least upper bound** for S, if b is an upper bound for S, and for every other upper bound c of $S, b \le c$.

If $S \subseteq \mathbf{F}$ has an upper bound, we say S is **bounded above**. The least upper bound of a set S is often referred to as the **supremum**, and we sometimes write **sup** S (or **l.u.b** S) to denote the supremum of S. If an upper bound of S is a member of S, then that upper bound is referred to as the **greatest member (maximum)** of S. In a similar manner, we can define **lower bound**, **greatest lower bound (infimum)**, abbreviated as **inf** (or **g.l.b**), **least member (minimum)** of S, and **bounded below**. Last, a set S is said to be **bounded** if it is bounded both above and below.

Discussion. Think about what the words 'upper bound for S' should mean. The definition should say that a number is an upper bound if it is larger than every member of the set. This is exactly what the definition does say, except we have allowed an upper bound to be equal to an element of the set. It follows from the Law of Trichotomy that to show b is not an upper bound for S, we must find a witness, $x \in S$, such that $b < x$.

What should 'b is the least upper bound of S' mean? It should mean that b is an upper bound and there is no upper bound that is smaller (less) than b — exactly what the definition asserts. Thus, to show that b is not the least upper bound of S, we must show that it is not an upper bound or find an upper bound that is less than b.

To further clarify these ideas, the reader may find it helpful to write out the definitions of upper bound and least upper bound in terms of the '*to the left of*' relation.

If b is the greatest member of S, then b must also be the supremum of S because for any $y \in S$, if $y < b$, then y is shown not to be an upper bound for S using b as the witness. However, a set, which is bounded above (below), need not possess a greatest (least) element. As an example, consider

$$S = \{x : 0 < x < 1 \text{ and } x \in \mathbf{R}\}.$$

Note 1 is the least upper bound for S, but $1 \notin S$ so S has no greatest element. $\square$

Example 1. Find upper bounds and the suprema of the following subsets of $\mathbf{R}$, if they exist:

$$
\begin{aligned}
H_1 &= \{x \in \mathbf{Q} \text{ and } x < 0\}; \\
H_2 &= \{0, 1, 2, 3\}; \\
H_3 &= \mathbf{R}.
\end{aligned}
$$

Solution. It is clear 4 (where the symbols $1, 2, 3, 4$ have their usual meaning) is an upper bound for both H_1 and H_2. The number 2, on the other hand, is an upper bound for H_1, but not H_2. The collection of all upper bounds for H_1 is $\{y : y \in \mathbf{R} \text{ and } 0 \le y\}$. To see that no real number c that is less than 0 can be an upper bound for H_1, we have only to find a rational number x, such that $c < x < 0$. The proof of this fact is left to Exercise 37. It is immediate 0 is the supremum of H_1. Note $0 \notin H_1$.

For H_2, 3 is seen to be an upper bound by directly verifying $x \le 3$ for each of the other elements $x \in H_2$, that is, by inspection. Since $3 \in H_2$, it is the supremum.

H_3 is the set of all real numbers, $\mathbf{R}$. Evidently, this set has no upper bounds **(WHY?)**.[7] Because $\mathbf{R}$ has no upper bounds, it has no least upper bound. $\square$

Discussion. The collection of rational numbers has no upper bounds, whence it has no least upper bound. This illustrates one of the reasons why a set can fail to have a supremum, namely, it has no upper bounds. The other reason a set can fail to have a supremum is that it has too many upper bounds. The question of when a set fails to have a supremum will be explored in Exercise 8. $\square$

We now prove the uniqueness of the supremum which will justify our use of the phrase '*the* supremum' in the above definition.

---

[7]When **(WHY?)**, or **(HOW?)**, appear in the text, they request the reader to supply a computation or recall a fact. In this case an appropriate mental calculation would be: consider $b \in \mathbf{R}$ a potential upper bound; then $b < b + 1 \in \mathbf{R}$, so b cannot be an upper bound for $\mathbf{R}$; hence $\mathbf{R}$ is not bounded above. As you read the proofs, whether there is a **(WHY?)** or not, if you are not sure of the reasoning, then you have to be prepared to fill in the details. It is only through this type of active participation with the material that you will develop understanding. As an aid, in the remainder of this chapter and the start of the next, we will supply the missing computations in discussions.

Theorem 0.4.1. *Let $H \subseteq \mathbf{F}$. If $\sup H$ exists, then it is unique.*

Proof. Let b and c be two suprema of H. Then both must be upper bounds for H. If they are not equal, then we may assume $b < c$ (**WHY?**). But this clearly means either b is not an upper bound for H or c is not the least upper bound. In either case, we contradict our assumptions about b and c. $\square$

Discussion. The straightforward proof, which is an indirect argument, uses **O1**, the trichotomy law, as well as the definition of supremum.

The reader should review the comments on indirect arguments that follow Theorem 0.2.2.

For the (**WHY?**), we may assume $b < c$, as opposed to $c < b$, because b and c are only names for the two suprema that have no other distinguishing features. We could not make this assumption if one of the suprema had a distinguishing property, for example, it was known to be a perfect square. In that situation we would have to directly argue both cases, namely, $b < c$ and $c < b$. $\square$

The next theorem gives a useful characterization of the supremum.

Theorem 0.4.2. *Let H be a nonempty subset of $\mathbf{F}$ bounded above by an element $b \in \mathbf{F}$. Then, b is the supremum of H if and only if for every positive ϵ (epsilon) there is an $x \in H$ such that $b - \epsilon < x \le b$.*

Proof. We prove the supremum has the property claimed; the converse is left to Exercise 3. Thus, let H be a nonempty subset of $\mathbf{F}$ and $b = \sup H$. Fix a positive ϵ. Now, $b - \epsilon < b$ (**WHY?**), so $b - \epsilon$ is not an upper bound for H. This means there is an $x \in H$ that will witness the fact that $b - \epsilon$ is not an upper bound. For such an x, we must have

$$b - \epsilon < x \le b,$$

where the last inequality is obtained by virtue of the fact that b is an upper bound for H. $\square$

Discussion. As a general rule, the symbol 'ϵ' always denotes a fixed small positive quantity, say, something smaller than 10^{-1000} or even smaller. With this in mind, we draw a picture. Notice that as ϵ is fixed close to 0, the quantity $b - \epsilon$ is fixed close to b on the left side (see Figure 0.4.2). The function of $b - \epsilon$ is to generate a witness to the fact that no number that is less than or equal to $b - \epsilon$ can be an upper bound for H. It does this by producing an x such that $x \in H$ and $b - \epsilon < x$. In the figure, $b - \epsilon$ has been chosen to witness that c, which lies to the left of $b - \epsilon$, is not an upper bound for H. Finally, as $b - \epsilon$ moves in on b, it kills the chance of any number to the left of b being an upper bound for H. Thus, there is no upper bound that is less than b, whence b is the supremum.

Figure 0.4.2 The guaranteed existence of x makes b the supremum of H.

The statement of Theorem 0.4.2 is an 'if and only if' statement. Thus, there are in fact two statements to be proved. We have proved only one of these and have left the other for the reader in Exercise 3.

Because this theorem is an 'if and only if' statement, it characterizes suprema and can be used as an alternative to the definition. If one wants to show a is the supremum of H, it is enough to show, first, a is an upper bound for H and, second, for every positive ϵ there is an $x \in H$ such that $a - \epsilon < x \le a$.

For the (**WHY?**), $\epsilon > 0$ implies $-\epsilon < 0$ implies $b - \epsilon < b$. $\square$

The concepts of upper bound and supremum are the tools we needed to state the last definition leading to the completeness axiom.

Definition. Let **F** be an ordered field. Then, **F** is **complete** (or **order-complete**) provided for every nonempty subset S of **F** that is bounded above, there is an element $x_S \in \mathbf{F}$ such that x_S is the supremum of S.

Completeness Axiom: *The real numbers are a complete ordered field.*

We now have our axioms for the real numbers, namely, **A1-A6**, **O1-O4**, and the Completeness Axiom. The first ten axioms, **A1-A6** and **O1-O4** permit many realizations. For example, **Q** is one such realization, while **R** is another. It is an amazing fact, although we will not prove it here, that this last axiom (Completeness) forces the real numbers to be unique! This result is a direct consequence of the Completeness Axiom, which has a very special nature, distinct from all the others. The first ten axioms discuss the behavior of *numbers*. The last axiom describes the behavior of *sets of numbers*. This is a key difference, and we will now explore some of its more remarkable consequences. However, we first note that as a consequence of the uniqueness of the real numbers, the completeness axiom can be restated as follows:

Supremum Principle: *Let S be a nonempty set of real numbers that is bounded above. Then there is a real number that is the supremum (least upper bound) of S.*

For the remainder of this book we are working in the structure $\mathbf{R} = \langle \mathbf{R}, +, \cdot, <, 0, 1 \rangle$ satisfying the 11 axioms above.

The approach we have taken has been nonconstructive; that is, we have laid down a set of 11 axioms and taken **R** to be any structure that satisfies these axioms. This leaves us with the question: Why should we believe a structure exists satisfying these axioms, much less a unique structure? One answer to this is that the real numbers, with which we are all familiar, satisfy these axioms and we are merely setting down the self-evident truths about **R** that we take as the place to start. While this deals with existence, it does not deal with uniqueness. Another approach is to obtain **R** constructively from a set of axioms known as Peano's Axioms. These axioms specify the structure of the natural numbers, **N**. While this approach may seem more satisfactory as a result of **N** being a simpler structure, one is still left with wondering why **N** should exist and be unique. Such an alternative development of **R** can be found in the book *Foundations of Analysis* by E. Landau or in other texts on advanced calculus.

As noted in Section 0.2, the main structures of interest in this text are **R**, **Q**, **Z**, and **N**. We have defined **R**. We turn now to the problem of identifying the positive integers (natural numbers) within the structure **R** and developing their essential properties.

Definition. A subset S of **R** is **inductively closed** provided that

(i) $1 \in S$.

(ii) If $x \in S$, then $x + 1 \in S$.

The collection of **natural numbers (positive integers)** is defined by

$$\mathbf{N} = \bigcap \{S : S \subseteq \mathbf{R} \text{ and } S \text{ is inductively closed}\}.$$

Discussion. The purpose of this definition is to select out for special consideration those numbers that one usually uses to count. At the same time, we want to capture the very powerful properties associated with the structure inherent in the natural numbers, namely, the power associated with the induction axiom that is used as a proof technique. To accomplish these two goals, we must make sure all of the counting numbers end up in **N** and, at the same time, ensure that no other numbers get into **N**. Putting 1 in **N**, and requiring closure under the 'unary' operation of 'addition by one' (usually referred to as the **successor operation**), will guarantee all the counting numbers are in **N**. However, there are many inductively closed sets that contain numbers other than the counting numbers (**give an example!**). To ensure we have only the counting numbers, we take an intersection over all possible such inductively closed sets. It is this intersection, then, that yields **N**. Thus, if we can prove **N** itself is inductively closed, then it will follow **N** is the smallest inductively closed subset of **R**. This is done in the next theorem. $\square$

Theorem 0.4.3. *The natural numbers, **N**, form an inductively closed subset of the real numbers **R**.*

Proof. We first claim **R**, itself, is inductively closed. But this is obvious from the field axioms. Now 1 belongs to every inductively closed set, S, and so is in the intersection. Let x be an arbitrary member of **N**. Then $x \in S$, where S is an arbitrary inductively closed set. Since S is inductively closed, $x + 1 \in S$. But S was arbitrary, so $x + 1$ belongs to the intersection of all the S's, namely, **N**. Thus, the natural numbers are an inductively closed set, as claimed. $\square$

Discussion. We point out that in general, if each member of a family of sets possesses a property, say, $\mathcal{P}$, it is not necessarily true their intersection will also inherit the property $\mathcal{P}$. In other words, the 'intersection' of all these sets need not be the smallest set with the property $\mathcal{P}$. For example, the intersection of a collection of nonempty sets may be empty. This is the reason why we need a proof to convince us **N** itself is inductively closed. $\square$

We shall use the standard notations for natural numbers; that is,

$$2 = 1 + 1; \quad 3 = 2 + 1; \quad 4 = 3 + 1; \quad \ldots.$$

Principle of Mathematical Induction: *Let M be any subset of the natural numbers that is inductively closed. Then $M = \mathbf{N}$.*

Proof. By definition, M is one of the sets forming the intersection that is $\mathbf{N}$. It is immediate $\mathbf{N}$ is a subset of M, and so equality follows. $\square$

The remainder of this section brings out the importance of the Principle of Mathematical Induction as an useful proof technique and also as a tool for recursive definitions.

Theorem 0.4.4. *The natural numbers have the following properties:*

(i) *Every natural number is greater than or equal to 1.*

(ii) *For every natural number, n, other than 1, there is a natural number m such that $n = m + 1$.*

(iii) *For every pair of natural numbers n and m, if $m < n$, then $m + 1 \leq n$.*

(iv) *Every nonempty subset of $\mathbf{N}$ has a least element.*

(v) *For every real number x, there exists a natural number n such that $x < n$.*

Proof. For (i), let
$$M = \{n : n \in \mathbf{N} \text{ and } 1 \leq n\}.$$
Observe $1 \in M$, and if $n \in M$, then $n + 1 \in M$, as is easily shown from the order axioms (**HOW?**). Thus, M is inductively closed, so $M = \mathbf{N}$.

For (ii), set
$$M = \{n : n \in \mathbf{N} \text{ .and. } n = 1 \text{ or for some } m \in \mathbf{N}, n = m + 1\}.$$
It is easily checked M is inductively closed, and so equals $\mathbf{N}$. Since (i) is established, it is clear $m \in \mathbf{N}$ implies $1 \neq m + 1$.

For (iii), let
$$M = \{n : n \in \mathbf{N} \text{ .and. for any } m \in \mathbf{N}, m < n \text{ implies } m + 1 \leq n\}.$$

Clearly, $1 \in M$ since there is no $m < 1$, so the statement '$m < n$ implies $m + 1 \leq n$' is always true.

Let $n \in M$ and consider the case for $n + 1$. Let $m \in \mathbf{N}$ and $m < n + 1$. If $m = 1$, then (i) asserts $m = 1 \leq n$, so $m + 1 \leq n + 1$. Otherwise, we have

$$\begin{aligned}
m &< n + 1 \\
m - 1 &< n \qquad \text{by } \mathbf{A4} \text{ and } \mathbf{O3}.
\end{aligned}$$

Note $m - 1 \in \mathbf{N}$ by (ii), since $m \neq 1$. Thus,

$$\begin{aligned}
(m - 1) + 1 &\leq n \qquad \text{since } n \in M; \\
m &\leq n \qquad \text{by } \mathbf{A2} \text{ and } \mathbf{A4}; \\
m + 1 &\leq n + 1 \quad \text{by } \mathbf{O3}.
\end{aligned}$$

Hence, $n + 1 \in M$. Then M is inductively closed and so equals $\mathbf{N}$.

To establish (iv), first note that we may assume $1 \notin H$, the given arbitrary nonempty subset of $\mathbf{N}$; otherwise, the result is established. We set

$$M = \{j : j \in \mathbf{N} \text{ and for all } n \in H, \; j < n\}.$$

By assumption, $1 \in M$. Further, if $j \in M$, and $k < j$ is a natural number, then $k \in M$ (**WHY?**). Evidently, $M \neq \mathbf{N}$, since H is a nonempty subset of $\mathbf{N}$, and $M \cap H = \emptyset$. It follows M is not inductively closed, whence there is an $x \in M$ such that $x+1 \notin M$. Let m witness this fact; that is, let $m \in M$ and $m+1 \notin M$. Then $m + 1 \in H$ (**WHY?**), and in fact is the least element of H, as desired.

For (v), we assume for the sake of argument there is a fixed real number x such that for every natural number n, $n \leq x$. Thus, $\mathbf{N}$ is bounded above, and hence by the supremum principle, $\mathbf{N}$ has a supremum; call it m. By Theorem 0.4.2, taking $\epsilon = \frac{1}{2}$, there is an $n \in \mathbf{N}$ such that

$$m - \frac{1}{2} < n \leq m,$$

whence by arithmetic

$$m < m + \frac{1}{2} < n + 1.$$

Since $n + 1$ is also a natural number, this contradicts our choice of m as the supremum of $\mathbf{N}$. Thus, our initial assumption of the existence of an upper bound for $\mathbf{N}$ was false. $\square$

Discussion. Some of these arguments are rather complicated, so we examine the argument for (iii) in detail. Rather than consider every pair of natural numbers n and m, we use induction upon n. For each n, we prove the statement 'for all m, if $m < n$ then $m + 1 \leq n$.' We define M to be the set of all n for which this statement is true.

To show $1 \in M$, we need only remark that there is no $m < 1$ (by (i)), and hence the condition 'if $m < 1$ then $m + 1 \leq 1$' is true by default.

Then we show $n \in M$ implies $n + 1 \in M$. We take $m < n + 1$ and show $m + 1 \leq n + 1$. We simply subtract 1 to get $m - 1 < n$, use the fact that $n \in M$ to get $m \leq n$, and add 1 to get $m + 1 \leq n + 1$. Because of the way M is defined, we must deal with the case $m = 1$ separately. Since $m - 1 = 0 \notin \mathbf{N}$, we cannot use the fact $n \in M$ to go from $m - 1 < n$ to $m \leq n$.

We have shown M is inductively closed, and by the Principle of Induction, $M = \mathbf{N}$. This completes the proof.

All these proofs have a common feature. We want to prove that a certain property holds for all natural numbers. The strategy is to define a set M that consists of all natural numbers enjoying that property. Once we establish M is inductively closed, we can conclude $M = \mathbf{N}$. The crucial steps are the correct formulation of the set M, and the ability to reason out that M is inductively closed.

For the (**HOW?**), observe from $0 \leq 1 \leq n$ we deduce $1 \leq 1 + 1 \leq n + 1$, whence $1 \leq n + 1$ by transitivity.

The property (iv) of Theorem 0.4.4 is called the **well-ordering property** of natural numbers. For the first (**WHY?**), consider that M is the collection of all natural numbers that are strict lower bounds for H. If $k \notin M$, then there must be $x \in H$ such that $x \leq k$, whence $j \notin M$ contrary to assumption. For the second **(WHY?)**, $m + 1 \in \mathbf{N}$, and there are no natural numbers between m and $m + 1$. Thus, if $m + 1 \notin H$, then $m + 1 \in M$, contrary to our assumption on m.

Property (v) is closely related to the **Archimedean property** that asserts that to every pair of positive real numbers, x and y, there is a natural number n such that $y < nx$ (Exercise 33). The Archimedean property is usually invoked to assert the existence of a natural number n such that $\frac{1}{n} < x$, where x is a fixed small positive number (Exercise 34). $\square$

The properties of the natural numbers stated in the above theorem are sometimes given a slightly different formulation known as **Peano's Axioms**. More precisely, these axioms state the set $\mathbf{N}$ of natural numbers possesses the following properties:

(i) $1 \in \mathbf{N}$;

(ii) Each $n \in \mathbf{N}$ possesses a unique **immediate successor**, $n' \in \mathbf{N}$;

(iii) $n' = m'$ implies $n = m$;

(iv) For each $n \in \mathbf{N}$, the statement '$n' = 1$' is false;

(v) If $M \subseteq \mathbf{N}$ is such that $1 \in S$, and the statement ($n \in M$ implies $n' \in M$) is true for each n, then $M = \mathbf{N}$.

In fact, $n' = n + 1$, in our familiar notation.

Discussion. We want to emphasize here that these are consequences of our axioms and that a key axiom in the development is the axiom of completeness.

The other feature in Theorem 0.4.4 of overwhelming importance is the use of the Principle of Mathematical Induction as a technique for proving theorems. This is also stated in Peano's Axiom (v) above. Suppose we are interested in proving a certain statement $S(n)$ involving natural numbers for each $n \in \mathbf{N}$. If we can verify the truth of $S(1)$, for the initial case, $n = 1$, and further, assuming the truth of $S(n)$, arrive at the truth of $S(n + 1)$ by suitable reasoning, then we have proved the truth of $S(n)$ for each $n \in \mathbf{N}$. Let

$$M = \{n \in \mathbf{N} : S(n) \text{ is true}\}.$$

Clearly, $1 \in M$, since $S(1)$ is true. If $n \in M$, then $S(n)$ is true, so $S(n + 1)$ is also true and, hence, $n + 1 \in M$. Thus, M is an inductively closed subset of $\mathbf{N}$ and so $M = \mathbf{N}$. In other words, $S(n)$ is true for each $n \in \mathbf{N}$. Such a proof is known as **proof by induction**.

An important fact to remember while using induction is that after checking the truth of the statement $S(n)$ for the initial case $n = 1$, we are trying to show the implication '$S(n)$ implies $S(n + 1)$' is always true. To do this, we proceed by assuming the truth of $S(n)$ and from there, derive, or investigate, the truth of $S(n+1)$. If we do not assume the truth of $S(n)$, the implication can never be false, and there is nothing to investigate. There is a common misunderstanding of this fact. When we say 'assume $S(n)$ is true', we are only using it as a tool for investigating the truth of the implication '$S(n)$ implies $S(n+1)$'. We are *not* asserting that $S(n)$ is indeed true. In fact, at this stage of the game, we do not know whether $S(n)$ is true for a specific value of n. The assumption that $S(n)$ is true is called the **induction hypothesis**.

A second point to remember is that we need both of these conditions

(i) $S(1)$ is true, and

(ii) the implication '$S(n)$ implies $S(n + 1)$' is true,

to conclude by mathematical induction that $S(n)$ holds for all n. Omitting one or the other may not prove what we want. These are illustrated in Exercises 18 and 19. Again, we need not start the induction with the initial case $n = 1$. We could, for example, prove a statement for $n \geq 3$, where the initial case to be checked will be for $n = 3$. $\square$

We will illustrate the use of the principle of induction, by proving the Binomial Theorem. But first, we require some definitions.

Definition. Let $n \in \{0\} \cup \mathbf{N}$, the **nonnegative integers,** and x a fixed real number. We define x^n by

(i) if $n = 0$ and $x \neq 0$, then $x^n = x^0 = 1$;

(ii) if $n = 1$, then $x^n = x^1 = x$;

(iii) if $n > 1$, then $x^n = x \cdot x^{n-1}$.

(Note that 0^0 is not defined.)

Definitions such as the one above are referred to as **recursive definitions**. To illustrate their use, let us calculate x^4 from this definition:

$$x^4 = x \cdot x^3 = x \cdot (x \cdot x^2) = x \cdot (x \cdot [x \cdot x]).$$

Definition. For a nonnegative integer, n, we define **n! (n factorial)** of n by

(i) $0! = 1$;

(ii) for $n \geq 1$, $n! = n \cdot (n - 1)!$

We then set
$$\left(\begin{array}{c} n \\ k \end{array} \right) = \frac{n!}{k!(n - k)!},$$
where $0 \leq k \leq n$, and refer to this quantity as the n**th binomial coefficient.** [The quantity $\binom{n}{k}$ is the same as that obtained by counting the number of combinations of n things taken k at a time. Thus, the coefficient is often referred to as 'n-choose-k'.]

In what follows, we shall be using Σ (**sigma**) notation, which is formally defined in Exercise 20.

Theorem 0.4.5 (Binomial Theorem). *For any nonzero x and y in* **R** *and any n in* **N**,

$$(x+y)^n = \sum_{k=0}^{n} \binom{n}{k} x^{n-k} y^k$$

(Note we have indicated the multiplication by juxtaposition.)

Proof. Let x and y be any pair of fixed real numbers. For this fixed pair, let M be the collection of natural numbers for which the theorem is true. We will be done if we can show $M = \mathbf{N}$. Clearly, the simplest way to do this is to show M is inductively closed. Now $1 \in M$ because

$$(x+y)^1 = x + y = \binom{1}{0} x^1 y^0 + \binom{1}{1} x^0 y^1.$$

Assume n is a fixed natural number belonging to M. Now

$$
\begin{aligned}
(x+y)^{n+1} &= (x+y) \cdot (x+y)^n & \text{by definition,} \\
&= (x+y) \cdot \left[\sum_{k=0}^{n} \binom{n}{k} x^{n-k} y^k \right] & \text{since } n \in M, \\
&= x \cdot \left[\sum_{k=0}^{n} \binom{n}{k} x^{n-k} y^k \right] + y \cdot \left[\sum_{k=0}^{n} \binom{n}{k} x^{n-k} y^k \right] & \text{by A6,} \\
&= \sum_{k=0}^{n} \binom{n}{k} x^{n+1-k} y^k + \sum_{k=0}^{n} \binom{n}{k} x^{n-k} y^{k+1}
\end{aligned}
$$

as a result of the generalized distributive law (see Exercise 21). We now want to rename the index variable in the second sum. We call the new index j and set $j = k + 1$. The second term of the sum now becomes

$$\sum_{j=1}^{n+1} \binom{n}{j-1} x^{n-(j-1)} y^j.$$

Notice $n - (j-1) = n + 1 - j$. Now we shift the index again (back from j to k), since we want the same index variable for both terms, this time leaving the range unchanged and changing only the name, to obtain the following formulation of the whole expression:

$$(x+y)^{n+1} = \sum_{k=0}^{n} \binom{n}{k} x^{n+1-k} y^k + \sum_{k=1}^{n+1} \binom{n}{k-1} x^{n+1-k} y^k.$$

Observe that the coefficient of the first term of the first sum satisfies

$$\binom{n}{0} = 1 = \binom{n+1}{0}$$

and the last term in the second sum satisfies

$$\binom{n}{n} = 1 = \binom{n+1}{n+1}.$$

The remaining terms occur in pairs, one for each sum, where the powers of x and y exactly match. Let us pick such a pair, with k arbitrary, but fixed and satisfying $1 \le k \le n$. We have

$$\binom{n}{k} x^{n+1-k}y^k + \binom{n}{k-1}x^{n+1-k}y^k = \left[\binom{n}{k} + \binom{n}{k-1}\right] x^{n+1-k}y^k.$$

Let us compute the coefficient:

$$
\begin{aligned}
\binom{n}{k} + \binom{n}{k-1} &= \frac{n!}{(n-k)!\,k!} + \frac{n!}{(n-(k-1))!\,(k-1)!} \\
&= \frac{n!\,[(n+1-k)!\,(k-1)! \;+\; (n-k)!\,k!]}{k!\,(n-k)!\,(n+1-k)!\,(k-1)!} \\
&= \frac{n!\,[(n+1-k) \;+\; k](n-k)!\,(k-1)!}{k!\,(n-k)!\,(n+1-k)!\,(k-1)!} \\
&= \binom{n+1}{k}.
\end{aligned}
$$

It is immediate from these calculations that

$$(x+y)^{n+1} = \sum_{k=0}^{n+1} \binom{n+1}{k} x^{n+1-k}y^k,$$

as desired. Thus $n+1 \in M$; hence, $M = \mathbf{N}$; and we are done. $\square$

Discussion. Aside from the fact that this is an important result that will be a useful tool in our future discussions, there are several points about the proof worth noting. The first is the use of the set M. Now the theorem is not really one single statement; rather, it is a collection of statements, one for each positive integer n. We use the set M to collect those n's for which the theorem is true (for that particular n). All proofs employing the principle of mathematical induction can be written in this form, and, indeed, the set M is there whether it is explicitly formulated or not. At this stage of your career, it will usually clarify your thinking to explicitly define M and use it appropriately. The other point is that the bulk of the proof is heavy algebraic manipulation. There is no escape from calculations of this type at almost any level in mathematics, and undoubtedly not in analysis. Algebra is certainly the most important tool in these discussions, and the reader would do well to sharpen his ability to manipulate expressions and be prepared to do so without the least fear. $\square$

Exercises

1. Define lower bound and greatest lower bound (infimum) for subsets of $\mathbf{R}$. Restate the Completeness axiom in terms of greatest lower bound, and prove the equivalence of the two forms.

2. Find two upper bounds, two lower bounds, the supremum, and the infimum for each of the following sets of real numbers (if such exist). You may assume Exercises 32–37, and 41, where needed.

(a) $\left\{-1, 4, \frac{1}{2}, \frac{9}{2}\right\}$;

(b) $\left\{-\frac{1}{n} : n \in \mathbf{N}\right\}$;

(c) $\{-\pi\}$;

(d) $\emptyset$;

(e) $\mathbf{Q}$;

(f) $(\mathbf{R} \sim \mathbf{Q}) \cup \mathbf{Z}$; [8]

(g) $\left\{\frac{1+(-1)^n}{2} : n \in \mathbf{N}\right\}$;

(h) $\left\{\frac{n+(-1)^n}{n} : n \in \mathbf{N}\right\}$;

(i) $\left\{n^{(-1)^n} : n \in \mathbf{N}\right\}$;

(j) $\left\{(-1)^n \left(\pi + \frac{1}{n}\right) : n \in \mathbf{N}\right\}$;

(k) $\left\{(-1)^n \left(\frac{1}{4} - \frac{8}{n}\right) : n \in \mathbf{N}\right\}$;

(l) $\{x : x \in \mathbf{Q} \text{ and } x \leq \sqrt{3}\}$;

(m) $\left\{x + \frac{1}{x} : x \in \mathbf{R} \sim \{0\}\right\}$;

(n) $\left\{x + \frac{1}{x} : \frac{1}{2} < x < 2\right\}$;

(o) $\left\{\frac{x}{1+x} : x > -1\right\}$;

(p) $\left\{\frac{3+(-1)^n}{2^{n+1}} : n \in \mathbf{N}\right\}$;

(q) $\left\{\frac{1}{3^n} + \frac{1}{5^{n-1}} : n \in \mathbf{N}\right\}$;

(r) $\left\{\frac{1}{m} + \frac{1}{n} : m, n \in \mathbf{N}\right\}$;

(s) $\left\{\frac{1}{2^m} + \frac{1}{3^n} : m, n \in \mathbf{N}\right\}$;

(t) $\{\sin x \cos x : x \in \mathbf{R}\}$;

(u) $\left\{\frac{1}{1+x^2} : x \in \mathbf{R}\right\}$;

(v) $\left\{\frac{x}{1+x^2} : x \in \mathbf{R}\right\}$;

(w) $\left\{n \sin \frac{n\pi}{2} + \frac{1}{n} \cos n\pi : n \in \mathbf{N}\right\}$[9];

(x) $\left\{\frac{x}{y} + \frac{y}{x} : x, y \in \mathbf{R} \sim \{0\}\right\}$;

(y) $\{x \in \mathbf{R} : (x-a)(x-b)(x-c)(x-d) < 0, \ a < b < c < d\}$;

(z) $\left\{(x+y+z)(\frac{1}{x} + \frac{1}{y} + \frac{1}{z}) : \ x, y, z \in \mathbf{R} \sim \{0\}\right\}$.

3. Supply the proof of the missing part of the 'if and only if' statement in Theorem 0.4.2 by emulating the discussion associated with Figure 0.4.2.

4. Prove the analogue of Theorem 0.4.2 for infimum: let H be a nonempty subset of $\mathbf{F}$ bounded below by $c \in \mathbf{F}$. Then, c is the infimum of H if and only if for every positive $\epsilon > 0$, there exists $x \in H$ satisfying $c \leq x < c + \epsilon$.

5. Fill in the missing details in the proof of Theorem 0.4.4.

6. Complete the algebraic manipulations in the proof of Theorem 0.4.5.

7. Show a nonempty finite subset of $\mathbf{R}$ has a supremum, infimum, greatest member, and a least member. Can you make a similar statement for infinite subsets?

8. Let $A \subseteq \mathbf{R}$. Show if the supremum of A does not exist, then $A = \emptyset$ or A is not bounded above.

9. If A is a nonempty bounded set and $B = \{y : y \text{ is an upper bound for } A\}$, prove $\inf B = \sup A$.

10. If $\sup A = \inf A$, what can you say about A?

11. If M is a nonempty, bounded set such that $\inf M > 0$, show $\sup\{\frac{1}{m} : m \in M\} = \frac{1}{\inf M}$.

12. If A and B are bounded sets and $A \subseteq B$, show $\sup A \leq \sup B$ and $\inf A \geq \inf B$. If $\sup A = \sup B$ and $\inf A = \inf B$, must A and B be identical?

13. If A_i is a family of subsets of $\mathbf{R}$, and if $m_i = \sup A_i$ for each i, show $\sup \cup A_i = \sup\{m_i\}$. Formulate and prove a similar statement for infimums.

14. For $A, B \subseteq \mathbf{R}$, recall the definitions of $-A$, $A + B$ and $A \cdot B$ from Exercise 0.2.17. Which of the following statements are true? Justify your answer.

---

[8] $A \sim B = \{x : x \in A \text{ and } x \notin B\}$.

[9] You may assume any required properties for the trigonometric functions.

(b) $\sup(-A) = -\inf A$;

(c) $\sup(-A) = -\sup A$;

(d) $\sup(A + B) = \sup A + \sup B$;

(e) $\inf(A + B) = \inf A + \inf B$;

(f) $\sup(A \cdot B) = (\sup A) \cdot (\sup B)$;

(g) $\inf(A \cdot B) = (\inf A) \cdot (\inf B)$;

(h) $A \subseteq B$ implies $\sup A \leq \sup B$;

(i) $A \subseteq B$ implies $\inf A \leq \inf B$;

(j) $\sup(A \cap B) \leq \min\{\sup A, \sup B\}$;

(k) $\inf(A \cap B) \geq \max\{\inf A, \inf B\}$.

15. Give two examples of inductively closed subsets of $\mathbf{R}$ different from $\mathbf{N}$ and $\mathbf{Q}$.

16. Use mathematical induction to prove the following statements:

(a) $\frac{1}{1 \cdot 2} + \frac{1}{2 \cdot 3} + \cdots + \frac{1}{n(n+1)} = \frac{n}{n+1}, (n \in \mathbf{N})$;

(b) $1 \cdot 2 \cdot 3 + 2 \cdot 3 \cdot 4 + \cdots + n(n+1)(n+2) = \frac{n(n+1)(n+2)(n+3)}{4}, (n \in \mathbf{N})$;

(c) $10^n - 3^n$ is always a multiple of 7;

(d) (**Bernoulli's inequality**): if $-1 \leq x, 1 + nx \leq (1+x)^n, (n \in \mathbf{N})$;

(e) $(1-x)^n \leq 1 - nx + \frac{n(n-1)}{2}x^2$ for $0 \leq x < 1$ and $n \geq 1$;

(f) $\frac{(2n)!}{(n!)^2} \leq 4^{n-1}$ for $n \geq 5$;

(g) $|\sin nx| \leq n|\sin x|$ for $n \in \mathbf{N}$ (assume standard properties of the sine function.);

(h) $a_1^2 + a_2^2 + \cdots + a_n^2 > 0$ unless $a_1 = a_2 = \cdots = a_n = 0$.

17. For each $n \in \mathbf{N}$, we define a real number a_n by
(i) $a_1 = \sqrt{2}$;
(ii) $a_{n+1} = \sqrt{2 + a_n}, (n \geq 1)$.
Show for every m, $a_m \leq a_{m+1} \leq 2$. (In this exercise, assume the existence of square roots of positive real numbers, with the usual defining property. This assumption is justified by Theorem 0.5.2.)

18. Consider the statement

$$P(n) : 1 \cdot 1! + 2 \cdot 2! + \cdots + n \cdot n! = (n+1)!$$

Show '$P(n)$ implies $P(n+1)$' is always true. Is $P(n)$ true for all n? Justify.

19. What is the fallacy in the following argument by induction?
Statement: All members of every group of n people have the same name.

Proof. Let $S(n)$ denote the given statement. Trivially, for $n = 1$, a group consisting of 1 person has the same name. Assume the truth of $S(n)$. Consider a group consisting of $(n + 1)$ people. Omitting a particular individual, say, X, we have a group of n people, so by induction hypothesis, all have the same name. Also, omitting a different individual Y, and adjoining X, we have a different group of n people, so all have the same name. So X and Y and, hence, all the $(n + 1)$ people have the same name. Thus, $S(n + 1)$ is valid, and hence by induction, $S(n)$ holds for all $n \in \mathbf{N}$. $\square$

20. Let m and n be integers with $m \leq n$ and f a function defined on the integers. Set

$$\sum_{k=m}^{n} f(k) = \begin{cases} f(m), & \text{if } n = m \\ f(n) + \sum_{k=m}^{n-1} f(k), & \text{if } m < n. \end{cases}$$

(a) Show $\sum_{k=1}^{n} a = n \cdot a$.

(b) Show $\sum_{k=1}^{n} k = \frac{n(n+1)}{2}$.

(c) Show $\sum_{k=1}^{n} f(k) = \sum_{k=0}^{n-1} f(k+1) = \sum_{k=1+m}^{n+m} f(k-m)$.

In the remaining exercises, the letters n, m, and k denote natural numbers.

21. Let $x_1, x_2, \ldots, x_n$ be n real numbers. Show for any fixed real number y,

$$y \cdot \sum_{k=1}^{n} x_k = \sum_{k=1}^{n} y \cdot x_k.$$

22. Let x be any fixed real number other than 1. Show

$$\sum_{k=0}^{n} x^k = \frac{1 - x^{n+1}}{1 - x}.$$

23. Show for any n and m and any real x, $x^n \cdot x^m = x^{n+m}$.

24. Let $x \neq 0$ be a real number and $n \in \mathbf{N}$. Define $x^{-n} = \frac{1}{x^n}$. Show $(x^n)^{-1} = (x^{-1})^n$, for every integer, n. Further, show if $x \neq 0$ and $m, n \in \mathbf{Z}$, then $x^m x^n = x^{m+n}$.

25. Prove there is a positive integer m such that $m \leq n$ implies $n^2 < 2^n$. [Hint: Apply the binomial theorem to the expression $(1 + 1)^n$.]

26. Prove there is a positive integer m such that $m \leq n$ implies $n^3 < 2^n$.

27. Prove there is a positive integer m such that $m \leq n$ implies $n^4 < 2^n$.

28. Let x be an arbitrary positive number. Show there is an $m \in \mathbf{N}$ such that $m \leq n$ implies $n^2 < (1 + x)^n$.

29. Prove the sum of two arbitrary positive integers is a positive integer.

30. Prove the product of two positive integers is a positive integer.

31. Define the set of **integers**, $\mathbf{Z}$, by $\mathbf{Z} = \{x : \ x = 0 \text{ or } x \in \mathbf{N} \text{ or } -x \in \mathbf{N}\}$. Prove $\mathbf{Z}$ is closed under sums and products.

32. Define $\mathbf{Q}$, the set of **rational numbers**, by $\mathbf{Q} = \{x : \ x = \frac{m}{n} \text{ for some } m, n \in \mathbf{Z} \text{ such that } n \neq 0\}$.

  (a) Prove $\mathbf{Q}$ is an ordered subfield of $\mathbf{R}$ that is contained in every subfield of $\mathbf{R}$.

  (b) Explain geometrically how you would find a rational number $\frac{m}{n}$ as a point on the real line.

  (c) Show further if we identify the integer $m \in \mathbf{Z}$ with the member $\frac{m}{1} \in \mathbf{Q}$, then we have the inclusions

$$\mathbf{N} \subset \mathbf{Z} \subset \mathbf{Q} \subset \mathbf{R}.$$

  (d) Let $\frac{p}{q} \in \mathbf{Q}$ where $p, q \geq 0$. We say $\frac{p}{q}$ is in **lowest terms** provided p and q have no common positive integer factor other than 1. Show every nonnegative rational number has a unique lowest terms representation. (You may use any required results from elementary number theory to establish this fact.) Further, explain how to extend the notion of lowest terms to negative rational numbers.

33. Prove the real numbers satisfy the **Archimedean Property**: for every pair of positive real numbers, x and y, there is an $n \in \mathbf{N}$ such that $y < n \cdot x$.

34. Prove for every positive real number, x, there is an $n \in \mathbf{N}$ such that $0 < \frac{1}{n} < x$.

35. For a nonempty subset A of $\mathbf{R}$, show the following statements are equivalent:
(i) $a = \sup A$;
(ii) For each $n \in \mathbf{N}$, $a - \frac{1}{n}$ is not an upper bound, but $a + \frac{1}{n}$ is always an upper bound for A.

36. Show for every pair of positive real numbers x and y, there exists an $n \in \mathbf{N}$ such that $0 < \frac{y}{n} < x$.

37. Prove there is a rational number between any two distinct real numbers.

38. Prove if $x \in \mathbf{R}$, then $\sup\{q \in \mathbf{Q} : q < x\} = x$.

39. Given $x \in \mathbf{R}$, prove there is a unique integer $n \in \mathbf{Z}$ satisfying $n \le x < n+1$. (The unique integer so determined is called the **integral part** of x and is denoted by $[x]$.)

40. Prove the sum of two rational numbers is rational. A real number is **irrational** if it is not rational. (For example, $\sqrt{2}$ is irrational; see Theorem 0.5.1.) What can be said about the sum of a rational with an irrational? sum of two irrationals? What about products in all four combinations?

41. Prove between any two distinct real numbers there is an irrational.

42. Prove between any two real numbers a and b, there exist an infinite number of rational numbers and an infinite number of irrational numbers.

43. Assume $x \ge 0$. Prove for each $n \ge 1$, there exists a finite decimal $r_n = x_0 \cdot x_1 x_2 \ldots x_n$, where $r_n \le x < r_n + 10^{-n}$.

44. Prove the Principle of Induction is equivalent to the well-ordering property of $\mathbf{N}$: every nonempty subset of $\mathbf{N}$ has a minimum [statement (iv) of Theorem 0.4.4.].

45. Prove the **second principle of mathematical induction**: Let S be a subset of $\mathbf{N}$ such that for each $n \in \mathbf{N}$, the inclusion $S_n \subseteq S$ implies $n \in S$, where $S_n = \{m : m \in \mathbf{N} \text{ and } m < n\}$. Then $S = \mathbf{N}$.
Use it to show if $a_1 = 1, a_2 = 2$ and for $n \ge 1$, $a_{n+2} = a_n + a_{n+1}$, then $a_n < \left(\frac{7}{4}\right)^n$ for each n.

46. The **Fibonacci sequence** $\{u_n\}$ is defined inductively by $u_1 = u_2 = 1$, and for $n \ge 2, u_n = u_{n-1} + u_{n-2}$. Prove the following:

(a) $\sum_{i=1}^{n} u_i = u_{n+2} - 1$;

(b) $\sum_{i=1}^{n} u_i^2 = u_n u_{n+1}$;

(c) $u_n = \frac{\alpha^n - \beta^n}{\alpha - \beta}$, where α, β are the roots of the quadratic equation $x^2 - x - 1 = 0$;

(d) $\alpha^{n-2} \le u_n \le \alpha^{n-1}$ for all n.

47. In Exercise 20(b), you were asked to prove the sum of the first n positive integers was given by $\frac{n(n+1)}{2}$. Consider the sum of the squares of the first n positive integers; show this sum is given by a polynomial in n of degree 3. [Hint: Assuming the existence of the polynomial, use values for the sum to find its coefficients; then use induction to establish the formula.]

48. Find a formula for the sum of the cubes of the first n positive integers.

49. Find a formula for the sum of the fourth powers of the first n positive integers.

50. Consider the field $\mathbf{F}$ of rational functions described in Exercise 0.3.16.
(a) Prove $\mathbf{F}$ does not possess the Archimedean Property.
(b) Show there is a member of $\mathbf{F}$ that exceeds every member of the form $\frac{n}{1}, n \in \mathbf{N}$.
(c) Exhibit a member $g \in \mathbf{F}, g > 0$ satisfying for all $n \in \mathbf{N}$, $0 < g \le \frac{1}{n}$.

(d) Show **F** does not satisfy the completeness axiom, by constructing a nonempty bounded set without a supremum.

51. Set $t_n = \left(1 + \frac{1}{n}\right)^n$. Use the Binomial Theorem to show

$$t_n = 1 + 1 + \frac{1}{2!}\left(1 - \frac{1}{n}\right) + \frac{1}{3!}\left(1 - \frac{1}{n}\right)\left(1 - \frac{2}{n}\right) + \cdots$$
$$+ \frac{1}{n!}\left(1 - \frac{1}{n}\right)\left(1 - \frac{2}{n}\right)\cdots\left(1 - \frac{n-1}{n}\right).$$

Conclude if $n \geq m$, then

$$t_n \geq 1 + 1 + \frac{1}{2}\left(1 - \frac{1}{n}\right) + \frac{1}{3!}\left(1 - \frac{1}{n}\right)\left(1 - \frac{2}{n}\right) + \cdots$$
$$+ \frac{1}{m!}\left(1 - \frac{1}{n}\right)\left(1 - \frac{2}{n}\right)\cdots\left(1 - \frac{m-1}{n}\right).$$

52. If C_k denotes the coefficient of x^k in the binomial expansion of $(1 + x)^n$, prove the following:

(a) $\sum_{k=0}^{n} C_k = 2^n$;   (b) $\sum_{k=0}^{n}(-1)^k C_k = 0$;

Compute: $\sum_{k=0}^{n} C_k^2$ and $\sum_{k=0}^{n} kC_k$.

53. Obtain the following factorizations:

(a) $x^3 + a^3 = (x + a)(x^2 - xa + a^2)$;

(b) $x^3 - a^3 = (x - a)(x^2 + xa + a^2)$;

(c) $x^4 - a^4 = (x - a)(x + a)(x^2 + a^2)$;

(d) $x^4 + a^4 = (x^2 - \sqrt{2}xa + a^2)(x^2 + \sqrt{2}xa + a^2)$;

(e) $x^n - a^n = (x - a)(x^{n-1} + ax^{n-2} + \cdots + a^{n-1})$;

(f) $x^n + a^n = (x + a)(x^{n-1} - ax^{n-2} + \cdots + a^{n-1})$ if n is odd.

54. Let $\mathbf{N} \times \mathbf{N} = \{(m, n) : m, n \in \mathbf{N}\}$. We prescribe the **dictionary ordering** on $\mathbf{N} \times \mathbf{N}$ (see Exercise 0.3.14). That is, $(k, l) > (m, n)$ if and only if either $k > m$ or else $k = m$ and $l > n$. Prove every nonempty subset of $\mathbf{N} \times \mathbf{N}$ possesses a least member with respect to the dictionary ordering.
Based on the above ordering, enunciate and prove a Principle of Induction on $\mathbf{N} \times \mathbf{N}$.
Mr X. formulates a principle of inductions on $\mathbf{N} \times \mathbf{N}$ as follows: let $S \subseteq \mathbf{N} \times \mathbf{N}$ be such that $(1, 1) \in S$ and $(k, l) \in S$ implies $(k+1, l+1) \in S$. Then $S = \mathbf{N} \times \mathbf{N}$. Compare the two formulations. Will your formulation be equivalent to that of Mr. X?

55. Prove for any nonzero real numbers a, b, c and $n \in \mathbf{N}$,

$$(a + b + c)^n = \sum_{i+j+k=n} \frac{n!}{i!\,j!\,k!}\, a^i b^j c^k, \quad 0 \leq i, j, k \leq n.$$

56. Define the product of n real numbers, $a_1 \cdot a_2 \cdots a_n$ (denoted by $\prod_{i=1}^{n} a_i$) recursively. Prove if $0 < a_i < b_i$, $\prod_{i=1}^{n} a_i < \prod_{i=1}^{n} b_i$.

57. Prove the following identity involving binomial coefficients:

$$\binom{p}{0}\binom{q}{n} + \binom{p}{1}\binom{q}{n-1} + \cdots + \binom{p}{n}\binom{q}{0} = \binom{p+q}{n}.$$

[Hint: Consider $(1 + x)^{p+q}$.]

0.5 Completeness: Further Consequences

In the last section, we used the Completeness Axiom to establish the basic properties of the positive integers–one of the most important being part (v) of Theorem 0.4.4 and its consequence, the Archimedean Property (see Exercises 0.4.33 and 0.4.34). Completeness has two other effects, one **algebraic** and the other **geometric**. It is these effects that we explore now.

The subfield of rational numbers, **Q**, was developed in the last section (see Exercise 0.4.32). It is straightforward that **Q** satisfies the order axioms. We show now the rational numbers are not **algebraically complete**, in the sense that there is an algebraic equation that has no rational number as a solution.

Theorem 0.5.1. *There does not exist a rational number that satisfies the equation*

$$x^2 = 2.$$

Proof. Let us assume, for the sake of argument, that there is a rational number, $\frac{m}{n}$, whose square is 2. First, note that we may take m and n positive and we may also assume m and n have no common factor (the reasons being implicit in Theorem 0.2.4 and the exercises following). The reader should prove (Exercise 3) if $2 \cdot k = i \cdot j$, i, j, and k positive integers, then there is a positive integer k' such that $2 \cdot k' = i$ or $2 \cdot k' = j$.[10] We now proceed. By choice of m and n,

$$\frac{m^2}{n^2} = 2,$$

whence

$$m^2 = 2 \cdot n^2.$$

Evidently, $m \cdot m = 2 \cdot j$, where $j = n^2$. Thus for some positive integer k, we have $m = 2 \cdot k$, so

$$(2k) \cdot (2k) = 2 \cdot n^2,$$

whence by cancellation,

$$2 \cdot k^2 = n^2.$$

It follows for some integer $k', n = 2 \cdot k'$. But this means m and n have a common factor of 2, contrary to the hypothesis! Thus, there can be no rational number whose square is 2. $\square$

Discussion. Geometrically, we say two line segments are **commensurable** provided their lengths are integral multiples of some fixed length d. As a consequence, when a line segment commensurable with the unit length is placed on the real line, with one end coinciding with the origin, the other end always lies on a rational point $\pm\frac{a}{b}$. To say that $\sqrt{2}$ is not a rational number amounts to saying that the diagonal of a square whose side is 1 unit is not commensurable with its side. Thus, if the diagonal is placed on the real line with one end coinciding with 0, the other end will not lie on a rational point. A geometrical proof of the above theorem using this concept can be found in *An Introduction to the Theory of Numbers* by G. H. Hardy and E. M. Wright. $\square$

---

[10]Proof of this fact can be found in any elementary text on number theory.

   The above result can be generalized to show that positive integers that are not already perfect squares, that is, the squares of other integers, do not have rational square roots. If we assume for a moment that there is a real number whose square is 2, then we can establish the connection between algebraic completeness and completeness in the sense of our axiom. Let us define a set

$$B = \{x : x \in \mathbf{Q} \text{ and } x^2 \leq 2\}.$$

The point is B is a nonempty set of rationals that does not have a rational number for a supremum (Exercise 2); thus, $\mathbf{Q}$ does not satisfy the completeness axiom. We will now use the completeness axiom to force the existence of real square roots for positive real numbers.

Theorem 0.5.2. *Let x be a positive real number; then there exists a positive real number y such that $y^2 = x$.*

Proof. Let x be a given positive real number. There are two cases to treat, $x < 1$ and $x \geq 1$; we treat the latter and leave the former to Exercise 4. Since $1 = 1^2$, we assume $x > 1$. Let

$$B = \{z : 0 < z^2 \leq x\}.$$

Evidently, $1 \in B$. Further, note if $z \in B$, then $z \leq x$, since $x < z$ implies

$$x < x^2 < x \cdot z < z^2 \qquad\qquad \text{by the order axioms,}$$

which contradicts $z \in B$. It follows B is a nonempty subset of $\mathbf{R}$ that is bounded above by x. Thus, B has a supremum; call it m. We want to show $m^2 = x$. To see this, suppose first $x < m^2$. Let $a = m^2 - x$. Choose $n \in \mathbf{N}$ such that

$$\frac{1}{n} < \frac{a}{2m}.$$

(**WHY** does such an n **exist?**) Then, $0 < \frac{2m}{n} < a$. Since m is the supremum of B, there exists a $z \in B$ such that

$$m - \frac{1}{n} < z < m; \qquad\qquad\qquad \textbf{(WHY?)}$$

evidently, for such a z, we have

$$x = m^2 - a < m^2 - \frac{2m}{n} < m^2 - \frac{2m}{n} + \frac{1}{n^2} = \left(m - \frac{1}{n}\right)^2 < z^2.$$

This contradicts $z \in B$; thus $m^2 \leq x$. Now suppose, on the other hand, that this last inequality is strict. Let $a = x - m^2$ and choose $n \in \mathbf{N}$ such that

$$\frac{1}{n} < \frac{a}{2m + 1}.$$

For such an n, we have

$$\left(m + \frac{1}{n}\right)^2 = m^2 + \frac{1}{n}\left(2m + \frac{1}{n}\right) \leq m^2 + \frac{1}{n}(2m + 1) < m^2 + a = x,$$

which contradicts the choice of m as $\sup B$. Thus, $m^2 = x$. $\square$

Discussion. The above proof exemplifies the type of calculations that occurs over and over again in elementary analysis. It should be understood that such calculations are not carried out in an *a priori* fashion; rather, once the set B has been written down, our intuition tells us m should have a certain desired property, in this case, $m^2 = x$. However, this intuition must be tested, and the test will in general be an algebraic manipulation employing the previously established facts. These facts must be in the head rather than in the textbook at one's side, if one is to hope for success in these endeavors. Further, one should expect to do considerable experimentation before the right calculation is found. In this case, the key is the recognition that if ϵ is 'small', then the quantity $(w + \epsilon)^2$ should be 'close to' w^2. The calculations only formalize this intuition by finding a particular number, in this case $m - \frac{1}{n}$ that will witness this intuitive belief. To find the correct value for $\frac{1}{n}$, we simply perform suitable algebraic experiments; in the case at hand, we square $\left(m - \frac{1}{n}\right)$ and ask how small does $\frac{1}{n}$ have to be so that the quantity

$$\frac{2m}{n} - \left(\frac{1}{n}\right)^2 < a.$$

When the answer has been determined, we are ready to write out the proof.

For the first **(WHY?)**, recall $\mathbf{N}$ is unbounded, so that given $a > 0$ we can choose $n > \frac{2m}{a}$ and thence $0 < \frac{1}{n} < \frac{a}{2m}$. For the second **(WHY?)**, we apply Theorem 0.4.2. The fact that $z < m$ instead of $z \le m$ follows from the choice of m in respect to x. $\square$

In principle, it might seem if we add to $\mathbf{Q}$ all the real solutions to polynomial equations, then we should have all the real numbers. This turns out not to be the case, since it is known there exist real numbers that do not satisfy any polynomial equation. An **algebraic number** is a real number that is a root of a polynomial equation with integer coefficients (see Exercise 10). Real numbers that are not algebraic numbers are called **transcendental numbers**. It turns out the collection of transcendental numbers is much larger than the collection of algebraic numbers. The numbers $e, \pi, e^\pi, 2^{\sqrt{2}}$ are some examples of transcendental numbers. The study of these numbers is a very fascinating one, since it is usually very difficult to establish the transcendental nature of some familiar numbers. The transcendentality of e, the base for natural logarithms, was demonstrated by Hermite in 1873. That π is transcendental was proved by F. Lindemann in 1882. Using his result, one can show $\frac{\pi}{2}$, $\pi + 1$, and $\sqrt{\pi}$ are transcendental. The first number that was proved transcendental was not e or π, but a number specifically constructed for this purpose by Liouville in 1844. The existence of a vast infinite supply of transcendentals was proved by Cantor in 1874. The difficulty of identifying a given number as algebraic or transcendental is illustrated by the fact that it is not yet known whether the numbers, π^π, $e \cdot \pi$, $e + \pi$ are algebraic or transcendental. The comforting result that any number of the form a^b is transcendental, where a is algebraic number different from 0 or 1, and b is an irrational algebraic number, was recently proved. This result is a culmination of a long effort to prove the so-called **Hilbert number**, $2^{\sqrt{2}}$, is transcendental.

In any case, it is clear completeness has important algebraic consequences. What about geometry?

If we think again of the real line being drawn as in Figure 0.3.1, then if we draw a line which intersects it, it divides the line into two parts that are almost disjoint; namely, those numbers to the left and those to the right (of course, there should be one in the middle). We can formalize this geometric idea as follows:

Definition. A **cut** (or more precisely, a **Dedekind cut**) of the real line is a pair of nonempty subsets L and R of $\mathbf{R}$ such that $L \cup R = \mathbf{R}$ and for every $x \in L$ and $y \in R, x < y$.

In terms of the geometry, L is the left-hand set and R is the right-hand set. Now the issue of substance here is whether there is a real number x that is in the middle. The answer is, of course, yes. But this is not a consequence of the field and order axioms, since it may happen that our field is $\mathbf{Q}$ and our cut 'passes through' $\sqrt{2}$. Thus, some form of the completeness axiom is required. We state now a theorem of Dedekind which is equivalent to our axiom of completeness, but that grows out of an alternative development of the real numbers via the Peano Axioms and Dedekind cuts.

Theorem 0.5.3. *Let L and R define a cut of the real line. Then there is one and only one real number a such that for every $x \in L$ and every $y \in R, x \leq a \leq y$.*

The proof of this theorem is left to Exercise 5. We want to emphasize that the historical development of the real numbers was from the positive integers to the rationals to the full reals, where the irrationals were obtained either as cuts of rationals (Dedekind) or as Cauchy sequences (Cantor) or as suprema of sets of rationals. These ideas are explored somewhat in the exercises but for a full treatment, we refer the reader to *Foundations of Analysis* by E. Landau. Historical perspective on these developments may be obtained by consulting *Mathematical Thought from Ancient to Modern Times* by M. Kline.

Exercises

1. Set $Q_a = \{x : x \in \mathbf{Q} \text{ and } x \leq a\}$, where a is fixed but an arbitrary real number. Prove the following:

    (a) $a = b$ if and only if $Q_a = Q_b$;

    (b) a is the supremum of Q_a;

    (c) when $a = \sqrt{2}$, Q_a has no largest element and $\mathbf{Q} \sim Q_a$ has no least element (Note $\mathbf{Q} \sim Q_a$ denotes the relative complement of $\mathbf{Q}_a$ in $\mathbf{Q}$.);

    (d) if $R_a = \{x : x \in \mathbf{R} \text{ and } x \leq a\}$, then for every $a \in \mathbf{R}$, $\sup R_a = \sup Q_a$.

2. Show the set $B = \{x : x \in \mathbf{Q} \text{ and } x^2 \leq 2\}$ does not possess a supremum in $\mathbf{Q}$, and the set $\mathbf{Q} \sim B$ has no infimum in $\mathbf{Q}$.

3. Supply the missing details in the proof of Theorem 0.5.1.

4. Complete the proof of Theorem 0.5.2.

5. Supply a proof of Theorem 0.5.3.

6. Show for any ordered field, Dedekind's Theorem (Theorem 0.5.3) is equivalent to the Completeness Axiom.

7. A natural number $p \neq 1$ is said to be a **prime number**, if it is divisible only by ± 1 and $\pm p$. Given any prime number p, show there exists a irrational real number x satisfying $x^2 = p$.

8. Prove the **Rational Root Theorem**: If the rational number $\frac{p}{q}$ is a root of the polynomial equation

$$a_n x^n + a_{n-1} x^{n-1} + \cdots + a_1 x + a_0 = 0,$$

where $n \geq 1$, $a_i (i = 1, 2, \ldots, n)$ are integers, and $a_n \neq 0$, then p divides a_0 and q divides a_n.

9. Use Exercise 8 to prove the following real numbers are not rational:

$$\sqrt{6}, (17)^{\frac{1}{5}}, \sqrt{\frac{3 + \sqrt{5}}{7}}, (3 + 5\sqrt{2})^{\frac{1}{3}}, \sqrt{1 + 2\sqrt{2} + 3\sqrt{3}}.$$

10. Prove every rational number is an algebraic number, but not conversely. [Recall an algebraic number is a root of a polynomial equation of the form

$$a_n x^n + a_{n-1} x^{n-1} + \cdots + a_1 x + a_0 = 0,$$

where $n \geq 1$, $a_i (i = 1, 2, \ldots, n)$ are integers and $a_n \neq 0$.] Also, prove the following numbers are algebraic numbers:

$$\sqrt{2 + 5^{\frac{1}{3}}}, \left(\frac{3 - 5\sqrt{6}}{7}\right)^{\frac{1}{3}}, (2 + 3\sqrt{6})^{\frac{2}{3}}, 2 + \sqrt{2} + \sqrt{3} + \sqrt{5}.$$

11. Let $0 \leq y$, a fixed real number. We say a nonnegative real number x is an nth **root** of y, (denoted $y^{\frac{1}{n}}$) provided $x^n = y$. Prove cube and fourth roots exist for any nonnegative real number y. Further show if $1 \leq y$, then

$$y^{\frac{1}{4}} \leq y^{\frac{1}{3}} \leq y^{\frac{1}{2}}.$$

12. State and prove the general result implicit in Exercise 11.

13. Show $0 < a < b$ implies $\sqrt{a} < \sqrt{b}$. State the general result applicable to all roots.

14. Let $x \geq 0$, $x \in \mathbf{R}$. For $m, n \in \mathbf{N}$, $n \neq 0$, define $x^{\frac{m}{n}} = \left(x^{\frac{1}{n}}\right)^m$. Show this definition makes sense for all rational powers of nonnegative real numbers except 0^0. Further show for all $m, n \in \mathbf{N}, n \neq 0$

$$\left(x^{\frac{1}{n}}\right)^m = (x^m)^{\frac{1}{n}}.$$

What can be said about order relations between the various rational powers?

15. If $a, b > 0$ and $\left(\frac{a}{b}\right)^2 < 2$, show
   (i) $2 < \left(\frac{a+2b}{a+b}\right)^2$, and
   (ii) $\left(\frac{a+2b}{a+b}\right)^2 - 2 < 2 - \left(\frac{a}{b}\right)^2$.
   (Thus, if $\frac{a}{b}$ is an approximation for $\sqrt{2}$, then, $\frac{a+2b}{a+b}$ is a better approximation.)

16. Prove Theorem 0.5.2 is false if the word 'real number' is replaced by 'rational number'. Where does the proof break down?

17. Prove there exists a real number $c > 0$ such that for all integers p and q, with $q \neq 0$, we have

$$\left| q\sqrt{2} - p \right| > \frac{c}{q}.$$

More generally, let a be a positive integer and let $\alpha = \sqrt{a}$ be irrational. Show there exists $c > 0$, such that for all integers p and q, with $q > 0$, we have

$$\left| q\alpha - p \right| > \frac{c}{q}.$$

0.6   Absolute Value

We begin this section with a brief treatment of notation. For real numbers a and b, define

$$[a, b] = \{x : a \leq x \leq b \text{ and } x \in \mathbf{R}\} \quad \text{to be the \textbf{closed interval}}$$
$$\text{with endpoints } a \text{ and } b;$$
$$(a, b) = \{x : a < x < b \text{ and } x \in \mathbf{R}\} \quad \text{to be the \textbf{open interval}}$$
$$\text{with endpoints } a \text{ and } b.$$

One can of course use combinations of the two such as $(a, b]$ or $[a, b)$ (called **half-open intervals**, open at the left endpoint a, the right endpoint b, respectively) with their obvious meanings. We also introduce two symbols $+\infty, -\infty$, (called **infinity** and **minus infinity**, respectively) with the property for each real number x, we have the order relation $-\infty < x < \infty$, and make the following abbreviations:

$$
\begin{aligned}
[a, \infty) &= \{x : a \leq x \text{ and } x \in \mathbf{R}\}; \\
(a, \infty) &= \{x : a < x \text{ and } x \in \mathbf{R}\}; \\
(-\infty, b) &= \{x : x < b \text{ and } x \in \mathbf{R}\}; \\
(-\infty, b] &= \{x : x \leq b \text{ and } x \in \mathbf{R}\}; \\
(-\infty, \infty) &= \{x : x \in \mathbf{R}\}.
\end{aligned}
$$

The reader should note ∞ and $-\infty$ are not real numbers, and are used here only for convenience. Thus, at this stage, it is meaningless to talk about expressions like $\infty \pm a, \infty \pm (-\infty), a \cdot \infty, \frac{\infty}{\infty}$, and so forth, where a is any real number.

Definition. Let x be an arbitrary real number. The quantity, **absolute value** of x, denoted by $|x|$, is defined by

$$|x| = \begin{cases} x, & \text{if } 0 \leq x \\ -x, & \text{if } x < 0. \end{cases}$$

Discussion. It is clear from the definition the absolute value is always nonnegative. One should think of the absolute value of x as a measure of the 'size' of x. Another way to think of it is as the **distance** from x to the origin, 0, on the real line. In the remainder of this book, many of our calculations will contain absolute values. Often the problem will be to get rid of the absolute value signs. This can be done most simply be dividing the problem into cases as specified by the definition. $\square$

Example 1. Find the set of all real numbers satisfying $|x - 6| < 3$.

Solution. According to the definition,

$$|x - 6| = \begin{cases} x - 6, & \text{if } 0 \leq x - 6 \\ -(x - 6), & \text{if } x - 6 < 0. \end{cases}$$

We may therefore divide the problem into two cases.

Case 1. $0 \leq x - 6$. Under this assumption we have $6 \leq x$ and the problem reduces to solving $x - 6 < 3$, which is clearly equivalent to $x < 9$. The solution set under case 1 is therefore $[6, 9)$.

Case 2. $x - 6 < 0$. The basic inequality is $-(x - 6) < 3$, which is equivalent to $6 - x < 3$. This is the same as $3 < x$, whence the solution set under this case is $(3, 6)$. Now the solution set for the whole inequality is the union of the two sets found under the two cases, which is $(3, 9)$. $\square$

Definition. For two real numbers x and y, the quantity $|x - y|$ is said to be the **distance between** x and y.

Discussion. As shown in Theorem 0.6.1, the quantity $|x - y|$ has all the right properties for a distance function. Thus, the absolute value is the algebraic tool for treating and manipulating that most fundamental geometric idea: distance. $\square$

Example 2. Give a geometric/graphical interpretation to the inequality $|x-6| < 3$.

Solution. The quantity $|x - 6|$ is the distance between x and 6. Hence, the function defined by

$$f(x) = |x - 6|,$$

which is defined for all $x \in \mathbf{R}$ returns as its value the distance from x to 6. A graph of f is presented in Figure 0.6.1. This distance between x and 6 is required to be less than 3, and in Figure 0.6.1 we have graphed the horizontal line defined by $g(x) = 3$. Now the question posed translates into: for what values of x is $f(x) \leq g(x)$? Inspecting Figure 0.6.1 reveals if $3 < x < 9$, then $f(x) < g(x) = 3$. The solution set can be described geometrically by measuring 3 units to the right and left of 6 to find the bounds, which numerically means compute $6 - 3$ and $6 + 3$. Then the inequality will be satisfied exactly if x is strictly between $6 - 3$ and $6 + 3$. $\square$

Discussion. If we wanted a function that returned distance from a point x on the x-axis to the fixed point a on the x-axis, what properties would it have to have? We would measure the distance from x to a and then plot as the function value the point having first coordinate x and second coordinate the measured

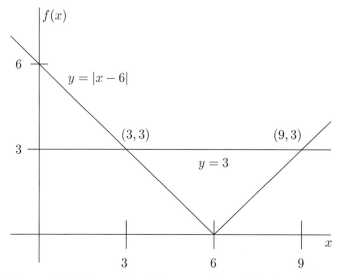

Figure 0.6.1 A graphical interpretation of the inequality, $|x - 6| < 3$. As discussed in the text, the solution set is $\{x : \ x \in \mathbf{R} \text{ and } 3 < x < 9\}$.

valued for the distance. Numerically, except for sign, the quantity $x - a$ gives the distance. We can ensure the value is nonnegative by taking the absolute value. This produces a graph with only one zero, namely, at $x = a$, which rises with a slope of 1 as x moves away from a to the right and rises with a slope of -1 as x moves away from a to the left. These facts are illustrated in the special case of $a = 6$ presented in Figure 0.6.1.

If we take any other number b then the solution set for the inequality $|x-a| < b$ will be the portion of the graph of $|x - a|$ that is strictly below the horizontal line defined by $y = b$. The boundaries of the solution are identified by the x coordinates of the point of intersection of the two graphs. All of these facts again are illustrated in Figure 0.6.1.

It is immediate that graphs and geometry can lead us directly to the solution set. Indeed, the pictures makes the solution so simple, one wonders why we would bother with the original solution at all! The answer is that while the geometry can make it transparent what the solution set ought to be, the argument based on geometry does not constitute a proof that the purported solution set is in fact the solution set. Only a sequence of statements, each of which is an axiom or a previously established theorem, can constitute a proof. *Statements about pictures cannot be part of proofs.* But they can be at the heart of our intuition. Thus, geometry must be used to show us plausible directions in which to proceed, but the rigor of proof must always be used to keep us from false paths. $\square$

We now establish the basic properties of distance.

Theorem 0.6.1. *Let x and y be arbitrary real numbers. Then*

(i) $|x - y| = 0$ *if and only if $x = y$; otherwise, $|x - y| > 0$;*

(ii) $|x - y| = |y - x|$;

(iii) *for any* z, $|x - y| \le |x - z| + |z - y|$.

Proof. By definition of absolute value, $0 \le |x - y|$. Evidently, this quantity is 0 exactly when $x = y$. For (ii), we may assume without loss of generality $x - y < 0$. But $-(x - y) = y - x > 0$, whence

$$|x - y| = -(x - y) = y - x = |y - x|$$

as desired. Now for (iii), fix an arbitrary z and consider

$$|x - y| = |(x - z) + (z - y)|.$$

We would like to split the term on the right in the obvious way while achieving the desired inequality. But the fact that

$$|(x - z) + (z - y)| \le |x - z| + |z - y|$$

is an immediate consequence of our next more general theorem. $\square$

Discussion. Part (iii) of this theorem is known as the **Triangle inequality**. The reason for this is the inequality has the geometric interpretation that says the distance between two points is less than or equal to the sum of the distances of the two points to any third point. $\square$

Theorem 0.6.2. *For any pair of real numbers* x *and* y, $|x + y| \le |x| + |y|$.

Proof. Fix x and y two arbitrary real numbers. We may assume $x \le y$. Now if $0 \le x$, or $y \le 0$, then we have equality between the two terms. Thus, the situation of interest is when $x < 0 \le y$. There are two possible cases under this assumption: $-x \le y$ and $y < -x$. For the former, we have $x < -x$, whence

$$0 \le x + y < -x + y = |x| + |y|.$$

Since $x + y$ is nonnegative, we are done under this assumption. The other case follows by observing

$$|x + y| = |-x + (-y)|$$

and using the fact just obtained. $\square$

Discussion. An alternative algebraic proof of the above follows from comparing $(|x + y|)^2$ with $(|x| + |y|)^2$ and using $|x|^2 = x^2$. We will ask the reader to explore this in Exercise 6. We have used case analysis in the above because such analysis can often be useful. $\square$

Theorem 0.6.3. *For any real numbers* x *and* y, $||x| - |y|| \le |x \pm y|$.

Theorem 0.6.4. *Let* x *and* y *be real numbers with* $0 < x$. *Then,*

$$|y| < x \text{ .if and only if. } -x < y < x.$$

The proofs of these two theorems are left to Exercises 1 and 2. Before closing this section we would like to make several comments on the methods for dealing with inequalities. Theorem 0.6.2 is an essential fact that will be employed repeatedly throughout this book. In many instances it will be used in exactly the manner of Theorem 0.6.1; namely, a situation will be created in which a number z is added and subtracted inside an absolute value and an inequality deduced:

$$|(x - z) + (z - y)| \leq |x - z| + |z - y|.$$

Of course, the 'z' that is added and subtracted is not chosen at random, but is determined by the situation. The point is the reader should be prepared for this to happen and ready with the question: Why that particular choice of 'z'?

Last, we remark on the content of Theorem 0.6.1. What is it telling us about distance? It says the distance between distinct points is positive, the distance between two points is independent of the direction in which it is measured, and finally the distance between two points is not more than the sum of the distances from the two to any third point. All these are what is required by experience!

Exercises

1. Prove Theorem 0.6.3.

2. Prove Theorem 0.6.4.

3. Generalize Theorem 0.6.2 to arbitrary finite sums of real numbers: For $n \in \mathbf{N}$ show $|\sum_{k=1}^{n} a_k| \leq \sum_{k=1}^{n} |a_k|$, where $a_1, a_2, \ldots, a_n$ are any real numbers.

4. Describe the following subsets of $\mathbf{R}$, their suprema and infima and, where appropriate, discuss the geometry of the inequality. The domain of n is the set of natural numbers.

   (a) $\{x : |x + 3| < 4\}$; (b) $\{x : |x^2 - 2| < 4\}$; (c) $\{x : |1 - 2x| < x + 1\}$;

   (d) $\{x : \left|\frac{x+2}{x-2}\right| < x\}$; (e) $\{x : |x - 1| < \frac{1}{x}\}$; (f) $\{x : \text{ for all } n, x < \frac{1}{n}\}$;

   (g) $\{x : x = \frac{n+1}{n}\}$; (h) $\{x : |5 - 2x| < |2x|\}$; (i) $\{x : |x + 2| + |3 - x| < 4\}$;

   (j) $\{x : |x^3| < 3\}$; (k) $\{x : \frac{1}{x + |x - 1|} < 2\}$.

5. Show $|x - a| < \epsilon$ .if and only if. $a - \epsilon < x < a + \epsilon$.

6. Give an algebraic proof of Theorem 0.6.2 along the lines indicated in the discussion.

7. Prove for all x, $-|x| \leq x \leq |x|$ and use this to establish Theorem 0.6.2.

8. Find the general solution set in $\mathbf{R}$ for the following inequalities ($a, b, c, d \in \mathbf{R}$):

   (a) $|ax + b| < c$; (b) $|ax + b| > c$;

   (c) $|x^2 - (a + b)x + ab| < c$; (d) $|x^2 + (a + b)x + ab| \geq c$;

   (e) $|(x - a)(x - b)(x - c)| \leq d$; (f) $|(x + a)(x - b)(x + c)| > d$.

9. For each of the subsets of real numbers in Exercise 4, there is a subset of rationals obtained by adding the condition $x \in \mathbf{Q}$. Repeat Exercise 4 for these sets.

10. Establish the following inequalities for any real numbers x, y, and z:

    (a) $|x \cdot y| = |x| \cdot |y|$; (b) $\left|\frac{x}{y}\right| = \frac{|x|}{|y|}$, if $y \neq 0$;

    (c) $\sqrt{x^2 + y^2} \leq x + y$, provided $x, y \geq 0$; (d) $\frac{|x|}{1 + |x|} + \frac{|y|}{1 + |y|} \geq \frac{|x+y|}{1 + |x+y|}$;

    (e) $\left|\frac{x}{y} + \frac{y}{x}\right| \geq 2$ provided $x, y \neq 0$; (f) $|x + y| + |x - y| \geq |x| + |y|$;

    (g) $2|xy| \leq z^2 a^2 + \frac{1}{z^2 b^2}$, if $z > 0$;

(h) $|x| + |y| + |z| + |x + y + z| \geq |x + y| + |y + z| + |z + x|$.

11. State and prove a result for products that is similar to that in Exercise 3.

12. Show any open interval of the form (a, b) can be expressed as $\{x : |x - c| < d\}$, where c and d are fixed real numbers determined by a and b.

13. Show a subset $S \subseteq \mathbf{R}$ is bounded if and only if there exists a real number $K > 0$, such that $|s| < K$ for all $s \in S$. Formulate a similar criterion for unboundedness.

14. By an **interval** I we mean a set of real numbers having the property that whenever $a, b \in I$, with $a < b$, then $(a, b) \subseteq I$. Show an interval must have one of the following forms:

$$[a, b], (a, b], [a, b), (a, b), (-\infty, a], (-\infty, a), [a, \infty), (a, \infty) \text{ or } \mathbf{R}.$$

15. Let x and y be arbitrary elements of an ordered field. The **maximum** and the **minimum** of the elements x and y are defined by

$$\max\{x, y\} = \begin{cases} x, & \text{if } y \leq x \\ y, & \text{if } x < y; \end{cases}$$

$$\min\{x, y\} = \begin{cases} x, & \text{if } x \leq y \\ y, & \text{if } y < x. \end{cases}$$

Prove the following:

(a) $\max\{x, -x\} = |x|$;

(b) $\min\{x, -x\} = -|x|$;

(c) $\max\{x, y\} = \frac{x + y + |x - y|}{2}$;

(d) $-\max\{x, y\} = \min\{-x, -y\}$;

(e) $\min\{x, y\} = \frac{x + y - |x - y|}{2}$;

(f) If $x < y$, $-\max\{|x|, |y|\} \leq x < y \leq \max\{|x|, |y|\}$.

16. The **positive part** x^+ of x is defined to be $\max\{0, x\}$ and the **negative part** x^- is $\min\{0, x\}$. Show $x = x^+ + x^-$ and $|x| = x^+ - x^-$.

17. The **signum** of a real number x is defined by

$$\operatorname{sgn} x = \begin{cases} 1, & x > 0 \\ 0, & x = 0 \\ -1, & x < 0. \end{cases}$$

Prove the following:

(a) $|x| = x \cdot \operatorname{sgn} x$;

(b) $\operatorname{sgn} x \cdot \operatorname{sgn} y = \operatorname{sgn}(xy)$;

(c) $\frac{\operatorname{sgn} x}{\operatorname{sgn} y} = \operatorname{sgn} \left\{ \frac{x}{y} \right\}$ if $y \neq 0$.

18. If A is a bounded set of real numbers and if $|a - b| < 1$ for all $a, b \in A$, show $\sup A - \inf A \leq 1$.

Chapter 1

Limits of Sequences

Our purpose now is to begin the careful study of analysis. At this level, we will be following the paths of many eighteenth- and nineteenth- century mathematicians who concerned themselves with trying to understand such concepts as limit, continuity, and derivative. Their efforts were hampered by the fact that they had an imperfect understanding of the heart of the subject, namely, the structure of the real number system. In fact, it was the existence of the problems associated with trying to correctly define the fundamental concepts that forced mathematicians to come to terms with the structure of the real numbers, and it was only after a deep understanding of the properties of real numbers had been achieved that a full treatment of analysis was obtained. For us, Chapter 0 contains the prerequisite knowledge of the real number system; the reader would do well to remember that the essential facts required in the sequel are to be found in our work on the fundamental properties of real numbers.

1.1   Sequences

The basic tool of analysis is the notion of a 'limit', and the simplest form of limit results from applying the concept to sequences.

Definition. A **sequence** is a function from the set $\mathbf{N}$ of all positive integers to a subset of the real numbers.

Discussion. A short treatment of the basic tools of set theory, including the definitions of ordered pair, function, domain, and range, may be found in the appendix. A reader who is not completely familiar with these should study the appendix with care.

When specifying sequences, instead of the usual functional notation, as for example,

$$\{(n, f(n)) : n \in \mathbf{N}\},$$

we will write

$$\{a_n\}_{n=1}^{\infty} \qquad \text{or, more simply,} \qquad \{a_n\},$$

where $a_n = f(n)$ is the value of the function at the natural number $n \in \mathbf{N}$. We will, therefore, often speak of the sequence

$$a_n = \text{some given functional expression in } n,$$

where this equation defines the nth **term** of the sequence or the term having **subscript** n. Thus, for example,

$$\left\{ \left(n, \frac{2n}{n^2 + 1} \right) : n \in \mathbf{N} \right\}, \qquad \left\{ \frac{2n}{n^2 + 1} \right\}_{n=1}^{\infty}, \qquad \left\{ \frac{2n}{n^2 + 1} \right\},$$

and

$$\text{the sequence whose } n\text{th term is given by } a_n = \frac{2n}{n^2 + 1}$$

are all notations specifying the same sequence, or function $f : \mathbf{N} \to \mathbf{R}$. In general, we prefer the simplest notation, namely, $\{a_n\}$, and will use this to specify sequences where no confusion will arise. Because sequences can be completely specified by defining the nth term, a sequence can also be easily defined recursively. That is, the first term a_1 is specified, and using induction, the term a_{n+1} is described using the knowledge of the previous term a_n (or perhaps, using some a_k, for $k \leq n$). For example, a sequence $\{a_n\}_{n=0}^{\infty}$ could be specified by the following recurrence rule:

$$a_0 = a_1 = 1, \ a_2 = 2, \text{ and for } n \geq 0, \ a_{n+3} = 2a_n - a_{n+1} + 3a_{n+2}.$$

The Fibonacci sequence in Exercise 0.4.46 is an example of a sequence that is defined recursively.

The reader will notice that this notation emphasizes our interest in the range of the function, rather than in the function itself. More particularly, the notation $\{a_n\}_{n=1}^{\infty}$ calls to mind moving through the discrete values of the range in a particular order, namely, according to the natural order of the integers. This notion will be a focus of our interest. $\square$

The next definition further establishes our identification of a sequence with the range of a function.

Definition. A sequence, $\{a_n\}_{n=1}^{\infty}$, is said to be **bounded** provided the set $\{a_n : n \in \mathbf{N}\}$ is bounded; that is, there exists a real number B such that for all $n \in \mathbf{N}$, $|a_n| \leq B$. A sequence that is not bounded is said to be **unbounded**.

Discussion. Notice that there is no mention of the domain of the function, which is obviously not a bounded set. The focus is on the range of the function, and the definition merely requires one to check whether the range is a bounded set. As the reader can see, a sequence $\{a_n\}$ is unbounded if for any given $K > 0$, there exists $n \in \mathbf{N}$ satisfying $|a_n| > K$. $\square$

Example 1. Graph the sequence whose nth term is given by $a_n = \frac{1}{n}$. Show this sequence is bounded.

Solution. A graph is shown in Figure 1.1.1(a). Alternately, we may draw a one-dimensional graph of the sequence as shown in Figure 1.1.1(b). If $x \in \{a_n : n \in \mathbf{N}\}$, then for some $n \in \mathbf{N}$, $x = \frac{1}{n}$. From the facts established in Chapter 0, we have $0 < \frac{1}{n} \leq 1$, whence $\{a_n\}$ is bounded. $\square$

Discussion. Consider Figure 1.1.1(a). If we allow our eye to move consecutively from $(1, 1)$ to $(2, \frac{1}{2})$ to $(3, \frac{1}{3})$ to $\cdots$, it becomes apparent that the points of the

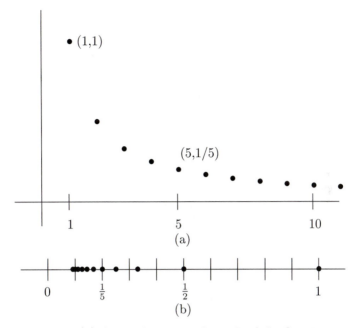

Figure 1.1.1(a) A two-dimensional graph of the first 11 terms of the sequence $\{\frac{1}{n}\}$.
(b) A one-dimensional plot of the same 11 terms of the sequence.

graph are getting closer and closer to the x-axis. Alternatively, consider the one-dimensional graph of the values of the sequence as shown in Figure 1.1.1(b). If we let our eye travel along the axis following the natural order in which the values of the sequence are formed, that is, starting on the right at 1, then continuing $\frac{1}{2}$, $\frac{1}{3}$, $\frac{1}{4}$, $\cdots$, and so forth, it is again apparent that the values of the sequence are approaching 0. Intuitively, we might think of the values of the sequence as a collection of stepping-stones forming a path that we are required to trace in a certain order. The issue is: Is there a fixed number to which this path leads? For the case above, the picture tells us that the path leads toward 0. However, there is a gap, so we are left with the question: Does the sequence get to 0? To be able to answer, yes, we would expect further terms of the sequence to fill in the gap in Figure 1.1.1(b), and this is indeed the case as the reader can check. □

The struggle to capture this notion of 'approaching to the limit' occupied many mathematicians in the nineteenth century. Their struggle yielded:

Definition. Let $\{a_n\}$ be a sequence of real numbers. We say that $\{a_n\}$ **has a limit** in case there exists a real number A, called the **limit** of the sequence, such that, for every positive real number ϵ, there exists a real number M such that

$$\text{if } n > M, \text{ then } |a_n - A| < \epsilon.$$

Discussion. The first thing that this definition does is to assert the existence of a number A. This number remains fixed for the rest of the definition. Thus, if a particular sequence is to satisfy the definition, one must try to find an appropriate candidate for A. The reader is likely familiar with many techniques for finding A from earlier courses in calculus and should feel free to employ these techniques to find the A's required. The most obvious such technique is simply to calculate various terms of the sequence and to see if they appear to be approaching a fixed number. Having found A, the goal becomes one of proving that the likely candidate does indeed satisfy the definition. The next number mentioned in the definition is ϵ (epsilon). Epsilon is an arbitrary positive real number. The purpose of ϵ is to test whether the sequence is eventually close to A and, further, to see if the sequence stays close to A. For a_n to be close to A, we would expect that $|a_n - A|$, which measures the distance from a_n to A, to be a small number; thus, if ϵ is going to be our test number, we would want ϵ to be small, in fact, very small. The way that ϵ tests whether a_n is close to A is by requiring us to find an M such that all the terms whose subscript is larger than M are within a distance of ϵ from A. Since terms with small subscripts are not likely to be very close to A, we would expect M to be a large number, and this will in general be the case. Just how large M will have to be depends on how small ϵ is. As a general rule, we can say that the smaller ϵ is, the larger M will have to be for the definition to be satisfied. Thus, the number M depends on ϵ. We can emphasize this fact (as some authors do) by writing $M = M(\epsilon)$, to stress that M is a function of ϵ and that as ϵ changes, the corresponding M will almost always also change.

We have allowed M to be any real number. There would be no loss of generality if M were required to be a positive integer because given $M \in \mathbf{R}$ there is an $M_1 \in \mathbf{N}$ such that $M < M_1$. In many cases we will require $M \in \mathbf{N}$ because we want to use M as a subscript to identify a particular term of the sequence.

Figure 1.1.2 illustrates the limit concept graphically. Let an arbitrary $\epsilon > 0$ be given and consider the horizontal strip of width 2ϵ generated by the lines, $y = A - \epsilon$ and $y = A + \epsilon$. A given term, a_n, of the sequence, $\{a_n\}$, lies inside this strip exactly if the inequality $|a_n - A| < \epsilon$ is satisfied. Thus, for the number, A, to be the limit of the sequence, we must be able to specify a point, M, on the x-axis, such that for all n lying to the right of M, the corresponding term, a_n, gets trapped within the horizontal strip. Figure 1.1.2 is two-dimensional. The reader should develop a one-dimensional analogue for Figure 1.1.2 (refer ro Exercise 1). $\square$

Notation. If $\{a_n\}$ is a sequence having the number A as a limit, we write

$$\lim_{n \to \infty} a_n = A$$

(or simply $\lim a_n = A$) and say **the limit as n tends to infinity** of a_n is A (or the sequence $\{a_n\}$ **converges** to the limit A).

Example 2. Show $\displaystyle\lim_{n \to \infty} \frac{n}{n+1} = 1$.

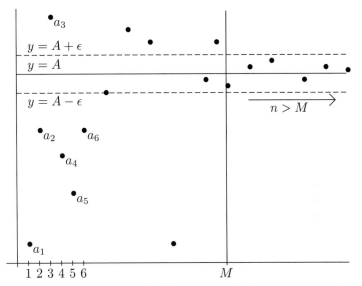

Figure 1.1.2 A geometric illustration of $\lim_{n \to \infty} a_n = A$. For $n > M$ all terms lie between $A - \epsilon$ and $A + \epsilon$.

Solution. Let $\epsilon > 0$ be fixed, but arbitrary. Then,

$$\left| \frac{n}{n+1} - 1 \right| = \left| \frac{n - (n+1)}{n+1} \right|$$

$$= \left| \frac{-1}{n+1} \right|$$

$$= \frac{1}{n+1}$$

by definition of the absolute value. Now, there is a positive integer M, such that $0 < \frac{1}{M} < \epsilon$. Fix M with this property. If $n > M$, then

$$\left| \frac{n}{n+1} - 1 \right| = \frac{1}{n+1} < \frac{1}{n} < \frac{1}{M} < \epsilon. \qquad \square$$

Discussion. The heart of the computation is finding M, once ϵ has been given. That an M with the desired properties exists was the subject of Exercise 0.4.34, which states a fundamental property of the real numbers, namely, that for any positive x, there is a positive integer N such that $0 < \frac{1}{N} < x$. Further, all the computations, together with the inequalities have also been established in the various exercises of Chapter 0. The example above is highly simplistic; however, it does illustrate the strategy involved in this type of problem. To begin with, the expression $|a_n - A|$ is simplified by algebraic means. The simplified expression is then replaced by other larger expressions, until one is found that can easily be made less than ϵ, provided only that n is sufficiently large. Figure 1.1.3 captures the geometry behind this example.

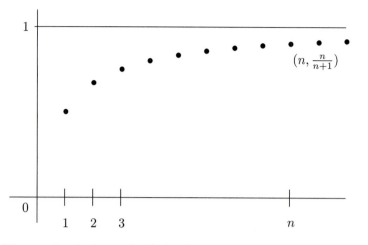

Figure 1.1.3 A graph of the first 11 terms of the sequence $\{\frac{n}{n+1}\}$ illustrating that the values increase to a limit of 1.

In this example, the number A, which is the limit of the sequence, was given. Had it not been, it would have to be found. One way of doing this is to calculate various terms of the sequence. For example, $a_{10} = \frac{10}{11}$, while $a_{100} = \frac{100}{101}$, and so forth. Alternatively, a little judicious algebra is also a useful tool, as we will continually reiterate throughout this book. In the present case, one observes that

$$\frac{n}{n+1} = \frac{\frac{n}{n}}{\frac{n}{n} + \frac{1}{n}} = \frac{1}{1 + \frac{1}{n}}.$$

Since the quantity $\frac{1}{n}$ becomes arbitrarily small as n becomes arbitrarily large, it is the limit must be 1. $\square$

Example 3. Show $\lim\limits_{n \to \infty} (\sqrt{n+1} - \sqrt{n}) = 0$.

Solution. Fix $\epsilon > 0$. Then

$$|\sqrt{n+1} - \sqrt{n}| = (\sqrt{n+1} - \sqrt{n}) \cdot \left(\frac{\sqrt{n+1} + \sqrt{n}}{\sqrt{n+1} + \sqrt{n}}\right)$$

$$= \frac{1}{\sqrt{n+1} + \sqrt{n}},$$

where the absolute value signs may be dispensed with, since n is a positive integer, whence all quantities are positive. It is straightforward that the last quantity is in fact less than $\frac{1}{2\sqrt{n}}$. Let $M = \frac{1}{\epsilon^2}$. Evidently, if $n > M$, then

$$\frac{1}{\epsilon} = \sqrt{M} < \sqrt{n} < 2\sqrt{n}.$$

Thus, for such an M and n, $|\sqrt{n+1} - \sqrt{n}| < \epsilon$, as desired. $\square$

Discussion. Again, note the strategy: use algebra to simplify the absolute value of the difference; replace the simplified quantity by another, larger quantity, which can easily be seen to be less than ϵ when n is sufficiently large. Evidently, the heart of the matter is an algebraic calculation; namely, we rationalize the numerator to get rid of the square roots.

The reader may wonder about the use of ϵ^2. The problem is that we must determine how large to choose M. If we start with

$$\frac{1}{2\sqrt{M}} < \epsilon,$$

this is equivalent to

$$\frac{1}{\epsilon} < 2\sqrt{M},$$

which in turn is equivalent to

$$\frac{1}{\epsilon^2} < 4M.$$

In the argument above, we have omitted the coefficient 4, since if the inequality is satisfied without the term 4, it will certainly be satisfied when the term 4 is present.

This same calculation, rationalizing, also finds the limit, had it not been given, since it is intuitively clear that $\frac{1}{\sqrt{n+1}+\sqrt{n}}$ must have 0 as a limit. $\square$

One might think that all sequences have limits. However, a little thought will convince one that a sequence that gets large without bound cannot possibly have a limit. Once one knows that some sequences have limits while others do not, it becomes clear that it would be a good idea to understand how a sequence can fail to have a limit, or **diverge**. As a first step in the process, we take the definition of **convergent** sequence (one having a limit) and form its negation.

Negation of the Limit Definition. A sequence $\{a_n\}$ does not have a limit if for every real number A, there exists a positive ϵ such that for every $M \in \mathbf{R}$, there exists an $n > M$ with $|a_n - A| \geq \epsilon$.

Discussion. The negation tells us that we must show that no real number can be the limit. Thus, an argument showing a sequence does not have a limit will start by picking an arbitrary real number that is then fixed and shown ultimately not to be the limit. Next, an ϵ is found that will witness the fact that the sequence does not *stay* close to this previously picked A. We emphasize *stay*. Some terms of the sequence may get closer and closer to A, but they cannot all stay close to A after a certain point. Thus, no matter how large M is taken, we must always find terms with subscript exceeding M whose distance from A is at least as large as ϵ. It seems intuitively clear that a sequence such as $\{n^2\}$ will not have a limit, because no matter what our concept of infinity is, the terms of this sequence are clearly headed there in a hurry. We consider two examples, now, that illustrate the two reasons why a sequence can fail to have a limit. $\square$

Example 4. Show the sequence $\{n^2\}$ has no limit.

Solution. Let A be a fixed but arbitrary real number. Let $\epsilon = 1$, and set

$$K = |A| + \epsilon = |A| + 1.$$

If $n \geq K$, then

$$
\begin{aligned}
|n^2 - A| &\geq \left| |n^2| - |A| \right| & \text{by Theorem 0.6.3} \\
&\geq \left| (|A| + 1) - |A| \right| & \textbf{(WHY?)} \\
&= 1.
\end{aligned}
$$

We have therefore shown that every term of sufficiently large subscript, $n \geq \max\{K, M\}$ for any choice of M, is not within ϵ distance of A. $\square$

Discussion. The terms of the sequence $\{n^2\}$ obviously get large without bound, so this sequence is unbounded. With this intuitive fact in our minds we set about trying to show that the negation of the limit definition is satisfied. From the negation, we see that the candidate for the limit, A, is arbitrary, which means that we can make no special assumptions about it. We can, however, choose ϵ, subject only to the requirement that ϵ is positive. In the solution, we have taken ϵ to be 1. It happens to be the case that for a sequence that gets large without bound, any positive real number will do as a choice for ϵ. (To see this, the reader should try carrying out the argument employing $\epsilon = 100$, or some other large number.) As stated, we have no control over A. This fact is accounted for in the argument by making the choice of K depend on the value of A, as well as the value of ϵ. So n must exceed both M and K, that is, the maximum, $\max\{M, K\}$.

The argument for unbounded sequences is inherently simpler than an argument showing that a bounded sequence does not have a limit, and this example should be thoroughly understood as a prerequisite to the next case. It also seems apparent that we should expect convergent sequences to be bounded.

Last, consider the **(WHY?)**. The observation required runs as follows. Because $n \in \mathbf{N}$,

$$|n^2| = n^2 \geq n \geq K = |A| + 1 > |A|,$$

whence

$$|n^2| - |A| \geq (|A| + 1) - |A| \geq 1,$$

where the last inequality permits the removal of the absolute value signs. $\square$

Example 5. Show the sequence $\left\{ (-1)^n \left(\frac{1}{2} - \frac{1}{n} \right) \right\}$ has no limit.

Solution. Fix A. Let $\epsilon = \frac{1}{3}$ and M be given. Set $n \geq \max\{M, K\}$, where $K = 6$. There are two possibilities concerning A: that A is nonnegative or that A is negative. Let us assume the former and also that $n = 2k+1$ for some $k \in \mathbf{N}$. Then,

$$
\begin{aligned}
\left| (-1)^n \left(\frac{1}{2} - \frac{1}{n} \right) - A \right| &= \left| (-1) \left[\left(\frac{1}{2} - \frac{1}{n} \right) + A \right] \right| \\
&\geq \left| \frac{1}{2} - \frac{1}{n} \right| & \textbf{(WHY?)} \\
&\geq \left| \frac{1}{2} - \frac{1}{6} \right| = \frac{1}{3}.
\end{aligned}
$$

The case when A is negative is treated similarly under the assumption that $n = 2k$ for some positive integer k (Exercise 2). □

Discussion. This sequence is bounded. To understand the intuitive reason this bounded sequence does not have a limit, we notice that there are two distinct numbers to which the terms of the sequence get close; namely, the odd terms are near $-\frac{1}{2}$, and the even terms are near $\frac{1}{2}$ (see Figure 1.1.4). (As well, it would be instructive to the reader to create a table of the odd and even terms for this sequence.) Notice also that the odd terms get as close as we please to $-\frac{1}{2}$ and in fact stay that close, while the even terms get as close as we please to $\frac{1}{2}$ and also stay that close. Thus, this sequence seems to 'wander' or 'oscillate'. A bounded sequence that has no limit will always 'wander' or 'oscillate' in the sense that there will be at least two distinct numbers to which its terms are sometimes close. Once these two numbers are known, a proof of divergence of the type given is easily constructed. We simply choose ϵ to be less than half the absolute value of the difference of the two numbers (in this case, $\frac{1}{3}$ is less than $\frac{1}{2} \cdot 1$) and use the fact that no matter what value of A is chosen, this value cannot be arbitrarily close to both the numbers simultaneously. This permits us to find a suitably large K with the property that for selected $n > K$, a_n will be close to one of the two numbers, and hence relatively far from the other (see Figure 1.1.4). In the example, after experimentation, it is found $K = 6$ will do the job. This strategy is important because it will work for all bounded sequences that have no limit.

The **(WHY?)** requires an explanation of how to eliminate the (-1) and the A from the previous term. The (-1) can be deleted because $|(-a)b| = |-a| \cdot |b|$; A can be deleted because both it and the quantity $\frac{1}{2} - \frac{1}{n}$ are positive. □

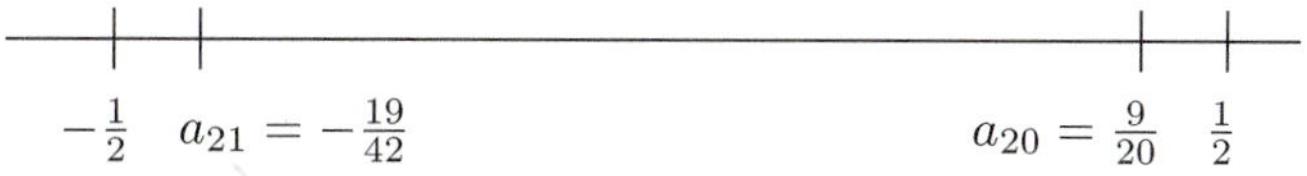

Figure 1.1.4 In Example 4, the odd terms get close to $-\frac{1}{2}$ while the even terms get close to $\frac{1}{2}$.

The problem of deciding whether or not a given sequence has a limit can be thought of as a game between two players, P1 and P2. The player P1 supplies A and M's, while the player P2 supplies ϵ's.

To see how this game works, first consider the sequence $\left\{ \frac{100}{n} \right\}$. The game starts by player P1 declaring A. In the case of the sequence $\left\{ \frac{100}{n} \right\}$, the smart player would set $A = 0$. P2 then names an ϵ, and it is up to P1 to then find an M such that the definition is satisfied. Let us suppose that P2 sets $\epsilon = 1$. P1 would then set $M = 100$ and would note that if $n > 100$, then $\left| \frac{100}{n} - 0 \right| < 1$. P2 would then reply that 1 was pretty large for ϵ and set $\epsilon = \frac{1}{100}$. P1 now has to find a new M; in this case he might set $M = 10,000$ and would observe that if $n > M$, then $\left| \frac{100}{n} - 0 \right| < \frac{1}{100}$. It is clear that for this sequence no matter how small P2 chooses ϵ, P1 can always find an M such that $n > M$ implies $\left| \frac{100}{n} - 0 \right| < \epsilon$. So in this case we say P1 wins the game, for the simple reason that P1 has only to choose $M \geq \frac{100}{\epsilon}$.

Now consider a second sequence, $\left\{\frac{(-1)^n}{2}\right\}$. For this sequence, P1 might start off by declaring $A = \frac{1}{2}$. P2 sets $\epsilon = \frac{1}{2}$ as well. P1 then takes $M = 200$. P2 then points out that $a_{501} = -\frac{1}{2}$ and that $\left|-\frac{1}{2} - \frac{1}{2}\right| \geq \frac{1}{2}$. P1 then tries a new M, say, $M = 1000$. But P2 can easily show, by using a_{1001}, that this choice does not work either. P1 then tries a new tack by changing A, say, setting $A = -\frac{1}{2}$. P2 is happy with $\epsilon = \frac{1}{2}$. P1 tries $M = 10,000$. P2 then points out that $a_{20,000} = \frac{1}{2}$ and that $\left|\frac{1}{2} - \frac{-1}{2}\right| > \frac{1}{2}$. It should be clear that for the second sequence, no matter how P1 picks A, by sticking with $\epsilon = \frac{1}{2}$, P2 can prevent P1 from finding a suitable M so that the conclusion of the definition is satisfied. Thus, in this case we say that P2 wins the game. Of course, there is a win for the first player exactly if the sequence has a limit and a win for the second player exactly if the sequence does not have a limit.

In Exercise 4, we ask the reader to obtain limits using the definition. It is only by doing these exercises that you, the student, can hope to become comfortable with the limit definition. It is essential that you do so, since the various limit definitions are the fundamental tools of the subject.

Exercises

1. Develop a one-dimensional analogue of Figure 1.1.3 and use it to explain the definition of limit of a sequence.

2. Complete the missing details in the solution to Example 5.

3. Given $\epsilon = 2.5$, find a suitable M such that for $n > M$,

$$\left|\frac{2n^3 + 5n}{3n^3 - 6} - \frac{2}{3}\right| < \epsilon.$$

   Repeat the exercise with $\epsilon = 0.5$, 0.005 and 0.0000005.

4. Use the definition of the limit to establish the existence or nonexistence of limits for the following sequences:

   (a) $\left\{\frac{n^2+1}{n+10}\right\}$;   (b) $\left\{\frac{n}{n^2+1}\right\}$;   (c) $\left\{(-1)^n\left(\frac{1}{10} - \frac{1}{n}\right)\right\}$;

   (d) $\left\{(-1)^n\left(\frac{1}{100} - \frac{1}{n}\right)\right\}$;   (e) $\left\{\frac{1+(-1)^n}{2}\right\}$;   (f) $\{(-1)^n n\}$;

   (g) $\left\{\frac{2n+1}{n+3}\right\}$;   (h) $\left\{\frac{1}{2n+5}\right\}$;   (i) $\left\{\frac{6n^2}{1-5n^2}\right\}$;

   (j) $\left\{\frac{(-1)^n 6n^3}{1+4n^3}\right\}$;   (k) $\left\{\sin\frac{n\pi}{2}\right\}$;   (l) $\left\{\left|\sin\frac{n\pi}{2}\right|\right\}$;

   (m) $\left\{\frac{\sin n}{n^2}\right\}$;   (n) $\left\{\frac{1}{2}\left(1 - \left(-\frac{1}{2}\right)^n\right)\right\}$;   (o) $\left\{\frac{2^n}{n^5}\right\}$;

   (p) $\left\{\frac{n^3}{2^n}\right\}$;   (q) $\left\{\frac{n^6}{3^n}\right\}$;   (r) $\left\{\frac{(-1)^n n^4}{2^n}\right\}$;

   (s) $\left\{\sqrt{\frac{n}{2n+1}}\right\}$;   (t) $\left\{\frac{1}{\sqrt[3]{2n-9}}\right\}$, $(n \geq 5)$;   (u) $\{\sqrt{n+3} - \sqrt{n}\}$;

   (v) $\left\{\frac{2^n}{n!}\right\}$;   (w) $\{\sqrt{2n} - \sqrt{n}\}$;   (x) $\left\{\frac{100^n}{n!}\right\}$;

   (y) $\left\{\frac{n!}{n^n}\right\}$;   (z) $\{(n+1)^{2/3} - n^{2/3}\}$;

   (aa) $\left\{\frac{2^n - (-2)^n}{n}\right\}$;   (bb) $\{\sqrt{4n^2 + n} - 2n\}$;

$$\text{(cc) } a_n = \begin{cases} \frac{n}{2n+1}, & \text{if } n \text{ is odd} \\ \frac{3n-5}{6n+1}, & \text{if } n \text{ is even;} \end{cases} \qquad \text{(dd) } a_n = \begin{cases} 2, & \text{if } n = 3k, k \in \mathbf{N} \\ \frac{2}{n^2}, & \text{if } n = 3k-1, k \in \mathbf{N} \\ 2^n, & \text{if } n = 3k-2, k \in \mathbf{N}. \end{cases}$$

5. Consider the two-player game described at the end of the section. For the sequence given by

$$a_n = \frac{(-1)^n(1+n)}{4n},$$

show that P2 has a win. What is an upper bound on ϵ, so that if P2 chooses ϵ less than this upper bound, P2 will be guaranteed a win?

6. Evaluate $\lim\limits_{n \to \infty} t_n$, where t_n is defined as follows:

(a) $t_n = \frac{1}{n^2} + \frac{2}{n^2} + \cdots + \frac{n}{n^2}$;

(b) $t_n = \left(1 - \frac{1}{2}\right)\left(1 - \frac{1}{3}\right)\left(1 - \frac{1}{4}\right)\cdots\left(1 - \frac{1}{n}\right)$;

(c) $t_n = \left(1 - \frac{1}{2^2}\right)\left(1 - \frac{1}{3^2}\right)\left(1 - \frac{1}{4^2}\right)\cdots\left(1 - \frac{1}{(n+1)^2}\right)$;

(d) $t_n = \frac{1}{1\cdot2} + \frac{1}{2\cdot3} + \cdots + \frac{1}{n(n+1)}$;

(e) $t_n = \dfrac{1 - 2 + 3 - 4 + \cdots - 2n}{\sqrt{n^2 + 1}}$;

(f) $t_n = \frac{s_{n+1}}{s_n}$, where $s_n = \frac{2^n n!}{n^n}$;

(g) $t_1 = \frac{1}{2}$, $t_{n+1} = t_n + \frac{1}{(n+1)(n+2)}$, $(n \geq 1)$;

(h) $t_n = \frac{r_n}{5}$, where r_n is the remainder when n is divided by 9.

7. If $a_n = a$ for all n, prove the constant sequence $\{a_n\}$ converges to the limit a.

8. Prove $\lim a_n = A$ if and only if $\lim b_n = 0$, where $b_n = a_n - A$ for each $n \in \mathbf{N}$.

9. Let $\{a_n\}$ and $\{b_n\}$ be two sequences. If there exists an integer k such that $a_n = b_n$ for $n > k$, show that either $\{a_n\}$ and $\{b_n\}$ have the same limit or else both fail to have limits.

10. Given a sequence $\{a_n\}$, define the sequence $\{b_n\}$ by setting $b_n = a_{n+k}$, where $k \in \mathbf{N}$ is fixed. Show that $\{b_n\}$ converges if and only if $\{a_n\}$ converges, and in that case, $\lim\limits_{n \to \infty} b_n = \lim\limits_{n \to \infty} a_n$.

11. If a real number b appears an infinite number of times as a term of a convergent sequence $\{a_n\}$, prove that $\lim\limits_{n \to \infty} a_n = b$.

12. If $\{a_n\}$ is convergent, and if $a_n \leq A$ for all n (after a certain stage), show that $\lim a_n \leq A$.

13. If $\{a_n\}$ and $\{b_n\}$ are two convergent sequences satisfying $a_n < b_n$ for all n, can you conclude that $\lim a_n < \lim b_n$?

14. If $\{a_n\}$ converges to a limit $L > 0$, show that there exists $M > 0$ such that $|a_n| > \frac{L}{2}$ for $n > M$.

15. Give examples of
(a) a sequence of rational numbers having an irrational number as limit;
(b) a sequence of irrational numbers having a rational number as limit.

16. If the sequence $\{a_n\}$ is such that the two sequences $\{a_{2n}\}$ and $\{a_{2n+1}\}$ formed by the even and the odd terms of $\{a_n\}$ both converge to L, prove that $\lim\limits_{n \to \infty} a_n = L$.

17. Let $f : \mathbf{N} \to \mathbf{N}$ be a one-to-one and onto function. Then the sequence $\{f(n)\}$ is a **rearrangement** of $\mathbf{N}$. A sequence $\{b_n\}$ is said to be a **rearrangement** of a sequence $\{a_n\}$ if there exists a rearrangement $\{f(n)\}$ of $\mathbf{N}$ such that $b_{f(n)} = a_n$. Prove $\{a_n\}$ converges to a limit L if and only if every rearrangement $\{b_n\}$ has a limit L.

18. Give a formal proof that for $b > 0$, $\lim\limits_{n \to \infty} \dfrac{[bn]}{bn} = 1$, where $[x]$ denotes the greatest integer not exceeding x. Hence, or otherwise, show for $a > 0, b > 0$, $\lim\limits_{n \to \infty} \dfrac{[bn]}{an} = \dfrac{b}{a}$, and $\lim\limits_{n \to \infty} \left[\dfrac{1}{an}\right] bn = 0$.

19. We say the sequence $\{a_n\}$ **diverges to** $+\infty$ (written $\lim a_n = +\infty$) provided given $K \in \mathbf{R}$ there exists $M \in \mathbf{R}$ satisfying $n > M$ implies $a_n \geq K$. Define '**diverges to** $-\infty$'. Which of the nonconvergent sequences in Exercise 2 diverge to $+\infty$ or $-\infty$?

20. If $\{a_n\}$ diverges to $+\infty$ and if there exists n_0 such that for $n > n_0$, $b_n \geq a_n$, prove $\{b_n\}$ diverges to $+\infty$.

21. Let $\{a_n\}$ be a sequence with $a_n > 0$ for all n and $\lim \dfrac{a_{n+1}}{a_n} = A > 1$. Prove $\lim a_n = +\infty$. What happens when $A = 1$?

22. Given an arbitrary sequence $\{a_n\}$, one can construct a new sequence $\{b_n\}$ by any of the following operations (to be made precise by the reader):

   (a) deleting a finite number of terms;
   (b) inserting a finite number of terms;
   (c) deleting an infinite number of terms;
   (d) deleting every second term;
   (e) inserting an infinite number of terms at random places;
   (f) inserting a 0 between every consecutive term;
   (g) altering a finite number of terms;
   (h) altering an infinite number of terms.

   Which of these operations will affect the convergence properties of the sequence $\{b_n\}$ in relation to the sequence $\{a_n\}$?

1.2   Basic Limit Theorems

In this section we establish the basic theorems that govern the behavior of limits of sequences. The reader should begin looking for patterns, because, as we shall see, many of these results are repeated over again for the different types of limits, and also because the methods that are used to obtain the results are limited in number. We shall try to emphasize this internal structure as we proceed.

Theorem 1.2.1. *The limit of a sequence, if it exists, is unique.*

Proof. Let $\{a_n\}$ be an arbitrary convergent sequence. Let A and B be any two real numbers both of which are limits of the sequence $\{a_n\}$. We assume for the sake of argument $A \neq B$, so that $d = |A - B| > 0$. Let $\epsilon = \frac{d}{3}$. Since A is a limit, there is a real number M_A, depending on A (as well as on ϵ), such that if $n \geq M_A$, then $|a_n - A| < \epsilon$. Similarly, since B is a limit, there is an $M_B \in \mathbf{R}$ such that if $n \geq M_B$, then $|a_n - B| < \epsilon$. Let M be the maximum of M_A and M_B. Then, if $n \geq M$,

$$|a_n - A| < \epsilon \text{ and } |a_n - B| < \epsilon.$$

Now, without loss of generality, we may assume $A < B$. Under this assumption, using Theorem 0.6.4 and observing that $B - A > 2\epsilon$, we have

$$a_n < A + \epsilon < B - \epsilon < a_n.$$

But this is absurd. Hence, we conclude $A = B$. $\square$

Discussion. In the solutions presented in the last section, we have seen that the calculations of arithmetic are the essential tools with which our arguments will be created. As can be seen above, these tools are also used in the creation of proofs. All of the necessary arithmetic was developed in Chapter 0. Moreover, a standard of proof was laid down there as well. In the arguments presented from this point forward, we shall employ the tools of arithmetic without explicit reference to which particular tool is required in a specific instance. Thus, for example, we have asserted above that $|A - B| > 0$; no reference is given as to which arithmetical fact would permit this conclusion; it is assumed the reader can recall that $A \neq B$ implies $|A - B| > 0$ is a theorem so that the argument given can be brought to the standard of proof given in Chapter 0. All our arguments, whether as proofs for theorems or as solutions to examples, must be able to meet this standard; this can be accomplished provided we are always able to supply the missing steps, which for the instance above means being able to immediately recall the content of Theorem 0.6.1(i). In this regard, no step may be taken as 'obvious' until the supporting justification is understood in its entirety!

Turning now to the details of the proof, notice the similarity of this proof to the technique employed in Example 1.1.5 that shows that 'oscillating' sequences do not have limits. The reader should also compare this proof with the proof of Theorem 0.4.1 that asserts that the supremum is unique. The geometry of the situation is made clear in Figure 1.2.1. The reader should make clear to himself why there is no loss of generality involved in making the assumption that $A < B$. Also, the details leading to the set of inequalities at the end of the proof that force the contradiction $a_n < a_n$ should be fully worked out.

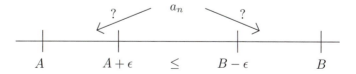

Figure 1.2.1 The figure illustrates the contradiction resulting from the assumption the limit is not unique (Theorem 1.2.1).

A contradiction could also be arrived at as follows: note that $d = |A - B| = |(A - a_n) + (a_n - B)| \leq |a_n - A| + |a_n - B| < \epsilon + \epsilon = 2\epsilon < d$. The process of 'adding' and 'subtracting' an equal quantity and using the Triangle inequality is a standard technique that the reader should grasp thoroughly. □

Theorem 1.2.2. *Let $\{a_n\}$, $\{b_n\}$, and $\{c_n\}$ be sequences such that for every $n \in \mathbf{N}$,*

$$a_n \leq b_n \leq c_n.$$

If $\{a_n\}$ and $\{c_n\}$ both converge to A, then $\{b_n\}$ also converges to A.

Proof. Let $\epsilon > 0$ be fixed. By hypothesis there is an M_1 such that if $n \geq M_1$, then $|a_n - A| < \epsilon$. Similarly, there is an M_2 such that $n \geq M_2$ implies $|c_n - A| < \epsilon$. Let M be the larger of M_1 and M_2. Now, if $n \geq M$, then

$$A - \epsilon < a_n \leq b_n \leq c_n < A + \epsilon.$$

It follows that

$$|b_n - A| < \epsilon,$$

as desired. Since ϵ was arbitrary, we are done. □

Discussion. Notice how the proof flows naturally from writing down the definition of convergence of a_n to A, followed by the definition of convergence of c_n to A, followed by choosing A so that the conditions are satisfied simultaneously. We then use Theorem 0.6.4 to manipulate the inequalities and obtain the desired form. Again, the geometry is clear and is shown in Figure 1.2.2. For obvious reasons, this theorem is often call the 'sandwich' theorem (or a 'squeeze' theorem) for sequences. □

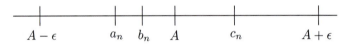

Figure 1.2.2 A geometric illustration of the 'sandwich theorem' with $a_n \leq b_n \leq c_n$.

The next theorem spells out an important property possessed by all convergent sequences.

Theorem 1.2.3. *Every convergent sequence is bounded.*

Proof. Let $\{a_n\}$ be an arbitrary convergent sequence. We must find a real number B such that for every $n \in \mathbf{N}$,

$$|a_n| \leq B.$$

To find B, let $\epsilon = 1$, and find M such that if $n \geq M$; then

$$|a_n - A| < \epsilon,$$

where A is the limit of the sequence. We may suppose that M is in fact a positive integer. Now the set

$$\{|a_n| : n \leq M\}$$

is a finite set and so has a maximum, say, B_1. Let B be the maximum of B_1 and $|A| + 1$. Evidently, for $n \leq M, |a_n| \leq B_1 \leq B$. For $n > M, |a_n - A| < 1$, whence $|a_n| \leq |A| + 1 \leq B$. Thus, for all $n, |a_n| \leq B$. $\square$

Discussion. The intuition leading to this theorem is quite simple. First, the definition of limit states that for A to be the limit of a sequence, all the terms of the sequence having large subscript must be 'close to' A. Since there are only a finite number of terms having small subscript, and every finite set of real numbers is bounded, we ought to be able to construct a bound for all the terms of the sequence. The proof only formalizes these intuitive reasonable ideas.

The contrapositive equivalent of the above theorem asserts that a sequence that is not bounded can never converge. This is a useful tool for showing certain sequences do not converge, because it may be easy to check that the sequence is unbounded (for example, the sequence $\{n^3\}$).

The statement of the above theorem is an implication, where the hypothesis is that the sequence is convergent and the conclusion is that it is bounded. The reader should ask the question: What about the converse? In this case, the converse turns out to be false. An easy counterexample would be the sequence discussed in Example 1.1.5. Then, the natural question would be: What additional conditions are needed to guarantee that the converse also holds? This is explored in the next section. $\square$

Our next several theorems concern the arithmetic of sequences. Just as one can think of adding two real numbers, so one could imagine adding together two sequences:

$$\{a_n\} + \{b_n\}.$$

The quantity displayed makes clear our intent, namely, to combine two sequences by an additive process; it does not, however, give any indication how this process might be performed. For sequences, we have settled on performing operations 'term by term'. This yields a defining equation

$$\{a_n\}_{n=1}^{\infty} + \{b_n\}_{n=1}^{\infty} = \{a_n + b_n\}_{n=1}^{\infty}.$$

In this equation, the operation on the left, which has been denoted by a standard $+$ sign, is defined in terms of a computation that we know how to perform and that is given on the right. Namely, we know how to add two numbers, and this is what the $+$ sign on the right-hand side requires us to do. This definition extends our usual definition of addition of numbers. Also, it is merely an example of the usual definition for addition of functions, with which the reader is almost certainly familiar.

Having defined addition of two sequences, we can immediately think of multiplying or dividing sequences, or of taking the absolute value of a sequence. In all cases, standard usage requires these operations to be performed, term by term. Thus,

$$\{a_n\} \cdot \{b_n\} = \{a_n \cdot b_n\},$$

and if $b_n \neq 0$ for every $n \in \mathbf{N}$, then, we define

$$\frac{\{a_n\}}{\{b_n\}} = \left\{ \frac{a_n}{b_n} \right\}.$$

Also, the absolute value of the sequence $\{a_n\}$ is given by $|\{a_n\}| = \{|a_n|\}$.

Having specified arithmetic operations for sequences, we are led to obvious questions, such as 'If $\{a_n\}$ converges to A, and $\{b_n\}$ converges to B, to what, if anything, will $\{a_n + b_n\}$ converge'? This question is addressed in the following theorems.

Theorem 1.2.4. *Let $\{a_n\}$ and $\{b_n\}$ be sequences having A and B as limits, respectively. Then $\{a_n + b_n\}$ converges to $A + B$.*

Proof. Fix $\epsilon > 0$. Let M_1 be chosen such that $n \geq M_1$ implies $|a_n - A| < \frac{\epsilon}{2}$. Let M_2 be chosen such that $n \geq M_2$ .implies. $|b_n - B| < \frac{\epsilon}{2}$. Set $M = \max\{M_1, M_2\}$. It follows that

$$
\begin{aligned}
|(a_n + b_n) - (A + B)| &= |(a_n - A) + (b_n - B)| \\
&\leq |a_n - A| + |b_n - B| \\
&< \frac{\epsilon}{2} + \frac{\epsilon}{2} = \epsilon,
\end{aligned}
$$

whenever $n \geq M$. $\square$

Discussion. From the sequence of statements in the proof, it appears as if one knows how to choose M_1 and M_2 before doing the calculation that splits the sum into its constituent parts by using the Triangle inequality (Theorem 0.6.2). This is not the case. In fact, we experiment with the calculation first, and then having found out how to choose M_1 and M_2 in terms of $\frac{\epsilon}{2}$, we write the proof accordingly. Thus, in the proof above, we preferred to choose M_1, M_2 such that the quantities $|a_n - A|, |b_n - B|$ are each less than $\frac{\epsilon}{2}$, rather than ϵ, because in the final calculation, we want $|(a_n + b_n) - (A + B)| < \epsilon$. The choice of $\frac{\epsilon}{2}$ is not obligatory, for if we choose $|a_n - A|$ and $|b_n - B|$ each less than ϵ, then, in the final step, $|(a_n + b_n) - (A + B)| < 2\epsilon$, and the conclusion will still be valid **(WHY?)**. The reader should note the use of Theorem 0.6.2, which will be a consistent feature of many of our elementary arguments. $\square$

Theorem 1.2.5. *Let $\{a_n\}$ and $\{b_n\}$ be sequences converging to A and B, respectively. Then the sequence $\{a_n \cdot b_n\}$ converges to $A \cdot B$.*

Proof. Fix $\epsilon > 0$. Consider

$$
\begin{aligned}
|a_n \cdot b_n - A \cdot B| &= |(a_n \cdot b_n - b_n \cdot A) + (b_n \cdot A - A \cdot B)| \\
&\leq |a_n \cdot b_n - b_n \cdot A| + |b_n \cdot A - A \cdot B| \\
&= |b_n| \cdot |a_n - A| + |A| \cdot |b_n - B|.
\end{aligned}
$$

Now $\{b_n\}$ is a convergent sequence, so by Theorem 1.2.3, there is a positive real number K, $(K > 1)$ such that $|b_n| < K$ for all $n \in \mathbf{N}$. Choose M_1 such that $n \geq M_1$ implies $|a_n - A| < \frac{\epsilon}{2K}$. Further, choose M_2 such that $n \geq M_2$ implies $|b_n - B| < \frac{\epsilon}{2(|A|+1)}$. Let $M = \max\{M_1, M_2\}$, whence for $n \geq M$ we have, starting from the inequalities above,

$$\begin{aligned}
|a_n \cdot b_n - A \cdot B| \;&\leq\; |b_n| \cdot |a_n - A| + |A| \cdot |b_n - B| \\
&<\; K \cdot \frac{\epsilon}{2K} + |A| \cdot \frac{\epsilon}{2(|A|+1)} \\
&<\; \frac{\epsilon}{2} + \frac{\epsilon}{2} = \epsilon. \qquad \square
\end{aligned}$$

Discussion. As noted earlier in the comments following Theorem 0.6.4, a feature of many of these proofs is the addition and subtraction of the same quantity inside the absolute value signs. The purpose in this case, as it will usually be, is to obtain quantities that we know can be made as small as we please. In this case, the terms are $|a_n - A|$ and $|b_n - B|$. Of course, there are likely to be leftover terms, and here the leftover terms are $|b_n|$ and $|A|$. But leftover terms are of no consequence as long as these terms are bounded and multiplied by other terms that can be made arbitrarily small. For the case at hand, since $|b_n| < K$, we can replace it by K, and then make $|a_n - A|$ so small that its product with K will be less than $\frac{\epsilon}{2}$. (This is why we need to be able to make $|a_n - A|$ as small as we please, and not merely less than ϵ.) With this in mind, the reader should now make sure she understands how the other term is dealt with. The reason for using $\frac{\epsilon}{2(|A|+1)}$, rather than $\frac{\epsilon}{2A}$, is to take care of the possibility that $A = 0$. Last, again note the use of Theorem 0.6.2. $\square$

Theorem 1.2.6. *Let $\{a_n\}$ be a sequence of nonzero terms that converges to A. If $A \neq 0$, then $\left\{\frac{1}{a_n}\right\}$ converges to $\frac{1}{A}$.*

Proof. Fix $\epsilon > 0$. Since $A \neq 0$, there is an M_1 such that $n > M_1$ implies $\frac{|A|}{2} < |a_n|$ **(WHY?)**. Then for $n > M_1$, we have

$$\begin{aligned}
\left| \frac{1}{a_n} - \frac{1}{A} \right| \;&=\; \frac{|A - a_n|}{|a_n| \cdot |A|} \\
&<\; \frac{2|a_n - A|}{|A|^2}.
\end{aligned}$$

From this calculation, it is clear that we should choose M_2 so that $n > M_2$ implies $|a_n - A| < \frac{A^2}{2}\epsilon$. Let $M = \max\{M_1, M_2\}$, whence for $n > M$ we have

$$\left| \frac{1}{a_n} - \frac{1}{A} \right| < \frac{2}{A^2} \cdot \frac{A^2}{2} \cdot \epsilon = \epsilon. \qquad \square$$

Discussion. The requirement that each of the terms be nonzero can be relaxed. The essential fact is that the limit be different from zero, for then all the terms will eventually be nonzero **(WHY?)**, and for those terms the limit of the inverses will be $\frac{1}{A}$. It is essential that the reader understand how and why we arrived at our choice of M_2. Respecting this, our comments following Theorem 1.2.5 should be reviewed. Note how we use the arithmetical fact that making the denominator of a fraction smaller increases the size of the fraction. Our calculations continually hinge on such basic elementary results from arithmetic! $\square$

The proofs of the following are left to Exercises 1, 2 and 4.

Theorem 1.2.7. *If $\{a_n\}$ converges to A, then $\{|a_n|\}$ converges to $|A|$.*

Theorem 1.2.8. *If $\{a_n\}$ and $\{b_n\}$ converge to A and B, respectively, and if $a_n \leq b_n$ for every $n \in \mathbf{N}$, then $A \leq B$.*

Theorem 1.2.9. *Let $B = \sup\{a_n : n \in \mathbf{N}\}$ and $b = \inf\{a_n : n \in \mathbf{N}\}$. If $\{a_n\}_{n=1}^{\infty}$ converges to A, then $b \leq A \leq B$.*

Definition. Let $f : \mathbf{N} \to \mathbf{N}$ be a strictly increasing function ($f(k) < f(k+1)$ for all $k \in \mathbf{N}$) with $f(k)$ denoted by n_k. If $\{a_n\}_{n=1}^{\infty}$ is any sequence, then

$$\{a_{n_k}\}_{k=1}^{\infty} = \{a_{n_k}\}$$

is called a **subsequence** of $\{a_n\}$.

Discussion. By a subsequence of a sequence, we should mean a new sequence constructed from terms of the old sequence taken in the same order as in the original. This is achieved by requiring the function f to be strictly increasing, and hence, one-to-one. Given a subsequence, the function f can be constructed simply by listing the terms of the subsequence in order and extracting the subscripts. From this observation, it follows that every infinite set of terms of the sequence can be realized as a subsequence. $\square$

Theorem 1.2.10. *A sequence $\{a_n\}$ converges if and only if every subsequence, $\{a_{n_k}\}$, converges to the same limit.*

Proof. We sketch the converse, leaving complete details of the proof to the reader as Exercise 18.

If some subsequence does not converge, we can immediately use that subsequence to construct a proof that the original sequence does not converge, along the lines used in Example 4. Thus, pick any subsequence and call its limit A. If some other subsequence has a limit, B, different from A, we can use A and B to construct a proof that the original sequence cannot converge following the lines of the proof of Example 5. $\square$

Exercises

1. Prove Theorem 1.2.7. Supply an example to show the converse of Theorem 1.2.7 is not in general true. However, if $\{|a_n|\}$ converges to 0, then prove $\{a_n\}$ also converges to 0.

2. Prove Theorem 1.2.8. Give an example that shows a_n may be strictly less than b_n for all n, but $A = B$.

3. For the following sequences $\{a_n\}$, use the limit theorems to establish convergence.

   (a) $a_n = \frac{2n-1}{3n+4-5^{1/n}}$;   (b) $a_n = \frac{n^2+n-1}{3n^2-4n}$;

   (c) $a_n = \frac{P(n+1)}{P(n)}$ where $P(x) = ax^3 + bx^2 + cx + d$, $a, b, c, d \in \mathbf{R}$;

   (d) $a_n = \frac{1^2+2^2+\cdots+n^2}{6n^3}$;   (e) $a_n = \frac{1^3+2^3+\cdots+n^3}{(2n+1)^4}$;

   (f) $a_n = \frac{1}{(n+1)^2} + \frac{1}{(n+2)^2} + \cdots + \frac{1}{(n+n)^2}$;

   (g) $a_n = 1 + x + x^2 + \cdots + x^n$, where $|x| < 1$.

4. Prove Theorem 1.2.9.

5. State and prove an appropriate limit theorem involving subtraction.

6. State and prove an appropriate theorem concerning the convergence of the ratio of two convergent sequences.

7. Without using Theorem 1.2.5, formally prove if $\lim_{n \to \infty} a_n = L$, then $\lim_{n \to \infty} a_n^2 = L^2$.

8. If $\left\{ \frac{a_n}{n} \right\}$ converges to $L \neq 0$, prove the sequence $\{a_n\}$ must be unbounded. Is the converse true?

9. Theorems 1.2.4 and 1.2.5 require $\{a_n\}$ and $\{b_n\}$ to be convergent. What can be said if this condition is dropped? Give examples that elucidate the situation.

10. Let $\{a_n\}$ converge to 0 and $\{b_n\}$ be bounded. Show $\{a_n b_n\}$ converges to 0.

11. Prove the following:
    (a) If $\lim a_n = +\infty$, then $\lim c a_n = +\infty$ if $c > 0$ and $\lim c a_n = -\infty$ if $c < 0$.
    (b) If $\lim a_n = +\infty$ and $\{b_n\}$ is bounded, then $\lim (a_n + b_n) = +\infty$.
    (c) If $\lim a_n = +\infty$, then $\lim(-a_n) = -\infty$.
    (d) If $a_n > 0$ for all n, $\{a_n\}$ diverges to $+\infty$ if and only if $\left\{ \frac{1}{a_n} \right\}$ converges to 0.

12. Show the term-by-term product of a bounded divergent sequence with a convergent sequence converges if and only if the limit of the convergent sequence is 0.

13. Prove or disprove the following:
    (a) If $\{a_n\}$ and $\{b_n\}$ diverge, $\{a_n + b_n\}$ diverges.
    (b) If $\{a_n\}$ converges and $\{b_n\}$ diverges, $\{a_n + b_n\}$ diverges.
    (c) If $\{a_n\}$ and $\{b_n\}$ diverge, $\{a_n - b_n\}$ diverges.
    (d) If $\{a_n\}$ and $\{b_n\}$ diverge, $\{a_n b_n\}$ diverges.
    (e) If $\{a_n\}$ and $\{a_n b_n\}$ converge, then $\{b_n\}$ converges.
    (f) If $\{a_n\}$ and $\{a_n b_n\}$ diverge, then $\{b_n\}$ diverges.

14. Let $\{a_n\}$ be a convergent sequence with limit A. Set

$$b_n = \frac{1}{n} \cdot \sum_{k=1}^{n} a_k,$$

that is, the **arithmetic average** of the first n terms of $\{a_n\}$. Prove $\{b_n\}$ is a convergent sequence. Does the fact that $\{b_n\}$ converges tell us anything about the convergence of $\{a_n\}$?

15. Repeat Exercise 15 with $b_n = (a_1 \cdot a_2 \ldots a_n)^{\frac{1}{n}}$, the **geometric average** of the first n terms of the first n terms of $\{a_n\}$.

16. Let $\{a_n\}$ and $\{b_n\}$ be sequences converging to A and B, respectively. Define

$$t_n = \frac{1}{n} \cdot \sum_{k=1}^{n} a_k b_{n-k}.$$

Show $\{t_n\}$ converges, and find its limit. What would happen if the term $\frac{1}{n}$ was omitted from the above expression for t_n?

17. Let $P(n)$ and $Q(n)$ be two nonzero polynomials in n. Let $\{a_n\}$ be the sequence obtained by forming the ratio of $P(n)$ over $Q(n)$. Show $\{a_n\}$ converges if and only if the degree of $Q(n)$ is at least as large as the degree of $P(n)$.

18. Prove the following: If $\{a_n\}$ converges, then every subsequence, $\{a_{n_k}\}$, converges to the same limit. Also complete the converse of Theorem 1.2.10 by filling in the missing details.

19. For each of the bounded divergent sequences in Exercise 1.1.3, find a convergent subsequence.

20. Let $\{a_n\}$ converge to a, where all numbers are nonnegative. Show $\{\sqrt{a_n}\}$ converges to $\sqrt{a}$.

21. Show for any convergent sequence of real numbers, the sequence obtained by taking the cube root of each term is convergent.

22. Let $a > 0$. Show $\{a^{\frac{1}{n}}\}$ converges to 1.

23. Show $\{n^{\frac{1}{n}}\}$ converges to 1. [Hint: Use the Binomial Theorem.]

24. Show $\lim\limits_{n\to\infty} \dfrac{a^n}{n!} = 0$.

25. Discuss the convergence or divergence of $\{(n!)^{\frac{1}{n}}\}$.

26. Discuss the convergence or divergence of the sequence $\{n^p a^n\}$ for various a and p.

27. Let

$$a_n = \sum_{k=1}^{n} \frac{1}{n^2 + k}.$$

Show $\{a_n\}$ is convergent. [Hint: Use Theorem 1.2.2.]

28. Repeat Exercise 28, where the sum is up to $2n$, instead of n.

29. Let $0 < a < 1$. Show the sequence having nth term $b_n = a^n$ converges to 0. What happens if $a > 1$?

30. If $0 < a \le b \le c$, show $\lim\limits_{n\to\infty} (a^n + b^n + c^n)^{1/n} = c$.

31. If $\{a_n + b_n\}$ converges to A, and $\{a_n - b_n\}$ converges to B, find the limit of $\{a_n b_n\}$.

32. If $a_0 + a_1 + \cdots + a_k = 0$, show

$$\lim\limits_{n\to\infty} [a_0\sqrt{n} + a_1\sqrt{n+1} + \cdots + a_k\sqrt{n+k}] = 0.$$

33. Let $a_n > 0$ for all n and set $b_n = \dfrac{a_{n+1}}{a_n}$ and $c_n = a_n^{1/n}$. If $\lim b_n = L$, prove $\lim c_n = L$. Is the converse true?

34. Let $\{a_n^{(i)}\}$ be a finite number of sequences, where $i = 1, 2, \ldots, k$. Generalize Theorems 1.2.4 and 1.2.5 for the sum (product) of these k sequences.

35. If $\lim b_n = 0$ and $|a_n - A| < Kb_n$ for some constant K, prove $\lim a_n = A$.

36. If $a_n > 0$ and if $\lim \dfrac{a_{n+1}}{a_n} = A < 1$, prove $\lim a_n = 0$. What happens when $A = 1$? Discuss the situation when $A > 1$. (Compare this with Exercise 1.1.19.)

1.3   Monotonicity and Its Consequences

We have defined the concept of limit for a sequence, but except for rather arduous calculations, as in the exercises at the end of Section 1.1, we have no means of knowing whether a given sequence has a limit. In this section we shall remedy that and in doing so will provide the first example of power of the completeness axiom as a tool of analysis.

   We have seen that a convergent sequence is always bounded, but not conversely. We are therefore led to look for conditions on a bounded sequence that would guarantee convergence. To this end, we study the following concept.

Definition.   A sequence $\{a_n\}$ is said to be **monotone increasing** if for all $n, m \in \mathbf{N}, n \leq m$ implies $a_n \leq a_m$.

   In a similar manner we can define **monotone decreasing** (the reader should do so). A sequence that is either monotone decreasing or increasing is said to be **monotone**. If the inequalities are strict ($n < m$ implies $a_n < a_m$), then the sequence is **strictly monotone** (increasing for the inequalities shown).

Discussion.   The definition of a monotonic increasing (decreasing) sequence conveys the simple idea that each term of the sequence is larger (smaller) than or equal to the preceding term. Monotonicity is a very useful concept, because it prevents the terms of a sequence from oscillating. Given a sequence $\{a_n\}$, there are several methods of testing whether it is monotonic increasing or not. One obvious way is to use the above definition. Sometimes, this can be hard. Other practical methods are

   (i) Check that the difference $(a_{n+1} - a_n) \geq 0$ for all n.

   (ii) Check that the ratio $\frac{a_{n+1}}{a_n} \geq 1$ for all n.

   (iii) Use induction on n.

   (iv) Using tools from elementary calculus, decide whether the function $f(n) = a_n$ is increasing, by using the first derivative test, namely, $f'(x) > 0$. (This is formally treated in Chapter 4, where we discuss differentiation.)

   One can write down similar conditions for monotonic decreasing sequences. Note that we need not require that the monotonicity definition is satisfied for all n. In many instances, it suffices to demand that the definition holds from a fixed $k \in \mathbf{N}$ onwards. $\square$

   The following theorem shows that in the presence of monotonicity, a bounded sequence will converge. It further tells us the value of the limit. It is a remarkable theorem on the convergence of sequences, and its proof employs the powerful tool of completeness of $\mathbf{R}$.

Theorem 1.3.1.   *Let $\{a_n\}_{n=1}^{\infty}$ be a monotone sequence. If $\{a_n\}$ is increasing and bounded above, then $\{a_n\}$ converges to $\sup\{a_n : n \in \mathbf{N}\}$. If $\{a_n\}$ is decreasing and bounded below, then $\{a_n\}$ converges to $\inf\{a_n : n \in \mathbf{N}\}$.*

Proof.   We will prove the statement for the case where $\{a_n\}$ is an increasing sequence. The decreasing case is left to Exercise 3. Since $\{a_n\}_{n=1}^{\infty}$ is a bounded sequence, $\{a_n : n \in \mathbf{N}\} \neq \emptyset$ is a bounded set. So let $A = \sup\{a_n : n \in \mathbf{N}\}$. Fix $\epsilon > 0$. By Theorem 0.4.2, there is a term of the sequence, say, a_M, such that

$$A - \epsilon < a_M \leq A.$$

Let $n \geq M$. Then by the monotonicity of $\{a_n\}_{n=1}^\infty$,

$$A - \epsilon < a_M \leq a_n \leq A,$$

whence

$$|a_n - A| < \epsilon. \qquad \square$$

Discussion. The geometry of the situation is summarized in the Figure 1.3.1(a) and (b) and is captured in the proof, an outline of which follows. Boundedness forces the supremum, A, to exist, and from the figure A should appear to the reader to be a natural candidate for the limit. The existence of the term of the sequence, a_M, satisfying $a_M \in (A - \epsilon, A]$ follows from Theorem 0.4.2. Once we have one term in the interval, monotonicity forces all the remaining terms to lie in the interval, completing the proof. We stress that in computing the supremum, we use the set, $\{a_n : n \in \mathbf{N}\}$, not the sequence, $\{a_n\}_{n=1}^\infty$.

This theorem states that for monotonic sequences, being convergent is equivalent to being bounded. In general, while every convergent sequence is necessarily bounded (Theorem 1.2.3), the converse is not generally true. But remarkably, for the monotonic sequences, the notions 'convergent' and 'bounded' coincide. This theorem is very powerful, and on many occasions, later in this book, we will invoke this theorem as a substitute for the Completeness Axiom. $\square$

Example 1. Show the sequence defined by $a_n = \left(1 + \frac{1}{n}\right)^n$ is convergent.

Solution. We use the Binomial Theorem (Theorem 0.4.5) to obtain

$$a_n = \left(1 + \frac{1}{n}\right)^n = \sum_{k=0}^n \binom{n}{k}\left(\frac{1}{n}\right)^k,$$

which is a sum of nonnegative terms. Now the $(k+1)$th term of the binomial expansion of a_n is

$$\frac{n!}{k!(n-k)!}\left(\frac{1}{n}\right)^k.$$

We can compare the $(k+1)$th term of a_n with the $(k+1)$th term of a_{n+1} by computing the ratio. Since both are positive, the latter will be larger (or equal to) the former if and only if the ratio of the former over the latter is less than or equal to 1. Thus,

$$\frac{\frac{n!}{k!(n-k)!}\left(\frac{1}{n}\right)^k}{\frac{(n+1)!}{k!(n+1-k)!}\left(\frac{1}{n+1}\right)^k} = \left(\frac{n+1-k}{n+1}\right)\left(\frac{n+1}{n}\right)^k,$$

and this quantity will be less than or equal to 1 if and only if

$$(n+1-k)(n+1)^{k-1} \leq n^k,$$

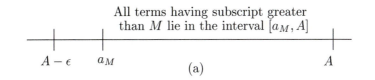

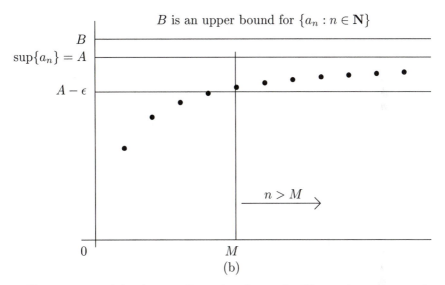

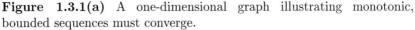

Figure 1.3.1(a) A one-dimensional graph illustrating monotonic, bounded sequences must converge.
(b) A two-dimensional graph illustrating the same fact.

for each $k \in \{0, 1, \ldots, n\}$. To establish this, we induct on k. For $k = 0$,

$$(n + 1 - 0)(n + 1)^{0-1} \leq n^0,$$

which is obviously true. Thus, assume the inequality for $k = m$, where $0 \leq m \leq n$. If $m = n$, we are done, otherwise we have

$$(n + 1 - m)(n + 1)^{m-1} \leq n^m,$$

whence

$$
\begin{aligned}
n^{m+1} &= n \cdot n^m \geq n(n + 1 - m)(n + 1)^{m-1} \\
&= [n \cdot (n + 1) - n \cdot m](n + 1)^{m-1} \\
&> [(n + 1) \cdot n - (n + 1) \cdot m](n + 1)^{m-1} \qquad \textbf{(WHY?)} \\
&= (n - m)(n + 1)^m \\
&= [(n + 1) - (m + 1)](n + 1)^m,
\end{aligned}
$$

as desired. It is immediate that

$$\sum_{k=0}^{n} \binom{n}{k} \left(\frac{1}{n}\right)^k \leq \sum_{k=0}^{n+1} \binom{n+1}{k} \left(\frac{1}{n+1}\right)^k,$$

since the $(k+1)$th term of the first sum is less than or equal to the $(k+1)$th term of the second. Thus, $\left(1 + \frac{1}{n}\right)^n$ is a monotone increasing sequence that will have a limit if we can show it is bounded above. Now, the reader can easily show by induction (Exercise 5) that

$$\frac{n!}{(n-k)!} \leq n^k,$$

for any $k \in \{0, 1, \ldots, n\}$. It follows that

$$\sum_{k=0}^{n} \binom{n}{k} \left(\frac{1}{n}\right)^k \leq \sum_{k=0}^{n} \frac{1}{k!} \leq 1 + \sum_{k=0}^{n} \left(\frac{1}{2}\right)^k.$$

The last sum is a finite geometric series that by Exercise 0.4.22 has sum less than 2. It follows that 3 is an upper bound for the sum, and hence $\{a_n\}$ converges to a limit $A \leq 3$. $\square$

Discussion. We see again, as we have already seen in the exercises, that the Binomial Theorem is a highly useful tool for doing analysis, a comment that applies to proof by induction, as well. This proof also employs facts about geometric series, facts that should be well known, but if not, they can be found in the exercises at the end of Section 0.4. At first glance, the computations perhaps seem arduous. While they are admittedly somewhat lengthy, they are completely straightforward and do not involve much more than the type of algebra learned in high school. The limit of the above sequence turns out to be the real number e (the base for the natural logarithm). The properties of e are more fully developed in Chapter 9. $\square$

Example 2. Find the limit of the sequence whose nth term is

$$\left(1 + \frac{1}{2n}\right)^n.$$

Solution. The sequence given can be rewritten as

$$\left(1 + \frac{1}{2n}\right)^n = \left[\left(1 + \frac{1}{2n}\right)^{2n}\right]^{1/2}.$$

Now the sequence $\left(1 + \frac{1}{2n}\right)^{2n}$ is clearly a subsequence of the sequence $\left(1 + \frac{1}{n}\right)^n$, which was just treated. The limit of $\left(1 + \frac{1}{n}\right)^n$ is e. Thus, since the limit of a subsequence is the limit of the sequence, the limit of $\left(1 + \frac{1}{2n}\right)^n$ is $\sqrt{e}$, by Exercise 1.2.18. $\square$

Example 3. Find the limit of the sequence defined recursively by

$$a_1 = \sqrt{2}; \qquad a_{n+1} = \sqrt{2 + a_n}, \quad n \geq 1.$$

Solution. This sequence has been shown to be monotone increasing with an upper bound of 2 (Exercise 0.4.17). Hence, $\lim a_n$ exists, by Theorem 1.3.1. Call this limit A. Consider the sequence $\{b_n\}_{n=1}^{\infty}$, where $b_n = a_{n+1}$. Since $\{b_n\}_{n=1}^{\infty}$ is a subsequence of $\{a_n\}_{n=1}^{\infty}$, it has the same limit, say, A. Now

$$\lim b_n \;=\; \lim \sqrt{2 + a_n}$$

from which we deduce (**HOW?**)

$$A \;=\; \sqrt{2 + A},$$

which in turn yields $A^2 - A - 2 = 0$. The solutions to this equation are -1 and 2. The initial conditions require $A > 0$, so the limit must be 2. $\square$

Discussion. To find the limit of the sequence, we first show that the limit of the sequence exists. Then, having found the limit, we use the recursion formula to generate an algebraic equation in the unknown limit. The equation is then solved, and the limit found. For the case at hand, the equation produces several solutions, only one of which can be the limit. Thus, the procedure can yield numbers that have no relation to the original problem. The point is that the recursion formula can be used to generate an equation that the limit of the sequence, if it exists, must satisfy. Knowledge of the sequence allows us to select the solution of the equation that is, in fact, the limit.

We note that a recursive formula can be used to generate an equation to be solved, even if the sequence being considered has no limit. For such a case, the calculations are nonsense. The generation of the equation which the limit must satisfy is meaningful only when it is known that the limit exists! $\square$

Exercises

1. Decide whether the following sequences $\{a_n\}$ are (i) monotone, (ii) bounded, (iii) convergent. Where possible, compute the limit.

   (a) $a_n = n - \frac{1}{n}$;

   (b) $a_n = n^2 + (-1)^n$;

   (c) $a_n = n^{1-n}$;

   (d) $a_n = \sin \frac{\pi}{2n}$;

   (e) $a_n = n + \sqrt{a + \frac{1}{n^2}}, a > 0$;

   (f) $a_n = \sqrt{n+1} - \sqrt{n}$;

   (g) $a_n = 3^{1/n}$;

   (h) $a_n = n\sqrt{1 + \frac{1}{n^2}}$;

   (i) $a_n = \frac{2}{\sqrt{n^2+1}-n}$;

   (j) $a_n = \sqrt{2n^2 + 3n - 4}$;

   (k) $a_n = \left(\frac{100n}{n^2}\right)$;

   (l) $a_n = \left(1 + \frac{1}{2} + \frac{1}{3} + \cdots + \frac{1}{n}\right)$;

   (m) $a_n = \frac{n^3}{\binom{2n}{n}}$;

   (n) $a_n = \left(1 + \frac{1}{2} + \frac{1}{4} + \cdots + \frac{1}{2^n}\right)$.

2. Find the limit of the following sequences $\{a_n\}$, where a_n is defined by

(a) $\left(1 + \frac{2}{n}\right)^n$;

(b) $\left(1 + \frac{1}{n}\right)^{3n}$;

(c) $\left(1 + \frac{1}{n+1}\right)^n$;

(d) $\left(1 + \frac{1}{n^2}\right)^n$;

(e) $\left(1 + \frac{1}{100+n}\right)^n$;

(f) $\left(1 + \frac{1}{2^n}\right)^{2^{2^n}}$;

(g) $\left(\frac{99}{100} + \frac{1}{n}\right)^{\sqrt{n}}$;

(h) $(\alpha a^n + \beta b^n)^{1/n}$, $0 < a < b$, $\alpha, \beta > 0$.

3. Prove a bounded decreasing sequence has a limit.

4. Must an unbounded sequence diverge to $+\infty$ or $-\infty$?

5. Fill in the missing induction argument in Example 1.

6. Define $\{a_n\}$ recursively by

$$a_1 = k, \ (k > 0); \qquad a_{n+1} = \sqrt{k + a_n}, \ (n \geq 1).$$

Show $\{a_n\}$ has a limit, and find it.

7. Define $\{a_n\}$ recursively by

$$a_1 = k \ (k > 0); \qquad a_{n+1} = \frac{k}{1 + a_n} \ (n \geq 1).$$

Show the sequence $\{a_n\}$ has a limit and find it.

8. Let x and y be positive real numbers. Define $\{a_n\}$ recursively by

$$a_0 = y, \qquad a_n = \frac{1}{2}\left(\frac{x}{a_{n-1}} + a_{n-1}\right) \ (n > 1).$$

Show $\{a_n\}$ is decreasing and converges to $\sqrt{x}$. Estimate $\sqrt{2}$ using a_4.

9. The sequence $\{a_n\}$ is defined recursively by

$$a_2 > a_1 > 0, \qquad a_{n+1} = \frac{a_n + a_{n-1}}{2} \ (n > 1).$$

Show the subsequences $\{a_{2n}\}$ and $\{a_{2n-1}\}$ are both monotonic, and converge to a common limit. Show $\lim a_n = \frac{2a_2 + a_1}{3}$.

10. Repeat Exercise 9 with $\{a_n\}$ defined by

$$a_2 > a_1 > 0, \qquad a_{n+1} = \sqrt{a_n a_{n-1}} \ (n > 1).$$

In this case, show $\lim a_n = \sqrt{a_2^2 a_1}$.

11. Let $a > b > 0$, $a_1 = \frac{a+b}{2}$, $b_1 = \sqrt{ab}$, and for $n \geq 1$,

$$a_{n+1} = \frac{a_n + b_n}{2}, \qquad b_{n+1} = \sqrt{a_n b_n}.$$

Show the sequences $\{a_n\}$ and $\{b_n\}$ converge to a common limit.

12. Let $\{a_n\}$ and $\{b_n\}$ be such that $a_1 > b_1 > 0$,

$$a_{n+1} = \frac{a_n + b_n}{2}, \qquad b_{n+1} = \frac{1}{2}\left(\frac{1}{a_n} + \frac{1}{b_n}\right) \ (n \geq 1)$$

Prove $\{a_n\}$ and $\{b_n\}$ both converge to $\sqrt{a_1 b_1}$.

13. A sequence $\{s_n\}$ is defined as follows:

$$s_1 = a > 0, \qquad s_{n+1} = \sqrt{\frac{s_n^2 + ab^2}{a+1}}, \ (n \geq 1), \text{ where } b > a.$$

Show $\{s_n\}$ is monotonic, and converges to b.

14. If $x_1 = 4$, $x_{n+1} = 3 - \frac{2}{x_n}$, $(n \geq 1)$, prove $\{x_n\}$ converges to a limit. Find the limit.

15. Let p be any positive integer other than 1. Let $\{a_n\}$ be any sequence of non-negative integers such that for every $n, 0 \leq a_n < p$. Define the sequence $\{b_n\}$ by

$$b_n = \sum_{k=1}^{n} \frac{a_k}{p^k}.$$

(a) Show the sequence $\{b_n\}$ converges to a real number in $[0, 1]$.
(b) Show if $x \in (0, 1)$, then there is exactly one sequence $\{a_n\}$ such that the sequence $\{b_n\}$ converges to x, unless $x = \frac{q}{p^m}$, for some positive integers m and q. In the latter case, show there are exactly two such sequences $\{a_n\}$.

16. A sequence $\{a_n\}$ is said to be **eventually monotone** if there is an M such that the subsequence consisting of all terms having subscript greater than or equal to M is monotone. Prove a convergence theorem about eventually monotone bounded sequences.

17. Let $\{a_n\}$ be a sequence having the property that there is an M and a q, where $0 < q < 1$, such that if $n \geq M$, then

$$\frac{|a_{n+1}|}{|a_n|} \leq q.$$

What can be said about the convergence of $\{a_n\}$?

18. Show every bounded sequence has a monotonic subsequence. Conclude every bounded sequence has a convergent subsequence.

19. Show the sequence $\left\{\frac{n^n}{n!e^n}\right\}$ tends to a limit.

20. Let f be any function from $\mathbf{N}$ into the collection of all sequences of digits from the set $\{0, 1, 2\}$. Show such a function can not be onto. Conclude the real numbers are 'uncountable' (see the appendix). [Hint: Assume that an onto f exists and construct a new sequence by changing the nth term of $f(n)$. The argument required here is the famous **diagonal argument** due to Cantor!]

21. Let ω (omega) denote the collection of all sequences of real numbers, and consider the following six mathematical sentences:

(a) $\exists A \forall \epsilon \exists M[\epsilon > 0 \text{ and } n > M \text{ .implies. } |a_n - A| < \epsilon]$,

(b) $\forall \epsilon \exists A \exists M[\epsilon > 0 \text{ and } n > M \text{ .implies. } |a_n - A| < \epsilon]$,

(c) $\forall \epsilon \exists M \exists A[\epsilon > 0 \text{ and } n > M \text{ .implies. } |a_n - A| < \epsilon]$,

(d) $\exists M \forall \epsilon \exists A[\epsilon > 0 \text{ and } n > M \text{ .implies. } |a_n - A| < \epsilon]$,

(e) $\exists A \exists M \forall \epsilon[\epsilon > 0 \text{ and } n > M \text{ .implies. } |a_n - A| < \epsilon]$,

(f) $\exists M \exists A \forall \epsilon[\epsilon > 0 \text{ and } n > M \text{ .implies. } |a_n - A| < \epsilon]$, where $\{a_n\}$ is an arbitrary sequence. Each of these formulae defines a subset of ω. For example, the subset consisting of all sequences satisfying the first formula is the usual collection of convergent sequences. Call the classes $C_1, C_2, \ldots, C_6$, respectively. Find all set containment relations that hold between these classes. Where possible, give a 'nice' description of the classes defined. (The symbols $\forall$ (for all) and $\exists$ (there exists) are the universal and existential quantifiers, respectively.)

22. Let $K > 0$. What class of sequences is defined by

$$\exists A \forall \epsilon \exists M [\epsilon > 0 \text{ and } n > M \text{ .implies. } |a_n - A| < k\epsilon]?$$

23. Consider the following classes of sequences:

$$
\begin{array}{rcl}
\omega & = & \text{the class of all real sequences;} \\
\mathbf{conv} & = & \text{the class of all convergent sequences;} \\
\mathbf{bdd} & = & \text{the class of all bounded sequences;} \\
\mathbf{div} & = & \text{the class of all divergent sequences;} \\
\mathbf{div}_\infty & = & \text{the class of all sequences that diverge to } +\infty; \\
\mathbf{div}_{-\infty} & = & \text{the class of all sequences that diverge to } -\infty; \\
\mathbf{osc} & = & \text{the class of all sequences that oscillate;} \\
\mathbf{mon} & = & \text{the class of all monotonic sequences.}
\end{array}
$$

Identify all possible containment relationships between these classes that you can think of from the development so far.

24. Show the convergence properties of a monotone sequence are completely determined by any of its subsequences.

25. Let $a_n = \left[\frac{n}{2}\right] - \frac{n^2}{2n+1}$. Find $\sup\{a_n\}, \inf\{a_n\}$. Does $\{a_n\}$ converge?

26. Let $0 < a < b \leq 1$.

    (a) Show $\frac{b^{n+1} - a^{n+1}}{b-a} < (n+1)b^n$.

    (b) Let $a_n = (1 + \frac{1}{n})^n$. Show $\{a_n\}$ is monotone increasing and bounded above by 4.

27. Show Theorem 1.3.1 can serve as a substitute for the supremum principle.

Chapter 2

Limits of Functions

Our purpose in this chapter is to study the limit concept as applied to functions. The notion of 'limit as x tends to a of a function f' is fundamental to all further ideas in real analysis. We shall develop it here and then use it to define 'continuous function' and to obtain the elementary properties of continuous functions.

In the last chapter, we began the study of the limit concept as applied to sequences that are special types of functions having for a domain the set of all natural numbers, $\mathbf{N}$. In this chapter, we apply the concept to functions having as domains, arbitrary subsets of the set $\mathbf{R}$ of real numbers.

2.1 Functions, Limits at Infinity

The concept of function is central to the rest of this book, and so we discuss it at some length. As can be seen from the treatment in the appendix, functions are collections of ordered pairs, where these collections satisfy certain constraints. Implicit in the definition is the notion of an **ordered pair**. These are defined in the appendix, but the property that is important, in fact essential, is that *ordered pairs are equal if and only if they are equal coordinatewise*, namely,

$$(a, b) = (c, d) \text{ .if and only if. } a = c \text{ and } b = d.$$

Here, a is called the **first coordinate** and b the **second coordinate**. We use this property to make the following definition:

Definition. A **function** is a set f of ordered pairs such that if x is the first coordinate of an ordered pair in f, then there is exactly one y such that $(x, y) \in f$.

Discussion. The trouble with this definition is it is too pat. The subtleties are hidden, and the motivation for the definition has long since disappeared. Analysis grew out of graphs and rules for computations and geometry, not out of rigorous discussions of ordered pairs. And so we should look to these for intuition (graphs, rules for computation, geometry). Let us examine these in turn.

Functions, as treated in the last century, were essentially rules for computation. By this we mean for any given number that was a proper input, one could obtain a unique output by applying the rule. When thinking about functions in this way, we are really identifying the function with a computing machine. For a concrete example of this idea, the reader should think of a calculator that will compute the square root of nonnegative numbers. Obviously, not every number is suitable for computation purposes by all machines. This gives rise to the idea of **domain**. (The domain of the 'root x' machine is the nonnegative reals.) Also, we notice that not all numbers will appear as outputs, which gives rise to the idea of **range**. (The range of 'root x' is the nonnegative reals.) The main features of this discussion are that the rule is **single-valued** (that is, it produces a unique output for each accepted input), has a domain, and has a range. The trouble with the machine concept is that it is not suitable for mathematical analysis. The important fact is that with each element of the domain we associate a unique element of the range. This can be captured by considering sets of ordered pairs, and the machine in the middle can be dispensed with. The notion of 'rule' has not disappeared, however. We often specify functions by giving the rule for computation, together with the domain; for example,

$$f(x) = 2x + 5, \ x \in \mathbf{R}.$$

In set notation, this type of specification for functions takes the form

$$\{(x, 2x + 5) : x \in \mathbf{R}\}.$$

Either type of specification gives us immediate access to an algebraic rule for computation. It is through judicious manipulation of this rule that we will be able to establish the interesting properties of a particular function. For concrete examples of this, the reader should review our computations with particular sequences. Manipulation of the rule was the method by which limits were established. The student should also note that regardless of the notation, a function f is a set of ordered pairs and that '$f(x)$' does not denote a function, but a particular output of the function f, called the **value of the function at** x.

We shall consider only ordered pairs of real numbers, so our functions are real valued functions of a real variable. We shall frequently use the arrow notation, namely, $f : D \to \mathbf{R}$, where the domain is the subset D of $\mathbf{R}$ and the range is contained in $\mathbf{R}$. We shall use Dmn f and Rng f throughout the book to denote the domain and the range of a function f. $\square$

Graphs provide us with pictures of functions. Pictures, in turn, lead naturally to questions about geometry. As such, pictures can be important aids to our intuition, and we should use them freely. To see how a graph can lead us to natural geometric questions, look at the graph of $f(x) = 2x$, $x \in \mathbf{R}$. The graph of this function (Figure 2.1.1) can be drawn by a single motion of the pencil across the page. The result is a continuous line, and mathematicians began to wonder what analytic properties of functions would lead to graphs that can be produced with a single stroke of a pen.

Graphs can also be misleading. For example, the graph of

$$f(x) = \left\{ \begin{array}{ll} x & x \in \mathbf{Q} \\ 1 - x & x \notin \mathbf{Q} \end{array} \right.$$

pictured in Figure 2.1.2 does not appear to represent a function because the rule appears to associate more than one y with a single value of x. However, examination of the rule shows that in fact there is a single unique y associated with each x, and so the rule is indeed single-valued, even though the picture cannot make it seem so.

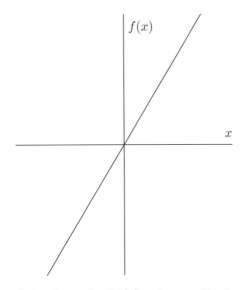

Figure 2.1.1 A graph of $f(x) = 2x$, $x \in \mathbf{R}$, showing that it can be drawn without lifting the pencil from paper.

In summary, we will consistently emphasize functions from the point of view of a domain, D, and a rule $f(x) = \cdots$, $x \in D$. We will draw graphs whenever these will aid our intuition, but we will always test our intuition by using the algebraic and analytic tools we have developed. Once again, we state that pictures are essential to the generation of intuition, but they cannot form a part of an analytic proof!

Unless explicitly stated otherwise, *all functions in this book will have both their domain and range contained in* $\mathbf{R}$. With respect to domains, the reader should keep in mind the standard domains of the elementary calculus, namely, open intervals (a, b), closed intervals $[a, b]$, and infinite intervals (a, ∞) or $(-\infty, a]$ to illustrate the various possibilities. It is essential that the reader first understand the content of our theorems and definitions for functions having intervals for their domains, before attempting to deal with the subtleties that arise for functions having domains that are arbitrary subsets of the real numbers. For this reason, functions having interval domains will be emphasized and many of the subtleties left to the exercises.

Definition. Let f be a function with domain D. We say that f **has a limit as** x **tends to** ∞ provided there exists an $A \in \mathbf{R}$ such that for every positive ϵ,

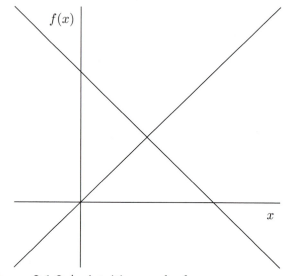

Figure 2.1.2 An intuitive graph of
$$f(x) = \begin{cases} x, & x \in \mathbf{Q} \\ 1 - x, & x \in \mathbf{R} \sim \mathbf{Q}. \end{cases}$$ Note: Both lines are part of the graph and both lines have infinitely many 'holes'.

there is an $M \in \mathbf{R}$ such that

$$x \in D \text{ and } x > M \text{ .implies. } |f(x) - A| < \epsilon.$$

In the case that a number A satisfying the definition exists, we say that A is the **limit** of $f(x)$ as $x \to \infty$, and we write

$$\lim_{x \to \infty} f(x) = A.$$

Notation. We will often use the symbol $+\infty$, instead of ∞, to emphasize we are looking at positive infinity, as opposed to negative infinity, which is denoted by $-\infty$.

Discussion. Let us compare this definition with that of the limit definition for sequences. Both require us to find a fixed real number A as a first step. Both use small values of ϵ as a test for closeness. Both require us to find an M, which depends on ϵ, such that the functional values (sequence values) are within a distance ϵ of A provided that x (or n, in the case of sequences) is at least as large as M. So what is the difference? Only that functions have a domain D that is an arbitrary subset of the real numbers, while for sequences the domain is the set of all natural numbers. In dealing with a sequence, we know that there are natural numbers—elements of the domain—that exceed any choice of M. For a function having an arbitrary domain, this is not necessarily the case; indeed, $(M, \infty) \cap D$ may very well be empty for some choices of M. Thus, we may expect that the differences between this definition of limit at $+\infty$ and the corresponding definition for sequences will arise for functions whose domain is bounded above. Such a difference is given in Exercise 3.

Let us discuss the existence of limit at infinity geometrically. If the limit, A, exists, then, given $\epsilon > 0$, we can construct an infinite horizontal strip around the line $y = A$, of width 2ϵ, bounded by the lines $y = A - \epsilon$ and $y = A + \epsilon$. Then, we come up with a number M such that for all x in the domain of f, to the right of the point M, the values $f(x)$ lie in this strip. In other words, we can arrive at a stage (determined by M), from which point onward, the graph of f lies entirely within this infinite strip. Figure 2.1.3 brings out these features, and is similar to Figure 1.1.3 for sequences. It is clear that if the limit at infinity did not exist, then for every number A, we can construct a horizontal strip of a suitable width, such that for every point M on the x-axis, we can always find points x to the right of M, where the graph shoots outside this strip. $\square$

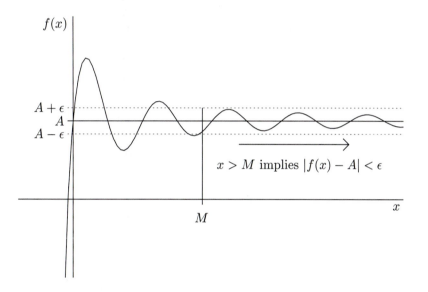

Figure 2.1.3 A graph illustrating $\lim\limits_{x \to \infty} f(x) = A$. Notice how for $x > M$, $f(x)$ lies in the ϵ-strip about A, that is, between the dashed lines.

Example 1. Show $f(x) = \frac{1}{x+1}$, $x \in \mathbf{Q} \cap [0, \infty)$ has a limit as x tends to ∞.

Solution. It is clear that we should take $A = 0$. Now let ϵ be a fixed positive real number. We have

$$\begin{aligned} |f(x) - 0| &= \left| \tfrac{1}{x+1} - 0 \right| \\ &= \tfrac{1}{x+1} \qquad \text{for } x \in \mathbf{Q} \cap [0, \infty). \end{aligned}$$

Let $M = \frac{1}{\epsilon}$. Now if $x > M$, then $x + 1 > x > \frac{1}{\epsilon}$, whence

$$|f(x) - 0| < \frac{1}{x} < \epsilon.$$

Thus, the limit is 0 as claimed. $\square$

Discussion. A graph of f is shown in Figure 2.1.4 and makes the conclusion intuitively obvious. The calculations are essentially the same as those we would

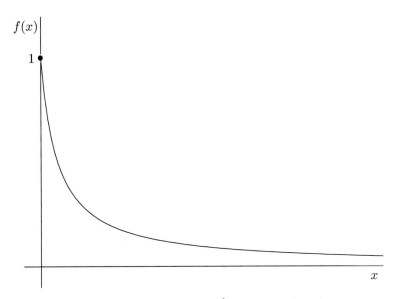

Figure 2.1.4 The graph of $f(x) = \frac{1}{x+1}$, $x \in \mathbf{Q} \cap [0, \infty)$. The graph has infinitely many 'holes'.

generate when considering the sequence $\left\{\frac{1}{n+1}\right\}$; indeed, it is these calculations that guided us in our choice of $A = 0$. Since the domain is unbounded, no special precautions need be taken beyond noting that all x's considered must come from the domain specified. $\square$

Negation of the Limit Definition. A function, f, having domain, D, will fail to have a limit as x tends to ∞ provided for every $A \in \mathbf{R}$, there exists $\epsilon > 0$ such that for every positive $M \in \mathbf{R}$, there exists $x \in D$ such that $x > M$, and

$$|f(x) - A| \geq \epsilon.$$

Discussion. We compare this negation with that of the analogous statement for sequences. Both begin with the phrase 'for every real number A, there exists $\epsilon > 0$ such that for every M', although in the sequence case there is no requirement that M be positive. The reader can check that the substance of the definition in the sequence case would not be changed by adding this requirement; nor would the substance in the function case be changed if we did not require M to be positive. In the case of a function, we are next required to find an x that is in the domain of f and greater than the previously chosen M. This x will witness the fact that the particular A cannot be the limit because for this x, $f(x)$ is not close to A. For sequences, since every positive integer n is in the domain of the sequence, we simply have to find an n greater than M with the property that a_n is not close to A. The two inequalities that validate 'is not close to' are identical.

The level of similarity between the two limit concepts is not remarkable because the definition for functions generalizes the concept for sequences by extending the definition to functions having more complicated domains.

It is implicit in the above that if a function fails to have a limit as x tends to ∞, then the domain, D, cannot be bounded above. This is due to the fact that establishing a falsity generally requires a counterexample, which involves witnesses. In the above, a witness, x, must be found, and three conditions must be satisfied:

(i) $x \in D$;

(ii) $x > M$;

(iii) $|f(x) - A| \geq \epsilon$.

The first two conditions apply directly to x. By virtue of M being arbitrary, they imply Dmn f is unbounded. The third condition asserts that x witnesses the fact A cannot be the limit. The point is, since a witness must be found, that is, shown to exist, we could not expect to prove that a limit did not exist by a vacuous argument based on a domain that was bounded above.

Last, in the discussion following the definition of limit at infinity, we have already pointed out the geometrical meaning of the nonexistence of limit at infinity. That geometrical discussion captures, in an intuitive sense, all the analytical points that are made above. $\square$

Example 2. Using any standard definition of the function $f(x) = \cos x$, $x \in \mathbf{R}$, show cos does not have a limit as x tends to ∞.

Solution. From any standard definition of $\cos x$ such as would appear in a calculus book, we know that $\cos 2n\pi = 1$, while $\cos(2n + 1)\pi = -1$, where $n \in \mathbf{N}$. Now set $\epsilon = 1$ and let $M > 0$ be arbitrary, but fixed. Let $n > M (n \in \mathbf{N})$ be chosen. Evidently, for any real number, A, whatsoever, the distance of A from one of $\cos n\pi$ and $\cos(n + 1)\pi$ must be at least 1 **(WHY?)**. Since $n\pi$ and $(n + 1)\pi$ are both members of D, we are done. $\square$

Discussion. A comment on our use of trigonometric functions is useful. In the early stages of our development, trigonometric functions will be used mostly as sources of counterexamples. The reason for this is that the development of the analytic properties of the trigonometric functions, such as those associated with computing various limits, requires a tractable analytic definition of the function. Suitable definitions will be developed later in Chapter 9, when an appropriate foundation has been laid. Until then, the reader, when asked to do problems involving trigonometric functions, should feel free to use any convenient definition and any properties that can be rigorously established from that definition. It is our belief that the properties employed above with respect to the cos function can be rigorously established from any of the usual definitions. To see this, the reader should take his favorite definition and go through the exercise of showing that $\cos 2n\pi = 1$, for all $n \in \mathbf{N}$. Similar remarks apply to the exponential and logarithmic functions.

Recalling our work with sequences, we found that a sequence failed to converge for one of two reasons, either it was unbounded, or it oscillated. Evidently, $\cos x$ oscillates as x tends to ∞, and this is the reason why it fails to converge to a limit as $x \to \infty$.

Last, we would reiterate that the definition of limits at infinity for functions is a generalization of the concept for sequences. Thus, each of the examples done for sequences can be reinterpreted in the context of limits of a function. To do this, one simply gives an analogous function definition based on the specification for the given sequence. The reader may obtain useful insights by reviewing the material on sequences, in this light. □

Given the relationship between the two limit definitions, we might expect that many theorems that are true about one type of limit are also true about the other type of limit. Indeed, this is one very good way of proceeding in mathematics. Namely, we know that certain things are true in a given situation; we ask which, if any, of these things remain true in the somewhat altered situation. For limits of functions, we might expect the following to be valid based on the results obtained for limits of sequences:

(i) The limit, if it exists, is unique.

(ii) The existence of the limit implies a boundedness condition on Rng f.

(iii) Theorems relating limits of functions to arithmetic operations may be valid (sum, product, absolute value).

(iv) Monotonicity and boundedness conditions may imply the limit of a function exists.

The reader should stop here and try to formulate suitable statements for the theorems indicated above, based on our previous work, and then try to prove those statements. In certain instances one will find that more hypotheses are needed to complete the proof. In other cases, the conclusion must be altered. By considering various examples, the industrious reader will be able to arrive at the correct statements of the theorems. It should be noted that the process of working out these statements will be empirical, and that proofs will be constructed only after this essentially experimental process has been completed.

Theorem 2.1.1. *Let f be a function with domain D that is not bounded above. If f has a limit at ∞, then this limit is unique.*

Proof. By assertion, f has a limit as x tends to infinity; call it A. Let B be any other limit. If $B \neq A$, then without loss of generality, we may assume that $B < A$, whence $A - B$ is positive. Set $\epsilon = \frac{A-B}{2}$. Since A is a limit, we can find a real number M_A, depending on A (as well as on ϵ), such that if $x > M_A$ and $x \in D$, then $|f(x) - A| < \epsilon$. Further, since B is also a limit, we can find M_B such that if $x > M_B$ and $x \in D$, then $|f(x) - B| < \epsilon$. Let $M = \max\{M_A, M_B\}$. Now D is not bounded above, so there is an $x_0 \in D$ with $x_0 > M$. For such an x_0 we have

$$f(x_0) < B + \epsilon = A - \epsilon < f(x_0),$$

which is absurd. We conclude $A = B$. □

Discussion. Compare this argument with that of Theorem 1.2.1. In essence the lines of reasoning, for both cases, are identical. The only significant difference between them is that at a key point we must use the fact that the domain of f is unbounded and contains members that are greater than M. This condition is automatically satisfied in the case of sequences. In Exercise 3 the reader will be asked to show that if a function has a bounded domain, then any real number will serve as a limit at infinity. In the above proof, one could also force a contradiction by writing $A - B = (A - f(x)) + (f(x) - B)$, and then applying the Triangle inequality. The reader should attempt such a proof. □

Theorem 2.1.2. *Let f have a domain D that is not bounded above. If f has a limit at ∞, then there is a real number B such that $\{f(x) : x \in D \text{ and } x > B\}$ is bounded.*

Proof. Let A be the limit at ∞ and $\epsilon = 1$. Then, by definition there is an M such that if $x > M$ and $x \in D$, then $|f(x) - A| < 1$. It is immediate that if we set $B = M$, the conclusion will follow. □

Discussion. This theorem differs from the comparable theorem on sequences (Theorem 1.2.3) in a fundamental way related to the notion of bounded. A function, g, is **bounded** if Rng g is a bounded set. Thus, for g to be bounded, there must be a fixed $K \in \mathbf{R}$ such that for all $x \in$ Dmn g, $|g(x)| < K$. For the f of the theorem, f is bounded only on the subset $D \cap (M, \infty)$ for some M, not necessarily on the whole of D. In contrast, for sequences we found that the existence of the limit forces the whole of the sequence to be bounded, that is, the sequence is bounded as a function, whence for every n, $|a_n| < K$. In Exercise 4 the reader will be asked to carefully explore this difference. □

Definition. Let f be a function with domain D. We say that f is **monotone increasing on** D provided that for every x and $y \in D$, $x < y$ implies $f(x) \leq f(y)$.

In the obvious way, we can define **monotone decreasing on** D (the reader should do so). A function will be called **monotone** if it is either monotone increasing or decreasing on its domain. It is **strictly monotone** if $x < y$ implies the strict inequality $f(x) < f(y)$ (for the increasing case). A similar definition can be given for the decreasing case.

Discussion. The definition of monotonicity for functions attempts to capture a basic property of graphs, namely, that some graphs go only 'up' as they go from left to right across the page, while others go only 'down'. The reader should ask what it means for a graph to go 'only up' and verify that the definition really does capture the intuitive ideas involved. □

Theorem 2.1.3. *If f be monotone on D and $\{f(x) : x \in D\}$ is bounded, then the limit at $\lim_{x \to \infty} f$ exists.*

Proof. Without loss of generality, we may assume that f is monotone increasing. There are two cases, namely, when D is bounded above and when D has no upper bound. The former is left to Exercise 3. Thus, we assume that D is not bounded above. Since the set $S = \{f(x) : x \in D\}$ is bounded above, we may set $A = \sup S$. Fix $\epsilon > 0$. By Theorem 0.4.2, we can find $y \in S$ such that $A - \epsilon < y \leq A$. But $y = f(x_0)$ for some $x_0 \in D$. Let $M = x_0$. Thus, if $x > M$ with $x \in D$, we have

$$A - \epsilon < y \leq f(x) \leq A,$$

whence $|f(x) - A| < \epsilon$. The decreasing case is left to Exercise 5. □

Discussion. Once again, we see that the crucial fact employed in the proof is the supremum principle. The reader should compare this theorem with Theorem

1.3.1. Note that the monotonicity was used in concluding $f(x) \geq f(x_0) = y$ from $x > M = x_0$. $\square$

These theorems illustrate the results that can be obtained for limits at ∞ and their similarity to those for sequences.

We conclude this section by defining the operations of addition, multiplication, and so on, as applied to functions. As in the case of sequences, this is done pointwise. Thus, for functions f and g with a common domain D, we define the **sum** $f + g$, the **product** $f \cdot g$, the **reciprocal** $\frac{1}{f}$, and the **absolute value function** f as follows:

$$
\begin{aligned}
(f + g)(x) &= f(x) + g(x), x \in D; \\
(f \cdot g)(x) &= f(x) \cdot g(x), x \in D; \\
\left(\frac{1}{f}\right)(x) &= \frac{1}{f(x)}, x \in D, f(x) \neq 0; \\
|f|(x) &= |f(x)|, x \in D.
\end{aligned}
$$

Obviously, $\dfrac{f}{g}$ stands for the function $f \cdot \dfrac{1}{g}$. The behavior of these functions as x approaches $\pm\infty$ is treated in Exercises 10—13.

Exercises

1. Define **limit as x tends to minus infinity** of f. Write a precise negation.

2. Prove theorems corresponding to Theorems 2.1.1—2.1.3 for limit as x tends to $-\infty$.

3. Let f have a bounded domain. Show any real number will serve as a limit at infinity; use this result to complete the proof of Theorem 2.1.3.

4. Give an example of a function that is not bounded, that is, has unbounded range but has a limit at infinity. Why is it impossible to construct a sequence with this property?

5. Fill in the missing details in the proof of Theorem 2.1.3.

6. Prove or disprove the following functions have limits at ∞ $(-\infty)$; find $\lim\limits_{x \to \infty} f(x)$ $(\lim\limits_{x \to -\infty} f(x))$ whenever it exists.

   (a) $f(x) = \frac{1}{x^2}$, $x \in \mathbf{R}^+ = \{x : x > 0 \text{ and } x \in \mathbf{R}\}$;   (b) $f(x) = \sin x$, $x \in \mathbf{R}$;

   (c) $f(x) = x^2$, $x \in [0, 10^{10000}]$;   (d) $f(x) = \frac{1 - x^2}{(1+x)^2}$, $x \in \mathbf{R}^+$;

   (e) $f(x) = \frac{x^3 + 1}{x^7 - 1}$, $x \in \mathbf{R} \sim \mathbf{Q}$;   (f) $f(x) = \frac{x - x^2}{1 + x^2}$, $x \in \mathbf{R}$;

   (g) $f(x) = \sqrt{2x} - \sqrt{x + 1}$, $x \in \mathbf{R}^+$;   (h) $f(x) = \frac{\cos x}{x}$, $x \in \mathbf{R}^+$;

(i) $f(x) = \sqrt{x + 10} - \sqrt{x + 2}$, $x \in \mathbf{R}^+$;

(j) $f(x) = \frac{2^x}{e^x}$, $x \in \mathbf{Q}$;

(k) $f(x) = \begin{cases} 1 - x, & x \in \mathbf{Q} \\ 1 + x, & x \notin \mathbf{Q}, \end{cases}$ $x \in \mathbf{R}$;

(l) $f(x) = \frac{-3x + 1}{\sqrt{x^2 + x}}$, $x \in \mathbf{R}^+$;

(m) $f(x) = \begin{cases} 1 - x, & x \leq 1 \\ 1 + x, & x > 1, \end{cases}$ $x \in \mathbf{R}$;

(n) $f(x) = \frac{\sqrt{x} - x}{\sqrt{x} + x}$, $x \in \mathbf{R}^+$;

(o) $f = \{(2, 3), (3, 4), (5, 7), (9, 11)\}$;

(p) $f(x) = \frac{3x - 2}{\sqrt{2x^2 + 1}}$, $x \in \mathbf{R}$;

(q) $f(x) = \sqrt{x - \sqrt{x}} - \sqrt{x + \sqrt{x}}$, $x \in [1, \infty)$.

7. Which of the following functions is monotone (strictly monotone) on their domains? Prove your answers.

(a) $f(x) = \cos x$, $x \in \mathbf{R}$;

(b) $f(x) = \frac{1}{x}$, $x \in \mathbf{R} \sim \mathbf{Q}$;

(c) $f(x) = \cos x$, $x \in \bigcup_{n \in \mathbf{N}} [2n\pi, (2n + 1)\pi]$;

(d) $f(x) = \frac{1}{x^2}$, $x \in \mathbf{R} \sim \mathbf{Q}$;

(e) $f(x) = |x|$, $x \in [-25, 0]$;

(f) $f(x) = \sqrt{x}$, $x \in \mathbf{R}^+$;

(g) $f(x) = 0$, $x \in \mathbf{R}$;

(h) $f(x) = 0$, $x \in \mathbf{Q}$;

(i) $f(x) = \begin{cases} 1 - 2x, & x < 1 \\ 2x, & x \geq 1, \end{cases}$ $x \in \mathbf{R}$;

(j) $f(x) = x^3$, $x \in \mathbf{R}$;

(k) $f(x) = \begin{cases} 1 - x, & x \in \mathbf{Q} \\ x, & x \in \mathbf{R} \sim \mathbf{Q} \end{cases}$;

(l) $f(x) = \frac{1}{x^2 + 1}$, $x \in \mathbf{R}$;

(m) $f(x) = \frac{x^2}{x^2 - 1}$, $x \in \mathbf{R} \sim \{-1, 1\}$;

(n) $f(x) = \frac{x}{x^2 + 1}$, $x \in \mathbf{R}$.

8. Which of the functions in Exercises 6 and 7 are bounded in their domains?

9. Let f and g be two functions that differ in value at a finite number of points in the common domain. Prove or disprove: $\lim_{x \to \infty} f(x) = \lim_{x \to \infty} g(x)$, provided it exists. What happens if f and g differ at an infinite number of points?

10. Let f and g be defined on D. Show

$$\lim_{x \to \infty} (f + g)(x) = \lim_{x \to \infty} f(x) + \lim_{x \to \infty} g(x)$$

assuming both limits on the right exist. Further, give an example to show the limit on the left may exist even though the limits on the right fail to exist.

11. Discuss the equation

$$\lim_{x \to \infty} (f \cdot g)(x) = \lim_{x \to \infty} f(x) \cdot \lim_{x \to \infty} g(x).$$

12. Show if $\lim_{x \to \infty} f(x) = A$, then $\lim_{x \to \infty} |f(x)| = |A|$, and $\lim_{x \to \infty} f^2(x) = A^2$. What about the converses?

13. Discuss $\lim_{x \to \infty} \frac{f(x)}{g(x)}$.

14. If f and g possess limits as x approaches ∞, and if $f(x) \leq g(x)$ throughout $\mathbf{R}^+$, show $\lim_{x \to \infty} f(x) \leq \lim_{x \to \infty} g(x)$. If $f(x) < g(x)$ throughout $\mathbf{R}^+$, can you conclude $\lim_{x \to \infty} f(x) < \lim_{x \to \infty} g(x)$?

15. We say f **diverges to** ∞ as x tends to ∞ provided for each $K > 0$, there exists a real number M such that $x \in \text{Dmn } f$ and $x > M$ .implies. $f(x) > K$. Formulate the concept of '**diverges to** $-\infty$'. Prove if $\lim_{x \to \infty} f = \infty$, then $\lim_{x \to \infty} \frac{1}{f} = 0$. Is the converse true?

16. Let $\lim\limits_{x\to\infty} \dfrac{f(x)}{g(x)} = A \neq 0$, where f and g are defined for $x > a \in \mathbf{R}$, and further
$g(x) > 0$ for $x > a$. Prove
(a) If $A > 0$, then $\lim\limits_{x\to\infty} f(x) = \infty$ if and only if $\lim\limits_{x\to\infty} g(x) = \infty$;
(b) If $A < 0$, then $\lim\limits_{x\to\infty} g(x) = -\infty$ if and only if $\lim\limits_{x\to\infty} g(x) = \infty$.

17. A function defined on $\mathbf{R}$ of the form

$$f(x) = \sum_{i=0}^{n} a_i x^i$$

where $n \in \mathbf{N}$, and the $a_i \in \mathbf{R}$ is called a **polynomial function**; if $a_n \neq 0$, we
say f is a **polynomial of degree** n. Let $P(x)$ and $Q(x)$ denote polynomials in
x, and set $f(x) = \dfrac{P(x)}{Q(x)}$ for $x \in D$, where D does not contain any of the zeros
of $Q(x)$. State and prove a theorem that completely describes the behavior of f
at ∞. Is this behavior eventually monotone? (See Exercise 1.3.16 for sequence
definition.) You may use results from elementary calculus and the Fundamental
Theorem of Algebra to settle the question of monotonicity.

18. Let f be defined on D which is not bounded above. Show $\lim\limits_{x\to\infty} f(x)$ exists, exactly
if for every sequence $\{a_n\}$ of elements of D that diverges to ∞, $\lim\limits_{n\to\infty} f(a_n)$ exists.

19. Show the function
$$f(x) = \frac{x \tan x + 2x - 1}{x + 1}$$

does not possess a limit as x approaches ∞, but if x ranges through a sequence
of values $x = n\pi + \frac{\pi}{4} (n = 0, 1, \dots)$, then $f(x)$ approaches a limit.

20. If $f : (a, \infty) \to \mathbf{R}$ is such that $\lim\limits_{x\to\infty} x f(x) = L \in \mathbf{R}$, show $\lim\limits_{x\to\infty} f(x) = 0$.

21. Obtain a set of sufficient conditions for the existence of $\lim\limits_{x\to\infty} (f \circ g)(x)$.

22. Show the concept of limit at infinity can be regarded as a two-person game.
Illustrate with examples.

2.2   Limit of a Function at a Real Number

The problem of formalizing the concept of limit was especially crucial for functions, since it is a prerequisite to any discussion of the geometry of the graphs
of functions. We might expect that the definition of the limit of a function at a
real number is quite similar to that for the limit at $+\infty$. To a degree this is the
case. However, at $+\infty$ we must worry about the intuitive idea of 'x traveling off
to $+\infty$' while now we want to capture the intuitive idea of 'x traveling toward
some fixed real number'.

Definition. Let f be defined on D and let $a \in \mathbf{R}$. We say that f has a **limit as
x tends to a,** provided that there exists an $A \in \mathbf{R}$ such that for every $\epsilon > 0$,

there exists $\delta > 0$ such that for every $x \in D$,

$$0 < |x - a| < \delta \text{ .implies. } |f(x) - A| < \epsilon.$$

Further, in the case that such a number A exists, we shall say that A is the **limit as x tends to** a of f or, more simply, as the **limit at** a of f and write

$$\lim_{x \to a} f(x) = A.$$

Discussion. If we compare the definition above with that of 'limit as x tends to $+\infty$', we find that they are the same until after the arbitrary ϵ is chosen. The definition above then requires us to find a positive δ, whereas the definition in the case of ∞ required us to find an M. Thus, the quantity δ depends on the ϵ, which was given in advance, just as the quantity M was a function of ϵ in the previous case. We may emphasize this dependence by writing $\delta = \delta(\epsilon)$. As was pointed out, M would generally be large since its purpose is to keep x close to $+\infty$. But δ will, in general, be very small since its purpose is to keep x close to a. It does this by requiring

$$|x - a| < \delta,$$

which restricts our attention to those members of the domain of f that are within a distance δ of a, if indeed there are any members of D that are close to a. In the case of x tending to ∞, x can travel only in one direction, while in the case of x tending to a real number a, the point a can be approached both from left and the right sides. The observant reader will have noticed that we have yet to comment on a significant part of the δ inequality, namely,

$$0 < |x - a|.$$

This part of the inequality has the effect of preventing x from actually taking the value a. Perhaps this seems strange, but remember that in the definition of the limit at ∞, x can never take the value ∞, since ∞ is not a number. Our interest is in what happens to the function values as x *tends to* ∞. In the same way, our interest is not what actually happens when x takes the value a, but rather what is happening when x *is close to* a, *but different from* a. That is why we require the distance from x to a to be positive.

The geometry of the situation can be made clear from a picture (see Figure 2.2.1). The picture has two parts, concerning a limit at a and another limit at b. In both cases we look at A as the potential limit. First, note that as soon as we are given an ϵ, this gives rise to an open interval, $(A - \epsilon, A + \epsilon)$, on the $f(x)$-axis (y-axis), which we shall call the ϵ-interval. This interval cuts out a horizontal strip that runs parallel to the x-axis. Similarly, the open interval $(a - \delta, a + \delta)$ generates a vertical strip of width 2δ, parallel to the y-axis. These two strips intersect in a rectangle. Now for the limit to exist, we must be able to find a δ such that the rectangle generated by the intersection of the δ-strip with the ϵ-strip completely contains the graph for all the domain values that lie inside the δ-interval (see Figure 2.2.1). For the δ-interval about a, the graph associated with domain values in the δ-interval is completely contained in the generated rectangle. However, for the δ-interval about b, the graph is not contained in the associated rectangle, that is, there are domain values inside the δ-interval whose function values lie outside the ϵ-interval. It can be seen from the figure that it is possible to choose a new δ_1-interval about b that is small enough to ensure that the portion of the graph in the vertical $2\delta_1$ strip also lies in the horizontal ϵ strip.

The reader should imagine ϵ shrinking, so that the interval about A shrinks. This will force the choice of δ to shrink if we are to generate a rectangle that will completely contain the part of the graph over its base. Thus, the δ we obtain depends on the ϵ, given in advance. If, as ϵ shrinks to 0, there is always a choice for δ, then the limit will exist. For the examples below, the reader should draw the appropriate picture to verify the geometry of the situation.

The limit of $f(x)$ as x approaches a is not to be confused with the value of the function f at the member a. The reader will have noted that we do not require a to be a member of the domain of f. As we shall see, this has as a consequence that we must always be aware whether or not a is a member of the domain of f, which adds a degree of complexity to our work. However, this complexity is a result of the fact that there are many naturally arising situations in which we want to compute a limit where a is not in the domain of f. For example, if $f(x) = \frac{1}{x}$, $x > 0$, we may ask, what is the limit of f as x tends to 0. So that we may deal with this and other natural questions, which may arise if the domain is not an interval, we have permitted this complexity to exist. This does not deny the fact that our main interest is in the case when a belongs to the domain of f. $\square$

Example 1. For $f(x) = x^2$, $x \in \mathbf{R}$, find the limit as x tends to 1.

Solution. We claim that $\lim_{x \to 1} f(x) = 1$. To see this, let us fix $\epsilon > 0$ and consider

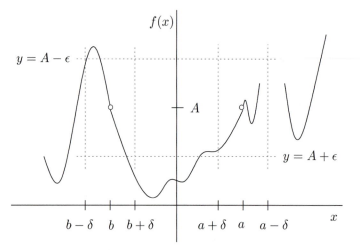

Figure 2.2.1 The figure depicts the geometry underlying the limit concept.

the quantity $|f(x) - 1|$. Now,

$$
\begin{aligned}
|f(x) - 1| &= |x^2 - 1| = |(x-1)(x+1)| \\
&= |x - 1| \cdot |x + 1|.
\end{aligned}
$$

Suppose $\delta \leq 1$ and $|x - 1| < \delta$. Then

$$|x - 1| < \delta \leq 1 \ \text{.implies.} \ 0 < |x + 1| < 3, \tag{$*$}$$

and, further,

$$|x - 1| < \delta \leq 1 \ \text{.implies.} \ |x + 1| \cdot |x - 1| < 3|x - 1|.$$

Let us therefore set $\delta = \min\left\{1, \frac{\epsilon}{3}\right\}$. Now, if $x \in D$ and $0 < |x - 1| < \delta$, we know

$$
\begin{aligned}
|f(x) - 1| &= |x^2 - 1| \\
&< 3|x - 1| \\
&< 3\delta \\
&\leq 3 \cdot \frac{\epsilon}{3} = \epsilon,
\end{aligned}
$$

which is the desired inequality. $\square$

Discussion. The first thing to note is that the primary tool for manipulating the expression $|f(x) - A|$ is algebra. The second is that there is a strategy, namely, to manipulate $|f(x) - A|$ into the form

$$|x - a| \cdot (\text{junk}),$$

where the 'junk' has the property that it is bounded provided that δ is sufficiently small. This is why we refer to this term as 'junk', since once we know it is bounded, we can overestimate it by a single number and we don't have to worry about it any more.

In the example above, we take $|x^2 - 1|$ and factor it to obtain

$$|x - 1| \cdot |x + 1|.$$

The term $|x - 1|$ is exactly the one that appears in the δ-inequality, and so it will be at least as small as δ. The 'junk' term in this case is $|x + 1|$. This term is bounded above by 3 as long as we require $\delta \leq 1$ because the condition $|x - 1| < \delta$ is equivalent to

$$1 - \delta < x < 1 + \delta.$$

In other words, by restricting δ, we force x to lie in the interval shown, and this permits us to assert $|x + 1| \leq 3$. Thus, if δ is small, that is, less than 1, then 3δ is also small, and in fact will be less than or equal to ϵ if δ is small enough.

The number 3 arises because of the line marked with a $(*)$ in the proof. If, alternatively, we had insisted that $\delta < 2$, then the reader can show that the 3 would be replaced by a 4, and the proof would be completed by setting $\delta = \min\{2, \frac{\epsilon}{4}\}$. Thus, there is nothing special about 3, other than it is determined by the restriction placed on δ, which, in turn, is useful for generating a bound on the 'junk' term.

In general δ *will depend on* ϵ. The nature of this dependence of δ on ϵ will be determined by the bound on the junk term, and perhaps other factors as further examples will show.

Figure 2.2.2 presents plots of $y = |x-1| = |x-a|$, $y = |x^2-1| = |f(x)-A|$, and $y = \epsilon$ on the same set of axes. Recall that the graph of $y = |x-1|$ is everywhere nonnegative and has exactly one zero, namely, at $x = 1$ (see Example 0.6.2). The graph of $y = |x^2 - 1|$ ($|f(x) - A|$) is also everywhere nonnegative. But it can have many zeros, namely, at each value of x for which $x^2 = 1$ ($f(x) = A$); in the present case there are two. One of the two zeros of $|x^2 - 1|$ also occurs at $x = 1$. This fact tells us that we have correctly chosen the value for the limit as $A = 1$. Any other choice for A will not produce a zero for $|x^2 - A|$ at $x = 1$ as the reader can verify with the aid of a graphing tool.

In Figure 2.2.3 an expansion of the graph in Figure 2.2.2 is presented centered at $x = 1$. In this figure the x values of the points of intersection between $y = |x^2-1|$ ($|f(x)-A|$) and $y = \epsilon$ have been identified as b and c and have the property that $b < a < c$. It is apparent from inspection of the figure that $|x^2 - 1| < \epsilon$ for $x \in (b, c)$. That such a b and c exist with these properties for arbitrary values of $\epsilon > 0$ is again entirely due to A being a root of $|f(x) - A|$. Evidently, a value for δ could be determined from the values for b and c. For example, the largest value that can be taken for δ in the present case is the minimum of $|b - a|$ and $|c - a|$. Though they appear to have the same value, they do not as the reader can check by solving the equation $|x^2 - 1| = \epsilon$ and finding b, c. The reader may wonder why our analytic approach is not tied to solving the equation $|f(x) - A| = \epsilon$. The answer is that the issue is not whether the equation $|f(x) - A| = \epsilon$ has a solution (which it will for almost all choices of A), rather the issue is whether $|f(x) - A|$ behaves as though it has a root at $x = a$. If $|f(x) - A|$ acts as though such a root exists, then reasoning in analogy to the Factor Theorem for Polynomials, we ought to be able to factor $|f(x) - A|$ into $|x - a|$ times a bounded nonzero term, and the general method discussed above will immediately apply.

The reader draw should an appropriate graph of $f(x) = x^2$ containing analogous elements to those presented in Figure 2.2.1. This will supply additional geometric intuition regarding the computation of the limit. $\square$

Example 2. Let $f(x) = \frac{1}{x+2}$, $x \in \mathbf{R} \sim \{-2\}$. Show $\lim\limits_{x \to -3} \dfrac{1}{x+2}$ exists.

Solution. Let $\epsilon > 0$ be arbitrary but fixed. Moreover, let us take as a candidate for A the number -1, which happens to be the value of the function at $x = -3$. Let us now consider the quantity $|f(x) - A|$. Algebra yields

$$
\begin{aligned}
\left| \frac{1}{x+2} - (-1) \right| &= \left| \frac{1 + (x+2)}{x+2} \right| \\
&= \left| \frac{x - (-3)}{x+2} \right| \\
&= |x - (-3)| \cdot \left| \frac{1}{x+2} \right|.
\end{aligned}
$$

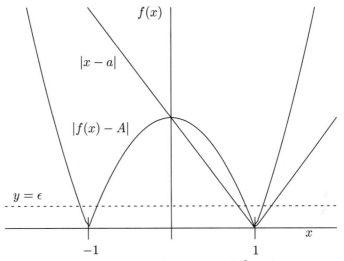

Figure 2.2.2 The graphs of $|x - 1|$ and $|x^2 - 1|$ showing the two apparent roots for $|x^2 - 1|$.

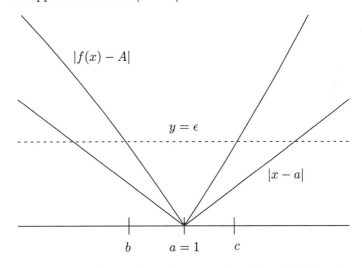

Figure 2.2.3 The graphs of $|x - a| = |x - 1|$ and $|f(x) - A| = |x^2 - 1|$ near $x = 1$. Notice $|x^2 - 1| < \epsilon$, provided $b < x < c$.

Further, if $\delta \leq \frac{1}{2}$, then

$$0 < |x - (-3)| < \frac{1}{2} \quad \text{.implies.} \quad x < -\frac{5}{2}$$

$$\text{.implies.} \quad x + 2 < -\frac{1}{2}$$

$$\text{.implies.} \quad \frac{1}{2} < |x + 2|$$

$$\text{.implies.} \quad \left| \frac{1}{x + 2} \right| < 2.$$

We now set $\delta = \min\left\{\frac{1}{2}, \frac{\epsilon}{2}\right\}$, whence

$$
\begin{aligned}
\left|\frac{1}{x+2} - (-1)\right| &= |x-(-3)| \cdot \left|\frac{1}{x+2}\right| \\
&< \delta \cdot \left|\frac{1}{x+2}\right| \\
&< \delta \cdot 2 \le \epsilon,
\end{aligned}
$$

as desired. $\square$

Discussion. The geometry of this situation is captured in Figure 2.2.4. Notice that the function is defined everywhere except at $x = -2$. Close to this value, the graph shoots up, or down, suggesting that problems may arise if x is permitted to be too close to -2. By requiring that $\delta < \frac{1}{2}$, we prevent x from getting arbitrarily close to -2.

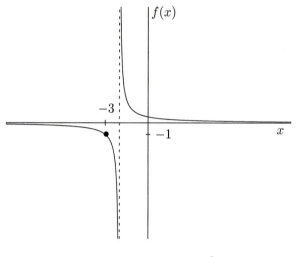

Figure 2.2.4 The graph of $f(x) = \frac{1}{x+2}$, $x \in \mathbf{R} \sim \{-2\}$ exhibiting $\lim\limits_{x \to -3} f(x) = -1$.

Algebraically, the problems at $x = -2$ show up in the term $\frac{1}{x+2}$, which is undefined when x takes -2 for a value. Moreover, since by using algebra, we have expressed $|f(x) - A|$ as a product of $|x - a|$ and $\left|\frac{1}{x+2}\right|$ $(a = -3)$, it is evident that we want to bound $\frac{1}{x+2}$ so that it can play the role of 'junk'. The restriction on δ is just what is required to accomplish this and yields an upper bound of 2. The solution is completed by setting δ to be the minimum of $\left\{\frac{1}{2}, \frac{\epsilon}{2}\right\}$. Note that $\frac{\epsilon}{2}$ was used instead of ϵ so that the added factor of 2 arising from the 'junk' term would disappear.

We have used a value of $\frac{1}{2}$ to prevent x from becoming arbitrarily close to -2. We could have used $\frac{3}{4}$, or any number that was less than 1. Had we used another number, an argument that was similar would have resulted. The reader would benefit from writing out one of these alternative arguments. In particular, it is essential that the reader understand why δ must be strictly less than 1. $\square$

Example 3. Let $f(x) = \frac{1}{x}$, $x \in (0, \frac{1}{10})$. Show f has a limit as x tends to $\frac{1}{10}$.

Solution. Fix a positive ϵ. We manipulate $|f(x) - A|$, where A has the value 10 to obtain

$$\left| \frac{1}{x} - 10 \right| = \left| \frac{1 - 10x}{x} \right| = \left| \frac{10}{x} \right| \cdot \left| x - \frac{1}{10} \right|.$$

Let $\delta = \min\left\{ \frac{1}{20}, \frac{\epsilon}{200} \right\}$; then

$$\left| \frac{1}{x} - 10 \right| = \left| \frac{10}{x} \right| \cdot \left| x - \frac{1}{10} \right|$$

$$< \quad 200 \cdot \left| x - \frac{1}{10} \right| \qquad \qquad \textbf{(WHY?)}$$

$$< \quad 200 \cdot \frac{\epsilon}{200} = \epsilon$$

whenever $x \in \left(0, \frac{1}{10} \right)$ and $0 < \left| x - \frac{1}{10} \right| < \delta$. □

Discussion. Again we follow the basic strategy of manipulating $|f(x) - A|$ to obtain a form $|x - a| \cdot (\text{junk})$, where there is a known bound on the size of 'junk'. In this case, the 'junk' turns out to be $\frac{10}{x}$, where $0 < x < \frac{1}{10}$. The problem with the quantity $\frac{10}{x}$ is that as x gets close to 0, $\frac{10}{x}$ becomes very large (see Figure 2.2.3) and, unfortunately, is unbounded (the reader should try some values to see that this does happen). But for $\frac{10}{x}$ to be really large, x must be really close to 0. Once we understand this fact, we see the solution to our problems is to prevent x from getting too close to 0. Thus, we require $\delta < \frac{1}{20}$, so that $\left| x - \frac{1}{10} \right| < \frac{1}{20}$ will force $x > \frac{1}{20}$. There is nothing special about $\frac{1}{20}$ other than the fact that it is less

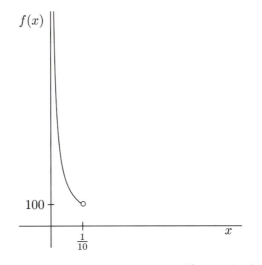

Figure 2.2.5 The graph of $f(x) = \frac{10}{x}$, $x \in \left(0, \frac{1}{10} \right)$.

than $\frac{1}{10}$ and greater than 0. Any fixed number satisfying this requirement may be used as an upper bound for δ and will lead to a bound on the 'junk' term; the reader might try $\frac{1}{19}$ as an upper bound for δ to see the effect on the calculations. Observe that the larger we allow the upper bound on δ to be, the closer x is permitted to be to 0 and the smaller we will ultimately have to choose δ to be to make the calculation work out. Thus, the ultimate smallness of δ depends not only on the smallness of ϵ but also on the bound for the 'junk' term. Last, we emphasize that $\frac{1}{10}$ is not in the domain of f as defined, and indeed the function is defined only on the left-hand side of $\frac{1}{10}$.

The reader should compare the arguments in the previous two examples. Both deal with similar functions. One difference between the two examples is in the fact that $\frac{1}{10}$ is considerably closer to 0 than -3 is to -2. The reader should identify how this fact changes the computations. $\square$

We turn our attention now to the problem of why a function may fail to have a limit at a given real number. As in the case of sequences, the first step in solving the problem is to write down precisely what is meant by a function failing to have a limit. This means negating the limit definition.

Negation of the Limit Definition. Let f be a function with domain D and let $a \in \mathbf{R}$. The limit as x tends to a of f does not exist provided that for every $A \in \mathbf{R}$ there exists an $\epsilon > 0$ such that for every $\delta > 0$ there is an $x \in D$ such that

$$0 < |x - a| < \delta \text{ .and. } |f(x) - A| \geq \epsilon.$$

Discussion. What this is telling us is that no matter how we pick A, we must be able to find a single fixed ϵ (the choice of ϵ will likely depend on the choice of A) such that no matter how small we choose δ there will always be an x in the domain of the function that is within a distance of δ from a (but not equal to a), and for this x, the distance from $f(x)$ to A will be at least as large as ϵ. Notice that a key fact here is that we must always be able to find an x that is in the domain and that satisfies $0 < |x - a| < \delta$. This means that for every choice of δ,

$$[(a - \delta, a) \cup (a, a + \delta)] \cap D \neq \emptyset.$$

As we shall see in the next section, this requirement means that a is a **limit point** of D. It follows that if for some particular choice of δ the above intersection is empty, then the limit as x tends to a of f will automatically exist, a fact pursued further in Exercise 8. $\square$

Our basic purpose is to understand why some functions should fail to have limits. Recall that we have already considered this question for sequences and we found that there were two fundamental reasons why a sequence should fail to have a limit. The first was that the sequence got large without bound as n approached ∞. The second was that the sequence 'wandered' or 'oscillated' as n tended to infinity without ever staying close to one fixed point. Thus, it seems reasonable that we might try these reasons, formulated in an appropriate manner, for functions.

Example 4. Let $f(x) = \frac{1}{x-2}$, $x \in D = \mathbf{R} \sim \mathbf{Q}$. Show f does not have a limit as x tends to 2.

Solution. Let A be any fixed real number and set $\epsilon = 1$. Let $\delta < \frac{1}{2}$. Now every interval of real numbers contains irrational numbers (Exercise 0.4.41), so there are members of the domain of f in $(2-\delta,2)$ and $(2,2+\delta)$. Let $x \in D \cap (2,2+\delta)$. An algebraic calculation will show $f(x) > 2$ (**HOW?**). Similarly, $x \in D \cap (2-\delta,2)$ implies $f(x) < -2$. Thus, no matter what the choice of A, either $|f(x) - A| \geq 1$ for $x \in D \cap (2, 2+\delta)$ or $|f(x) - A| \geq 1$ for $x \in D \cap (2-\delta, 2)$. Thus, f has no limit at 2 as claimed. $\square$

Discussion. The graph of f is shown in Figure 2.2.6. The picture illustrates the fact that if x is close to 2 on the right, then $f(x)$ is a very large positive number, while if x is close to 2 on the left, then $f(x)$ is a very large negative number. This fact is employed in the proof by choosing A fixed, either nonnegative or negative, and then choosing an x on the side of 2 that produces function values of opposite sign to witness $|f(x) - A| \geq \epsilon$. $\square$

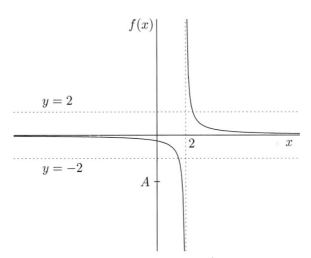

Figure 2.2.6 The graph of $f(x) = \frac{1}{x-2}$ showing non-existence of limit at 2. For A as shown, choosing $x \in (2, 2+\delta)$ guarantees $|f(x) - A| \geq 1$.

Example 5. Let $f(x) = \begin{cases} x, & \text{if } x \leq 1 \\ 2x, & \text{if } x > 1. \end{cases}$ Show f does not have a limit as x tends to 1.

Solution. Fix $A \in \mathbf{R}$ and set $\epsilon = \frac{1}{2}$. Let $\delta > 0$ be chosen. If $A \leq \frac{3}{2}$, then for any $x \in (1, 1+\delta)$, we have

$$|f(x) - A| = |2x - A| > |2 - A| \geq \left|2 - \frac{3}{2}\right| = \frac{1}{2}.$$

If $A > \frac{3}{2}$, then for any $x \in (1-\delta, 1)$,

$$|f(x) - A| = |x - A| > |1 - A| > \left|1 - \frac{3}{2}\right| = \frac{1}{2}.$$

In either case, A cannot be the limit. $\square$

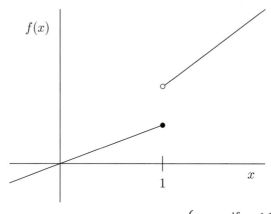

Figure 2.2.7 The graph of $f(x) = \begin{cases} x, & \text{if } x \leq 1 \\ 2x, & \text{if } x > 1 \end{cases}$

showing the 'jump' at $x = 1$.

Discussion. The graph of f is presented in Figure 2.2.7. Inspection of the figure reveals a 'jump' in the graph at $x = 1$. The reader will recall from previous remarks that for f to have a limit at 1, we must be able to choose A such that $|f(x) - A|$ behaves as though it has a root at 1. By using a graphing utility, the reader can easily verify that $|f(x) - A|$ does not have a root at $x = 1$, no matter what value of A is chosen.

Example 6. Let $f(x) = \sin \frac{1}{x}$, $x \in \mathbf{R} \sim \{0\}$. Show f does not have a limit as x tends to 0.

Solution. Let A be any fixed real number, and let $\epsilon = 1$. We take it as an established fact that

$$\sin \frac{n\pi}{2} = \begin{cases} 1, & \text{if } n = 4k+1, & k \in \mathbf{N} \\ 0, & \text{if } n = 2k, & k \in \mathbf{N} \\ -1, & \text{if } n = 4k+3, & k \in \mathbf{N}. \end{cases}$$

Let $\delta > 0$ be chosen. We can find $k \in \mathbf{N}$ such that

$$\frac{2}{(4k+1)\pi} < \delta.$$

For such a value of k, we know that either

$$\left| \sin \frac{(4k+1)\pi}{2} - A \right| \geq 1$$

or

$$\left| \sin \frac{(4k+3)\pi}{2} - A \right| \geq 1.$$

Since for any choice of k, these numbers are in D, we are done. $\square$

Discussion. The graph of $\sin \frac{1}{x}$ near 0 is shown in Figure 2.2.8. This is an example of a function that oscillates near a given point. In fact, as x moves in toward 0 from either side, the function moves up and down an infinite number of times. It is clear that a function with such behavior at a given point cannot have a limit at that point. Notice the proof that the limit fails to exist does not use all the domain near 0. Rather, we have selected a sequence of domain points that converges to 0 and then used the fact that the associated sequence of range points does not have a limit to show the function does not have a limit. (The reader will have an opportunity to explore this idea further in Exercise 9.) It does not matter that if, for example, A were 0, the function goes through this point infinitely often. To have a limit the function values must get *close to a fixed A and stay there* as the domain points get close to a. The reader should carefully compare these remarks with those concerning why sequences fail to have limits. $\square$

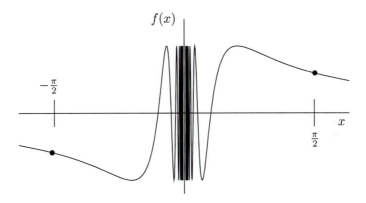

Figure 2.2.8 The graph of $f(x) = \sin \frac{1}{x}$, $x \in \mathbf{R} \sim \{0\}$, showing the behavior of the function near 0.

In summary, our previous three examples show that a function will fail to have a limit at $a \in \mathbf{R}$ if f is either unbounded near a, or f has a jump at a, or f oscillates near a. While we are not in a position now to show these are the only reasons why a function can fail to have a limit at a, we will do so later.

Exercises

1. Given $0 < \delta < 1$, verify the following estimates:

   (a) $|x - 2| < \delta$ .implies. $|x^2 - 4| < 5\delta$;

   (b) $|x - 2| < \delta$ .implies. $\left|\frac{x-2}{x+3}\right| < \frac{1}{4}\delta$;

   (c) $|x - 3| < \delta$ .implies. $|2x^2 - 11x + 15| < 3\delta$;

   (d) $|x + 2| < \delta$ .implies. $|x^2 + 3x + 2| < 2\delta$;

   (e) $|x + 1| < \delta$ .implies. $|x^3 + 1| < 7\delta$.

If in (a), $|x - 2| < \delta$ is replaced by $|x - 100| < \delta$, is the estimate still valid? What about analogous changes in (b)–(e)?

2. For each of the following functions with the given domains, use the definition to decide whether the limit exists at the point given.

(a) $f(x) = c$; $x \in \mathbf{R}$, $(c \in \mathbf{R})$, $a = 2$;   (b) $f(x) = c$; $x \in \mathbf{Q}$, $(c \in \mathbf{R})$, $a = 2$;

(c) $f(x) = 3x + 5$; $x \in \mathbf{R}$, $a = 2$;   (d) $f(x) = 4x - 3$; $x \in \mathbf{R}$, $a = -\frac{1}{2}$;

(e) $f(x) = \sqrt{x}$; $x \in \mathbf{R}^+$, $a = 0$;   (f) $f(x) = x^{\frac{1}{3}}$; $x \in \mathbf{R}$, $a = 2$;

(g) $f(x) = \frac{x+3}{2x}$; $x \in \mathbf{R} \sim \{0\}$, $a = -2$;   (h) $f(x) = \frac{x^2+2x}{2x-3}$; $x \in \mathbf{R} \sim \{\frac{3}{2}\}$, $a = 3$;

(i) $f(x) = \frac{1}{x^2}$; $x \in \mathbf{R} \sim \mathbf{Q}$, $a = 0$;   (j) $f(x) = \frac{1}{x^2}$; $x \in \mathbf{R} \sim \mathbf{Q}$, $a = \frac{1}{100}$;

(k) $f(x) = x^4 - 2x$; $x \in \mathbf{R}$, $a = 3$;   (l) $f(x) = \frac{1}{1+x^2}$; $x \in \mathbf{Q}$, $a = 0$;

(m) $f(x) = \frac{1}{1+x^3}$; $x \in [0, 1]$, $a = 1$;   (n) $f(x) = \frac{x^2+2x}{x-1}$; $x \in \mathbf{R} \sim \{1\}$, $a = 3$;

(o) $f(x) = \cos \frac{1}{x}$; $x \in (0, 1)$, $a = 0$;   (p) $f(x) = \frac{\sqrt{x-2}-2}{x-6}$; $x \in \mathbf{R} \sim \{6\}$, $a = 6$;

(q) $f(x) = \frac{x^3-27}{x-3}$; $x \in \mathbf{R} \sim \{3\}$, $a = 3$;   (r) $(x) = \frac{x^2-x-2}{x^2-2x}$; $x \in \mathbf{R} \sim \{0, 2\}$, $a = 2$;

(s) $f(x) = \sin \frac{1}{x^2}$; $x \in (-1, 0)$, $a = 0$;   (t) $f(x) = x \sin \frac{1}{x}$; $x \in (0, 1)$, $a = 0$;

(u) $f(x) = \frac{1}{x} \sin \frac{1}{x}$; $x \in (0, 1)$, $a = 0$;   (v) $f(x) = 2^x$; $x \in \mathbf{Q}$, $a = 0$;

(w) $f(x) = e^x$; $x \in \mathbf{Q}$, $a = 1$;   (x) $f(x) = [x]$; $x \in \mathbf{R}$, $a \in \mathbf{Z}$; $a \notin \mathbf{Z}$;

(y) $f(x) = x \left[\frac{1}{x}\right]$; $x \in \mathbf{R} \sim \{0\}$, $a = 0$;   (z) $f(x) = x + \frac{x}{|x|}$; $x \in \mathbf{R} \sim \{0\}$, $a = 0$;

(a') $f(x) = \begin{cases} 1, & x > 1 \\ \frac{1}{n}, & \frac{1}{n} < x \leq \frac{1}{n-1} \quad n \in \mathbf{N},\ n > 1 \\ 0, & x \leq 0, \end{cases} \qquad a = 0$;

(b') $f = \{(0, 2), (2, 3), (5, 8), (6, 6), (10, 0)\}$, $a = 0$, $a = 3$;

(c') $f(x) = \begin{cases} 3 + x, & x \leq 1 \\ 3 - x, & x > 1, \end{cases} \qquad x \in \mathbf{R}, a = 1$;

(d') $f(x) = \begin{cases} 2x, & x \in \mathbf{Q} \\ 1 - 2x, & x \notin \mathbf{Q}, \end{cases} \qquad a = \frac{1}{2}, a = \frac{1}{4}$;

(e') $f(x) = \dfrac{\sqrt{1+x} - \sqrt{1+x^2}}{\sqrt{1-x^2} - \sqrt{1-x}}$; $x \in (-1, 1) \sim \{0\}$, $a = 0$.

(f') $f(x) = \lim_{n \to \infty} \dfrac{\ln(2+x) - x^{2n} \sin x}{1 + x^{2n}}$; $x \in (-2, \infty)$, $a = 1$.

3. Let $f(x) = x^2$, $x \in (0, 1)$. Show the limit exists as x tends to -1. Is this limit unique?

4. Prove $\lim_{x \to 0} \dfrac{x+2}{x+1} = 2$. Given $\epsilon = 5, 1, \frac{1}{10}, \frac{1}{100}$, find the corresponding δ for which $0 < |x| < \delta$ witnesses $\left| \frac{x+2}{x+1} - 2 \right| < \epsilon$

5. Prove $\lim_{x \to a} f(x) = A$ if and only if $\lim_{x \to a} g(x) = 0$, where $g(x) = f(x) - A$.

6. Let f be defined on D. We say f is **unbounded at** $a \in \mathbf{R}$ provided for every M and every positive δ there is an $x \in (a - \delta, a + \delta) \cap D$ such that $x \neq a$ and $|f(x)| \geq M$. Show if f is unbounded at a, then the limit as x tends to a does not exist.

7. Let f be defined on D. We say f **oscillates at** $a \in \mathbf{R}$ if there exist $A, B \in \mathbf{R}$ such that $A < B$ and for every positive δ there exist $x, y \in (a - \delta, a + \delta) \cap D$, but distinct from a, such that $f(x) \le A$ and $f(y) \ge B$. Show if f oscillates at a, then the limit as x tends to a of f does not exist.

8. Let f be defined on D and suppose for some $a \in \mathbf{R}$ and some $\delta > 0$ that $D \cap [(a - \delta, a) \cup (a, a + \delta)]$ is empty. Show any real number will serve as the limit as x tends to a of f.

9. Let f be defined on D. Suppose there is a sequence $\{a_n\}_{n=1}^{\infty} \subseteq D \sim \{a\}$ such that $\{a_n\}$ converges to a and the sequence $\{f(a_n)\}$ has no limit. Show the limit as x tends to a of f does not exist. Note we have excluded a from being a value of $\{a_n\}$. Give an example to show why this is necessary.

10. State the contrapositive of the result in Exercise 9.

11. Let f be defined on D such that every sequence of domain points that converges to a gives rise to a sequence of range points having the same limit, A. Show the limit as x tends to a of f is A.

12. Let $f(x) = \begin{cases} x - 1, & \text{if } x \le 1 \\ x^3, & \text{if } x > 1. \end{cases}$ Show f does not have a limit at 1. Further use a graphing utility to study the graphs of $|f(x) - A|$ near $x = 1$ and develop a graphical argument as to why the limit cannot exist.

13. Let $f : \mathbf{R} \to \mathbf{R}$ satisfy the relationship $f(x + y) = f(x) + f(y)$ for all $x, y \in \mathbf{R}$. If f possesses a limit at $x = 0$, show it must be 0. Further prove f has a limit at every real number a.

14. Let f, g be defined on D and a be a limit point of D. Show if both f and $f + g$ have limits at $x = a$, then g has a limit at $x = a$. Is the same conclusion valid if we consider fg instead of $f + g$?

15. Show how the problem of whether $\lim\limits_{x \to a} f(x)$ exists can be interpreted as a two-person game. Illustrate with examples.

16. Prove if $\lim\limits_{x \to a} f(x) = A > 0$, there exists a positive δ such that $f(x) > 0$ for $0 < |x - a| < \delta$. Is the result true when $A = 0$?

17. If $f(x) \le g(x)$ for each x in their common domain, prove $\lim\limits_{x \to a} f(x) \le \lim\limits_{x \to a} g(x)$. If $f(x) < g(x)$ for each x, does it follow that $\lim\limits_{x \to a} f(x) < \lim\limits_{x \to a} g(x)$?

18. Prove the following 'squeeze theorem': If $f \le h \le g$, and if $\lim\limits_{x \to a} f = \lim\limits_{x \to a} g = L$, then $\lim\limits_{x \to a} h = L$.

19. Prove or disprove:
    (a) $\lim\limits_{x \to a} f(x) = A$ if and only for each $c \in \mathbf{R}$, $\lim\limits_{x \to a - c} f(x + c) = A$.;
    (b) $\lim\limits_{x \to a} f(x) = A$ if and only for each $c \ne 0$, $\lim\limits_{x \to \frac{a}{c}} f(cx) = A$.

20. Formulate the concept of $\lim\limits_{x \to a} f = +\infty$ (respectively, $-\infty$). Prove if $\lim\limits_{x \to a} f = +\infty$ and $\lim\limits_{x \to a} g = A$, then, $\lim\limits_{x \to a} (f + g) = +\infty$. Further, if $A \ne 0$, $\lim\limits_{x \to a} (fg) = \infty$. Discuss the situation when $A = 0$.

21. Give examples of functions f and g such that both f and g have no limit at a, but $f + g$ (respectively, fg) has a limit at a.

22. Show if $\lim\limits_{x \to a} f(x) = \infty$, then, $\lim\limits_{x \to a} \dfrac{1}{f(x)} = 0$. Also, prove if $0 < f(x) < \infty$ for all $x \in \text{Dmn } f$ and $\lim\limits_{x \to a} \dfrac{1}{f(x)} = 0$, then $\lim\limits_{x \to a} f(x) = \infty$.

2.3   Basic Limit Theorems

As with the previous types of limits, we want to set down a collection of theorems that describe the fundamental relations governing the behavior of limits of functions at a, and also the consequences of the existence of the limit at a for the function f. Before proceeding, it will be useful to have one definition of a topological concept and a generalization of bounded function.

Definition. Let $D \subseteq \mathbf{R}$ and $a \in \mathbf{R}$. Then, a is said to be a **limit point of** D provided for every positive δ,

$$\{x : 0 < |x - a| < \delta\} \cap D \neq \emptyset.$$

Discussion. The effect of the assertion that a is a limit point of D is to ensure every open interval that contains a also contains points of D other than a. In fact, it can be shown every such open interval contains an infinite number of points of D (see Exercise 13). The concept of limit point arises in topology and will be explored in greater detail in Chapter 3. $\square$

Definition. Let f be defined on D. Then, f is said to be **bounded on** D, provided the set $\{f(x) : x \in D\}$ is a bounded subset of $\mathbf{R}$.

Discussion. This generalizes our previous definition because D does not have to coincide with $\mathrm{Dmn}\, f$. One easily shows f is bounded on D if and only if there exists a positive real number K such that $|f(x)| < K$ for every $x \in D$. This definition should be compared with the equivalent definition of a bounded sequence. $\square$

Theorem 2.3.1. *Let a be a limit point of D, the domain of f, and suppose the limit as x tends to a of f exists. Then*

(i) *the limit at a is unique,*

(ii) *f is bounded in the interval $(a - \delta, a + \delta) \cap D$ for some positive δ.*

Proof. To establish (i), let us assume for the sake of argument there are two distinct limits at a, say, A and B, with $A < B$. Let us set $\epsilon = \frac{B-A}{2}$. Since the limit as x tends to a of f is A, we can find δ_A such that for all $x \in D$,

$$0 < |x - a| < \delta_A \text{ .implies. } |f(x) - A| < \epsilon.$$

Similarly, we can find δ_B such that for all $x \in D$,

$$0 < |x - a| < \delta_B \text{ .implies. } |f(x) - B| < \epsilon.$$

Let $\delta = \min\{\delta_A, \delta_B\}$. Since a is a limit point of D, there is an $x_0 \in D$ such that $0 < |x_0 - a| < \delta$. For such an x_0, we have

$$f(x_0) < A + \epsilon = B - \epsilon < f(x_0).$$

But this is contradiction, whence $A = B$ as desired.

To prove (ii), let A be the limit as x tends to a of f and set $\epsilon = 1$. By assumption, there exists a positive δ such that

$$0 < |x - a| < \delta \text{ .implies. } |f(x) - A| < 1.$$

Evidently, this implies for all $x \in D$,

$$0 < |x - a| < \delta \text{ .implies. } |f(x)| < |A| + 1.$$

Now if $a \notin D$, let $B = |A| + 1$. Otherwise, set $B = \max\{|f(a)|, |A| + 1\}$. In either case, we have found a bound, B, for f on $(a - \delta, a + \delta)$. □

Discussion. Let us compare this theorem with our earlier results. For (i), the theorems of interest are 1.2.1 and 2.1.1. In all cases the problem is to establish the uniqueness of the limit. In each instance we proceed by assuming there are distinct limits A and B. And in every case the proof continues by forcing the existence of a number (sequence value or function value) that must be close to A and at the same time close to B. This is done by choosing ϵ to measure the closeness and then finding either M or δ to obtain the a_n or $f(x)$. Note in the two earlier proofs we chose the maximum of M_A and M_B, while in this proof we choose the minimum of δ_A and δ_B. The reader can follow this thread even further by comparing the arguments to that for the uniqueness of the supremum (Theorem 0.4.1). Finally, notice that finding the required small value of δ is not enough. We must find a particular x_0 that satisfies

$$x_0 \in (a - \delta, a + \delta) \cap D,$$

since it is $f(x_0)$ that will act to witness the contradiction in the argument.

Another way to obtain a contradiction is to write

$$|A - B| = |(A - f(x)) + (f(x) - B)| \leq |A - f(x)| + |f(x) - B|$$

and then conclude that one of $|A - f(x)|$ and $|f(x) - B|$ must be greater than or equal to $\frac{|B-A|}{2}$.

Turning to the second part of the theorem, we suggest the reader compare this proof with the arguments in Theorems 1.2.3 and 2.1.2. Note that while the fact a is a limit point of D is essential to the proof of (i) **(WHY?)**, it is unnecessary for the result in (ii). (See Exercise 8.) □

Example 1. Let $f(x) = x$, $x \in [0, 1]$. Show the limit as x tends to 2 exists, but is not unique.

Solution. Let A be a fixed, but arbitrary real number. Let $\epsilon > 0$ be arbitrary and $\delta = 1$. Evidently, if $0 < |x - 2| < \delta$, then for each x, $x \in D$ .implies. $|f(x) - A| < \epsilon$. Since A was arbitrary, the limit cannot be unique. □

Discussion. The essential fact on which the argument depends is there are no x's in the domain that are within one unit of distance from the point 2. For this reason, the crucial implication is satisfied vacuously because its hypothesis is always false. Notice the argument is still valid even if 2 is added to the domain of f. The key fact is 2 would still not be a limit point of the domain. □

The theorem and example together suggest that as a general rule, we should require the point at which we are computing a limit be a limit point of the domain of the function. In the future, unless otherwise stated, *we will assume the limits are computed only at limit points of the domain of the function in question.* This will ensure if the limit exists, then we can discuss the limit as a unique real number.

As in the case of limits of sequence and limits at ∞, there are natural questions about how the arithmetic of functions interact with the limiting process for the case of limits at a. We present theorems dealing with products and reciprocals of functions, leaving other theorems to the exercises.

Theorem 2.3.2. *Let f and g be defined on D, and let a be a limit point of D. If the limit of f and g both exist as x tends to a, then*

$$\lim_{x \to a} (f \cdot g)(x) = \lim_{x \to a} f(x) \cdot \lim_{x \to a} g(x).$$

Proof. Let $\epsilon > 0$ be fixed and consider the quantity $|(f \cdot g)(x) - A \cdot B|$, where A and B are the limits at a of f and g, respectively. Now

$$
\begin{aligned}
|(f \cdot g)(x) - A \cdot B| &= |f(x) \cdot g(x) - A \cdot B| \\
&= |f(x) \cdot g(x) + (f(x) \cdot B - f(x) \cdot B) - A \cdot B| \\
&= |(f(x) \cdot g(x) - f(x) \cdot B) + (f(x) \cdot B - A \cdot B)| \\
&\leq |f(x) \cdot g(x) - f(x) \cdot B| + |f(x) \cdot B - A \cdot B| \\
&= |f(x)| \cdot |g(x) - B| + |f(x) - A| \cdot |B|.
\end{aligned}
$$

By Theorem 2.3.1, there is an $M > 0$ and a positive δ_1 such that if $x \in (a - \delta_1, a + \delta_1) \cap D$, then $|f(x)| < M$. Now, let $C = \max\{|B|, 1\}$. We choose δ_2 such that for all $x \in D$,

$$0 < |x - a| < \delta_2 \text{ .implies. } |g(x) - B| < \frac{\epsilon}{2M}.$$

Also, we choose δ_3 such that for all $x \in D$,

$$0 < |x - a| < \delta_3 \text{ .implies. } |f(x) - A| < \frac{\epsilon}{2C}.$$

It is immediate that if $\delta = \min\{\delta_1, \delta_2, \delta_3\}$, then for all $x \in D$,

$$0 < |x - a| < \delta \text{ .implies. } |(f \cdot g)(x) - A \cdot B| < \epsilon. \qquad \textbf{(WHY?)}$$

Thus, the limit of the product exists as claimed is the product of the limits. $\square$

Discussion. In the hypothesis of this theorem, we have required a be a limit point of D. The reason for this is we want

$$\lim_{x \to a} (f \cdot g)(x)$$

to be a unique quantity. For this to happen, the quantities on the right-hand side must be unique, and this requires a to be a limit point of D.

The reader should compare the argument here with that of Theorem 1.2.5. Especially, the reader should observe the similarities in the manipulation of the term $|(f \cdot g)(x) - A \cdot B|$ with $|a_n \cdot b_n - A \cdot B|$. In both cases, the first step consists of adding a form of 0, the precise form being determined by the context, followed by an application of the Triangle inequality. This sequence of steps is common and should be thoroughly understood. The reader may wonder why we choose C as above. The purpose is to avoid any worry about the case $B = 0$. Last, note we used Theorem 2.3.1 (ii), where the hypothesis contains no assumption regarding a being a limit point of D. See our remarks following Theorem 2.3.1. $\square$

Theorem 2.3.3. *Let f be defined and nonzero on D. If the limit as x tends to a of f exists and is not zero, then the limit as x tends to a of $\frac{1}{f}$ exists and equals*
$$\frac{1}{\lim\limits_{x \to a} f}.$$

Proof. The proof of the case when a is not a limit point of D is left to Exercise 9. Thus, we assume that a is a limit point of D and the limit as x tends to a of f is $A \neq 0$. It follows that $\frac{|A|}{2} > 0$. From the fact that the limit exists, we can find a positive δ_1 such that

$$0 < |x - a| < \delta_1 \text{ and } x \in D \text{ .implies. } \frac{|A|}{2} < |f(x)|. \qquad \textbf{(WHY?)}$$

Now

$$\left| \frac{1}{f(x)} - \frac{1}{A} \right| = |f(x) - A| \cdot \frac{1}{|f(x)| \cdot |A|}.$$

As a consequence of the inequality containing δ_1, we have

$$\frac{1}{|f(x)| \cdot |A|} < \frac{2}{A^2}.$$

Now let $\epsilon > 0$ be fixed but arbitrary. Since the limit at a exists, choose δ_2 such that $0 < |x - a| < \delta_2$ and $x \in D$ .implies. $|f(x) - A| < \frac{\epsilon \cdot A^2}{2}$. Let $\delta = \min\{\delta_1, \delta_2\}$. For such a choice of δ, we have

$$x \in D \text{ and } 0 < |x - a| < \delta \text{ .implies. } \left| \frac{1}{f(x)} - \frac{1}{A} \right| < \epsilon. \qquad \square$$

Discussion. The reader should compare the argument above with that for Theorem 1.2.6. They are almost identical. $\square$

In the exercises at the end of the previous section, we had the reader use the limit definition to establish whether various functions had limits. The theorems generated in this section simplify many of those calculations, as the following example illustrates.

Example 2. Let $f(x) = x^3$, $x \in \mathbf{R}$. Show $\lim\limits_{x \to 2} x^3$ exists.

Solution. We first check that 2 is a limit point of D. To do this, first let $\delta > 0$ be arbitrary. Since $2 + \delta \in D$, and δ is arbitrary, 2 is a limit point of D as required. Now,

$$\lim_{x \to 2} x^3 = \lim_{x \to 2} x \times \lim_{x \to 2} x \times \lim_{x \to 2} x = 2 \times 2 \times 2 = 8,$$

and the limit is computed. $\square$

Discussion. To make full use of these limit theorems, a set of basic functions whose limits are known must be established. Then, a large class of new functions is constructed using the arithmetic of functions; new limits are then computed as applications of the theorems. For the algebraic functions, polynomials, and their ratios, this only requires that we be able to compute

$$\lim_{x \to a} x,$$

which has the effect of reducing the problem of computing limits for polynomials to trivialities. $\square$

Exercises

1. Discuss the equations:

   (a) $\lim\limits_{x \to a} (f \pm g)(x) = \lim\limits_{x \to a} f(x) \pm \lim\limits_{x \to a} g(x);$ (b) $\lim\limits_{x \to a} |f|(x) = |\lim\limits_{x \to a} f(x)|;$

   (c) $\lim\limits_{x \to a} f^2(x) = (\lim\limits_{x \to a} f(x))^2.$

2. Where applicable, use the results of this section to do Exercise 2.2.2.

3. Some of the problems in Exercise 2.2.2 cannot be attacked using the theorems developed. Formulate limit theorems that would be useful for (e) and (f) of that exercise.

4. State and prove a general result that would be useful for computing limits as x tends to a of polynomial functions.

5. To what extent can the result of Exercise 4 be extended to rational forms of polynomials? State and prove a suitable result.

6. State and prove a result for functions comparable to Theorem 1.2.7.

7. State and prove a result for functions comparable to Theorem 1.2.8.

8. Show Theorem 2.3.1(ii) remains true even if a is not a limit point of D.

9. Complete the proof of Theorem 2.3.3.

10. Let f be defined on D and have the property that the limit as x tends to a of f is A and further $A > B$. Show there is a δ-interval $(a - \delta, a + \delta)$ such that if x is a domain point other than a that is in the interval, then $f(x) > \frac{A+B}{2}.$

11. Use examples to give a complete discussion of the outcome of the product of f with g when one or both of the limits on the right in the equation of Theorem 2.3.2 do not exist.

12. Let g be bounded in a δ-interval of a. Give a sufficient condition on f so the product of f with g may have a limit at a.

13. Show a is a limit point of D provided every open interval that contains a also contains an infinite number of points of D.

14. Let f and g be defined on $\mathbf{R}^+$. Discuss some sufficient conditions for the existence of $\lim\limits_{x \to a} (f \circ g)(x)$ of the composite function $f \circ g$.

15. Let a be a limit point of the domain D of the function f. Prove or disprove: $\lim\limits_{x \to a} f$ exists if and only if given $\epsilon > 0$, there exists $\delta > 0$ such that $|f(x_1) - f(x_2)| < \epsilon$ whenever $x_1, x_2 \in D$, $0 < |x_1 - a| < \delta$ and $0 < |x_2 - a| < \delta$.

16. Let f be defined on D, and let $a \in D$. We say f is **locally bounded at** a, provided there exists $\delta > 0$ such that the set $\{f(x) : x \in (a - \delta, a + \delta)\}$ is bounded. We also say f is **locally bounded** on the domain D, provided f is (locally) bounded at each member of D. Give an example to show a function that is locally bounded on D can fail to be bounded on D.

2.4   Monotone Functions

For sequences, we saw the property of being monotone had a powerful effect on the existence of a limit. As well, with functions in the case of limits at infinity, we saw monotonicity again played a role. We should therefore examine how monotonicity influences the existence of a limit for functions as x tends to a particular point, a. Recall that every bounded monotonic sequence has a limit. Based on this, it would seem reasonable to try to prove if f is monotone on D and bounded in a δ-interval about a, then the limit as x tends to a exists. The reader who tries to prove this will encounter certain difficulties (Go ahead and try!), that stem from the fact that the statement is not true. But more than that, they arise from the fact that we did not carefully model our earlier results. In the definition of limit at infinity for functions, x must approach $+\infty$ from values that are less than $+\infty$, since there are no real numbers that are greater than $+\infty$. Similarly, in the definition of limit at $-\infty$, x must approach through reals, all of which are greater than $-\infty$. In short, x can approach from only one side. With the definition of limit as x tends to a, x can approach from either direction (left or right), and this is the rub! The cure for this is simple: the notion of a **one-sided limit**.

Definition. Let f be defined on D. We say that the **limit as** x **tends to** a **from the left** exists if there exists $A \in \mathbf{R}$ such that for every positive ϵ there exists a positive δ such that $x \in D \cap (a - \delta, a)$ .implies. $|f(x) - A| < \epsilon$. If the limit exists, we write

$$\lim_{x \to a^-} f(x) = A.$$

This limit is often referred to as the **left-hand limit at** a. In a similar manner we invite the reader to define the **right-hand limit at** a [**limit as** x **tends to** a **from the right**, with the notation $\lim_{x \to a^+} f(x)$].

Discussion. If we compare this definition with the definition of 'limit as x tends to a of f', we see the condition

$$x \in D \quad \text{and} \quad 0 < |x - a| < \delta$$

has been replaced by

$$x \in D \cap (a - \delta, a).$$

The effect of this change is to restrict the values of x that can be substituted into the term $|f(x) - A|$ to those that lie on the left-hand side of x, as shown in Figure 2.4.1. Since we are not allowed to use x's on the right-hand side of a to show the definition does not hold, we would expect more functions to have a left-hand limit at a than have a limit at a, and this is indeed the case. Similar remarks apply to the right-hand limits. In Exercise 3, we ask the reader to show limit at a exists if and only if both the left-hand and right-hand limits at a exist and are equal.

It is important that the reader distinguish between the concept being captured with one-sided limits and that of Example 2.2.3. In the case of one-sided limits, points of D may exist on both sides of the point a at which the one-sided limit is being computed. In the case of the example, the domain of the function is $(0, \frac{1}{10})$, and the point of interest is $\frac{1}{10}$, which has domain points only on one side. In the former case, the restriction to consideration on only one side of a is by design of the definition; in the latter case it is because a is the right-hand endpoint of the domain. $\square$

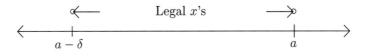

Figure 2.4.1 The δ-interval to be considered when finding the left-hand limit at a.

Example 1. Let
$$f(x) = \begin{cases} 1, & x \in (-1, 0) \\ \sin \frac{1}{x}, & x \in (0, 1). \end{cases}$$

Find the left-hand limit at 0.

Solution. From inspection of the definition of f, we conclude $A = 1$. Let $\epsilon > 0$ be fixed. If we set $\delta = 1$, the desired inequalities will follow. $\square$

Discussion. Functions, constant to the left of a point (such as f of this example on the left of 0), provide one of the few examples where the choice of δ is completely independent of the choice of ϵ. The reader can show the limit at 0 does not exist, and further the right-hand limit as x tends to 0 does not exist. The graph of this function is pictured in Figure 2.4.2. Other examples can be constructed that show both one-sided limits can exist, while the limit does not exist. Such examples appear in Exercise 1 and Example 2 (below). $\square$

Example 2. Let
$$f(x) = \begin{cases} \frac{x}{2}, & x \le 1 \\ 2x + 1, & x > 1. \end{cases}$$

Find the left- and right-hand limits at $x = 1$.

Solution. We shall show that the right-hand limit is 3. Let $\epsilon > 0$ be given, and

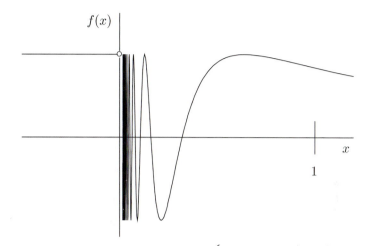

Figure 2.4.2 The graph of $f(x) = \begin{cases} 1, & x \in (-1,0) \\ \sin \frac{1}{x}, & x \in (0,1). \end{cases}$ The
limit from the left at $x = 0$ exists; the limit from the right does
not.

set $\delta = \frac{\epsilon}{2}$. If $x \in (1, 1+\delta)$, clearly, $|x - 1| < \delta$. Now

$$\begin{aligned} |f(x) - 3| &= |(2x + 1) - 3| \\ &= 2|x - 1| \\ &< 2\delta = \epsilon, \end{aligned}$$

proving our claim. Taking $\delta = 2\epsilon$, and repeating the above argument for the
interval $(1 - \delta, 1)$, it is readily seen that the left-side limit is $\frac{1}{2}$. $\square$

Discussion. The graph of $f(x)$ is shown in Figure 2.4.3. As shown, the graph is
composed of two pieces, one on either side of the point 1. The left-hand branch
is the part $f(x) = \frac{x}{2}$ that approaches the limit $\frac{1}{2}$ as x approaches 1 from the left
side. The right-hand branch is defined by $f(x) = 2x + 1$ and approaches 3 as
x approaches 1 from right. Since these two limits are different, the limit as x
approaches 1 does not exist. The geometrical meaning of this situation is that
there is a 'jump' of $\frac{5}{2}$ in the graph at the point 1. Therefore, for any real number
A, we can choose $\epsilon < \frac{5}{4}$ so that whatever the choice of $\delta > 0$, there will be points
in the interval $(1 - \delta, 1 + \delta)$ for which the function values, $f(x)$'s, will lie outside
the horizontal strip generated by $y = A - \epsilon$ and $y = A + \epsilon$. $\square$

In the above example, the function considered was monotonic increasing and
it had both one-sided limits. It is plausible that every monotonic function will
have one-sided limits. The next theorem states the best possible monotonicity
theorem for functions.

Theorem 2.4.1. *Let f be monotone on D. If $b, c \in D$ with $b < c$, then for every
a such that $b < a < c$, the one-sided limits as x tends to a of f exist.*

Proof. We will assume that f is monotone decreasing on D. Let a be given. We
assume that a is a limit point of $D \cap (a - 1, a)$ and compute the left-hand limit

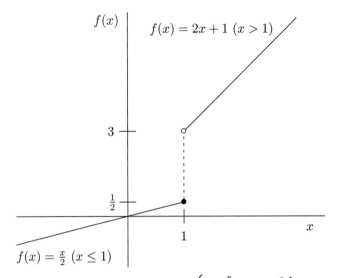

Figure 2.4.3 The graph of $f(x) = \begin{cases} \frac{x}{2}, & x \le 1 \\ 2x+1, & x > 1, \end{cases}$ showing the left- and right-hand limits at $x = 1$ are different.

at a. The remaining cases are left to Exercise 5. To continue, let

$$S = \{y : x < a \text{ and } y = f(x)\}.$$

Now, S is not empty since $b \in D$ and $b < a$. Further, since f is monotone decreasing and $a < c$ and $c \in D$, we have that $f(c)$ is a lower bound for S. Let $A = \inf S$, where the existence of A is obtained from the supremum principle. For a fixed $\epsilon > 0$, there is a $y \in S$ such that $A \le y < A + \epsilon$. Thus, $y = f(x)$ for some $x \in D$, whence we let $\delta = a - x$, and for this δ we have that $x \in D \cap (a-\delta, a)$ .implies. $|f(x) - A| < \epsilon$ (**WHY?**), as required. $\square$

Discussion. Even though this proof is written for decreasing functions. It is similar in its basic structure to Theorem 2.1.3 and, following the thread further, to Theorem 1.3.1. A particular item the reader should note is in the definition of S we have required $x < a$. If it happened that f was defined at a, one would be tempted to use less than or equal, and thus permit $a \in S$. Had we proceeded in that manner, the proof would have broken down. For the proof above to really imitate those at infinity, the inequality must be strict. Also, the strictness of the inequality emphasizes the fact that we have no interest in the value of the function at a, or even if the function is defined at a.

Note the theorem requires $b < a < c$. In the case $D = [b, c]$, it can also be shown that $\lim_{x \to b+} f$ and $\lim_{x \to c-} f$ exist. $\square$

Exercises

1. Graph the functions and find the one-sided limits indicated if they exist.

(a) $f(x) = \begin{cases} 3x, & x < 1 \\ x+1, & x > 1, \end{cases}$ limit at 1^+ and 1^-;

(b) $f(x) = \begin{cases} x^2, & x > 0 \\ x^3, & x < 0, \end{cases}$ limit at 0^+ and 0^-;

(c) $f(x) = \begin{cases} \frac{1}{x}, & x > 0 \\ 0, & x < 0, \end{cases}$ limit at 0^+;

(d) $f(x) = \begin{cases} 1-x, & x < 2 \\ 2x+3, & x \geq 2, \end{cases}$ limit at 2^+;

(e) $f(x) = \begin{cases} x^2+3, & x < 0 \\ 3(x-1), & x > 0, \end{cases}$ limit at 0^+;

(f) $f(x) = \begin{cases} \frac{1}{x}, & x \in \mathbf{Q}, x > 0 \\ \frac{1}{x^2}, & x \notin \mathbf{Q}, x > 0, \end{cases}$ limit at 2^-;

(g) $f(x) = \begin{cases} x, & x \in \mathbf{Q} \\ 1-x, & x \notin \mathbf{Q}, \end{cases}$ limit at 0^+;

(h) $f(x) = \begin{cases} x, & x \in \mathbf{Q} \\ 1-x, & x \notin \mathbf{Q}, \end{cases}$ limit at $(\frac{1}{2})^-$;

(i) $f(x) = \begin{cases} x, & x \in \mathbf{Q} \\ -x, & x \notin \mathbf{Q}, \end{cases}$ limit at 1^+;

(j) $f(x) = \left| \sin \frac{1}{x} \right|, \ x \in \mathbf{R} \sim \mathbf{Q}$ limit at 0^-;

(k) $f(x) = \begin{cases} \frac{1}{x^2}, & x \in \mathbf{Q} \sim \{0\} \\ \frac{1}{x^3}, & x \notin \mathbf{Q}, \end{cases}$ limit at 0^+;

(l) $f(x) = \begin{cases} x^2, & x \in \mathbf{Q} \\ 0, & x \in \mathbf{R} \sim \mathbf{Q}, \end{cases}$ limit at 0^+;

(m) $f(x) = 2^x, \ x \in \mathbf{Q}$,   limit at 1^-;

(n) $f(x) = e^x, \ x \in \mathbf{Q}$,   limit at 0^+.

2. Show $\displaystyle\lim_{x \to 0+} \frac{x}{a} \left[\frac{b}{x}\right] = \frac{b}{a}$, where $[x] = $ greatest integer $\leq x$.

3. Show if a is a limit point of D, the domain of f, and if the limit as x tends to a of f exists, then both one-sided limits at a exist and are equal.

4. Let a be a limit point of D, the domain of f. Find conditions on the one-sided limits that will guarantee the existence of the limit at a.

5. Fill in the missing details in the proof of Theorem 2.4.1.

6. Show if f is strictly monotone on D, then f is one-to-one. Is the converse true?

7. Show if f is strictly monotone on D, then f has an inverse, f^{-1}, which is also monotone.

8. Let f be strictly monotone on D and let the limit as x tends to a of f exist. What, if any, information does this give about the existence of limits for f^{-1} in each of the following cases?

    (a) D is an open interval and a is an interior point of D;

    (b) D is an arbitrary subset of the real numbers, and a is any limit point of D.

9. Give an example of a monotone function for which the limit as x tends to 1 does not exist.

10. What can be said about the general shape of the graph of a monotone function?

11. Let I denote the interval $(-\infty, z]$ (or $(a, z]$ or $[a, z]$), and let J denote the interval $[z, \infty)$ (or $[z, b)$ or $[z, b]$). Prove if f is monotonic on I and J, then f is monotonic on $I \cup J$.

12. Let f be defined on $[a, b]$. Suppose for every $c \in [a, b]$ there is a δ-interval about c such that f is monotone increasing over this δ-interval (this is called **local monotonicity**). Show f is monotone increasing on $[a, b]$. (See Exercise 2.6.19.)

13. In Exercise 12, can $[a, b]$ be replaced by an arbitrary subset of **R**? If not, describe the most general sets for which the theorem is true.

14. If f is locally monotone on D, what can be said about its one-sided limits?

15. Prove or disprove: If $g(x) = f(\frac{1}{x})$ where f is defined for $x > 0$, then, $\lim\limits_{x \to \infty} f(x) = A$ if and only if $\lim\limits_{x \to 0+} g(x) = A$.

2.5   Continuity

Suppose we have a collection of functions defined on an interval. If we draw the graph of such functions, we will observe that some can be drawn in one smooth 'continuous' sweep of our pen, while others have many breaks or jumps.

Example 1. Sketch the graph of the following functions:

(a) $f(x) = x^2$, $x \in [-2, 2]$;

(b) $f(x) = \begin{cases} \frac{1}{x}, & x \in [-2, 2] \text{ and } x \neq 0 \\ 0, & x = 0; \end{cases}$

(c) $f(x) = \begin{cases} x, & x \in [-2, 1] \\ -x, & x \in (1, 2]. \end{cases}$

Solution. The graphs are shown in Figure 2.5.1. $\square$

Discussion. The reader will note that while the graph of the first function can be drawn in one 'continuous' motion, without lifting one's pen from the paper, the other graphs cannot be drawn in this manner. This is a very striking property of the first graph, and it is natural to wonder if it can be captured by a suitable mathematical definition. In fact, mathematicians of the past several centuries did confront this question, namely,

> Is there a way to specify those curves that can be drawn with a single stroke of one's pen?

Their answer is our next definition. $\square$

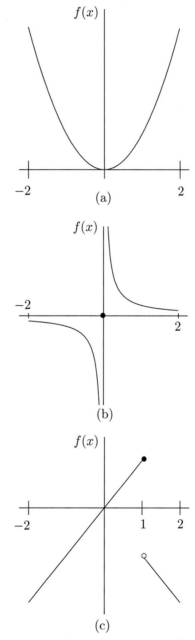

Figure 2.5.1 Three illustrations of continuous
and discontinuous functions. Function definitions
are given in the text.

Historical Note. Discussions about the fundamental nature of function and/or continuous function took place in the mathematics community from 1750 to 1840.[1] Among the notables involved were Bolzano, Cauchy, Dirichlet, Euler, Fourier, Gauss, and LaGrange.

During much of the period under discussion, there was great argument as to the very nature of function. Indeed some of the mathematicians listed would not have accepted the function in (c) above as a function and would have limited their considerations to functions having single analytic definitions such as the function in (a) above. Driving much of the discussion was the existence of physical processes for which no apparent analytical description existed. (Such a process was the flow of heat.) The modern definition of function defined on an interval was given in a paper by Dirichlet in 1837, and in an earlier paper he considered pathological functions like that of Example 5 below (Kline, 1972, p. 950).

Associated with the definition of function is the definition of continuous function. These considerations focused on defining a continuous function on an interval. Indeed, the first correct definition (see Kline, 1972, p. 951) was given in a paper by Bolzano in 1817 and is essentially the definition we give below. Cauchy also considered the problem and formulated a definition using infinitesimals; again the focus was on a process occurring over a domain.

It is interesting that the correct definition of continuity precedes the correct definition of function and that both discussions above were based on an incomplete understanding of **R**. □

Definition. Let f be defined on D. We say that f is **continuous on** D, provided that for every $y \in D$ and every positive ϵ, there is a positive δ such that for all $x \in D$

$$0 < |x - y| < \delta \text{ .implies. } |f(x) - f(y)| < \epsilon.$$

Further, if for a particular $a \in D$ we know that for every positive ϵ, there is a positive δ such that for all $x \in D$

$$0 < |x - a| < \delta \text{ .implies. } |f(x) - f(a)| < \epsilon$$

we will say that f **is continuous at** a.

Discussion. First, the definition given above defines continuity for functions on an arbitrary domain, not simply closed or open intervals. Second, the definition of continuity focuses entirely on points in the domain of f. This is in contrast to the case of computing a general limit where the point of interest, a, could be outside the domain of f.

To pass from the definition of being continuous on D to being continuous at a single point a of D, one simply replaces y in the domain definition by a and eliminates the universal quantification on y (now a). Evidently, then, a function is continuous on its domain exactly if it is continuous at each point of its domain. Most modern treatments give primacy to continuity at a point and from this definition proceed to continuity on D. Following the historical development

---

[1]This discussion is based on the work of M. Kline. The reader seeking more detail should examine pp. 949 ff. of Kline's book listed in the bibliography.

and because our treatment leads naturally to consideration of uniform continuity issues, we have given primacy to continuity on D.

Inspection of the definition for continuity at a leads to the procedure for checking continuity at a found in most calculus books, namely, the following three items must be established:

(i) f must be defined at a;

(ii) the limit as x tends to a must exist (that is, $\lim_{x \to a^+} f = \lim_{x \to a^-} f$);

(iii) the limit must be $f(a)$.

If all three are satisfied, then the function is indeed continuous at a. If any one of the above fails, the function is not continuous at a.

There is another feature of the definition that should be emphasized. Recall that in the definition of limit, the value of δ was dependent on the choice of ϵ. Quite obviously δ still depends on ϵ. However, there is an additional dependence. Notice that y is chosen before δ; that is, an argument will proceed by fixing an arbitrary $y \in D$. It seems plausible that the value that is finally arrived at for δ will depend on the value of y or, more generally, where in the domain of f we happen to be working. To further clarify this point, consider Example 2.2.1. Notice that the final choice of δ arises out of the line

$$|x - 1| < 1 \text{ .implies. } 0 < |x + 1| < 3. \qquad (*)$$

Thus, δ generally depends on the value of a in problems related to finding a limit. (Similar dependencies arise in Examples 2.2.2 and 2.2.3 and should be reviewed by the reader.) As a result, we can expect that the dependence of δ on y will introduce additional complexities for which our proofs must account.

The thrust of the discussion above relates to the analytic properties of the definition. Our initial impetus grew out of geometry. Thus, the question of how well this definition captures the intuitive idea 'single stroke of the pen' remains to be seen. This will be studied in further examples. $\square$

Example 2. Let $f(x) = x^2$, $x \in [-2, 2]$. Show f is continuous on this interval.

Solution. Let $y \in [-2, 2]$ be arbitrary but fixed. We show that f is continuous at y. Applying algebra to $|f(x) - f(y)|$, we get

$$
\begin{aligned}
|x^2 - y^2| &= |x - y| \cdot |x + y| \\
&\leq |x - y| \cdot 2 \cdot \max\{|x|, |y|\} \qquad (*) \\
&\leq |x - y| \cdot 4.
\end{aligned}
$$

Fix a positive ϵ, and let $\delta = \frac{\epsilon}{4}$. It is immediate that for all $x \in [-2, 2]$

$$0 < |x - y| < \delta \text{ .implies. } |f(x) - f(y)| < \epsilon.$$

Since y was an arbitrary member of D, the function is continuous on D. $\square$

Discussion. The reader should compare these computations with those in Example 2.2.1 and realize that continuity arguments are extensions of limit arguments. Indeed, a continuity argument merely establishes for every $a \in D$,

$$\lim_{x \to a} f(x) = f(a).$$

For this reason, the similarities should be substantial, as can be seen from the fact that in its broad outline the argument above is the same as that for Example 2.2.1. In particular, the main feature is to express $|f(x) - f(y)|$ as a product, one factor of which is $|x - y|$ and the other factor of which is a 'junk' term that is bounded. Once this is accomplished, the argument follows.

The value of δ will depend on the value of y, or the part of the domain in which we are interested. In the present case, the key inequality is established in the line marked with an asterisk $(*)$. It appears we have eliminated the dependence on y, but the dependence is only hidden. To see this, the reader should work the example again, but this time assuming that $D = [-10, 10]$. The point is to fully understand the way in which the nature of the domain affects the formulation of the argument.

As a last remark, the graph of this function can be drawn with a single stroke of the pen, as shown in Figure 2.5.1(a). So, we see for the first case, the definition agrees with the concept we were trying to capture. $\square$

Example 3. Let $f(x) = \frac{1}{x}$, $x \in [-2, 0) \cup (0, 2]$. Show f is continuous on its domain D.

Solution. Let $y \in D$ be fixed, and a positive ϵ given. Then, $y \neq 0$, whence $0 < \frac{|y|}{2}$. We may assume that any choice of δ will satisfy $\delta < \frac{|y|}{2}$. Under this assumption, if $|x - y| < \delta$, then

$$
\begin{aligned}
\left| \tfrac{1}{x} - \tfrac{1}{y} \right| &= |x - y| \cdot \tfrac{1}{|x| \cdot |y|} \\
&\leq |x - y| \cdot \tfrac{2}{y^2}. \qquad \textbf{(WHY?)}
\end{aligned}
\qquad (*)
$$

Let us therefore set $\delta = \min\left\{ \frac{|y|}{2}, \frac{\epsilon}{2} y^2 \right\}$. It is immediate that $x \in D$ and

$$
0 < |x - y| < \delta \text{ .implies. } \left| \frac{1}{x} - \frac{1}{y} \right| < \epsilon,
$$

as desired. Thus, $f(x) = \frac{1}{x}$ is continuous on $[-2, 0) \cup (0, 2]$. $\square$

Discussion. The calculations in this example are essentially the same as those done in Examples 2.2.2 and 2.2.3. In the present example, the reader should carefully note the role of y in obtaining the value of δ. The critical inequality that delineates this dependence is marked with an asterisk. There is also another source of dependence, namely, the assumption, $\delta < \frac{|y|}{2}$, that acts to prevent x from being arbitrarily close to 0.

The example has additional interest. Looking back at our initial comments regarding geometry, this example seems in direct conflict to our intuition, or else our claim that we have really captured the 'heart and soul' of continuity is false. The reason for the seeming conflict is the fact that the graph cannot be drawn with a single stroke of the pen. To see this, the reader should observe that the graph of this function coincides with the graph presented in Figure 2.5.1(b), except that no point would be plotted at $x = 0$, since f is not defined at 0 in the present example. Since the function graphed in Figure 2.5.1(b) was used as an example of a function that was not continuous, we appear to have something of a problem. However, there is a resolution.

Consider the function $g(x) = x^2$ with $D = [-2, -1) \cup (1, 2]$. The graph of the function g is in two pieces (see Figure 2.5.2), but this is due to the fact that the domain is in two pieces! Moreover, the break in the graph of g, where we are forced to pick up our pen, occurs at the points that are missing from the domain of g, namely, points in the interval $[-1, 1]$. Since we cannot move our pen across the domain without lifting it from the paper, it seems reasonable to believe that we should not be able to draw the graph at a single stroke of the pen. For this reason, it would seem that the function, g, which is the same as that presented in Example 2 except for the missing points in the domain, should be considered to be continuous, and this is the approach mathematicians have taken. The method they have adopted for pursuing this tack is to restrict continuity considerations to points that are in the domain of the function. Points not belonging to the domain are not relevant.

Returning now to the present example, the function is continuous exactly because the point $x = 0$ is missing from its domain. Notice that any attempt to add $x = 0$ to D will force us to define $f(0)$, and this will immediately cause the new function, which is now defined on the whole of $[-2, 2]$, to be discontinuous. In summary, we expect our geometric intuition to be borne out in full only for continuous functions whose domain is composed of a single interval.

There is still one other important fact to be gleaned from these examples. The function, $g(x) = x^2$, $x \in [-2, -1) \cup (1, 2]$ can be extended to a continuous function whose domain is all of $[-2, 2]$. (The function h with domain D' **extends** the function f with domain D, if $D \subseteq D'$ and for all $x \in D, f(x) = h(x)$.) As mentioned above, the function $f(x) = \frac{1}{x}$, which is defined on $[-2, 0) \cup (0, 2]$ does not have a continuous extension to all of $[-2, 2]$. The reason that no such extension exists is that the one-sided limits at 0 do not exist, so there is no way to choose a value for $f(0)$ that will lead to a continuous function. These ideas are explored further in the next example. $\square$

Definition. A function g is said to be an **extension** of a function f provided Dmn $f \subset$ Dmn g and for all $x \in$ Dmn f, $f(x) = g(x)$. If g is continuous, then the extension is said to be a **continuous extension**.

Example 4. Let

$$f(x) = \begin{cases} x^2, & x \in [-2, 0) \\ 1, & x \in (0, 2]. \end{cases}$$

Show this function is continuous, but has no continuous extension to $[-2, 2]$.

Solution. We first deal with the problem of establishing continuity. Fix $y \in D$ and a positive ϵ. Now $y \neq 0$, whence we may assume that any choice of δ satisfies $\delta < \frac{|y|}{2}$. The reader can check that $\delta = \min \left\{ \frac{|y|}{2}, \frac{\epsilon}{4} \right\}$ will do the job. (**WHY** is the restriction of δ absolutely essential to the argument?)

To see that no continuous extension exists, we observe (the reader can show) that

$$\lim_{x \to 0^+} f(x) = 1,$$

while

$$\lim_{x \to 0^-} f(x) = 0.$$

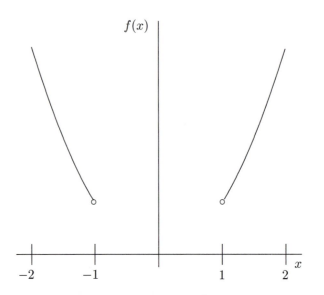

Figure 2.5.2 The graph of $f(x) = x^2$, $x \in [-2, -1) \cup (1, 2]$.

Thus, as shown in Exercise 2.4.3, the limit as x tends to 0 of f does not exist and so could not be equal to $g(0)$, for any extension g of f. $\square$

Discussion. There are several important points here. The first is 0 is a limit point of the domain of f. The second is if there is a continuous extension of f to $[-2, 2]$, then the value of the extension, g, at 0 will have to be given by

$$g(0) = \lim_{x \to 0} g(x) = \lim_{x \to 0} f(x).$$

Since this limit does not exist in the present case, as illustrated in Figure 2.5.3, no continuous extension to all of $[-2, 2]$ is possible. However, if the limit had existed, then the value assigned to $g(0)$ would be determined by this limit. In particular, if the value of the limit exists and is unique, then only one possible extension is available. These ideas are explored further in the exercises. $\square$

The notion of extension is an important tool. Consider the computation indicated by 2^x, for x an arbitrary nonnegative real number. Work in Chapter 0 established the existence of real numbers corresponding to $2^{1/n}$, for $n \in \mathbf{N}$ (or $\mathbf{Z}$). For this reason, we can carry out the computation for $2^{p/q}$, $p, q \in \mathbf{N}$. However, we are left to wonder about the outcome of computations such as that indicated by 2^π or other computations involving irrational exponents. The way around this problem is to use continuity. It can be established, with the tools developed, that the function $f(x) = 2^x$ with domain $\mathbf{Q}$ is continuous. It is reasonable to believe that any extension of 2^x to all of $\mathbf{R}$ that purports to capture exponentiation should also be continuous. Thus, the solution to the problem is to extend 2^x to all of $\mathbf{R}$ by continuity. This is a simple example of the use of extension by continuity. The general power x^y is formally done in Chapter 9. Other examples occur in many areas of mathematics.

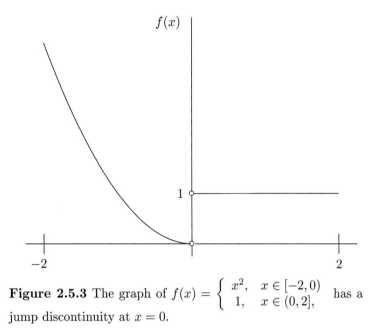

Figure 2.5.3 The graph of $f(x) = \begin{cases} x^2, & x \in [-2, 0) \\ 1, & x \in (0, 2], \end{cases}$ has a jump discontinuity at $x = 0$.

We turn now to the question of why a function fails to be continuous. As always, the first step is to write down the negation of the definition of continuity.

Negation of the Definition of Continuity. A function f is **not continuous** on its domain D, if there exists a $y \in D$ such that for some $\epsilon > 0$ and every $\delta > 0$, there exists an $x \in D$ such that

$$0 < |x - y| < \delta \quad \text{and} \quad |f(x) - f(y)| \geq \epsilon.$$

Discussion. Thus, f is not continuous on D if it is not continuous at some particular point of D. This means we are required to find $a \in D$ such that either

(i) $\lim_{x \to a} f(x)$ does not exist, or

(ii) $\lim_{x \to a} f(x)$ does exist, but is not equal to $f(a)$.

In either case, to be a candidate to witness the discontinuity of f, the point a must be a limit point of D. $\square$

A function f that is not continuous at $a \in D$ is said to be **discontinuous at** a or to have a **discontinuity at** a. If for every $y \in D$, f is discontinuous at y, we say that f is **totally discontinuous** on D or **nowhere continuous** on D. Functions that are totally discontinuous are not often encountered but are by no means rare. We give an example.

Example 5. Show the function given by

$$f(x) = \begin{cases} 1, & x \in \mathbf{R} \sim \mathbf{Q} \\ 0, & x \in \mathbf{Q} \end{cases}$$

is totally discontinuous.

Solution. Let y be an arbitrary, but fixed member of $\mathbf{R}$. Let $\epsilon = \frac{1}{2}$, and let $\delta > 0$ be fixed. Now the interval defined by $|x - y| < \delta$ contains both rationals and irrationals. If y is rational, pick x in the interval to be irrational. If y is irrational pick x in the interval to be rational. In either case,

$$0 < |x - y| < \delta \quad \text{and} \quad |f(x) - f(y)| \geq \epsilon = \frac{1}{2}.$$

Thus, f is not continuous at y. Since y was an arbitrary member of D, we are done. $\square$

Discussion. The graph of the function in Example 5 looks like two horizontal lines, one passing through $y = 1$ and the other being the x-axis. The reason for this is that the points on both parts of the graph are so close together that it is impossible to detect any holes. The reader should settle in his own mind how to picture this function.

It is instructive to think about how a function can get to be so discontinuous. For example, how would you build a really discontinuous function? You might start with a function like $f(x) = x$, $x \in \mathbf{R}$ and change values of the function at various domain points. But quickly you would find that this is a very inefficient way of producing discontinuities. To really produce discontinuities, you have to get your hands on lots of points in the domain, as in the above example, that can be thought of as arising from a constant function whose values were shifted on the rationals. This type of thinking suggests trying to classify the various types of discontinuities that may arise as discussed below. $\square$

If we think of discontinuities of functions as representing 'pathology', then we can classify discontinuities in terms of the degree of pathology they represent. The least pathological of all the discontinuities are the **removable discontinuities**. We say that the discontinuity at a is **removable**, provided that the limit as x tends to a exists and that

$$\lim_{x \to a} f(x) \neq f(a).$$

Removable discontinuities can be 'removed' simply by changing the function value at a to make it agree with the limit at x tends to a. For this reason, a function with removable discontinuities can be thought of as being 'almost' continuous. A simple example of a function having a removable discontinuity is

$$f(x) = \begin{cases} x^2, & x \in [-2, 0) \cup (0, 2] \\ 1, & x = 0. \end{cases}$$

To remove this discontinuity, we merely set $f(0) = 0$. Thus, the 'repair' can be accomplished by moving only one point on the graph. The graph of the function in this example is presented in Figure 2.5.4.

All other types of discontinuities require that

$$\lim_{x \to a} f(x) \text{ does not exist.}$$

Recall that even though this limit does not exist, both one-sided limits as x tends to a of f may exist. An example of this type was given in Example

4, and its graph is presented in Figure 2.5.3. Such a situation, where both one-sided limits exist but are different, represents the next step on the scale of bad behavior. If f is discontinuous at a, but both one-sided limits at a exist, we say that f has a **simple discontinuity at** a. Simple discontinuities include removable discontinuities. However, they are not limited to removable discontinuities. Inspection of Figure 2.5.3 shows the discontinuity to be of a more serious nature than a removable discontinuity. It cannot be fixed by changing the value of the function at a single point, since there is a big 'jump' in the graph. For this reason, discontinuities where the one-sided limits exist but disagree are referred to as **jump** discontinuities.

Simple discontinuities are often referred to as **discontinuities of the first kind**. All other types of discontinuities are **discontinuities of the second kind**. Discontinuities of the second kind involve the greatest degree of pathology, since they require the nonexistence of a one-sided limit at some point in the domain. As we have already remarked, this means that the function is unbounded or that it oscillates in a 'bad way'. Figure 2.5.1(b), which presents the graph of $f(x) = \frac{1}{x}$, with $f(0) = 0$, contains an example of a discontinuity of the second kind due to the function being unbounded near 0. The ultimate in 'gross' behavior is perhaps shown by $\sin\frac{1}{x}$, as discussed in Example 2.4.1. The problem is that the function oscillates an infinite number of times in any neighborhood of 0, as can be seen from inspection of Figure 2.4.2.

The classification scheme in respect to continuity presented above is not the only possible scheme for studying the behavior of functions. The importance of such schemes lies in the fact that they provide a methodological base for dealing with generic questions of the form:

How bad can a function be and still be integrable?

'Integrable' can be replaced by 'differentiable', or any other mathematical property of interest. We will from time to time approach questions of this type in the remainder of this text, and continuity will play a major role in the discussions.

Natural questions arise concerning the arithmetic of continuous functions. The basic arithmetical properties of continuous functions are summarized in the following theorem.

Theorem 2.5.1. *Let f and g be continuous on D, then*

  (i) *$f \pm g$ is continuous;*

  (ii) *$f \cdot g$ is continuous;*

  (iii) *if $g(x) \neq 0$ for all $x \in D$, then $\dfrac{f}{g}$ is continuous;*

  (iv) *$|f|$ is continuous.*

Proof. Exercise 4. $\square$

Theorem 2.5.2. *Let f be continuous on D with $\operatorname{Rng} f \subseteq E$. If g is continuous on E, then $g \circ f$ is continuous on D.*

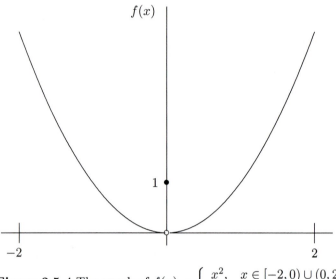

Figure 2.5.4 The graph of $f(x) = \begin{cases} x^2, & x \in [-2,0) \cup (0,2] \\ 1, & x = 0, \end{cases}$
has a removable discontinuity at $x = 0$.

Proof. Let $y \in D$ and let $\epsilon > 0$ be fixed. Set $w = f(y)$; then $w \in E$. Since g is continuous on E, there is a $\delta_1 > 0$ such that $u \in E$ and $|u - w| < \delta_1$ imply $|g(u) - g(w)| < \epsilon$. Since f is continuous on D and $\delta_1 > 0$, we can find $\delta > 0$ such that $x \in D$ and $|x - y| < \delta$ imply $|f(x) - f(y)| < \delta_1$. But this means that

$$|(g \circ f)(x) - (g \circ f)(y)| < \epsilon \qquad\qquad \textbf{(WHY?)}$$

whenever $x \in D$ and $|x - y| < \delta$. $\square$

Discussion. The intuition of the proof is really quite simple. Once we fix our y, we find its image in the range of g by computing $(g \circ f)(y) = g(w)$, where w is in the range of f. Since g is continuous, the positive ϵ-interval centered at $g(w)$ gives rise to a positive δ_1-interval centered at w. This δ_1-interval about w is interpreted as an ϵ-interval; we then apply the continuity of f to obtain the required δ-interval about y.

   To illustrate the utility of this theorem, consider the sequence discussed in Example 1.3.3. In the solution we considered the equation

$$\lim_{n \to \infty} b_n = \lim_{n \to \infty} \sqrt{2 + a_n},$$

where it was known that $\lim b_n = \lim a_n = A$. We then argued that

$$A = \lim_{n \to \infty} b_n = \lim_{n \to \infty} \sqrt{2 + a_n} = \sqrt{2 + A},$$

with the reader left to fill in the steps. A suitable calculation could be based on the product theorem for limits of sequences. However, given that that $f(x) = \sqrt{x}$ is continuous on its domain, Theorem 2.5.2 can be used to obtain the result. This idea is more fully explored in Exercise 13. $\square$

Before proceeding to the exercises, we want to draw the reader's attention to the function that appears in Exercise 1(h). This function is known as the **ruler function**. It has rather odd continuity properties; indeed, its behavior could only be termed pathological. We toyed with the idea of including this as an example. We decided not to include it on the grounds that this function provides an essential test of the student's understanding of many of the ideas which have been developed in the text. In short, if you, the student, can do this exercise, then you have gone a long way on the road to understanding analysis. If you cannot do it straightaway, then never fear, neither could we. The point is, keep trying until you can!

Exercises

1. Give a complete discussion of the continuity properties of the following functions, including types of discontinuities. Prove your answer, using definitions only.

   (a) $f(x) = x^3$, $x \in \mathbf{R}$;

   (b) $f(x) = 1 - x^2$, $x \in \mathbf{Q}$;

   (c) $f(x) = \frac{1}{x^2}$, $x \in [-3, 0) \cup (0, 2]$;

   (d) $f(x) = \frac{1}{1+x^2}$, $x \in \mathbf{R}$;

   (e) $f(x) = \begin{cases} x^2, & x \in [0, 1] \\ x^3, & x \in (1, 5]; \end{cases}$

   (f) $f(x) = \begin{cases} x^2, & x \in \mathbf{Q} \\ -x^2, & x \in \mathbf{R} \sim \mathbf{Q}; \end{cases}$

   (g) $f(x) = \begin{cases} x^4, & x \in [0, 1] \\ x + 1, & x \in (1, 3); \end{cases}$

   (h) $f(x) = x - [x]$, $x \in \mathbf{R}$;

   (i) $f(x) = \begin{cases} x, & x \in \{\frac{1}{n} : n \in \mathbf{N}\} \\ 0, & x \in [-1, 0); \end{cases}$

   (j) $f(x) = 2^x$, $x \in \mathbf{Q}$;

   (k) $f(x) = e^x$, $x \in \mathbf{Q}$;

   (l) $f(x) = \begin{cases} x, & x \in \mathbf{Q} \\ 1 - x, & x \in \mathbf{R} \sim \mathbf{Q}; \end{cases}$

   (m) $f(x) = \frac{1}{2} - x + \frac{1}{2}[2x] - \frac{1}{2}[1 - 2x]$, $x \in [0, 1]$;

   (n) $f(x) = \begin{cases} 0, & x \in \mathbf{R} \sim \mathbf{Q} \\ \frac{1}{q}, & x = \frac{p}{q} \in \mathbf{Q} \text{ in lowest terms}; \end{cases}$

   (o) $f(x) = \begin{cases} 0, & x \in \mathbf{R} \sim \mathbf{Q} \\ (-1)^p \cdot q, & x = \frac{p}{q} \in \mathbf{Q} \text{ in lowest terms}. \end{cases}$

   For the remaining problems, only consider continuity at the indicated point. You may assume knowledge of the domains and ranges of the sin, log, and exp functions.

   (p) $f(x) = \begin{cases} x \sin \frac{1}{x}, & x \neq 0 \\ 0, & x = 0, \end{cases}$ at $y = 0$;

   (q) $f(x) = \begin{cases} e^{1/x}, & x \neq 0 \\ 0, & x = 0, \end{cases}$ at $y = 0$;

   (r) $f(x) = \begin{cases} x^2 \sin^2 \frac{1}{x}, & x \neq 0 \\ 0, & x = 0, \end{cases}$ at $y = 0$;

   (s) $f(x) = \lim_{n \to \infty} \frac{\ln(2 + x) - x^{2n} \sin x}{1 + x^{2n}}$, $x \in (-2, \infty)$, at $y = 1$;

(t) $f(x) = \lim\limits_{n \to \infty} \dfrac{x^n \left(A + \sin \frac{1}{x-1}\right) + B + \sin \frac{1}{x-1}}{x^n} + 1;\ x \notin \{0, 1\},\ A, B \in \mathbf{R}$, at $y = 1$.

2. Discuss, from the definition, the continuity, or lack thereof, and properties at 0 for the following functions:

(a) $f(x) = \begin{cases} \sin \frac{1}{x^2}, & x \in \mathbf{R} \sim \{0\} \\ 0, & x = 0; \end{cases}$
   (b) $f(x) = \begin{cases} x \sin \frac{1}{x}, & x \in \mathbf{R} \sim \{0\} \\ 1, & x = 0. \end{cases}$

3. Define $f(0)$ so as to make the following functions continuous at the point $x = 0$:

(a) $f(x) = \frac{\sin(\sin x)}{x}$;   (b) $f(x) = \frac{\sin(\sin x^2)}{x}$;   (c) $f(x) = \frac{\sin(\sin(\sin x))}{x}$.

4. Prove Theorem 2.5.1.

5. Let $f : D \to \mathbf{R}$ be continuous at c and $f(c) > 0$. Show there exists an $\epsilon > 0$ such that $f(x) > 0$ for $x \in D \cap (c - \epsilon, c + \epsilon)$.

6. Let $f : D \to \mathbf{R}$ be continuous at c. Prove there exist $\delta, K > 0$ such that $|f(x)| < K$, for $x \in D \cap (c - \delta, c + \delta)$.

7. Let f be continuous on $[a, b] \cup [c, d]$ where $b < c$. Show if y is an arbitrary real number and e is any fixed element of (b, c), then there is a continuous function g with domain $[a, d]$ which extends f such that $f(e) = y$.

8. Let f be continuous on D, and let $a \notin D$ be a limit point of D. Show if g is a continuous extension of f to $D \cup \{a\}$, then g is unique. Further, find conditions on f which guarantee such a g will exist.

9. Let f be continuous on $\mathbf{Q}$ and g be a continuous extension of f to $\mathbf{R}$. Is g unique?

10. Develop the basic properties of the function $f(x) = e^x$ on $\mathbf{R}$, including the extension of the laws of exponents to arbitrary real exponents.

11. Let f be a polynomial function on $\mathbf{R}$. Show f is continuous on all of $\mathbf{R}$.

12. Let f and g be a polynomial functions on $\mathbf{R}$. Consider the rational function defined by $\frac{f}{g}$ and having domain $D = \{x : g(x) \neq 0\}$. Give a complete discussion of the continuity properties for $\frac{f}{g}$.

13. Let f be continuous on D and $\{a_n\}$ be a sequence of elements of D which converges to $a \in D$. Show $\{f(a_n)\}$ is a convergent sequence. Can we drop the restriction $a \in D$? What about the converse?

14. Let f and g be continuous on D. Show $\max\{f, g\}$ defined by

$$\max\{f, g\}(x) = \max\{f(x), g(x)\}$$

is continuous on D. Do the same for $\min\{f, g\}$, defined in the obvious way.

15. A function f is said to be **additive** provided $f(x + y) = f(x) + f(y)$ for each x and y in its domain. Show if f is additive and continuous on $\mathbf{R}$, then f is defined by $f(x) = cx$ for some $c \in \mathbf{R}$. (It can be shown that not every additive function from $\mathbf{R}$ into $\mathbf{R}$ is continuous.)

16. Let f be defined on $\mathbf{R}$ and continuous. Suppose $f(x) = x^2$ whenever x is rational. What can be said about $f(x)$ for x irrational?

17. Suppose f is defined and continuous on $[0, 1]$. If $f(x)$ is always rational and $f(1) = 2$, what is $f(0)$?

18. Let f be monotone on $[a, b]$ and suppose the image of $[a, b]$ under f is an interval; that is, $f([a, b])$ is an interval. Show f is continuous on $[a, b]$.

19. A function f is **subadditive** if $f(x + y) \leq f(x) + f(y)$ for $x, y \in \mathbf{R}$. Show if f is subadditive, $f(0) = 0$, and if f is continuous at 0, then f is continuous on $\mathbf{R}$. Is the condition $f(0) = 0$ necessary?

20. If f and g are continuous on $\mathbf{R}$, and if $f(x) = g(x)$ for each $x \in \mathbf{Q}$, show f and g are identical. What crucial property of $\mathbf{Q}$ is needed in the proof?

21. If $f : \mathbf{R} \to \mathbf{R}$ is continuous and $f\left(\frac{m}{2^n}\right) = 0$ for each $m \in \mathbf{Z}, n \in \mathbf{N}$, prove $f = 0$.

22. Identify and classify the discontinuities of the functions in Exercise 1.

23. A function g is said to be the **restriction** of the function f to a domain D' provided $D' \subset \mathrm{Dmn}\, f$ and $f = g$ on D'. Prove if f is continuous on D, and $D' \subset D$, then the restriction of f to D' is also continuous. Show by an example the converse need not be true.

24. Give an example of a function f, which is strictly increasing on D, but f^{-1} is not continuous on $f(D)$.

25. Let f be increasing in the interval $[a, b]$. Show

$$\lim_{x \to b^-} f(x) - \lim_{x \to a^+} f(x) \geq \sum_{k=1}^{n} \left[\lim_{x \to x_k^+} f(x) - \lim_{x \to x_k^-} f(x) \right],$$

where $a < x_1 < \cdots < x_n < b$. Hence, conclude
(a) the set of discontinuities of a monotonic function is at most countable;
(b) there is a point of continuity of f in every open subinterval of $[a, b]$.

2.6 Properties of Continuous Functions

Let us return now to our model continuous function, namely, a function defined on an interval whose graph can be drawn with a single continuous sweep of the pen. With this in mind, consider the following situation:

> The function f is defined on an interval I, $a, b \in I$ with $a < b$ and $f(a) < f(b)$.

This situation is pictured in Figure 2.6.1. Since f is continuous, the points $(a, f(a))$ and $(b, f(b))$ can apparently be connected by a single unbroken curve. This process will generate a set of y values, namely,

$$\{y : \text{ there is an } x \in [a, b] \text{ such that } f(x) = y\}.$$

A little experimentation should quickly convince the reader every real number in the interval $[f(a), f(b)]$ should belong to the set of generated y values. Since what we have just asserted is so obviously 'true', we should try to formalize it as a theorem.

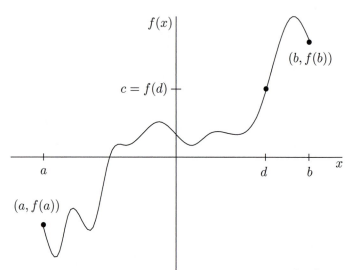

Figure 2.6.1 The graph of f defined on an interval $[a, b]$ with $a < b$ and $f(a) < f(b)$. Given $c \in (f(a), f(c))$, a point of the domain, d, is found in $[a, b]$, with $f(d) = c$, thereby illustrating the intermediate value theorem.

Theorem 2.6.1. *Let f be continuous on an interval I with $a < b$ and $a, b \in I$. If $f(a) < c < f(b)$, then there is an $x \in (a, b)$ such that $f(x) = c$.*

Proof. Consider the set

$$A = \{x : x \in [a, b] \text{ and } f(x) < c\}.$$

This set is nonempty, since $a \in A$; it is bounded above, since $x \in A$ implies $x \leq b$. Thus, it has a supremum that we call d. Moreover, the continuity of f on $[a, b] \subseteq I$ together with $f(a) < c < f(b)$ imply $a < d < b$. We claim $f(d) = c$. To establish the claim, we eliminate the other possibilities. First, suppose $c > f(d)$. Let $\epsilon = c - f(d)$. By continuity of f at d, there is a $\delta > 0$ such that

$$d < x < d + \delta \text{ .implies. } f(x) < f(d) + \epsilon = c.$$

Since $(d, d + \delta) \cap [a, b] \neq \emptyset$, this contradicts our choice of d as $\sup A$ (**WHY?**); moreover, it follows $d \neq a$. On the other hand, suppose $f(d) > c$. Let $\epsilon = f(d) - c$. Since f is continuous at d, there is a $\delta > 0$ such that

$$d - \delta < x < d \text{ .implies. } c = f(d) - \epsilon < f(x).$$

The existence of an $x \in (d - \delta, d] \cap A$ is guaranteed by Theorem 0.4.1, but such an x cannot be a member of A. This again contradicts our choice of $d = \sup A$. It follows $f(d) = c$ as desired. $\square$

Corollary. *Let f be continuous on I, an interval. Then, the range of f is an interval.*

Proof. Let $a, b \in \mathrm{Rng}\, f$, with $a < b$. We must show (see Exercise 0.6.12) if $c \in (a, b)$, then $c \in \mathrm{Rng}\, f$. But this is the content of Theorem 2.6.1. $\square$

Discussion. Theorem 2.6.1 is often referred to as the **Intermediate Value Theorem**. This is a good name, since it describes what the theorem does. An alternative way to think of the theorem is in terms of the corollary, which spells out the fact that continuous functions map intervals to intervals. A portion of the hypothesis of Theorem 2.6.1 is that the domain of f is an interval. This requirement on the domain cannot be relaxed, as can be seen from the examples of the last section. Thus, this theorem is not simply a theorem about continuous functions, rather it is a theorem about continuous functions and intervals.

In the proof, the reader should pay close attention to the use of the supremum principle. This is the heart of the argument, and any attempt to prove the result without using this principle, in some form or other, must fail. The key is the definition of the set to which we apply the principle, namely,

$$\{x : x \in [a, b] \text{ and } f(x) < c\}.$$

The choice of this set should appear natural in the sense that if we apply f to d, the supremum, then it should be intuitively obvious that $f(d) = c$. We emphasize that this basic intuition can only come about by sitting down and experimenting with examples in the presence of the question: how could one generate an x such that $f(x) = c$? Last, note while the supremum principle generates the candidate value, d, for us, it is continuity that forces d to have the desired properties. $\square$

Theorem 2.6.1 results from an interaction between the notion of continuity and a property of the domain of the function. The reader may wonder whether other such interactions exist. Such interactions do exist, and we study them in the next two theorems.

Theorem 2.6.2. *Let f be continuous on a closed interval $[a, b]$. Then, the range of f is bounded.*

Proof. Define the set S by

$$S = \{c : c \in [a, b] \text{ and } f \text{ is bounded on } [a, c]\}.$$

S is not empty since $a \in S$. Further, if $b \in S$, we are through; otherwise, S is bounded above by b. Thus, S has a supremum. Call it d. We claim $d = b$. Suppose not. Then, $d < b$. Evidently, since f is continuous, there is a $\delta > 0$ such that $|d - x| < \delta$ .implies. $|f(x)| < |f(d)| + 1$. Since δ is positive, this contradicts d being the supremum of S, so $b = d$ and continuity of f at b ensures $b \in S$ **(WHY?)**. Thus, f is bounded on $[a, b]$, as claimed. $\square$

Theorem 2.6.3. *Let f be defined and continuous on $[a, b]$. Then, f attains the supremum of its range.*

Proof. Let $C = \text{Rng } f$. By the preceding theorem, C is bounded above. Since $f(a) \in C$, C is nonempty and has a supremum that we denote by c. We claim $c = f(x)$ for some $x \in [a, b]$. To see this, set $g(x) = c - f(x)$, $x \in [a, b]$. Either $g(x) = 0$ for some $x \in [a, b]$, or $g(x) > 0$ for all $x \in [a, b]$. If the former, we are done. If the latter, then g is continuous and nonzero on $[a, b]$, whence $\frac{1}{g}$ is continuous on $[a, b]$ by Theorem 2.5.1(iii). Hence $\frac{1}{g}$ must be bounded on $[a, b]$, again by Theorem 2.6.2. But given $\epsilon > 0$, because c is the supremum of C, we can find $x \in [a, b]$ such that $0 < g(x) < \epsilon$. Thus, $\frac{1}{g}$ cannot be bounded contradicting Theorem 2.6.2. $\square$

Corollary 1. *Let f be defined and continuous on $[a, b]$. Then, the range of f is a closed interval.*

Corollary 2. *Let f be defined and continuous on $[a, b]$. Then, there is a point $x_0 \in [a, b]$ such that for all $x \in [a, b]$, $f(x) \le f(x_0)$.*

Discussion. These theorems should be thought of as being about what continuous functions do to various types of domains. Theorem 2.6.1 says if we apply a continuous function to an interval, the result will be an interval. Theorems 2.6.2 and 2.6.3 together say if the interval is closed, then the result will also be a closed interval. Thus, continuous functions have the property that they preserve certain properties of subsets of **R**. This fact suggests all kinds of questions, such as exactly what properties are preserved by continuous functions, or are there other classes of functions that preserve other properties of domain sets, and so on. Such questions have been asked by mathematicians. We will provide some of the answers in later chapters. Nevertheless, the reader should explore some of the possibilities, or at least formulate additional questions. $\square$

Theorem 2.6.4. *Let f be continuous and strictly increasing on an interval I. Then, f^{-1} is continuous and strictly increasing on Rng f.*

Proof. It has already been shown in Exercise 2.4.7 that f^{-1} exists and is strictly increasing. Hence, let $y \in \text{Rng } f$ and $\epsilon > 0$ be fixed. Let $w = f^{-1}(y)$. There are several possibilities regarding where w is in I. If, for example, $I = [w, w]$, then we are done since $f^{-1} = \{(y, w)\}$, which is continuous (**WHY?**). We will treat in detail the case where w is the left-hand endpoint of I. With no loss of generality, we may assume $w + \epsilon \in I$. Let $\delta = f(w + \epsilon) - y$. It is immediate from the monotonicity conditions that if $x \in \text{Rng } f$ and $|x - y| < \delta$, then $|f^{-1}(x) - f^{-1}(y)| < \epsilon$, as desired. The proof of the other cases is analogous and left to Exercise 2. $\square$

Discussion. The intuition of the situation is best obtained from a picture (Figure 2.6.2). The reader should study how the condition that the domain is an interval is used, since the theorem fails without this condition. Prime examples of a pair of functions that satisfy the conditions of the theorem are the logarithmic and exponential functions; these functions are developed in Chapter 9. $\square$

Exercises

1. Complete the proof of Theorem 2.6.3.

2. Complete the proof of Theorem 2.6.4.

3. Find an example of a strictly increasing function that is continuous on $[0, 2] \cup (3, 5]$ whose inverse is not continuous.

4. Let f be strictly increasing and continuous on A. What can be said about f^{-1}?

5. Show $f(x) = x^8 - 17x^6 + x^3 + 1$ has a root between 0 and 1.

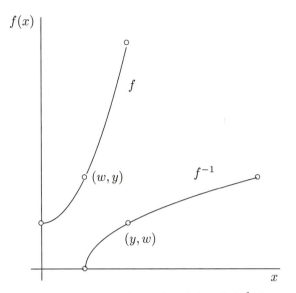

Figure 2.6.2 Typical graphs of f and f^{-1} showing their relationship. The cases in the proof of Theorem 2.6.4 are identified by circles.

6. Give an example of a continuous function with domain $\mathbf{Q}$ that does not satisfy the Intermediate Value Theorem.
   (In this problem and the next, the intervals being discussed are intervals on $\mathbf{Q}$.)

7. Let f be continuous on $\mathbf{Q}$ and suppose f does satisfy the Intermediate Value Theorem. What can be said about f?

8. Let f be continuous on $[a, b]$ with $x_1, x_2, \ldots, x_n \in [a, b]$ and $g_1, g_2, \ldots, g_n$ real numbers all of one sign. Show

$$\sum_{i=1}^{n} f(x_i)g_i = f(c) \sum_{i=1}^{n} g_i$$

for some $c \in [a, b]$.

9. Let f be defined on $[0, 2]$ and have at most one point of discontinuity on this interval. If $f(x)$ is rational for each $x \in [0, 1)$ and irrational for each $x \in (1, 2]$, prove f has exactly one point of discontinuity. What is that point?

10. Let f and g be continuous on $\mathbf{R}$ and suppose $f(x) = g(x)$ for each $x \in \mathbf{Q}$. What conclusion can be drawn about f and g?

11. Let f be one-to-one and continuous on $[a, b]$. Show f is strictly monotone there.

12. Let $f(x) = a_0 + a_1 x + \cdots + a_n x^n$, where n is odd and $x \in \mathbf{R}$. Show the range of f is all of $\mathbf{R}$, whence f must have a real root.

13. Let f be as in Exercise 12, except that n is even. Show if $a_0 a_n < 0$, then f has at least two real roots.

14. Let f be increasing and have the intermediate value property on $[a, b]$. Show f is continuous.

15. Let $f : [a, b] \to [a, b]$ be continuous. Prove there exists $c \in [a, b]$ such that $f(c) = c$. (Such a point is called a **fixed point** for the function f.) Does this result hold if we replaces $[a, b]$ by either $[0, \infty)$ or $(0, 1]$?

16. Let $f : \mathbf{Q} \to \mathbf{R}$ be continuous and satisfy the condition that for each $M > 0$ there exist $x, y \in \mathbf{Q}$ such that $f(x) < -M$ and $f(y) > M$. Show the extension of f to $\mathbf{R}$ by continuity is onto $\mathbf{R}$.

17. If f is continuous in $[a, b]$ and $\frac{\lambda}{\mu}$ is positive, show for some $\xi \in [a, b]$

$$f(\xi) = \frac{\lambda f(a) + \mu f(b)}{\lambda + \mu}.$$

18. Given

$$f(x) = \lim_{n \to \infty} \frac{\ln(2 + x) - x^{2n} \sin x}{1 + x^{2n}}$$

explain why $f(x)$ does not vanish in $\left[0, \frac{\pi}{2}\right]$ even though $f(0)$ and $f\left(\frac{\pi}{2}\right)$ differ in sign.

19. Let f be defined on $[a, b]$. f is said to be **locally monotone increasing** on $[a, b]$ if for each x there is a $\delta > 0$ such that f is monotone increasing on $(x - \delta, x + \delta)$ (see Exercise 2.4.12). Show if f is locally monotone on $[a, b]$, then f is monotone on $[a, b]$.

20. Give an example of a continuous, strictly increasing function f, defined in $[0, 1] \cup [2, 3]$, whose inverse is not continuous.

21. Let $f : [a, b] \to \mathbf{R}$ be continuous and $f(x) > 0$ for all $x \in [a, b]$. Show there exists a constant $C > 0$ such that $f(x) \geq C > 0$ for $x \in (a, b)$. Give an example to show the result need not be true if $[a, b]$ is replaced by either $[0, \infty)$ or $(0, 1]$.

22. Let $f : [a, b] \to \mathbf{R}$ be continuous and $g(x) = \sup \{f(t) : a \leq t \leq x\}$. Show $g(a) = f(a)$, g is increasing and continuous at a.

23. Use Theorem 2.6.1 to show for each real number $y > 0$ there exists a positive mth root for $m \in \mathbf{N}$. [Hint: Consider x^m.]

24. If f and g are continuous on $[a, b]$, $f(a) \geq g(a)$, and $f(b) \leq g(b)$, prove there exists $c \in [a, b]$ such that $f(c) = g(c)$.

25. If f is continuous on $\mathbf{R}$ and for some $a, b \in \mathbf{R}$, $f(a) \cdot f(b) < 0$, prove there exists $x \in (a, b)$ such that $f(x) = 0$.

26. Give examples of functions that are not continuous but still possess the intermediate value property.

27. Construct a two-to-one function from $[0, 1]$ to $\mathbf{R}$ [that is, $\{x : f(x) = y\}$ is either empty, or consists of exactly two members.]. Show such a function cannot be continuous.

Chapter 3

A Little Topology

In this chapter, we develop the basic topological properties of the real line. The reader might wonder: what is topology? One answer is that it is the study of those properties of spaces that are invariant under homeomorphisms. True as the last statement may be, it is hardly an answer for the uninitiate. To understand what this answer means, we must come to grips with the notion of 'topological property'.

Consider for a moment the concept of function. One way to think of functions is to focus on collections of ordered pairs, and to consider what properties such collections might have. Evidently, the function concept itself can be treated this way, as in the definition of function that appears in the appendix on set theory. This approach is abstract and elegant. However, it loses much of the *raison d'être* of functions. Functions exist to 'get you from one place to another place'. This is probably why one of the synonyms for function is 'map'. If you are in a given place and have a map, the map tells you how to get to other places. If you have a given element of a domain, and you have a function on that domain, the function tells you how to find a given element of the range. Thinking of functions in this way makes one focus on three things: domain, transfer process, and range.

Now let us think of our domain and range as being arbitrary sets. As sets, they have no internal structure. We would not care, for example, what type of entities make up the members of the set. However, if we know something about the nature of the transfer process between the two sets, we can say something about the relationship of the sets to one another. For example, suppose the domain set consists of 10 marbles. If the transfer process satisfies the conditions of a function, we know the range of the function can contain at most 10 elements, although we cannot say what type of items these elements might be. If, as well, we know the function is one-to-one, while again we cannot say what the range set is composed of, we can say it includes 10 items.

Evidently, then, transfer processes that are one-to-one functions have a very interesting property. They guarantee that an attribute of the domain is also present in the range. The attribute to which we refer is that of 'size' of the domain set. The key fact we want to highlight is that a select group of functions, namely, one-to-one functions, can guarantee to preserve certain attributes of their domains. Indeed, in the case of one-to-one functions and the size attribute, the preservation is in both directions; that is, the domain has 10 members if and only if the range has 10 members.

Consider now the case of continuous functions. The motivation for defining this class of functions was an attempt to collect together all those functions whose graphs could be drawn with a single stroke of the pen. After a definition of continuity was given, it was discovered that for the graph of the continuous function to have the required property, the domain to which the function was applied also had to have a certain property. Further, work with the continuity concept led to several theorems of the form

> If f is continuous on D and D has property $\mathcal{P}$, then the range of f has property $\mathcal{P}$.

Obviously then, continuous functions are special in that they preserve attributes, or structure, of their domains as one uses the function to move to the range.

Still another way of looking at these ideas is the following. Consider the interval $A = [0, 1]$ and the set $B = [0, 1] \cup [2, 3]$. It is impossible to find a continuous function that maps A onto B. The conclusion follows because the image of a closed bounded interval under a continuous function must be a closed bounded interval and B is not an interval. Thus, even though there are many functions that will map A onto B, none of these functions will be continuous. What this means is that a continuous function must somehow be inextricably wedded to its domain, and the very notion of continuity must be tied up with the structure of $\mathbf{R}$.

On the other hand, if we consider B as the domain and A as the range, we see there are continuous functions mapping B onto A (the reader should construct one). Thus, the property of being continuous for a function is not enough to guarantee the domain, and range will have equivalent properties. To ensure this, we will need more, namely, that both the function and its inverse are continuous.

In topology, a one-to-one, continuous function, whose inverse is also continuous, is called a **homeomorphism**. In the remainder of this chapter, we will be looking for **topological properties**, namely, those properties of $\mathbf{R}$, or of subsets of $\mathbf{R}$, that are **invariant** (do not change) upon application of such a map.

In seeking such properties, the focus will be on a comparison of the attributes of the domain of a continuous function with that of the range of the function. The function itself will be of little interest, other than to specify its continuity properties. The identification process will consist of showing that a given property, $\mathcal{P}$, is preserved by one-to-one continuous functions, since if continuity guarantees that an attribute passes from the domain to the range of a function, the continuity of the inverse, for the case of homeomorphisms, will reverse the passage.

The identification of topological properties will tell us a great deal about the internal structure of $\mathbf{R}$, providing information about such questions as: how closely the points of $\mathbf{R}$ are packed together; is $\mathbf{R}$ connected, or does it have holes? With this in mind, let us begin.

3.1 Basic Topological Concepts

The basic objects for study in topology are 'spaces' (or topological spaces). Exactly what a space should be was a matter of contention for a goodly length of time during the latter part of the last century and the first part of this century. The end result of these deliberations was that a space was a nonempty set with a certain collection of subsets selected out and termed 'open sets'. For us, our set is $\mathbf{R}$, the set of all real numbers. We already have a concept of 'open interval', and with a little thought we can generalize it to 'open set'.

Think for a moment about the open interval (a, b), with $a < b$. For any fixed $c \in (a, b)$, there is an open interval containing c that is completely contained in (a, b). This is an utter triviality, since we may choose this interval to be (a, b). To see this is not true for an arbitrary subset X of $\mathbf{R}$, note that $a \in [a, b]$, but there is no open interval contained in $[a, b]$ that also has a as a member. As the reader can show, $(a, b) \cap \mathbf{Q}$ does not contain any open interval of positive length.

Definition. A subset S of $\mathbf{R}$ is **open** provided that for every $c \in S$, there is an open interval (a, b) such that $c \in (a, b) \subseteq S$.

Discussion. It is immediate from the definition that each set of the form (x, y), $x, y \in \mathbf{R}$, which we have been calling an open interval is, in fact, an open subset of $\mathbf{R}$. It might be thought that the only open sets are the open intervals; however, this is not the case since the following subsets of $\mathbf{R}$ are all open:

$$\mathbf{R}, \quad \emptyset, \quad (2, \infty), \quad (3, 5) \cup (7, 10), \quad (-\infty, -1) \cup (1, 4).$$

While the last two examples show the open subsets of $\mathbf{R}$ consist of more than the open intervals, the definition suggests that the open subsets must be very closely related to open intervals. The reason for this is that whenever a point c is in an open set, the set must also include an open interval surrounding c. Thus, it can be seen that every open subset can be obtained as a union of open intervals (**HOW?**). Recall, in Section 2.6, we showed the property of being an interval was a topological property, although it was not stated in this way. This suggests that the property of being an open set might also be a topological property. Before stating the appropriate theorem, we need a definition. $\square$

Definition. Let f be a function having domain, D, and let $A \subseteq D$. The **image** of A under f (denoted by $f(A)$) is defined by

$$f(A) = \{y : \text{there exists } x \in A \text{ and } y = f(x)\}.$$

Theorem 3.1.1. *Let f be a one-to-one continuous function having domain D, and suppose $A \subseteq D$ is an open subset of $\mathbf{R}$. Then $f(A)$ is an open subset of $\mathbf{R}$.*

Proof. Let $y \in f(A)$. We must find an open interval about y that is a subset of $f(A)$. Now $y = f(x)$ for some $x \in A$. Since A is open, there exists an open interval (c, d) such that $x \in (c, d) \subseteq A$. We may assume that $c, d \in A$ (**WHY?**). Since f is one-to-one, $f(c)$, $f(d)$, and $f(x)$ are three distinct points in $f(A)$. Without loss of generality, we may assume that $f(c) < f(x)$. Since f is continuous and one-to-one, $f(d) \le f(x)$ would yield a contradiction. To see this, first note that $f(d) = f(x)$ contradicts the fact that f is one-to-one. Thus, suppose $f(c) < f(d) < f(x)$. Evidently, by the Intermediate Value Theorem there exists $z \in (c, x)$ such that $f(z) = f(d)$, whence f is again not one-to-one. A similar argument takes care of the case when $f(d) \leq f(c)$. Thus, by the Intermediate Value Theorem and its consequences,

$$y = f(x) \in (f(c), f(d)) \subseteq f(A)$$

follows. Thus, $f(A)$ is open. $\square$

Discussion. The proof that the attribute of being an open set is a topological property employs two essential facts, namely, (1) the function is one-to-one, and (2) the image of an interval is again an interval, which is a consequence of continuity.

To see that (1) cannot be relaxed, consider $f(x) = x^2$, where $x \in \mathbf{R}$. It is easily seen that $f((-1, 1)) = [0, 1)$, which is not an open interval, and is not an open set because $0 \in [0, 1)$, but there is no open interval containing 0 that is a subset of $[0, 1)$. $\square$

Example 1. Show the attribute of being bounded is not a topological property.

Solution. Let $f(x) = \frac{1}{x}$ be defined on $D = (0, 1)$. Since f is strictly decreasing on D, it is one-to-one. That it is continuous follows from Example 2.5.3. But Rng $f = (1, \infty)$, which is an unbounded set. $\square$

Discussion. The proof that a particular attribute of a set is not topological requires a counterexample. In this case the function $f(x) = \frac{1}{x}$ together with its domain serve to show the image of a bounded set is not necessarily bounded. The reader should compare this result with Theorem 2.6.3 and its corollaries. $\square$

Definition. Let S be a subset of $\mathbf{R}$ and $c \in S$. Then c is an **interior point** of S provided there is an open interval (a, b) such that $c \in (a, b) \subseteq S$. If c is not an interior point of S, c is said to be a **boundary point**.

Discussion. If the reader thinks of what an interior point should be according to the usual meaning of interior, then this usage is completely consistent with the intuitive meaning. Any set can have interior points, not just open sets. But the concepts are related by the following: A set is open if and only if all of its points are interior points (see Exercise 11).

If $c \in S$ is not an interior point, we have called c a boundary point of S. This should make sense intuitively, since the boundary usually refers to an edge. Thus, if c is not surrounded by points of S, then there must be points which are not in S that are close to c. If one thinks about this further, one sees that there may also be points that are not in S that one would want to call boundary points. This idea will be discussed further in Example 3 and Exercises 31 and 32. $\square$

Example 2. Find the interior and boundary points of

$$\bigcap \left\{ \left(-\frac{1}{n}, 1 + \frac{1}{n} \right) : n \in \mathbf{N} \right\}.$$

Solution. Since $-\frac{1}{n} < 0$ and $1 < 1 + \frac{1}{n}$ for every $n \in \mathbf{N}$, it follows that $[0, 1]$ is contained in the intersection. On the other hand, if $a \notin [0, 1]$, two cases ensue.

First assume $a < 0$. By the Archimedean Property, there exists $N \in \mathbf{N}$ such that $a < -\frac{1}{N}$, whence a does not belong to the intersection. On the other hand, if $a > 1$, a similar argument shows that at least one member of the intersection misses the point a. Hence, we conclude that the intersection is precisely the closed interval $[0, 1]$. If $c \in (0, 1)$, then c is clearly an interior point of $[0, 1]$. Since there is no open interval containing 0 and contained in $[0, 1]$, we see that 0 is not an interior point. Similarly, 1 is also not an interior point. Thus, the set of interior points of the given set is $(0, 1)$. This argument also tells us what the boundary points are, namely, all points in $[0, 1]$ that are not interior points. Thus, the set of boundary points is $\{0, 1\}$. $\square$

Discussion. The intuition for what the intersection should be comes from

$$\lim_{n \to \infty} \left(-\frac{1}{n} \right) = 0 \quad \text{and} \quad \lim_{n \to \infty} \left(1 + \frac{1}{n} \right) = 1. \ \square$$

Example 3. Find all interior and boundary points of $A = \left\{ \frac{1}{n} : n \in \mathbf{N} \right\}$ and its complement.

Solution. The set, A, contains no open interval (**WHY?**), whence it can have no interior points. Thus, every point of A is a boundary point of A. Now, consider A', the complement of A. The reader can verify

$$A' = (-\infty, 0] \cup (1, \infty) \cup \left(\bigcup \left(\frac{1}{n+1}, \frac{1}{n} \right) : n \in \mathbf{N} \right).$$

Every point of A', except 0, is an interior point. Thus, 0 is a boundary point of A'. $\square$

Discussion. The key fact used in this example is that for a set to have interior points, it must contain an open interval. The essential conclusion of the example is the complement of a set can have boundary that which do not belong to the original set and these points also should be thought of as being boundary points of the original set. This suggests the boundary of a set should consist of the boundary points of the given set together with the boundary points of the complement (see Exercises 31 and 32). $\square$

Theorem 3.1.2. *Let A be a collection of open sets. Then $\bigcup A$ is open. If A is a finite collection, then $\bigcap A$ is open.*

Proof. Let $c \in A$. Then there exists $B \in A$ such that $c \in B$. Now B is open, so there is an open interval (a, b) such that $c \in (a, b) \subseteq B \subseteq \bigcup A$ and we are done. Now to complete the proof, assume that A is finite, say, $\{A_1, \ldots, A_n\}$. Let $c \in \bigcap A$. For each $i = 1, 2, \ldots, n$, we can choose a positive δ_i such that

$$(c - \delta_i, c + \delta_i) \subseteq A_i.$$

Let $\delta = \min\{\delta_1, \ldots, \delta_n\}$, whence $(c - \delta, c + \delta) \subseteq \bigcap A$. $\square$

Discussion. Unions and intersections are the arithmetic operations on sets, as plus and times are the arithmetic operations on numbers. This theorem answers the question of the degree to which the attribute of being open is preserved by these arithmetic operations on sets.

The only significant feature of the argument is the use of the finiteness of A to ensure that $\delta > 0$. It is possible for the infimum of an infinite collection of positive real numbers is zero. For example, $\inf \left\{ \frac{1}{n} : n \in \mathbf{N} \right\} = 0$. The reader should note this carefully and create an example of an infinite collection of open sets whose intersection is not open (see Exercise 7).

On a more general note, the branch of mathematics known as set-theoretic topology (general topology) begins by defining a **topology** on a nonempty set X as a collection of subsets (called **open sets**) that have the property that arbitrary unions of members of the collection and finite intersections of members of the collection belong to the collection, as well as the empty set and X. As the reader can readily see by a comparison of the general definition with the content of the theorem above, such a definition is not abstractly and artificially coined, but stems from basic properties of open sets on the real line. Similarly, many of the abstract definitions and ideas in mathematics are carved out of well-known elementary properties of the real numbers and are motivated by a desire to obtain deeper insight into the nature of the real number system. $\square$

We have seen that the concept of open interval can be generalized to that of open set. The reader may be wondering whether the concept of closed interval also has a generalization. The answer is yes. Again, the generalization is obtained by looking at the relationship of the set as a whole to particular points. Recall the definition of a limit point of a set: a is a **limit point** of $A \subseteq \mathbf{R}$ if for every positive δ, $[(a - \delta, a) \cup (a, a + \delta)] \cap A \neq \emptyset$.

Definition. A subset S of $\mathbf{R}$ is **closed** if S contains all its limit points.

Discussion. It is easily checked that every closed interval is closed. The following are also examples of closed subsets of $\mathbf{R}$:

$$\mathbf{R}, \quad \emptyset, \quad [2,3] \cup [-1,0], \quad [1,1], \quad \{0\} \cup \left\{ \frac{1}{n} : n \in \mathbf{N} \right\}.$$

The fact that arbitrary intersections and finite unions of closed sets are again closed is pursued in Exercises 12 and 13. The reader is invited to supply appropriate examples demonstrating that an arbitrary union of closed sets need not be closed. $\square$

The next theorem expresses the fundamental relationship between open and closed sets. It could also be used as the definition of a closed set, from which one could arrive at the above definition of a closed set as a theorem.

Theorem 3.1.3. *Let $A \subseteq \mathbf{R}$. Then A is open if and only if the complement, A', of A is closed.*

Proof. Let A be open. Fix $a \in \mathbf{R}$, such that a is a limit point of A'. If $a \in A'$, then we are done; hence for the sake of argument we assume $a \in A$. But then, there is a neighborhood of a (any open set containing a), say, $(a - \delta, a + \delta)$, that is completely contained in A. Since $A \cap A' = \emptyset$, a cannot be a limit point of A'. Thus, we have a contradiction, whence $a \in A'$. It follows A' is closed.

Conversely, suppose A' is closed for a fixed A. Let $a \in A$, and so not in A'. Thus, a is not a limit point of A', whence there is a neighborhood of a, say,

$(a - \delta, a + \delta)$, such that the intersection of this neighborhood with A' is empty. It is immediate that $(a - \delta, a + \delta) \subseteq A$, whence A is open. $\square$

Discussion. Note the use of an indirect argument to show the complement of an open set is closed. This type of reasoning is often very useful in elementary topological proofs. Also, both proofs involve little more than manipulation of the relevant definitions. What is essential is that the reader have these at the 'tip of the tongue', since otherwise it is impossible to draw together the relevant facts.

Theorem 3.1.3 also tells us that by taking set-theoretic complements, any statement known to be true about open sets can be translated into a statement about closed sets that will also be true. For example, we have shown the union of an arbitrary collection of open sets is open. Taking the union of a collection of open sets corresponds to taking the intersection of the complements of these same sets. Thus, the intersection of an arbitrary collection of closed sets must be closed. The reader will be asked to give a direct proof of this fact in Exercise 13. $\square$

One of the interesting features of the topology of the reals is that all open sets can be generated using open intervals having rational endpoints.

Theorem 3.1.4. *Let A be open. Then A is a countable union of open intervals with rational end points.*

Proof. Fix $c \in A$. Then there is an open interval (a, b) such that $c \in (a, b) \subseteq A$ with a and b both rational. If we label the interval as $(a, b)_c$, then $A = \bigcup\{(a, b)_c : c \in A\}$. Since $\mathbf{Q}$ is countable, this collection of such intervals must be countable. $\square$

Discussion. As defined in the appendix, a set is countable if it can be put in one-to-one correspondence with $\mathbf{N}$. A critical fact required for this proof is that $\mathbf{Q}$ is countable (see the appendix). The reader should give a detailed explanation as to why an interval with rational endpoints exists. This should be followed by a detailed explanation as to why the collection of intervals is countable, since A in general will be uncountable. $\square$

Theorem 3.1.5. *Let A be open. Then A is a countable disjoint union of open intervals.*

Proof. For $x, y \in A$, we write $x \cong y$ if and only if $x, y \in (a, b) \subseteq A$ for some $a, b \in \mathbf{R}$. It is a straightforward matter to check that $\cong$ defines an equivalence relation (see the appendix) on A. It follows the equivalence classes are a collection of disjoint intervals. Further, each equivalence class contains at least two points **(WHY?)** and so, since it is an interval, contains a rational number. Thus, the collection of intervals is countable. To complete the proof, we need only check that an interval bounded above (below) does not contain its supremum (infimum). Thus, consider an interval, I, which is bounded above, and call its supremum b. If $b \in A'$, we are done. To see why $b \in A'$ must be so, consider $b < c$, where c is arbitrary. Then there is a c' such that

$$b < c' \leq c \quad \text{and} \quad c' \in A',$$

since otherwise $b \cong c$ and b could not be an upper bound for I. It follows b is a limit point of A', and so is a member of A', since A' is the complement of an open set. Thus, $b \in A'$ as claimed. □

Discussion. The use of equivalence relations to get a decomposition of A into intervals is a nice application of equivalence relations as a tool. Since what we are trying to establish is that A is a union of a collection of disjoint open intervals, the properties of equivalence classes make the proof a triviality. However, the proof based on equivalence relations is not nearly so self-motivating as a direct construction of the required intervals. The constructive process also acts to motivate the definition of $\cong$, since the interval $I \subseteq A$ that contains $a \in A$ is obtained by

$$I_a = \bigcup \{(b, c) : a \in (b, c) \subseteq A \text{ and } b, c \in \mathbf{R}\}.$$

Once one has the collection of intervals I_a, one must deal with the problem that many of the intervals are the same. The natural way to deal with problems of this type is to generate a suitable equivalence relation that identifies all those objects that are the same except for their name. □

The development above establishes that all the basic topological properties of the real numbers have at their heart the order and completeness axioms. Indeed, the very definition of open set has as its foundation the concept of 'open interval', which itself arises out of the order properties of $\mathbf{R}$. It is clear then why this topology (collection of open sets) is called the **order topology** on $\mathbf{R}$.

Definition. Let A be a subset of $\mathbf{R}$. Set

$$A_{lm} = \{x : x \text{ is a limit point of } A\}.$$

A_{lm} is referred to as the **derived set** of A.

Theorem 3.1.6. *Let $A \subseteq \mathbf{R}$. Then the derived set of A is closed.*

Proof. Let x be a limit point of A_{lm}. We must show x is a limit point of A. Thus, fix $\delta > 0$. Since x is a limit point of A_{lm}, there exists $y \in A_{lm}$ such that $0 < |y - x| < \frac{\delta}{2}$. Since y is a limit point of A, there exists $z \in A$ such that $0 < |y - z| < \frac{\delta}{2}$. It follows

$$0 < |x - z| \leq |x - y| + |y - z| < \delta.$$

Since δ was arbitrary, x is a limit point of A. □

Discussion. This proof should be reminiscent of many of our earlier proofs with limits. In essence, it makes the following points. To be a limit point of A_{lm}, a point x must have points of A_{lm} arbitrarily close to it. These points, in turn, by virtue of the fact that they belong to A_{lm}, must have points of A that are arbitrarily close to them. Thus, some points of A must end up being close to the point x, which makes x a limit point of A. The proof simply formalizes this intuition. □

Theorem 3.1.7. *Let $A \subseteq \mathbf{R}$. Then $A \cup A_{lm}$ is closed. Moreover, if $B \subseteq \mathbf{R}$ is closed and $A \subseteq B$, then $A \cup A_{lm} \subseteq B$.*

Proof. Let x be a limit point of $A \cup A_{lm}$. Then either x is a limit point of A, or x is a limit point of A_{lm}. In the latter case, x is a limit point of A by our last theorem. Thus, $A \cup A_{lm}$ is closed. The remainder of the proof is left to Exercise 14. □

Example 4. Find the set of limit points of **Q**.

Solution. Let x be an arbitrary member of **R** and fix $\delta > 0$. Then there is a rational number q such that $0 < |q - x| < \delta$. Thus, every real number is a limit point of **Q**. Thus, $\mathbf{Q}_{lm} = \mathbf{R}$. □

Discussion. This example shows the rationals are distributed throughout all the real numbers, and no matter how we might try, we can never find a real number that is very far from all rational numbers. It also tells us every real number can be obtained as a limit of a sequence of rational numbers. A set, A, with this property—that is, every member of **R** can be obtained as a limit point of A—is said to be **dense** in **R**. □

The main thrust of this section has been to develop criteria for distinguishing various types of sets of real numbers. The principal type among these is the collection of open sets. If the reader thinks about it, it will be clear that there are connections between the concept of open set and that of continuous function. One reason for suggesting this is the fact that the open interval concept plays such an important role in the definitions of limit and continuity. The next two theorems delineate the connection between the concept of continuous function and the concepts of open and closed set.

Theorem 3.1.8. *Let f be defined on **R**. Then f is continuous on **R** if and only if for every open set $O \subseteq \mathbf{R}$, $f^{-1}(O)$ is an open subset of **R**.*

Proof. Let f be continuous, $O \subseteq \mathbf{R}$ be open and $x \in f^{-1}(O)$. Now $f(x) = y \in O$, whence there is a positive ϵ such that $|y - z| < \epsilon$ implies $z \in O$. Since f is continuous on **R**, f is continuous at x, so there is a $\delta > 0$ such that $|x - w| < \delta$ implies $|f(x) - f(w)| < \epsilon$. But this means that $f(w) \in O$, whence $w \in f^{-1}(O)$. Thus, $f^{-1}(O)$ is open as desired.

Conversely, suppose that for every open set $O \subseteq \mathbf{R}$, $f^{-1}(O)$ is open, and fix $x \in \mathbf{R}$. We must show that f is continuous at x. Set $y = f(x)$ and fix $\epsilon > 0$. Evidently, $O = (y - \epsilon, y + \epsilon)$ is an open set. Thus, by hypothesis, $f^{-1}(O)$ is open. Since $x \in f^{-1}(O)$, we can choose $\delta > 0$ such that $|x - z| < \delta$ implies $z \in f^{-1}(O)$, which, in turn, implies $|f(x) - f(z)| < \epsilon$. Since ϵ was arbitrary, f is continuous at x. Since x was arbitrary, we are done. □

Discussion. This theorem provides a means for characterizing continuous functions that is independent of the context of real numbers. Moreover, the proof is relatively straightforward.

The reader may wonder why we have required the domain of the function to be the totality of the real numbers. The reason is most simply explained by considering an example. Let $f(x) = x + 3$, $x \in [0, 1]$. Evidently, this is a continuous function having $[3, 4]$ for its range. If we let $O = (2, 5)$, we can ask about the nature of $f^{-1}(O)$, that is, is this set open? In answering this question, consider the point $4 \in O$. We know $f(1) = 4$ and $|z - 1| < \frac{1}{2}$ and $z \in [0, 1]$ will imply that $f(z) \in O$. Thus, the argument presented as the proof of the theorem still has validity. But, on the other hand, we know that $f^{-1}(O) = [0, 1]$, which is a closed subset of **R**. The reason for the breakdown is that not every z that satisfies $|z - 1| < \frac{1}{2}$ also satisfies $z \in [0, 1]$.

The resolution of this problem relates to the fact that we want to characterize continuous functions *on a domain*. The flaw arises because we consider $D = [0, 1]$ as a subset of $\mathbf{R}$ and relate the definition of open and closed sets to $\mathbf{R}$ instead of to D. Once this problem is recognized, the solution is easy. We simply define the concept of an open set **relative to** D (see Exercises 36 and 37). $\square$

We close this section with the following theorem that establishes the relationship between closed sets and continuous functions. Given that this theorem is completely analogous to Theorem 3.1.8 and that a set is closed exactly if its complement is open, the reader may wonder about an analogue to Theorem 3.1.1 that states that open sets are preserved by one-to-one continuous functions. The reader is asked to find a counterexample in Exercise 18. A more limited version of the analogous theorem is proved in Section 3.6.

Theorem 3.1.9. *Let f be defined on $\mathbf{R}$. Then f is continuous on $\mathbf{R}$ if and only if for every closed set $C \subseteq \mathbf{R}$, $f^{-1}(C)$ is a closed subset of $\mathbf{R}$.*

Proof. Exercise 17. $\square$

Exercises

1. Find the interior and the limit points of each of the following sets:
   (a) $\emptyset$;   (b) $\mathbf{R}$;   (c) $(0, 1)$;   (d) $\mathbf{N}$;   (e) $\left\{ \frac{1}{n} : n \in \mathbf{N} \right\}$;
   (f) $(-1, 3) \cap \mathbf{Q}$;   (g) $\left\{ \frac{1}{m} + \frac{1}{n} : m, n \in \mathbf{N} \right\}$;   (h) $(-2, 4) \cap (\mathbf{R} \sim \mathbf{Q})$;
   (i) $\left\{ \frac{1}{n} : n \in \mathbf{N} \right\} \cup (2, 3) \cup (3, 4) \cup \{5\} \cup [6, 7] \cup (\mathbf{Q} \cap [8, 9])$.

2. Which of the sets in Exercise 1 are open? Which are closed?

3. Let A be an open subset of $\mathbf{R}$. If one point is deleted, is A still open? If a finite number of points are deleted, is A still open? If A is infinite, and a countably infinite set of points is removed, is A still open? What can you say if a set that in its entirety constitutes a strictly monotone sequence is deleted from A?

4. Let A be a closed subset of R. If one point is deleted, is A still closed? If a finite number of points is deleted, is A still closed? If A is infinite, and a countably infinite set of points is removed, is A still closed? What can you say if a set that in its entirety constitutes a strictly monotone sequence is deleted from A?

5. Let A be an open subset of R. If one point is added, is A still open? If a finite number of points is added, is A still open? If A is infinite, and a countably infinite set of points is added, is A still open? What can you say if a set that in its entirety constitutes a strictly monotone sequence is added from A?

6. Let A be a closed subset of R. If one point is added, is A still closed? If a finite number of points is added, is A still closed? If A is infinite, and a countably infinite set of points is added, is A still closed? What can you say if a set that in its entirety constitutes a strictly monotone sequence is added from A?

7. Give an example of a collection of open sets whose intersection is not open.

8. Show every finite set is closed and has no limit points.

9. Let $A \subseteq \mathbf{R}$ be a nonempty, closed bounded set. Show $\sup A \in A$.

10. (a) Construct a set of real numbers with exactly three limit points.
  (b) Construct a subset of $[0, 1]$ with a countably infinite number of limit points.

11. Show a set is open if and only if it is composed entirely of interior points.

12. Show a finite union of closed subsets of $\mathbf{R}$ is closed. Show this cannot be extended to infinite unions.

13. Give a direct proof of the fact that an arbitrary intersection of closed sets is closed.

14. Let $A \subseteq \mathbf{R}$. Show there is a unique smallest closed set, $\overline{A}$, that contains A. Show this set $\overline{A} = A \cup A_{lm}$. This set is called the **closure** of A.

15. Establish the following properties of closure:
  (a) A is closed if and only if $A = \overline{A}$; (b) $\overline{\emptyset} = \emptyset$;
  (c) for every $A \subseteq \mathbf{R}, \overline{A} = \overline{\overline{A}}$; (d) for every $A, B \subseteq \mathbf{R}, \overline{A} \cup \overline{B} = \overline{(A \cup B)}$.

16. Prove or disprove
  (a) $\overline{\bigcup A} = \bigcup \overline{A}$; (b) $\overline{\bigcap A} = \bigcap \overline{A}$;
  where A is a family of arbitrary subsets of $\mathbf{R}$.

17. Prove Theorem 3.1.9.

18. Give an example of a one-to-one continuous function from $\mathbf{R}$ into $\mathbf{R}$ for which there is a closed set A such that $f(A)$ is not closed.

19. Let $A \subseteq \mathbf{R}$. Then, $a \in A$ is called an **isolated point** of A provided there is an open interval (b, c) such that $(b, c) \cap A = \{a\}$. Find the set of isolated points of the subsets in Exercise 1.

20. Show

  (a) if A is open, then A has no isolated points;

  (b) if A is closed, then for each $a \in A, a$ is either an isolated point of A or a limit point of A;

  Prove or disprove: If every point of A is isolated, then A is closed.

21. Is it true that if every point of A is a limit point of A, then A is closed? What about if no point of A is a limit point?

22. A set $A \subseteq \mathbf{R}$ is **clopen** if it is both open and closed. Give an example of a clopen subset of $\mathbf{R}$. Find all clopen subsets of $\mathbf{R}$.

23. Let $f : \mathbf{R} \to \mathbf{R}$ be one-to-one and continuous. Let (a, b) be an open interval. Show $f((a, b))$ is also an open interval and give a precise description of the image interval. What can be said about the image of a closed interval?

24. Let $f : \mathbf{R} \to \mathbf{R}$ be one-to-one, continuous. Show f must either be strictly increasing or strictly decreasing. Does this still hold if the domain of f is an arbitrary interval? What about if Dmn f is an arbitrary subset of $\mathbf{R}$?

25. Let $f : \mathbf{R} \to \mathbf{R}$ be a homeomorphism (that is, f is one-to-one and continuous and f^{-1} is also continuous). Let $A \subseteq \mathbf{R}$. Show each of the following statements is true about A if and only if it is also true about $f(A)$.
  (a) A has an isolated point; (b) A is clopen;
  (c) A is an interval; (d) A is a union of two disjoint intervals;
  (e) A has exactly two limit points; (f) A has a nonempty interior.

26. Show $A = \mathbf{R}$ is not a topological property.

27. Show there is a continuous one-to-one function that maps $[0,1] \cup (2,3)$ onto $[0,1)$. Does there exist a function that will do the same job for $[0,1] \cup [2,3)$? Interpret these results in the context of Exercise 25(d).

28. Let A°, $\overline{A}$, and A' denote, respectively, the interior, the closure, and the complement of A. Show for any $A \subseteq \mathbf{R}$,
   (a) $A^{\circ\circ} = A^\circ$;   (b) $(A \cap B)^\circ = A^\circ \cap B^\circ$ (Can $\cap$ be replaced by $\cup$?);
   (c) $(A')^\circ = (\overline{A})'$.

29. Let A be a subset of $\mathbf{R}$. Show there is a unique largest open set that is a subset of A. Show this set is in fact A°.

30. For a fixed $A \subseteq \mathbf{R}$, show there are at most 14 distinct sets that can be obtained from A by successive applications of the operations of complement, interior, and closure. Further, show there is in fact a set that will yield 14 distinct sets via these operations.

31. For a subset $A \subseteq \mathbf{R}$, define the **boundary**, $b(A) = \overline{A} \cap \overline{(A')}$. Prove the following:
   (a) $\overline{A} = A \cup b(A)$;   (b) $A^\circ = A \sim b(A)$;   (c) $\mathbf{R} = A^\circ \cup b(A) \cup (A')^\circ$.
Compute the boundary of the sets described in Exercise 1.

32. Show the boundary of A as defined in Exercise 31 consists of the union of the boundary points of A and the boundary points of A'.

33. Define the **exterior** of A, Ext (A), to be the complement of the closure of A. Prove or disprove the following: Ext $(A \cup B) =$ Ext $(A) \cap$ Ext (B).

34. Find necessary and sufficient conditions such that $(A_{lm})_{lm} = A_{lm}$. Give an example showing $(A_{lm})_{lm} \neq A_{lm}$.

35. Theorem 3.1.5 provides an powerful characterization of open subsets of $\mathbf{R}$ in terms of a countable union of disjoint open intervals. Does a similar characterization of closed subsets of $\mathbf{R}$ exist?

36. Let $D \subseteq \mathbf{R}$. We define the **relative topology** on D by calling a subset $U \subseteq D$ open (relative to D) if and only if $U = A \cap D$ for some open subset $A \subseteq \mathbf{R}$. We also say U is **relatively open** in D. Prove the following results for $A \subseteq D$:

   (a) A point $c \in D$ is a limit point of A with respect to the relative topology on D if and only if it is a limit point of A in $\mathbf{R}$.

   (b) The closure of a subset $A \subseteq D$ with respect to the relative topology on D is the set $\overline{A} \cap D$.

   (c) The set $A \subseteq D$ is closed in the relative topology on D if and only if $A = X \cap D$, where X is a closed set in $\mathbf{R}$.

37. Let f map D into $\mathbf{R}$. Show f is continuous on D if and only if for every open set $O \subseteq \mathbf{R}$, $f^{-1}(O)$ is open in the relative topology on D.

38. Let f map D into $\mathbf{R}$. Show f is continuous on D if and only if for every closed set $C \subseteq \mathbf{R}$, $f^{-1}(C)$ is closed in the relative topology on D.

39. A subset S of $\mathbf{R}$ is **connected** provided there do not exist open sets O_1 and O_2 such that $O_1 \cap O_2 = \emptyset$, $O_i \cap S \neq \emptyset (i = 1, 2)$, and $S \subseteq O_1 \cup O_2$. Show S is connected if and only if S is an interval. Can a set that consists entirely of isolated points ever be connected?

40. A set, A, is called **path connected** if for every $x, y \in A$, there exists a continuous function, ϕ from $[0,1]$ into A such that $\phi(0) = x$ and $\phi(1) = y$. Show every connected subset of $\mathbf{R}$ is path connected.

3.2 Properties of Closed, Bounded Subsets of R

In Sections 2.6 and 3.1, we developed several facts about the images of subsets of **R** under one-to-one, continuous functions. It was shown that the image of an interval was an interval, and the image of a closed bounded interval was again a closed bounded interval. It was also demonstrated that the image of a bounded subset of **R** was not necessarily bounded. Thus, closed bounded intervals behave nicely under continuous functions. Since we have generalized the notion of a closed interval, we may wonder to what extent the above-mentioned theorem about the image of a closed bounded interval under a continuous function can be generalized. The complete answer to this question will be given in Section 3.6. In this section we develop the important properties of closed bounded subsets of **R**.

Since the generalized definition of a closed set is based on the notion of limit point, it is natural to expect that limit points must play a key role. Thus, the first result in this section is the Bolzano-Weierstrass Theorem on the existence of limit points for infinite bounded sets. The reader should not be surprised by the fact that all the results of this section have the completeness of **R** at the heart of their proofs.

Theorem 3.2.1 (Bolzano-Weierstrass). *Let A be an infinite subset of* **R** *that is bounded. Then A has a limit point.*

Proof. Since A is bounded, there is an $M > 0$ such that for all $a \in A, -M < a < M$. We now define a collection of closed intervals for each $n \in \mathbf{N}$:

$$I_0 = [-M, M];$$
$$I_n^i = \left[-M + 2M\frac{(i-1)}{2^n}, -M + 2M\frac{i}{2^n} \right] \text{ for } 1 \leq i \leq 2^n.$$

The collection $I_n^i \ 1 \leq i \leq 2^n$ has union $[-M, M]$ and divides the interval into 2^n subintervals of equal length. From each collection, we pick a subinterval called J_n and having the property that $J_n \cap A$ is infinite. We define J_n inductively as follows:

$$J_0 = I_0,$$

Further, suppose $J_1, J_2, \ldots, J_m$ have been defined so that

$$J_m \subseteq J_{m-1} \subseteq \cdots \subseteq J_0; \quad J_k \cap A \text{ is infinite for } 0 \leq k \leq m$$

and each J_k is chosen from the collection I_k^i where $1 \leq i \leq 2^k$. By assumption there is a fixed s such that

$$J_m = \left[-M + \frac{2M(s-1)}{2^m}, -M + \frac{2Ms}{2^m} \right].$$

If $I_{m+1}^{2s-1} \cap A$ is infinite, set $J_{m+1} = I_{m+1}^{2s-1}$. Otherwise, set $J_{m+1} = I_{m+1}^{2s}$. It is clear $J_{m+1} \subseteq J_m$ and $J_{m+1} \cap A$ is infinite. Last, J_{m+1} is chosen from among the I_{m+1}^i as required. Let a_n denote the left-hand endpoint of J_n, b_n the right-hand endpoint. Then $\{a_n\}$ is monotone increasing and $\{b_n\}$ is monotone decreasing. Further, $a_{n+k} \leq b_n$ for all $k \in \mathbf{N}$. It is immediate $\{a_n\}$ has a limit, call it b. To see that b is a limit point of the set A, let a positive ϵ be given and choose n so that

$$J_n \subseteq [b - \epsilon, b + \epsilon].$$

Such a choice is possible since $b_n - a_n = \frac{1}{2^n}$, for any given n. Since $J_n \cap A$ is infinite, we have that every deleted interval about b will contain points of A. Thus, b is the required limit point of A. $\square$

Discussion. The intuitive idea behind this proof is simple. We start with an interval that contains an infinite number of points of A. We divided it into two equal parts. Surely one of the halves must contain an infinite number of points of A. If the lower half has an infinite number of points from A, we select it for attention; if not, we select the upper half. But now we are looking at an interval that has half the length, and we repeat the procedure. This process generates a sequence of intervals, each of which is contained in its predecessor, and each has half the length of its predecessor. The endpoints of the intervals form two monotone sequences. At the appropriate moment we haul out the supremum principle in the form of the theorem on bounded monotone sequences and use it to generate a limit point. The reader will note that without the Completeness Axiom, the proof could not be accomplished. The reader will be asked to discuss this fact further in Exercise 12. $\square$

Theorem 3.2.1 has an immediate corollary that states the first important property of closed bounded subsets of **R**.

Corollary. *Let A be a closed bounded subset of* **R**. *Then every infinite subset of A has a limit point in A.*

Discussion. This corollary is in the form of an implication. Evidently, an immediate question of interest would be whether the converse is true. What would be particularly interesting about this question is that the condition—'every infinite subset of A has a limit point in A'—makes no mention of 'bounded'. The problem with the concept of boundedness is that it requires a notion of distance as a prerequisite. Thus, a converse could supply an important tool for generalizing the notion of 'closed and bounded' to situations where no distance function was available, provided we could obtain a definition of limit point that did not depend on distance. Such a definition exists (see Exercise 4), whence a converse would supply a more general notion of closed bounded sets.

Of course, all of the above is a mere pipe dream, unless a proof of the converse could be generated. It seems apparent that the condition at least guarantees that A is closed (**WHY?**). However, to show that A is also bounded is a horse of a different color. The industrious reader may wish to try. Otherwise, read on, Macduff. $\square$

Theorem 3.2.2. *Let A be a subset of* **R**. *Then A is closed and bounded exactly if every infinite subset of A has a limit point in A.*

Proof. Let A satisfy the condition that every infinite subset of A has a limit point in A. Let x be an arbitrary limit point of A. Let $O_n = \left(x - \frac{1}{n}, x + \frac{1}{n}\right)$. Evidently, $\bigcap O_n = \{x\}$. Since x is a limit point of A, there is a sequence of distinct points $\{a_n\}$, such that $a_n \in A \cap O_n$. This sequence has a limit that, not surprisingly, is x. Now the points of the sequence constitute an infinite collection of points of A. Such a set has a limit point. Since a convergent sequence has only one limit point, and that limit point is x, we conclude that $x \in A$ as required. Thus, A is closed. Now, consider the collection of sets $[-n, n]$, where $n \in \mathbf{N}$. If A is unbounded, we may pick points $\{a_n\}$ such that $a_n \notin [-n, n]$. Such a sequence does not have a limit point in **R**, much less in A. Thus, A must be bounded. The reverse implication is contained in the corollary above. $\square$

Definition. A sequence $\{J_n\}$ of intervals is called **nested** provided that $J_{n+1} \subseteq J_n$ for each $n \in \mathbf{N}$.

Discussion. If one thinks of what 'nested' should mean, then a sequence of nested intervals has the right properties, in that each successive interval is found inside its successor. Note that if the length of the first interval is finite, then the lengths of a sequence of nested intervals form a monotonic decreasing sequence. $\square$

Theorem 3.2.3. *Let* $\{J_n\}$ *be a sequence of closed, bounded, nonempty, nested intervals. Then* $\bigcap J_n \neq \emptyset$. *Further, if the length of the intervals has limit 0 as* n *tends to* ∞, *then the intersection consists exactly of a single point.*

Proof. Because J_n is a closed, bounded, nonempty interval, $J_n = [a_n, b_n]$ with $a_n \leq b_n$. Thus, consider $\{a_n\}_{n=1}^{\infty}$ and $\{b_n\}_{n=1}^{\infty}$. Note that $\{a_n\}$ is monotone increasing and $\{b_n\}$ is monotone decreasing. If a and b denote the respective limits of $\{a_n\}$ and $\{b_n\}$ that must exist, then $a \leq b$ (**WHY**?). It is immediate that $[a, b] \subseteq J_n$ for each n. Further, if $c < a$, then there is an n such that $c < a_n$, and so $c \notin J_n$ and so not in the intersection. Thus,

$$\bigcap J_n = [a, b] \neq \emptyset,$$

the last since $a \leq b$. To complete the proof, if the limit of the lengths of the J_n's is 0, then $b_n - a_n$ must tend to 0 as n tends to infinity. This forces $a = b$. $\square$

Discussion. Again note that the ultimate force that takes care of the argument is the Completeness Axiom. We cannot overemphasize the importance of this axiom for the structural properties of the reals. $\square$

Example 1. Show by example that an analogous theorem about a nested sequence of open intervals is not valid.

Solution. Let $O_n = \left(0, \frac{1}{n}\right)$. If $x \in \bigcap O_n$, then $x > 0$. On the other hand, it is also the case that $x < \frac{1}{n}$ for every $n \in \mathbf{N}$, whence $x \leq 0$. Thus, $\bigcap O_n$ is empty. $\square$

Discussion. The point that one would like to have in $\bigcap O_n$ is 0. It is not there because it is the 'missing' limit point from all the sets. This is why the sets must all be closed in the hypothesis of Theorem 3.2.3. In Exercise 3 we ask the reader to show that bounded is also a necessary part of the hypothesis. $\square$

The next theorem is also sometimes referred to as the Bolzano-Weierstrass Theorem.

Theorem 3.2.4. *Let* $\{a_n\}$ *denote a bounded sequence of real numbers. Then* $\{a_n\}$ *contains a convergent subsequence.*

Proof.   There are two cases, namely, that $\{a_n\}$ takes finitely many distinct values and that $\{a_n\}$ takes infinitely many distinct values. For the first case, we note that there must be a subset $M \subseteq \mathbf{N}$ such that M is infinite and such that $m, n \in M$ implies $a_n = a_m$. But now $\{a_m : m \in M\}$ is a convergent subsequence since it is a constant sequence. Thus, the first case is established. For the second, $\{a_n\}$ denotes an infinite bounded set of real numbers. It follows that we can apply Theorem 3.2.1 to obtain a limit point. It is left to Exercise 5 to show that there is a subsequence that converges to the limit point. □

Discussion.   The basic fact being employed in the proof of the first case is that if we divide up an infinite set, in this case $\mathbf{N}$, into a finite number of disjoint subsets, then at least one of the subsets must be infinite. Another way of stating this fact is to say that the union of a finite collection of finite sets is finite, which is a basic theorem of 'cardinal arithmetic', although a nontrivial one. In any case, it is the heart of the argument for the first case. For the second, we employ Theorem 3.2.1, which hides the use of the Completeness Axiom, but the reader should not forget its presence or its necessity to the argument.

Consider for a moment what these theorems are telling us about the convergence of sequences in $\mathbf{R}$. Specifically, suppose we want to construct a sequence of real numbers that has no limit. In past discussions we have asserted that such a sequence must either be unbounded or it must oscillate in the sense that there are at least two distinct real numbers to which terms of the sequence are close to infinitely often. Theorem 3.2.4 supplys the means to prove this assertion. Thus, suppose $\{a_n\}$ is any bounded nonconvergent sequence. We claim that that $\{a_n\}$ contains two convergent subsequences having distinct limits. The proof of this fact is left as Exercise 7. But what this means is the reason all of our examples of nonconvergent sequences failed to converge for one of the two named reasons is not that we carefully selected the examples to support this assertion, but rather because these are the only possible ways in which a sequence of real numbers can fail to be convergent. □

Definition.   Let $S \subseteq \mathbf{R}$. A collection U of open sets is called an **open cover** of S provided $S \subseteq \bigcup U$.

Theorem 3.2.5 (Heine-Borel).   *Let $S \subseteq \mathbf{R}$. Then S is closed and bounded if and only if every open cover of S contains a finite subcollection that is also an open cover of S.*

Proof.   Let us assume that S is closed and bounded. Then there is an $M, 0 \le M$, such that $S \subseteq [-M, M]$. With this in mind, we can define the collection of intervals I_n^i, $1 \le i \le 2^n$, as in Theorem 3.2.1. Let U be any fixed open cover of S, and let us suppose for the sake of argument that no finite subcollection of U will cover S. Then we can choose a sequence of intervals, $\{J_n\}$, $n \in \mathbf{N} \cup \{0\}$, such that

(i) J_n is chosen from among I_n^i, $1 \le i \le 2^n$;
(ii) $J_{n+1} \subseteq J_n$ for each $n \in \mathbf{N}$;
(iii) no finite subcollection of U will cover $J_n \cap S$ for each $n \in \mathbf{N}$.

Let a denote the unique member of $\bigcap \{J_n : n \in \mathbf{N}\}$. Then a is a limit point of S. Since S is closed, $s \in S$, whence there is a $P \in U$ such that $a \in P$. Since P is open, there is an interval about a of positive length, say, $(a - c, a + c)$, such that this interval is completely contained in P. But now we can choose n sufficiently large that $J_n \subseteq (a - c, a + c)$. For this n, J_n is covered by a finite subcollection of U, since $J_n \subseteq P$. This contradicts our choice of J_n, whence our assumption that no finite subcollection of U covers S must be false.

To complete the proof, we must show that the requirement that S be closed and bounded is necessary. We prove the contrapositive. Thus, let us assume that S is not closed and bounded, whence either S is not closed or S is not bounded. If S is not closed, then S has a limit point a that is not an element of S. Consider the sets U_n, defined by

$$U_n = \left\{ x : \left(x < a - \frac{1}{n} \text{ .or. } a + \frac{1}{n} < x \right), \text{ where } n \in \mathbf{N} \right\}.$$

Let U consist of the sets $U_n, n \in \mathbf{N}$. Then U is a cover of S, but no finite subcollection will cover S(**WHY?**). If, on the other hand, S is not bounded, then consider the collection of open intervals $(-n, n)$, for $n \in \mathbf{N}$. Clearly this collection covers S, but no finite subcollection will cover S. This completes the proof. $\square$

Discussion. The proofs given are really sketches of the proof, and many important details have been left out. In Exercise 6, the reader will be asked to completely flesh out the details of these arguments. It should be clear that the same technique has been used to obtain each of the results in this section. A nested sequence of sets J_n with certain properties is constructed. The main feature is that the length of the intervals shrinks to 0, and so by completeness we can find a single point on which to focus our attention. The properties of this point are in part determined by constraints on the J_n's, and once we have this point to fix upon, we can get a contradiction as in the last argument or use its properties directly as in Theorem 3.2.1. $\square$

Recall that Theorem 3.2.2 supplied us with a characterization of closed and bounded sets that did not depend on the notion of bounded. Also, Theorem 3.2.5 tells us closed bounded subsets of **R** can be characterized in terms of the property that every open cover has a finite **subcover** (a subcollection that is also a cover). Thus, the concept of closed and bounded can be rephrased in terms relating only to the concept of open set. While the notion of open set as it relates to **R** is intimately related to the notion of distance, the theorem above permits us to develop a more general notion of 'closed and bounded', where the primary concept is of open set and is unrelated to distance.

Definition. Let $A \subseteq \mathbf{R}$. Then A is **compact,** provided every open cover of A contains a finite subcover.

Discussion. Examination shows this definition does not contain any features that are dependent upon **R**, whence it can be generalized to arbitrary topological spaces, which as we have mentioned are composed of a nonempty set, X, together with a collection of distinguished subsets of X that are called the open sets. The generalized notion of open cover remains identical to the existing notion and quite clearly reduces to the present notion for **R**. In those more general situations where a concept of boundedness makes sense, it is natural to ask for an analogue of Theorem 3.2.5, and such theorems will in almost all cases exist.

A finite set is obviously compact, but there there exist infinite sets that are not compact. Theorem 3.2.5 lists all compact subsets of $\mathbf{R}$, namely, they are precisely all the sets that are both closed and bounded. The word compact conveys a notion of small and its meaning led Hermann Weyl to quip: A city is 'compact' if it can be guarded by a finite number of arbitrarily nearsighted policemen. □

The theorems discussed in this section were all proved during the height of the study of elementary analysis that took place in the last century and many carry the names of some of the most eminent mathematicians of the period, for example, Bolzano, Borel, Heine, and Weierstrass.

Exercises

1. Show the sets I_n^i generated in Section 3.2.1 have all the properties claimed.

2. Show the infinite sequence generated in the proof of Theorem 3.2.2 that is asserted not to have a limit point in fact has no limit point.

3. Show at least one of the intervals mentioned in Theorem 3.2.3 must be bounded for the conclusion to hold.

4. We will say x is a **limit point** of A provided every open set O for which $x \in O$ contains points of A other than x. Show this definition of limit point is equivalent to our previous definition of limit point.

5. Complete the details of Theorem 3.2.4.

6. Complete the details of Theorem 3.2.5.

7. Let $\{a_n\}$ be a bounded, nonconvergent sequence of distinct real numbers. Show $\{a_n\}$ has at least two distinct limit points. Further, show any bounded, nonconvergent sequence must have at least two subsequences that converge to distinct limits.

8. Let f be defined on $[a, b]$, but unbounded there. Show there is a $c \in [a, b]$ such that f is unbounded on $(c - \delta, c + \delta) \cap [a, b]$ for every positive δ.

9. Recall (Exercise 2.3.16) that a function f defined on $[a, b]$ is said to be **locally bounded** provided for every $c \in [a, b]$ there is a positive δ such that f is bounded on $(c - \delta, c + \delta)$. Show a function that is locally bounded on $[a, b]$ is bounded on $[a, b]$.

10. Consider the field $\mathbf{Q}$ of rational numbers. A subset S of $\mathbf{Q}$ is open if there exists an open subset, O, of $\mathbf{R}$ such that $S = O \cap \mathbf{Q}$. A subset C of $\mathbf{Q}$ is closed provided it is the complement of an open subset of $\mathbf{Q}$. (This is just the relative topology on $\mathbf{Q}$ of Exercise 3.1.36.) Show the open subsets of $\mathbf{Q}$ satisfy the basic theorems on open sets contained in Section 3.1.

11. Show $[0, 1] \cap \mathbf{Q}$ is not compact in the topology of Exercise 10. Find an infinite subset of $\mathbf{Q}$ that is compact in this topology.

12. Prove or disprove Theorem 3.2.1 for $\mathbf{Q}$.

13. Prove or disprove Theorem 3.2.2 for $\mathbf{Q}$.

14. Prove or disprove Theorem 3.2.3 for **Q**.

15. Can you characterize the compact subsets of **Q**?

16. Let $\{a_n\}$ be a sequence having a subsequence that converges to a. Then a is called a **cluster point** (or **cluster value**) for the sequence. Show every bounded nonconvergent sequence has at least two cluster points. What is the relationship between cluster points and limit points of a sequence?

17. Find all cluster points for the following sequences whose nth term is given by a_n. Which of the cluster points are limit points?

    (a) $a_n = 1 + \frac{(-1)^n}{n}$; (b) $a_n = (-1)^n$; (c) $a_n = \sin\frac{n\pi}{16}$;

    (d) $a_n = (-1)^n n$; (e) $a_n = \frac{1}{n}\cos\frac{n\pi}{5}$.

18. Let $\{a_n\}$ be any sequence having $\mathbf{Q} \cap [0,1]$ for its range. Find all the limit points for such a sequence.

19. Let f be a function and a a limit point of the domain of f. Suppose f is bounded on an open set containing a, but the limit as x tends to a of f does not exist. Show f oscillates at a (see Exercise 2.2.7).

20. Let f be a function defined on (a,b) and that is not continuous at $c \in (a,b)$. Show either f is unbounded on an open set containing c or f oscillates at c.

21. Let $\{a_n\}$ be a sequence. Show it has a limit a if and only if every subsequence has a as a cluster point.

22. Give an example of a sequence having

    (a) no limit points; (b) exactly one limit point;
    (c) exactly two limit points; (d) exactly five limit points;
    (e) a countably infinite collection of limit points.
    Is it possible for a sequence of real numbers to have an uncountable collection of limit points?

23. Let $A \subseteq \mathbf{R}$. Prove A is compact if and only if every sequence of points from A has a cluster point in A.

24. If $\{a_n\}$ has at least one limit point, can we conclude it is bounded?

25. Let $A \subseteq \mathbf{R}$ and let $\mathcal{C}$ be a collection of closed subsets of A. $\mathcal{C}$ is said to satisfy the **finite intersection property** if every finite subcollection of $\mathcal{C}$ has a nonempty intersection. Prove A is compact if and only if for each collection $\mathcal{C}$ of closed subsets of A that satisfies the finite intersection property, $\bigcap \mathcal{C} \neq \emptyset$.

26. Let A be an infinite closed bounded subset of **Q**, say, $[0,1] \cap \mathbf{Q}$. Find a collection $\mathcal{C}$ of subsets of A such that the intersection of any finite subcollection of subsets of $\mathcal{C}$ is nonempty, but $\bigcap \mathcal{C} = \emptyset$.

27. We say X is **dense in** Y if every point of Y is a member of X or a limit point of X. Which of the following are true?

    (a) $\mathbf{R} \sim \mathbf{Q}$ is dense in **R**; (b) $\mathbf{R} \sim \mathbf{Q}$ is dense in **Q**;
    (c) **Q** is dense in **Q**; (d) **N** is dense in **R**;
    (e) **N** is dense in **N**; (f) **Q** is dense in **N**;
    (g) **Q** is dense in **R**; (h) (a,b) is dense in **R**;
    (i) **R** is dense in **Q**; (j) (a,b) is dense in $[a,b]$;
    (k) $[a,b]$ is dense in (a,b); (l) $\{\frac{1}{n} : n \in \mathbf{N}\}$ is dense in $(0,1)$.

28. Show Y is dense in X exactly if $X \subseteq \overline{Y}$.

29. Let $f : \mathbf{Q} \to [0,1]$, such that $f(\mathbf{Q})$ is dense in $[0,1]$. Show $f(\mathbf{Q})$ is not closed.

30. If $f : A \to \mathbf{R}$ is continuous and A is compact, prove $f(A)$ is compact. Show the result is no longer true if the word 'compact' is replaced by 'closed' or 'open'.

31. Show the following subsets of $\mathbf{R}$ are not compact by actually exhibiting an open cover that fails to produce a finite subcover:

   (a) $\mathbf{N}$;   (b) $(0, 1)$;   (c) $\mathbf{R} \sim \mathbf{Q}$;   (d) $\left\{ \frac{1}{n} : n \in \mathbf{N} \right\}$.

32. Show a compact subset of $\mathbf{R}$ is closed. What about the converse?

33. Show compactness is a topological property.

34. Let f and g be two functions that agree on a dense subset of Dmn f = Dmn g. Find a condition that will guarantee $f = g$.

3.3   The Cauchy Criterion

In the last two sections we have seen the importance of the concept of 'limit point'. First, we saw how this concept was used as the foundation of the definition of a closed set. Second, we saw that limit points play an essential role in the characterization of closed bounded sets. This suggests that limit points, and related topics, may be worthy of further investigation.

The connection between limit points and limits of sequences are many. One connection is the alternate form of the Bolzano-Weierstrass Theorem asserting that a bounded sequence must have a cluster point. Another is that a is a limit point of S exactly if there is a sequence of distinct points from S that converges to a. Going the other way, we notice that a sequence of distinct points has a limit exactly if, when considered as a set, it has a unique limit point. Given this close relationship between limit points and limits of sequences, we may wonder whether it is also possible to characterize convergent sequences purely in terms of the sequence and with no reference to the limit. That there is hope is suggested by the fact that a bounded monotone sequence must have a limit, a statement that refers only to properties of the sequence.

Definition. Let $\{a_n\}$ be a sequence of real numbers. We say that $\{a_n\}$ **converges in the sense of Cauchy** (is **Cauchy convergent** or, simply, is a **Cauchy sequence**) provided for every $\epsilon > 0$, there is an $N = N(\epsilon)$ such that

$$n > N \text{ and } m > N \text{ .implies. } |a_n - a_m| < \epsilon.$$

Discussion. If we compare this definition with the usual definition of convergence given in Section 1.1, we see the basic change is to replace A by a_m in the expression

$$|a_n - a_m| < \epsilon.$$

As in the original definition, $N = N(\epsilon)$, depends on the value of ϵ, which is given in advance. The intuition behind this change is simple. For sequences converging in the usual sense, we notice that as soon as $n > N$, we have

$$|a_n - A| < \epsilon$$

and that this means that *all* the terms with subscript larger than M must be close to A. If all are close to A, then all must be close together; in fact we get

$$|a_n - a_m| < |a_n - A| + |a_m - A| < 2\epsilon.$$

Further, it is intuitively reasonable that if all the terms of a sequence with sufficiently large subscript can be forced into an arbitrarily small interval, then the sequence should be convergent in the usual sense. □

The intuitive discussion given above is formalized in the following theorem.

Theorem 3.3.1. *Let $\{a_n\}$ be a sequence of real numbers. Then $\{a_n\}$ has a limit if and only if $\{a_n\}$ is Cauchy convergent.*

Proof. First, assume $\{a_n\}$ has a limit, A. Let $\epsilon > 0$ be given, and find N such that $n \geq N$ implies $|a_n - A| < \frac{\epsilon}{2}$. If $m, n > N$, then

$$|a_n - a_m| \leq |a_n - A| + |a_m - A| < \frac{\epsilon}{2} + \frac{\epsilon}{2} = \epsilon.$$

Thus, $\{a_n\}$ is Cauchy convergent as claimed.

Conversely, suppose $\{a_n\}$ is a Cauchy convergent sequence. Choosing $\epsilon = 1$, find $N(1)$. Fix $n > N(1)$, and let $M = |a_n| + 2$. Since there are only a finite number of terms whose subscript does not exceed $N(1)$, and since $m > N(1)$ implies $|a_m| < M$, $\{a_n\}$ is bounded. By Theorem 3.2.4 $\{a_n\}$ has a convergent subsequence. Let the limit be denoted by A. Fix $N(\epsilon)$, where $\epsilon > 0$ is arbitrary, then we can find an $n > N(\epsilon)$, such that $a_n \in (A - \epsilon, A + \epsilon)$ (**WHY?**). Now for any $m > N(\epsilon)$, we have

$$|a_m - A| \leq |a_m - a_n| + |a_n - A| < 2\epsilon.$$

Since ϵ was arbitrary, $\{a_n\}$ converges to A. □

Discussion. The heart of the argument that every convergent sequence is Cauchy convergent is the straightforward implementation of the Triangle inequality. The proof of converse can be summarized in three steps. First, a Cauchy sequence is bounded. Second, the completeness of the underlying field guarantees the existence of a subsequence converging to a limit. Third, the Cauchy property ensures the limit of the convergent subsequence has to be the limit of the sequence.

The use of the Completeness Axiom in the second step is vital. For example, if $\mathbf{R}$ is replaced by $\mathbf{Q}$ in the statement of the theorem, the result is no longer true. □

A principal use of the Cauchy criterion is in establishing that a given sequence has a limit without having to find the limit.

Example 1. Consider the sequence $\{a_n\}$ defined by $a_n = \sum_{i=1}^{n} \frac{1}{i^2}$. Show $\{a_n\}$ is a convergent sequence.

Solution. To apply the Cauchy criterion, we must show that provided n, m are sufficiently large, the quantity $|a_n - a_m|$ will be arbitrarily small. Thus, consider the quantity A_k defined by

$$A_k = \sum_{i=2^k}^{2^{k+1}-1} \frac{1}{i^2}.$$

Direct computation establishes $A_k \leq 2^{-k}$ (see Exercise 1). Now, fix $\epsilon > 0$. Since the A_k's are bounded by terms of a geometric series, as an application of the formula established in Exercise 0.4.22, we have for all $s, t \in \mathbf{N}$ such that $s \leq t$

$$\sum_{k=s}^{t} A_k \leq 2^{-(s-1)}.$$

To complete the argument, let s be chosen so that $2^{-s} < \epsilon$. Now for any $n, m \geq s$, there exists a $j \in \mathbf{N}$ such that

$$|a_n - a_m| \leq \sum_{k=s}^{s+j} A_k < \epsilon.$$

Thus, $\{a_n\}$ is indeed a Cauchy sequence, and so converges. $\square$

Discussion. It happens that the limit of the sequence given in Example 1 is $\frac{\pi^2}{6}$. We suggest that even given this information, the reader would find it a difficult task to prove this sequence has a limit using the definition of Chapter 1. This is not to suggest inadequacy on the part of the reader. Rather, it is to suggest that establishing that a particular real number is the limit of an arbitrary sequence can be extremely difficult, in part because it is impossible to know the true value of most real numbers. For example, most of us think we know what π is. But this is only because we have given π a name, and not because we know the value of all its decimal places. Suffice it to say, most real numbers do not have names and thus are totally unknown. $\square$

Example 1 and the discussion provide evidence of the utility of the Cauchy convergence concept. However, these are not the only values of Cauchy convergence. To expand on this idea, consider a sequence $\{q_n\}$ of rational numbers that converges to $\sqrt{2}$. Viewed as a sequence of real numbers, we have a real number, $\sqrt{2}$, that is the limit of this sequence, and so the sequence satisfies the definition of convergence. However, if we think of the sequence as a sequence of rational numbers, and if our definition of convergence requires us to find a rational number that is the limit, then $\{q_n\}$ is no longer convergent since $\sqrt{2} \notin \mathbf{Q}$. This lack of convergence is particularly disturbing, since we have not altered the sequence, itself, in any way. We have merely changed the context in which we are considering the sequence by changing the underlying field (topological space) in which the sequence is found. But changing the context in which the sequence is considered has no effect on whether the sequence is Cauchy, since the property of being Cauchy refers only to properties of the sequence. This fact that makes the Cauchy concept a tool of such power that it can even serve as the heart of a replacement for the Supremum Principle (see Exercise 29).

Exercises

1. Let $\{A_k\}$ be defined as in Example 1. Show $A_k \leq 2^{-k}$.

2. Show directly from the definition of a Cauchy sequence that the sum of two Cauchy sequences must again be a Cauchy sequence.

3. Show directly from the definition of a Cauchy sequence that the product of two Cauchy sequences must again be a Cauchy sequence.

4. Give a direct proof that every Cauchy sequence is bounded. Is the converse true?

5. Give examples of sequences that are monotone but not Cauchy and Cauchy but not monotone.

6. Show every subsequence of a Cauchy sequence is Cauchy. Now consider sequences of elements of $\mathbf{Q}$. Suppose a subsequence of a Cauchy sequence has a limit in $\mathbf{Q}$. What can be said about the original sequence?

7. Show if $\{a_n\}$ is Cauchy, then $\{|a_n|\}$ is Cauchy. Is the converse true?

8. If $\{a_n\}$ is Cauchy and for all n, $a_n > 0$, will $\{\frac{1}{a_n}\}$ be Cauchy?

9. Let $I_n = [a_n, b_n]$ be such that $I_{n+1} \subseteq I_n$ and $\lim_{n \to \infty} (b_n - a_n) = 0$. Show $\{a_n\}$ and $\{b_n\}$ are Cauchy and conclude $\bigcap I_n$ contains exactly one point.

10. Let $\{a_n\}$ be any sequence. Which of the following statements about $\{a_n\}$ are equivalent to the assertion $\{a_n\}$ is a Cauchy sequence?

    (a) For every $\epsilon > 0$, there is an $N \in \mathbf{N}$ such that for every $p, q \in \mathbf{N}$, $|a_{N+p} - a_{N+q}| < \epsilon$.

    (b) For every $\epsilon > 0$, there is an $N \in \mathbf{N}$, such that for every $n \in \mathbf{N}$, $|a_N - a_{N+n}| < \epsilon$.

    (c) For every $\epsilon > 0$ and $p \in \mathbf{N}$, there is an $N \in \mathbf{N}$, such that $|a_N - a_{N+p}| < \epsilon$.

    (d) For every $\epsilon > 0$ and $p \in \mathbf{N}$, there is an $N \in \mathbf{N}$ such that $|a_N - a_{N \cdot p}| < \epsilon$.

    (e) For every $p \in \mathbf{N}$, $\lim_{n \to \infty} (a_n - a_{n+p}) = 0$.

11. Can there exist a function $f : \mathbf{N} \to \mathbf{N}$ such that for an arbitrary sequence $\{a_n\}$, $\{a_n\}$ is Cauchy if and only if for every positive ϵ there is an $N \in \mathbf{N}$ such that $n > N$ implies $|a_n - a_{f(n)}| < \epsilon$.

12. Let f map $\mathbf{N} \times \mathbf{N}$ into $\mathbf{N}$ and suppose the sequence $\{a_n\}$ satisfies $|a_n - a_m| < f(n, m)$. Will any of the following f's guarantee $\{a_n\}$ is Cauchy?

    (a) $f(n, m) = \frac{1}{n+m}$;   (b) $f(n, m) = \frac{mn}{n+m}$;   (c) $f(n, m) = \frac{mn}{n^2+m^2}$.

13. Let $x_1 < x_2$ be any two real numbers. For $n > 2$, set $x_n = w_1 x_{n-1} + w_2 x_{n-2}$, where w_1, w_2 are nonnegative real numbers whose sum is 1. Show $\{x_n\}$ is Cauchy. Find its limit.

14. Let $a_n = \sqrt{n}$. Show for n sufficiently large $|a_{n+k} - a_n| < \frac{k}{2\sqrt{n}}$. Why does this not establish $\{a_n\}$ is Cauchy?

15. Let $\{x_n\}$ be a Cauchy sequence such that for all n, $x_n \in \mathbf{N}$. Show $\{x_n\}$ must eventually be constant.

16. Let $\{x_n\}$ be a Cauchy sequence whose values all lie in A. If A is an infinite collection of isolated points, must $\{x_n\}$ eventually become constant?

17. Suppose $\{a_n\}$ satisfies the condition $|a_{n+1} - a_n| \leq \frac{1}{2^n}$. Show $\{a_n\}$ is a Cauchy sequence.

18. Let $\{a_n\}$ be a sequence for which there exist constants r and c, $0 < r < 1$, $c > 0$ such that $|a_{n+1} - a_n| < cr^n$. What can be said about the convergence of $\{a_n\}$? Will your conclusions be altered if instead, you assume the condition $\lim(a_{n+1} - a_n) = 0$?

19. Let $\{a_n\}$ be any sequence and set

$$s_n = \sum_{i=1}^{n} a_i; \qquad t_n = \sum_{i=1}^{n} |a_i|.$$

Show if $\{t_n\}$ is Cauchy, then $\{s_n\}$ will also be Cauchy. Is the converse true?

20. Which of the following sequences $\{s_n\}$ is Cauchy?

(a) $s_n = \sum_{i=1}^{n} \frac{1}{i!}$; (b) $s_n = \sum_{i=1}^{n} \frac{1}{i}$; (c) $s_n = 1 + \frac{(-1)^n}{5} + \frac{1}{n}$;

(d) $s_n = \sum_{i=1}^{n} \frac{(-1)^i}{i}$; (e) $s_n = \sum_{i=1}^{n} (-1)^i$; (f) $s_n = \sum_{i=1}^{n} \frac{(-1)^i}{i!}$.

21. Let $\{x_n\}$, $\{y_n\}$, and $\{z_n\}$ be three sequences of real numbers such that $z_n = x_n$ if n is odd and y_n otherwise. Under what conditions will $\{z_n\}$ be Cauchy?

22. Let f be continuous on D, and suppose $\{a_n\}$ is a Cauchy sequence of elements of D. Show if $\{a_n\}$ has a limit in D, then the sequence $\{f(a_n)\}$ will also be Cauchy, but if $\{a_n\}$ has a limit that is not in D, then the sequence $\{f(a_n)\}$ need not be Cauchy.

23. Let f be continuous on $D \subseteq \mathbf{R}$ and $\{a_n\}$ a Cauchy sequence of elements of D. If $D = \mathbf{R}$, must $\{f(a_n)\}$ be Cauchy? If D is open, must $\{f(a_n)\}$ be Cauchy? If D is closed, must $\{f(a_n)\}$ be Cauchy?

24. Suppose f is defined on $[a, b]$ and f maps Cauchy sequences to Cauchy sequences. Must f be continuous?

25. A sequence $\{a_n\}$ is **contractive** if there exists a constant k, $0 < k < 1$, such that

$$|a_{n+2} - a_{n+1}| \leq k|a_{n+1} - a_n|.$$

Show every contractive sequence is Cauchy. Further, show if a is the limit of such a sequence, then

$$|a - a_n| \leq \frac{k}{1-k}|a_n - a_{n-1}|.$$

What can be said when $k = 1$?

26. The polynomial equation $x^3 - 5x + 1 = 0$ has a root, r, with $0 < r < 1$. Use an appropriate contractive sequence to compute r to within 10^{-4}.

27. A function $f : \mathbf{R} \to \mathbf{R}$ is a **contraction map** if there is k, $0 < k < 1$, such that for all $x, y \in \mathbf{R}$,

$$|f(x) - f(y)| \leq k|x - y|.$$

Show a contraction map, f, always has a unique fixed point, that is, a point t satisfying the equation $f(t) = t$.

28. State and prove a theorem for $\mathbf{Q}$ analogous to Theorem 3.2.4.

29. Let $\mathbf{Q}^*$ denote the collection of all Cauchy convergent sequences of elements of $\mathbf{Q}$. For $\{a_n\}, \{b_n\} \in \mathbf{Q}^*$, define $\{a_n\} \cong \{b_n\}$ if and only if $\lim_{n \to \infty} (a_n - b_n) = 0$. Show $\cong$ is an equivalence relation on $\mathbf{Q}^*$. Show the equivalence classes so obtained form an ordered field under suitable field operations. Further show this field is complete and can be identified in a natural way, with $\mathbf{R}$.

30. A set S is **linearly ordered** by $\le$ if $\le \subseteq S \times S$ and $\le$ is reflexive, antisymmetric, transitive (see the appendix), and has the property that for all x, y either $(x, y) \in \le$ or $(y, x) \in \le$. A linear order of S is **dense** if for all x, y such that $x < y$ there exists z such that $x < z < y$. Show if S is countable and $\le_1$ and $\le_2$ are two dense linear orders of S having no upper or lower bounds, then there is a mapping $f : S \to S$ such that f is one-to-one and onto and for all $x, y \in S$, $x \le y$ if and only if $f(x) \le f(y)$.

31. A function, $\sigma(n, m)$, mapping $\mathbf{N} \times \mathbf{N}$ into $\mathbf{R}$ is called a **double sequence**. A double sequence, $\{\sigma(n, m)\}$, is said to have a **limit** provided there exists $A \in \mathbf{R}$ such that for every positive ϵ there exists $N \in \mathbf{N}$ such that

$$|\sigma(n, m) - A| < \epsilon \text{ whenever } n, m > N.$$

In the event the limit A exists, we write

$$\lim_{n, m \to \infty} \sigma(n, m) = A.$$

Develop a Cauchy criterion for double sequences.

32. Let $\{\sigma(n, m)\}$ be a double sequence having a limit, A. Further, suppose for each fixed M, $\lim_{n \to \infty} \sigma(n, M)$ exists and for each fixed N, $\lim_{m \to \infty} \sigma(N, m)$ exists. These limits are called **partial limits**. . Show the two **iterated limits** satisfy

$$\lim_{M \to \infty} \lim_{n \to \infty} \sigma(n, M) = \lim_{N \to \infty} \lim_{m \to \infty} \sigma(N, m) = A.$$

Show by an example there is a convergent double sequence such that none of the individual partial limits, $\lim_{n \to \infty} \sigma(n, M)$ or $\lim_{m \to \infty} \sigma(N, m)$, exists.

33. Let $\sigma(n, m) = \frac{n}{n+m}$ and consider the double sequence $\{\sigma(n, m)\}$. Show all partial limits exist, the two iterated limits exist, but the double sequence does not have a limit.

3.4 Limit Superior and Limit Inferior

In the last section we saw it was possible to specify the class of convergent sequences of real numbers by focusing entirely on properties of the sequence and omitting any reference to the limit. In this section, we further investigate the properties of sequences, but the focus is on a more general concept related to limit points.

Definition. Let $\{a_n\}$ be a sequence. A number $x \in \mathbf{R}$ is called a **cluster point** of the sequence, provided there is a subsequence $\{a_{n_k}\}$ of $\{a_n\}$ that converges to x.

Discussion. Cluster points (values) were already introduced in Exercise 3.2.16. Cluster points and limit points are obviously related. This is a conclusion of the version of the Bolzano-Weierstrass Theorem for sequences. That they are not the same can be seen from an examination of the proof of the Bolzano-Weierstrass Theorem (Theorem 3.2.4), which falls into two cases. The first case assumes that the sequence only takes a finite number of distinct values. Thus, as a set of real numbers, $\{a_n : n \in \mathbf{N}\}$ is finite, and a finite set has no limit points (see Exercise 3.1.8). On the other hand, if the sequence takes on infinitely many distinct values, then the sequence forms an infinite bounded set of real numbers that has a limit point. This limit point must, in turn, be the limit of a subsequence converging to it. From this, it is easy to see every limit point of a sequence is a cluster point, but not the other way around. And, indeed, it is also clear why cluster point is a right concept, if we want an analogous concept to limit point in the context of sequences. $\square$

By the Bolzano-Weierstrass Theorem, we know a bounded sequence will have cluster points. In general, then, an arbitrary sequence is likely to have lots of cluster points. The problem is always one of being able to get one's hands on a particular cluster point. Thus, one would like a mechanical process for manipulating a sequence that will be guaranteed to produce a cluster point.

Consider then the collection of cluster points of a sequence. A given real number can become a cluster point in one of two ways. Either it is a limit point of the sequence, or it is a value for terms of the sequence that is repeated an infinite number of times. Now, we know the collection of limit points of a set is closed (see Theorem 3.1.6). It therefore seems plausible the collection of cluster points of a sequence is also closed, and indeed this is the case (see Exercise 3). It follows if the collection of cluster points of a sequence is bounded above, then the supremum of this collection will be a cluster point of the sequence (see Exercise 3.1.9). Thus, there are two natural cluster points to try to find, the least cluster point and the greatest cluster point. What we would like is a method for picking out these cluster points that depends only on the terms of the sequence and not on the set of cluster points. This leads to the following definition.

Definition. Let $\{a_n\}_{n=1}^{\infty}$ be a sequence and for each $n \in \mathbf{N}$, set

$$b_n = \sup\{a_k : k \geq n\}.$$

The **limit superior** of $\{a_n\}$, denoted by $\overline{\lim}\{a_n\}$, is defined by

$$\overline{\lim}\{a_n\} = \inf\{b_n : n \in \mathbf{N}\} = \inf\{x : x = \sup\{a_k : k \geq n\} \text{ for some } n \in \mathbf{N}\}$$

provided the quantity on the right exists. Similarly, we define the **limit inferior** by

$$\underline{\lim}\{a_n\} = \sup\{x : x = \inf\{a_k : k \geq n\} \text{ for some } n \in \mathbf{N}\}.$$

We will usually simplify the notation by writing

$$\overline{\lim} a_n \text{ and } \underline{\lim} a_n$$

to denote the limit superior and inferior, respectively. (We may also use the terminology **lim sup**, and **lim inf**, respectively.)

Discussion. Let us determine the conditions under which the limit superior will, or will not, exist. Because the limit superior is an infimum, by using the axiom of completeness we can guarantee the existence of the limit superior provided the set of real numbers over which this infimum is calculated is nonempty and bounded below. If either of these conditions fails, the limit superior will not exist. Specifically, the limit superior will fail to exist if

$$A = \{x : x = \sup\{a_k : k \geq n\} \text{ for some } n \in \mathbf{N}\}$$

is either empty or not bounded below.

Consider how A could be empty. To become a member of A, an element must be a supremum of a set, A_n, of the form

$$A_n = \{a_k : k \geq n\}.$$

For all values of n, the sets A_n are never empty, since each A_n consists of a tail of the sequence. For example, for the sequence whose mth term is given by $\frac{1}{m}$, we have

$$A_n = \left\{\frac{1}{n}, \frac{1}{n+1}, \frac{1}{n+2}, \cdots\right\}.$$

Thus, for a particular value of n, the supremum of A_n can fail to exist only if A_n is not bounded above. This can happen only if the sequence $\{a_n\}$ is itself not bounded above as would happen if the mth term of the sequence were m^2, to take another example. But, whether any particular tail of a sequence is bounded is now a general property of the sequence. That is, if the supremum of a single A_n fails to exist, then for every value of n, the supremum of A_n will fail to exist, and A will be empty. In summary, A will be empty exactly if $\{a_n\}$ is not bounded above.

The second reason A can fail to have an infimum is it is not bounded below. Under this assumption, A must be nonempty, whence $\{a_n\}$ is bounded above. As in the definition, set $b_n = \sup A_n$. The sequence $\{b_n\}$ can be shown to be monotone decreasing (see Exercise 1). Since $b_n \in A$, and these are the only members of A, we see A is not bounded below exactly if the sequence $\{b_n\}$ diverges to $-\infty$. But this implies that $\{a_n\}$ satisfies the condition, for all $m \in \mathbf{N}$, there is an $N \in \mathbf{N}$ such that $n > N$ implies $a_n < -m$. Evidently, this is a strong condition, since it requires that all remaining terms of the sequence be below any potential lower bound, and hence the sequence diverges to $-\infty$.

In summary, then, the limit superior of a sequence will fail to exist for sequences that are not bounded above and those that diverge to $-\infty$. As well, for a sequence that is bounded above, if the limit superior does not exist, then the sequence will not be bounded below, whence the limit inferior will not exist either. $\square$

Example 1. Find the limit superior and the limit inferior of the sequence $\{a_n\}$, where $a_n = 1 + (-1)^n + \frac{1}{2^n}$.

Solution. For any fixed n, $a_n = 2 + \frac{1}{2^n}$, if n is even and $\frac{1}{2^n}$ when n is odd. Thus, $b_n = \sup\{a_k : k \geq n\} = 2 + \frac{1}{2^n}$ if n is even, and $2 + \frac{1}{2^{n+1}}$ when n is odd. Hence

$$\overline{\lim}\,\{a_n\} = \inf\{b_n : n \in \mathbf{N}\} = 2.$$

A similar calculation shows $\underline{\lim}\,\{a_n\} = 0$. $\square$

Discussion. The calculations are straightforward from the definition. The reader should sketch a graph of the sequence. From this she will discover the sequence oscillates between small intervals around 0 and 2. $\square$

Example 2. Find the limit superior and limit inferior of $\{a_n\}$, where $a_n = 2^n$.

Solution. Because $\{a_n\}$ is not bounded above, the limit superior does not exist. For the limit inferior, consider

$$c_n = \inf\{a_k : k \geq n\}.$$

Evidently, $c_n = a_n = 2^n$, since $\{a_n\}$ is monotone increasing and c_n diverges to $+\infty$. Thus, the supremum over $\{c_n : n \in \mathbf{N}\}$ does not exist, whence the limit inferior does not exist, even though the sequence $\{a_n\}$ is bounded below. $\square$

Discussion. The argument presented is a direct application of the ideas presented in the discussion following the definition of limit superior. $\square$

Recall the motivation for the limit superior notion was to provide a means for finding the greatest cluster point of a sequence.

Theorem 3.4.1. *The limit superior of a sequence is the greatest cluster point of the sequence.*

Proof. Let $\{a_n\}$ be a sequence having a as its limit superior. We want to generate a subsequence that converges to a. Consider b_n as specified in the definition of limit superior. Because a exists, b_n exists for all $n \in \mathbf{N}$. Now, either

$$b_n \in \{a_k : k \geq n\} \quad \text{for all } n \in \mathbf{N}$$

or for some $n_0 \in \mathbf{N}$

$$b_{n_0} \notin \{a_k : k \geq n_0\}.$$

If the former, then the subsequence defined by taking the b_n's in order and avoiding repetitions of the same a_k will converge to a (see Exercise 1). If the latter, then $b_{n_0} > a_k$ for all $k \geq n_0$. However, because it is the supremum of this set, it must be a limit point of the set and hence a cluster point of the sequence (see Exercise 2). Note also $b_{n_0} = b_n$ for all $n > n_0$, so that $b_n \to a$. To complete the proof, let a_{n_k} be any convergent subsequence. Call its limit b. Observe, $a_n \leq b_n$ for all $n \in \mathbf{N}$, whence $a_{n_k} \leq b_{n_k}$ for any pair of subsequences generated identically. Since $\{b_n\}$ is a monotone decreasing sequence converging to a, $b \leq a$ follows, and we are done. $\square$

Discussion. The key idea in this proof is that if the supremum (infimum) of a set is not a member of that set, then it must be a limit point of the set. It arises as an application of Theorem 0.4.2 and enables us to construct the required subsequences.

A similar proof reveals the limit inferior of a sequence is the least cluster point of the sequence. The reader is asked to write out a detailed proof of this fact as Exercise 9. Thus, the sequence $\{a_n\}$ converges to a limit if and only if $\overline{\lim}\, a_n = \underline{\lim}\, a_n$.

The reader should note that the proof and statement of the theorem assume the existence of the limit superior. Thus, the theorem does not assert that the greatest cluster point will always be the limit superior. Examples of sequences having a greatest cluster point but no limit superior may be found in the exercises. $\square$

Theorem 3.4.2. *Let $\{a_n\}$ be a sequence for which either the limit superior or the limit inferior exists. Then $\{a_n\}$ has a cluster point.*

Proof. Immediate from Theorem 3.4.1. $\square$

Discussion. For sequences of distinct points, this result generalizes that of Theorem 3.2.1. Comparison of arguments reveals the cluster point generated in Theorem 3.2.1 is merely shown to exist, whereas the argument of 3.4.2 is completely constructive, provided of course we view the supremum of a set as an essentially 'constructed' number. $\square$

Theorem 3.4.3. *Let $\{a_n\}$ be a sequence. Then $\{a_n\}$ has a limit if and only if the limit superior and the limit inferior exist and are equal.*

Proof. Exercise 4. $\square$.

In previous cases where we have formulated a limiting process, we have immediately raised questions about its behavior with respect to the usual operations of arithmetic. Similar question can be asked about the limit superior and inferior. One relationship that is known to exist is

$$\overline{\lim}\{a_n + b_n\} \le \overline{\lim}\{a_n\} + \overline{\lim}\{b_n\}.$$

The reader is asked to explore and develop these types of relationships in Exercise 20.

Exercises

1. Let $\{a_n\}$ satisfy the hypothesis of Theorem 3.4.1. Show the sequence $\{b_n\}$ (see proof of 3.4.1) is monotone decreasing and converges to a, the limit superior. Further, show if $b_n \in \{a_m : m \in \mathbf{N}\}$ for all $n \in \mathbf{N}$, then $\{a_n\}$ has a monotone decreasing subsequence that converges to a.

2. Let A be a nonempty bounded set such that $\sup A \notin A$. Show A cannot be finite and $\sup A$ is a limit point of A. Use these facts to construct a sequence of points from A converging to $\sup A$.

3. Show the collection of cluster points of a sequence is a closed set.

4. Let $\{a_n\}$ be a sequence and consider the subsequence defined by a_{n+m} for a fixed $m \in \mathbf{N}$. Show the subsequence has the same set of cluster points as the original sequence.

5. Show $\underline{\lim} a_n \le \overline{\lim} a_n$.

6. Prove Theorem 3.4.3.

7. Recall the sequence $\{b_n\}$ as given in the definition of limit superior. Find $b_1, \ldots, b_6$ inclusive for each of the following sequences:

   (a) $\{(-1)^n n\}$; (b) $\left\{\frac{n^4+5}{3^n}\right\}$; (c) $\{-n^2\}$.

8. Find the limit inferior, the limit superior, and all cluster points for each of the sequences, $\{a_n\}$, where a_n is given below:

   (a) $(-1)^n$;

   (b) $\frac{n+(-1)^n(2n+1)}{n}$;

   (c) $(-1)^n\left(1-\frac{1}{n}\right)+(-1)^{n+1}\left(1+\frac{1}{n}\right)$; (d) $(-1)^{n+1}+\sin\frac{n\pi}{4}$;

   (e) $\frac{1+\cos\frac{n\pi}{2}}{(-1)^n n^2}$;

   (f) $n^{\sin\frac{n\pi}{2}}$;

   (g) $3\sin\frac{n\pi}{2}+(-1)^n\left(2+\frac{1}{n}\right)$; (h) $\frac{1}{2^n}+(-1)^n\cos\frac{n\pi}{4}+\sin\frac{n\pi}{2}$;

   (i) $\left(1+\frac{1}{n}\right)\left(1+\sin\frac{n\pi}{2}\right)^{\frac{1}{n}}$; (j) $\left(1+\frac{2^n}{e^n}\right)\sin\frac{n\pi}{2}$;

   (k) $2^{(-1)^n}\left(1+\frac{1}{n^2}\right)+3^{(-1)^{n+1}}$; (l) $(-1)^n\left(1-e^n|\cos\frac{n\pi}{2}|\right)$;

   (m) $n\cos\left(\frac{n\pi}{2}\right)$; (n) $\frac{2}{3},\frac{1}{3},\frac{3}{4},\frac{1}{4},\frac{4}{5},\frac{1}{5},\frac{5}{6},\ldots$;

   (o) $(1+(-1)^n)n^{(-1)^{n+1}}+\cos\frac{n\pi}{6}$; (p) $\frac{3}{2},\frac{-1}{2},\frac{4}{3},\frac{-1}{3},\frac{5}{4},\frac{-1}{4},\frac{6}{5},\ldots$;

   (q) $a_1=0,a_{2n}=\frac{a_{2n-1}}{2},a_{2n+1}=\frac{1}{2}+a_{2n}$; (r) $\left(1+\frac{(-1)^n}{n}\right)^n$;

   (s) $-\frac{n}{4}+\left[\frac{n}{4}\right]+(-1)^n$, where $[x]$ is the greatest integer function;

   (t) $r_n+r_{9n}+(-1)^n$, where r_n is the remainder when n is divided by 3;

   (u) $\frac{r_{2n+1}}{3}+r_{4n}+\frac{1}{2^{n+1}}$, where r_n is the remainder when n is divided by 9;

   (v) $a_k=\frac{j}{k+1}$, for $n=k^2+j, j=1,2,\ldots,2k+1, k=0,1,\ldots$.

9. Show the limit inferior is the least cluster point of a sequence.

10. Discuss the limit inferior, with particular reference to the conditions for its nonexistence.

11. Let $\{a_n\}$ be a sequence. Show $\{a_n\}$ has a limit a if and only if every subsequence has a as a cluster point.

12. Give an example of a sequence having
    (a) no cluster points; (b) exactly one cluster point;
    (c) exactly two cluster points; (d) exactly five cluster points;
    (e) a countably infinite collection of cluster points.
    Is it possible for a sequence of real numbers to have an uncountable collection of cluster points?

13. Recall the distinction between cluster points and limit points. Give an example of a sequence having

    (a) exactly one cluster point and no limit points;

    (b) exactly two cluster points and no limit points;

    (c) exactly five cluster points and no limit points;

    (d) a countably infinite collection of cluster points and no limits.

    Is it possible for a sequence of real numbers to have an uncountable collection of cluster points and no limit points?

14. Let $A\subseteq\mathbf{R}$. Prove A is compact if and only if every sequence of points from A has a cluster point in A.

15. Suppose $\{a_n\}$ has a cluster point. What can be said about the existence of the limit superior and the limit inferior?

16. Find a simple condition on a sequence that is equivalent to the existence of both the limit superior and the limit inferior.

17. Let $f : \mathbf{N} \to \mathbf{Q} \cap (0,1)$ be an onto function. Thinking of f as the sequence defined by $a_n = f(n)$, find the limit superior, the limit inferior, and all the cluster points.

18. Suppose $\overline{\lim} \, a_n = a$. Show for every positive ϵ there is an $N \in \mathbf{N}$ such that $n > N$ implies $a_n < a + \epsilon$. Formulate a similar result for the limit inferior.

19. Find a condition on the limit superior and inferior that is equivalent to the existence of a limit for a sequence.

20. Prove or disprove; where a counterexample is obtained, find the strongest possible relationship, or show none can hold.
    (a) $\overline{\lim} \, (s_n + t_n) = \overline{\lim} \, s_n + \overline{\lim} \, t_n;$ (b) $\underline{\lim} \, (s_n + t_n) = \underline{\lim} \, s_n + \underline{\lim} \, t_n;$
    (c) $\overline{\lim} \, (s_n t_n) = \overline{\lim} \, s_n \cdot \overline{\lim} \, t_n;$ (d) $\underline{\lim} \, (s_n t_n) = \underline{\lim} \, s_n \cdot \underline{\lim} \, t_n;$

21. If $\{a_n\}$ converges to a and $\{b_n\}$ is bounded, show

$$\overline{\lim} \, (a_n b_n) = a \cdot \overline{\lim} \, b_n.$$

22. Show $a = \overline{\lim} \, a_n$ if and only if the following two conditions hold:

    (i) for every $\epsilon > 0$, there is an $N \in \mathbf{N}$ such that $n > N$ implies $a_n < a + \epsilon$;

    (ii) for every $\epsilon > 0$ and every $N \in \mathbf{N}$, there is an $n \in \mathbf{N}$ such that $n > N$ and $a - \epsilon < a_n$.

23. State a result (similar to Exercise 19) for limit inferiors and prove it.

24. Let s_n be any sequence and define t_n by

$$t_n = \sum_{i=1}^{n} \frac{s_i}{n}.$$

Find any order relations ($\leq$) between

$$\overline{\lim} \, s_n, \quad \overline{\lim} \, t_n, \quad \underline{\lim} \, s_n, \quad \text{and} \quad \underline{\lim} \, t_n.$$

25. Let $\{a_n\}$ be a sequence of nonnegative reals. Define $s_n = \frac{a_{n+1}}{a_n}$ and $t_n = (a_n)^{\frac{1}{n}}$. Show

$$\underline{\lim} \, s_n \leq \underline{\lim} \, t_n \leq \overline{\lim} \, t_n \leq \overline{\lim} \, s_n.$$

Moreover, give examples to show these inequalities may be strict.

3.5   Uniform Continuity

We began this chapter by pointing out the study of topology had to do with properties of sets that were preserved by one-to-one continuous functions. In Section 3.2 we focused on closed bounded subsets of $\mathbf{R}$ and showed such sets had very nice properties. The reader may already suspect that the image of a closed bounded set under a continuous function is again a closed bounded set. A limited version of this result was already established in Section 2.6, where it was shown the image of a closed interval was again a closed interval. Thus, the question of whether a continuous function takes closed bounded sets to closed bounded sets begs for an answer.

We can, however, approach things from another point of view. Consider the class of continuous functions having a given domain, $D \subseteq \mathbf{R}$. We might wonder whether the fact that D was a closed bounded subset of $\mathbf{R}$ would have any effect on the properties of the function, f. Looking at things from this direction provides a totally different perspective and quite conceivably could lead to new and interesting results. This approach is fruitful and will be explored in this section.

We want to give the definition of the concept of uniform continuity. As a lead-in to this definition, recall the definition of continuity. Consider then a function, f, that is defined on an open interval, (a, b). (While we could deal with a more general domain, this setting is sufficiently rich to meet our requirements.) Pick a particular $c \in (a, b)$ and ask whether f is continuous at c. To answer yes to this question, we must compute $\lim_{x \to c} f(x)$. This computation involves first setting a value for ϵ and then finding δ, meeting the requirements of the definition of continuity.

Recall that the value of δ in general depends both on the value of ϵ and also on the value of c. This dependence could be expressed functionally by

$$\delta = \delta(\epsilon, c).$$

The nature of dependence of δ on c can best be understood by considering some examples. Thus, consider the open interval $(0, 10)$. For the function defined by $f(x) = x$, there is no dependence, that is no matter what value of $c \in (0, 10)$ is selected, setting $\delta = \epsilon$ will ensure

$$|f(x) - f(c)| = |x - c| < \epsilon \text{ whenever } |x - c| < \delta.$$

This is because the only occurrence of c in the expression $|f(x) - f(c)|$ is in the the quantity $x - c$.

As a second example, consider the function $f(x) = x^2$. The computation that arises requires working with the quantity

$$|x^2 - c^2| = |x + c| \cdot |x - c|,$$

which must be made less than ϵ merely by making δ small (see Example 2.5.2 for further discussion of this function). The size of the term $|x + c|$ depends on where in the interval the point c resides. If it is near 0, then this term could almost be ignored in the calculation. If this term is near 10, then $|x + c|$ must be accounted for by making δ smaller. Because $|x + c|$ is bounded by 20 on the interval $(0, 10)$, if we take $\delta = \frac{\epsilon}{20}$, this will guarantee

$$|x^2 - c^2| < \epsilon \text{ whenever } |x - c| < \delta,$$

irrespective of the value of c. For this example, a functional dependence of δ that will ensure the continuity requirement can be written as $\delta(\epsilon, c) = \min\{\epsilon, \frac{\epsilon}{c+10}\}$. But the dependence of this expression on c can be eliminated because c is required to lie in the interval, $(0, 10)$; this is what led to $\delta = \frac{\epsilon}{20}$.

As a third example, consider $f(x) = \frac{1}{x}$. If we pick $c \in (0, 10)$, we end up working with

$$\left| \frac{1}{x} - \frac{1}{c} \right| = \left| \frac{c - x}{c \cdot x} \right| \le \frac{\delta}{|c| \cdot |x|}.$$

No matter what we do, we cannot avoid the dependence of δ on c, because as c gets close to 0, the quantity $\frac{1}{cx}$ becomes arbitrarily large. To cope with this, we must take δ much smaller than c, as well as being smaller than ϵ (see Example 2.5.3 for previous discussion).

A concept of continuity that avoided the dependence of δ on the choice of c would have utility. Such a concept exists, and we now present its definition.

Definition. A function f on $D \subseteq \mathbf{R}$ is said to be **uniformly continuous** on D, provided that for every $\epsilon > 0$, there is a $\delta > 0$ such that for every $x, y \in D$,

$$|f(x) - f(y)| < \epsilon \text{ whenever } |x - y| < \delta.$$

Discussion. Functions are uniformly continuous on domains, not at single points. Thus, while a function is continuous on its domain if it is continuous at each point of its domain, no comparable statement can exist for uniform continuity. With this in mind, let us compare the definition of continuity on D with that of uniform continuity on D.

The definition of continuity on D asserts that for every $y \in D$ and every $\epsilon > 0$, there is a $\delta > 0$ such that for all $x \in D$, $|x - y| < \delta$ implies that $|f(x) - f(y)| < \epsilon$.

To change the continuity definition into the uniform continuity definition, one merely changes the order of quantification of the variables! Specifically, one moves the 'for all y' quantification inside of (to the right of) the 'there exists δ'. While the change itself is small, the effect of this change is substantial.

In the usual definition of continuity, we are first handed a y and an ϵ and then told to look for a δ. As discussed above, the choice of δ depends on both the size of ϵ and the particular value of y that we have been given.

In the uniform continuity definition, we are first handed an ϵ. No mention is made of any members of the domain. Instead, we are immediately asked to start a search for δ. Further, after we have a candidate for δ, this candidate must be checked against every $y \in D$ before it can be accepted. In other words, it must be the case for every pair $x, y \in D$,

$$|f(x) - f(y)| < \epsilon \text{ whenever } |x - y| < \delta.$$

All that is required to reject a candidate δ is that we be able to find one single y with an x sufficiently close to it such that

$$x, y \in D, \qquad |x - y| < \delta \qquad \text{and} \qquad |f(x) - f(y)| \ge \epsilon.$$

Thus, the choice of δ can depend only on the choice of ϵ. It must be independent of where we choose x and y from the domain.

To summarize, for continuity on D, given ϵ, we search point by point for a δ, depending on both ϵ and the point y in question, whereas for uniform continuity on D, we look for a single δ that works for all points in that domain. $\square$

Example 1. Let $f(x) = \sqrt{x}$, $x \in [0, \infty)$. Show f is uniformly continuous.

Solution. Let $\epsilon > 0$ be given. Choose $\delta = \epsilon^2$ and observe if $x, y \in [0, \infty)$, then either both $x, y \in [0, \epsilon^2)$ or at least one of x, y is greater or equal to ϵ^2. In the first case, we have $\sqrt{x}, \sqrt{y} \in [0, \epsilon)$, and therefore,

$$|f(x) - f(y)| = |\sqrt{x} - \sqrt{y}| < |\epsilon - 0| = \epsilon.$$

In the latter case, $\sqrt{x} + \sqrt{y} \geq \sqrt{\epsilon^2} = \epsilon$, whence

$$
\begin{aligned}
|f(x) - f(y)| &= |\sqrt{x} - \sqrt{y}| \\
&= \frac{|x - y|}{\sqrt{x} + \sqrt{y}} \leq \frac{|x - y|}{\epsilon} < \frac{\delta}{\epsilon} = \frac{\epsilon^2}{\epsilon} = \epsilon
\end{aligned}
$$

whenever $|x - y| < \delta$. Hence, for every $x, y \in [0, \infty)$, we have

$$|x - y| < \delta \text{ .implies. } |f(x) - f(y)| < \epsilon.$$

Since δ depends only on ϵ, the function $f(x) = \sqrt{x}$ is indeed uniformly continuous on $[0, \infty)$. $\square$

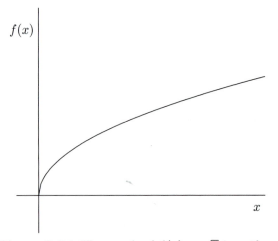

Figure 3.5.1 The graph of $f(x) = \sqrt{x}$ is uniformly continuous in $[0, \infty)$, because, intuitively, the rate of increase is bounded as x gets large.

Discussion. Let us compare the solution above with that of Example 2.5.3, which shows $g(x) = \frac{1}{x}$ is continuous on $[-2, 0) \cup (0, 2]$. (Note that we have renamed the function for the purpose of discussion.)

The first step in showing g is continuous is to pick a $y \in \text{Dmn } g$ followed by an ϵ. For the remainder of that argument, the value of y is kept fixed. The argument then proceeds by restricting the choice of δ. This restriction is based on the choice of y, which had been previously fixed. The quantity $|g(x) - g(y)|$ is then examined, and a final choice for δ made. Again we note the final form of the choice of δ depends both on y and δ. Contrast this with our argument above for f.

First, an ϵ is chosen. The quantity $|f(x) - f(y)|$ is then manipulated with no assumptions as to where in the domain the quantities x and y might reside. Out of this examination, after considerable experimentation, a choice for δ is made that depends only on ϵ. Actually, we have mildly overstated our case. In fact we did use information about the position of y, but since we had only two cases (any finite number would do), we were able to find one value of δ that would work independently of where y was located in the domain. We referred to experimentation. What we mean is that we rewrote $f(x) - f(y)$ and then made two key observations. The first was that as long as $x, y < \epsilon^2$, we could see directly that

$$|f(x) - f(y)| < \epsilon \text{ whenever } |x - y| < \delta = \epsilon^2.$$

On the other hand, if $x, y > \epsilon^2$, only simple algebra is needed to achieve the desired conclusion. Thus, the problem is resolved by recognizing that only two computational situations need be treated and performing the required algebra.

Still, this does not fully explain the intuition used in constructing the argument. Consider the graph of f that is presented in Figure 3.5.1. Starting from $(0, 0)$, the graph rises sharply at first and then ever more slowly. Consider what this geometry means computationally. Let us suppose x and y are a fixed distance apart; that is, the value of $|x - y|$ is fixed, although x and y are allowed to slide along the horizontal axis. As x, y slide to the right, what happens to $|f(x) - f(y)|$? It gets smaller. Indeed, it goes to 0 as is evident from the computations presented in the example. But what is really important is if we slide x, y to the left, there is a maximum value for $|f(x) - f(y)|$. In fact, the maximum is $\sqrt{|x - y|}$, which accounts for the computations presented.

On the other hand, if we repeat these geometric considerations for g, we come to rather different conclusions. Again, let $|x - y|$ be fixed. If we slide x, y to the right toward the point, 2, $|g(x) - g(y)|$ gets smaller, and if we could slide it far enough, by extending the domain, the difference would become arbitrarily small. However, as we slide x, y to the left, bringing them close to 0, the quantity $|g(x) - g(y)|$ becomes arbitrarily large. This is the heart of the problem, and why for g it is impossible to find a δ that is independent of the position of y in the domain of g. $\square$

Example 2. Show $f(x) = x^2$ is uniformly continuous on $[0, 8]$.

Solution. Let $\epsilon > 0$ be fixed. Let $x, y \in [0, 8]$ and consider

$$
\begin{aligned}
|x^2 - y^2| &= |x - y||x + y| \\
&\leq |x - y| \cdot 16.
\end{aligned}
$$

Let $\delta = \frac{\epsilon}{16}$. For this choice of δ, we have

$$|x - y| < \delta \text{ implies } |x^2 - y^2| < \epsilon,$$

as desired. □

Discussion. Computationally, the argument is similar to previous arguments for continuity (see Example 2.5.2). The key point is that a bound of 16 for the term $|x + y|$ can be found that is completely independent of where x and y are in the given interval.

The intuition behind the argument lies in the following. Consider x, y that are a fixed distance apart, and again ask what happens to $|x^2 - y^2|$ as we slide x, y along the interval comprising the domain of the function. Slide it to the left, and the functional difference becomes smaller. Slide it to the right, and the functional difference becomes larger. However, it cannot become arbitrarily large for $x, y \in [0, 8]$. Thus, to find the required dependence on δ, we consider that part of the domain where the difference in functional values can be largest. This occurs near 8, which yields the factor of 16. □

To further illustrate the concept of uniform continuity, we turn to the question of how a function on a given domain can fail to be uniformly continuous. Again, the first step in this process is to negate the definition, and we suggest the reader, herself, do this before reading further.

Negation of the Definition of Uniform Continuity. A function f is not uniformly continuous on a domain D if there exists an $\epsilon > 0$ such that for every $\delta > 0$ there exist $x, y \in D$ such that

$$|x - y| < \delta \text{ .and. } |f(x) - f(y)| \geq \epsilon.$$

Discussion. To satisfy this statement for a function defined on D, we must be able to find a single fixed value for ϵ such that no matter how small we choose δ, there will always exist a pair of points in D having the property that the δ-inequality holds while the ϵ-inequality fails for this pair of points. A little thought should convince the reader that a function that is not continuous on its domain cannot be uniformly continuous on that domain. Since we have already detailed the reasons why a function may fail to be continuous on a domain, our interest here should focus on the question of why a function that is continuous on a domain D can fail to be uniformly continuous there. It is a remarkable fact, which we will prove later, that if the domain for the function is closed and bounded, then continuity is enough to ensure uniform continuity. Thus, we are interested in functions that are continuous on sets that are not closed and bounded (not compact). Our next three examples will cover all the possibilities. However, we already have the key intuition.

Suppose we fix $|x - y|$. If we can find a place in the domain that has the property that when we slide x, y close to this place, the change in $|f(x) - f(y)|$ becomes arbitrarily large, then there will be no hope that f could be uniformly continuous on its domain. Evidently, such a situation arises whenever the function is unbounded at a point. It would be nice if this type of situation were the only situation where uniform continuity cannot be achieved. However, two other situations also arise as the examples below show. □

Example 3. Show $f(x) = \frac{1}{x}, x \in (0, 1]$ is not uniformly continuous.

Solution. Again, let $\epsilon = \frac{1}{2}$, and $1 \geq \delta > 0$ be fixed. Let $x, y \in D$, where

$|x - y| = \frac{\delta}{2}$ and $0 < x, y < \delta$. Clearly, such an x and y exist. For this choice, we have

$$\left| \frac{1}{x} - \frac{1}{y} \right| = |x - y| \left(\frac{1}{xy} \right) \geq \left(\frac{\delta}{2} \right) \left(\frac{1}{\delta^2} \right)$$

$$= \frac{1}{2\delta} \geq \frac{1}{2},$$

as required. $\square$

Discussion. The reader should not be surprised this function is not uniformly continuous on the interval mentioned. After all, this function has been continually referred to in previous discussion as a function for which we cannot eliminate the dependence of δ on the choice of y.

It is important to note the value chosen for ϵ was one of convenience. Any value would have worked, although the larger the value for ϵ, the closer to 0 we may have to choose y in order to ensure $\left| \frac{1}{x} - \frac{1}{y} \right|$ is greater than ϵ. $\square$

Example 4. Show $f(x) = x^2, x \in [0, \infty)$ is not uniformly continuous.

Solution. Let $\epsilon = 1$ and let $1 \geq \delta > 0$ be fixed. Clearly, we can choose an x and y satisfying $|x - y| = \frac{\delta}{2}$ and $x, y > \frac{2}{\delta}$. For such a choice of x and y, we have $x, y \in D, |x - y| < \delta$, and

$$|x^2 - y^2| = |x - y| \cdot |x + y| \geq \left(\frac{\delta}{2} \right) \left(\frac{2}{\delta} + \frac{2}{\delta} \right)$$

$$= 2 \geq \epsilon,$$

as desired. $\square$

Discussion. In the example above, the domain of the function is closed but unbounded. Further, the function is continuous on its entire domain and, in consequence, has no point at which the function is unbounded.

In the argument, the choice of ϵ appears magical. Actually it is not; any value for ϵ would have done. To see why, consider the computations with $|x^2 - y^2|$. Notice that the factor $|x + y|$ depends only on the values of x and y and not on the distance between x and y. Thus, as x and y get large without bound, the term $|x + y|$ also gets large without bound. As a result, to make the product small, the term $|x - y|$ (which measures the closeness of x and y) must be small enough to compensate for the growth of $|x + y|$. It is clear that no fixed amount of 'smallness' can do the job for the unbounded values of x and y generated as x and y go to infinity. Thus, we must choose ever-smaller values for δ as we choose larger values for x and y, and it is this bit of intuition that is the heart of the argument. $\square$

In Examples 3 and 4, it is possible, by moving x, y, to make the value of $|f(x) - f(y)|$ as large as we please. One might think this is a requirement for uniform continuity to fail, and if we could show, for example, that for each value of δ, the value of $|f(x) - f(y)|$ was bounded, provided only $|x - y| < \delta$, then uniform continuity would hold. Such is not the case, as the oscillating example below shows.

Example 5. Show $f(x) = \sin \frac{1}{x}, x \in (0, 1)$ is not uniformly continuous.

Solution. Let $\epsilon = 1$ and $1 > \delta > 0$ be fixed. We can choose $n \in \mathbf{N}$ such that $n > \frac{1}{\delta}$. If we let $x = \frac{2}{\pi(2n+1)}$ and $y = \frac{2}{\pi(2n+3)}$, then $0 < x, y < \delta$, whence

$$|x - y| < \frac{1}{n^2} < \delta \ .\text{and.}\ \left|\sin\frac{1}{x} - \sin\frac{1}{y}\right| = 2 \geq 1,$$

as required. $\square$

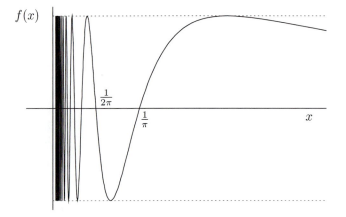

Figure 3.5.2 The graph of $f(x) = \sin\frac{1}{x}$, $x \in (0,1)$ is not uniformly continuous, although it is continuous on this interval.

Discussion. The properties of the sine function required for this example are no more than those presented in Example 2.2.5 in which it was shown that the function $\sin\frac{1}{x}$ had no limit as x tended to 0. The intuition behind the argument can be grasped by thinking of the graph of the sine function stretching out to minus infinity. Now take this graph and push it in from left to right starting at minus infinity and keeping the point $(1, \sin 1)$ fixed. We could think of this as compressing a spring that had been infinitely long to a finite length (see Figure 3.5.2). Thus, we trap an infinite number of oscillations between 0 and 1. Since these oscillations reach a uniform height of 1 on each side of the line $y = 0$, and a complete oscillation exists in any δ-interval about 0, we have all that we need to destroy any hope for uniform continuity. Last, to see $\sin\frac{1}{x}$ is continuous, apply Theorem 2.5.2 to the functions $\sin x$ and $\frac{1}{x}$. That the former is continuous will be established in Chapter 10.

In terms of the intuitive approach we have been developing, consider a fixed value for $|x - y|$. If we slide x, y along the axis of the independent variable, then we see that $|f(x) - f(y)|$ never exceeds 2. However, no matter how small we take δ, we see we can always find values for x, y such that $|f(x) - f(y)|$ will be close to 2. Thus, even though $|f(x) - f(y)|$ does not become arbitrarily large, as in the previous examples, we still do not have uniform continuity.

To complete our intuitive discussion, consider the following. Suppose f is continuous on D, which we may assume is an interval. If we fix a $y \in D$ and a

$\delta > 0$, then we can define the set

$$A_\delta^y = \{z : \text{ there exists } x \in (y - \delta, y + \delta) \cap D \text{ and } |f(x) - f(y)| = z\}.$$

Intuitively, numbers get into A_δ^y provided they are obtainable as a distance between $f(x)$ and $f(y)$ for an x in the δ-interval about y. Now, if A_δ is the union of all the A_δ^ys, for $y \in D$, then the supremum of A_δ will be the smallest ϵ (assuming the supremum exists) that for this choice of δ will yield a valid implication of the form

$$|x - y| < \delta \text{ implies } |f(x) - f(y)| \leq \epsilon.$$

The formation of the set A_δ amounts to looking at the possible values of $|f(x) - f(y)|$ that can be obtained by sliding x, y around in D. What we have observed is the supremum of A_δ must exist for each choice of δ if we are to have any hope of achieving uniform continuity. Further, if we are to avoid the type of misbehavior displayed in Example 5, it must be the case that by making δ small, we can make the supremum over A_δ small.

Still another way of phrasing the question being posed here is: For a given size change in the independent variable, what is the largest change in the dependent variable that can result? Uniform continuity tells us that small changes will produce small changes. A more complete treatment of this approach is spelled out in Exercise 4. □

A useful characterization of the uniform continuity concept in terms of sequences is the following:

Theorem 3.5.1. *Let f be defined on $D\subseteq\mathbf{R}$. Then f is uniformly continuous on D if and only if for every pair of sequences $\{a_n\}_{n=1}^\infty$ and $\{b_n\}_{n=1}^\infty$ of elements of D,*

$$\lim(a_n - b_n) = 0 \qquad implies \qquad \lim(f(a_n) - f(b_n)) = 0.$$

Proof. We prove the condition ensures the uniform continuity of f and leave the main implication to Exercise 20. Thus, suppose f is defined on D and for every pair of sequences $\{a_n\}_{n=1}^\infty$ and $\{b_n\}_{n=1}^\infty$ of elements of D,

$$\lim(a_n - b_n) = 0 \quad .\text{implies.} \quad \lim(f(a_n) - f(b_n)) = 0.$$

Suppose f is not uniformly continuous and choose $\epsilon > 0$ that will witness this fact. Now let $\delta_n = \frac{1}{n}$. Since f is not uniformly continuous with ϵ as witness, we may choose a_n and b_n such that $|a_n - b_n| < \delta_n$ and for which

$$|f(a_n) - f(b_n)| \geq \epsilon > 0.$$

The sequences $\{a_n\}_{n=1}^\infty$ and $\{b_n\}_{n=1}^\infty$ derived by this process satisfy $\lim(a_n - b_n) = 0$, so we have a contradiction. □

Discussion. A related result states that functions that are uniformly continuous preserve Cauchy sequences, Exercise 19. A natural question is whether a function that preserves Cauchy sequences must be uniformly continuous. We leave that question to the reader as part of Exercise 19. □

Exercises

1. Use the definition of uniform continuity to show which of the following functions are uniformly continuous. You may assume sin is continuous with the usual domain and range.

   (a) $f(x) = 2x + 1$, $x \in \mathbf{R}$;
   (b) $f(x) = 1 - x$, $x \in \mathbf{Q}$;
   (c) $f(x) = x^2$, $x \in \mathbf{N}$;
   (d) $f(x) = x^3$, $x \in \mathbf{Q}$;
   (e) $f(x) = \frac{1}{x^2}$, $x \in (.0001, 1]$;
   (f) $f(x) = \frac{1}{x^2}$, $x \in (0, 1)$;
   (g) $f(x) = \frac{1}{1+x^2}$, $x \in \mathbf{R}$;
   (h) $f(x) = \frac{10}{1+x^2}$, $x \in \mathbf{Q}$;
   (i) $f(x) = x^{1/3}$, $x \in \mathbf{R}$;
   (j) $f(x) = \begin{cases} 1, & x > 0 \\ 0, & x < 0, \end{cases}$ $x \in \mathbf{R} \sim \{0\}$;
   (k) $f(x) = \sqrt{x+1}$, $x \geq -1$;
   (l) $f(x) = \frac{x^2+1}{x-3}$, $x \in \mathbf{R} \sim \{3\}$;
   (m) $f(x) = x^{\frac{1}{5}}$, $x \in \mathbf{R}$;
   (n) $f(x) = \frac{x^2+1}{x^2-3}$, $x \in [-2, 1]$, $x \neq -\sqrt{3}$;
   (o) $f(x) = \frac{x}{x^2+1}$, $x \in \mathbf{R}$;
   (p) $f(x) = \sin x^2$, $x \in \mathbf{R}$;
   (q) $f(x) = x \sin \frac{1}{x}$, $x \in (0, \pi)$;
   (r) $f(x) = \sqrt{x} \sin \frac{1}{x}$, $x \in (0, \pi)$;
   (s) $f(x) = x \sin \frac{1}{x^2}$, $x \in (0, \pi)$.

2. Let $P(x)$ and $Q(x)$ be two polynomials with integer coefficients. Let $f(x) = \frac{P(x)}{Q(x)}$, and have as its domain all real numbers that are not zeros for $Q(x)$. Find necessary and sufficient conditions for f to be uniformly continuous on its domain.

3. Discuss the uniform continuity properties of $f(x) = x^{\frac{1}{n}}$, $n \in \mathbf{N}$, on its domain.

4. Let f be defined on an interval I. Define A_δ^y and A_δ as at the end of the section. Define the functions g and h by

$$g(y, \delta) = \begin{cases} \sup A_\delta^y, & \text{if this exists} \\ \text{undefined, otherwise;} \end{cases}$$

$$h(\delta) = \begin{cases} \sup A_\delta, & \text{if this exists} \\ \text{undefined, otherwise.} \end{cases}$$

   (a) Show f is continuous on D if and only if for every $y \in D$, $\lim\limits_{\delta \to 0} g(y, \delta) = 0$ for $\delta > 0$.

   (b) Show f is uniformly continuous on D if and only if $\lim\limits_{\delta \to 0} h(\delta) = 0$ for $\delta > 0$.

5. Let $f(x) = x^2$, $x \in \mathbf{R}$, with g and h as in Exercise 4. Give explicit descriptions for g and h. Do the same for $f(x) = 1 - 2x$, $x \in \mathbf{R}$, and $f(x) = \frac{1}{x}$, $x \in (0.1, 1)$.

6. Let f and g be uniformly continuous on a bounded set, D. What can be said about the sum $f + g$ and product fg? Do your conclusions alter if D is an arbitrary interval, in particular, $D = \mathbf{R}$?

7. Let f be uniformly continuous on (a, b). Show f is bounded on (a, b).

8. Prove the Cauchy criterion for functions: $\lim\limits_{x \to a} f(x)$ exists and is finite if and only if for every $\epsilon > 0$ there exists $\delta > 0$ such that for every $x, y \in \text{Dmn } f$, $0 < |x - a| < \delta$, and $0 < |y - a| < \delta$ imply $|f(x) - f(y)| < \epsilon$.

9. Let f be uniformly continuous on (a, b). Show the left-hand limit at b exists.

10. Give examples that show the results in Exercises 7 and 9 are not true if uniformly continuous is replaced by continuous.

11. Suppose f is defined on (a, b) and for each $x \in (a, b)$ there is a $\delta > 0$ such that f is uniformly continuous on $(x - \delta, x + \delta)$, that is, f is **locally uniformly continuous**. Is f uniformly continuous on (a, b)? Would the answer change if the δ-intervals were closed?

12. Let $f(x) = \sqrt{1 - x^2}, |x| < 1$. Is f uniformly continuous?

13. Let f be uniformly continuous on an interval I. Show f is uniformly continuous on every subinterval of I.

14. Let f be monotone increasing, continuous, and bounded on an interval I. Show f is uniformly continuous on I.

15. Let I_1 and I_2 be two closed intervals. Show if f is uniformly continuous on I_1 and I_2, then f is uniformly continuous on $I_1 \cup I_2$.

16. Can the result of Exercise 15 be extended to countably infinite collections of closed intervals?

17. Let D be bounded, a be a limit point of D that is not in D, and f be continuous on D but not having a continuous extension to $D \cup \{a\}$. Is it true f is not uniformly continuous?

18. Show the function $f : \mathbf{Q} \to \mathbf{Q}$ defined by $f(x) = \frac{1}{2 - x^2}$ is continuous but not uniformly continuous on its domain.

19. Suppose f is uniformly continuous on a domain $D \subseteq \mathbf{R}$, and $\{x_n\}$ is a Cauchy sequence in D. Show $\{f(x_n)\}$ is a Cauchy sequence in $\mathbf{R}$. Is the converse true?

20. If $f : D \to \mathbf{R}$ is uniformly continuous, and for sequences $\{a_n\}$, $\{b_n\}$ in D, if $\lim_{n \to \infty} (a_n - b_n) = 0$, show $\lim_{n \to \infty} (f(a_n) - f(b_n)) = 0$. Give an example to show mere continuity is not enough to ensure the validity of this result.

21. A function $f : D \to \mathbf{R}$ is **Lipschitz**, provided there is a constant $K > 0$ such that

$$|f(x) - f(y)| \le K|x - y|$$

for $x, y \in D$. Show a Lipschitz map is uniformly continuous.

22. For each of the functions given below, either find a value of K proving the function to be Lipschitz on its domain or show no such K can exist.

(a) $f(x) = 2x + 1$, $x \in \mathbf{R}$;
(c) $f(x) = x^3$, $x \in [-8, 8]$;
(e) $f(x) = x^3$, $x \in \mathbf{R}$;
(g) $f(x) = \frac{1}{x^2}$, $x \in (0, 1)$;
(i) $f(x) = \sqrt{x}$, $x \in [0, \infty)$;
(k) $f(x) = \sin x$, $x \in \mathbf{R}$;
(m) $f(x) = \sin \frac{1}{x}$, $x \in (0, \pi)$;

(b) $f(x) = x^2$, $x \in [-2, 2]$;
(d) $f(x) = x^2$, $x \in \mathbf{R}$;
(f) $f(x) = \frac{1}{x^2}$, $x \in (.0001, 1]$;
(h) $f(x) = \frac{1}{1 + x^2}$, $x \in \mathbf{R}$;
(j) $f(x) = x^{\frac{1}{3}}$, $x \in \mathbf{R}$;
(l) $f(x) = \sqrt{x + 1}$, $x \ge -1$;
(n) $f(x) = x \sin \frac{1}{x}$, $x \in (0, \pi)$.

23. Let $f(x) = x^{\frac{p}{q}}$, where $\frac{p}{q} \in \mathbf{Q}$ and f is defined on D. Find conditions under which f will be Lipschitz and explicitly describe the dependence of these conditions on D.

24. Let f be defined on (a, b) and $y \in (a, b)$. Define the **limit superior of f at y** by

$$\overline{\lim}_{x \to y} f(x) = \inf_{\delta > 0} \sup_{0 < |x - y| < \delta} f(x).$$

(a) Define the limit inferior, $\underline{\lim}$, analogously.

(b) Show $\overline{\lim}_{x \to y} f(x) \leq A$ if and only if for every positive ϵ, there exists a positive δ such that for every x with $0 < |x - y| < \delta$ we have $f(x) \leq A + \epsilon$.

(c) Show $\overline{\lim}_{x \to y} f(x) \geq A$ if and only if for every positive ϵ and positive δ there exists x with $0 < |x - y| < \delta$ we have $f(x) \geq A - \epsilon$.

(d) Show $\underline{\lim}_{x \to y} f(x) \leq \overline{\lim}_{x \to y} f(x)$ with equality if and only if the limit exists.

25. Find the $\overline{\lim}$ and $\underline{\lim}$ for each of the following functions at 0.

(a) $f(x) = x^{1/3}$, $x \in \mathbf{R}$;

(b) $f(x) = \begin{cases} 1, & x \geq 0, \\ 0, & x < 0; \end{cases}$

(c) $f(x) = \begin{cases} \frac{1}{x}, & x \neq 0, \\ 0, & x = 0; \end{cases}$

(d) $f(x) = \begin{cases} \frac{1 - \cos x}{x}, & x \in [-\pi, 0) \cup (0, 2] \\ 0, & x = 0; \end{cases}$

(e) $f(x) = \sin x^2$, $x \in \mathbf{R}$;

(f) $f(x) = x \sin \frac{1}{x}$, $x \in (0, \pi)$;

(g) $f(x) = \sqrt{x} \sin \frac{1}{x}$, $x \in (0, \pi)$.

26. Let f be defined on (a, b) and $y \in (a, b)$; f is said to be **lower semicontinuous** at y provided $f(y) \leq \underline{\lim}_{x \to y} f(x)$. Similarly, one can define **upper semicontinuous** at y. A function is said to be **lower (upper) semicontinuous on** (a, b) if it is lower (upper) semicontinuous for each $y \in (a, b)$.

(a) Show f is lower semicontinuous at y if and only if for each positive ϵ there exists a positive δ such that $f(y) \leq f(x) + \epsilon$ whenever $|x - y| < \delta$.

(b) Show f is continuous at y exactly if it is both upper and lower semicontinuous at y.

(c) Show f is lower semicontinuous on (a, b) if and only if the set $\{x : f(x) > a\}$ is open for each $a \in \mathbf{R}$.

(d) Suppose f has a jump discontinuity at y. Under what conditions will f be lower semicontinuous at y?

27. Determine the semicontinuity properties at 0 for each of the functions in Exercise 25.

3.6   Continuous Functions on Closed Bounded Sets

At the beginning of the last section, we stated that it was our intention to answer the question of whether the attribute of being closed and bounded was a topological property. We have already answered this question for the case of closed bounded intervals in Section 2.6. The first theorem of this section produces a proof for the general case.

Theorem 3.6.1. *Let f be defined and continuous on a closed bounded subset of $\mathbf{R}$. Then the range of f is a closed bounded subset of $\mathbf{R}$.*

Proof. Let D be the domain of f. Equivalently, it suffices to show $f(D)$ is closed and bounded. To establish this, it is sufficient to show $f(D)$ is compact. Thus, let $\{O_\alpha\}$ be an arbitrary collection of open sets covering $f(D)$ and fix $y \in f(D)$. Then $y \in O_{\alpha(y)}$ for some $\alpha(y)$ that we fix. Since $O_{\alpha(y)}$ is open, there is an $\epsilon_{y,\alpha(y)} > 0$ such that if $|z - y| < \epsilon_{y,\alpha(y)}$, then $z \in O_{\alpha(y)}$. Also, since f is continuous, for each $x \in D$ such that $f(x) = y$, there is a $\delta_{x,\alpha(y)}$ such that $|x - w| < \delta_{x,\alpha(y)}$ and $w \in D$ imply $|f(w) - f(x)| < \epsilon_{y,\alpha(y)}$. Now set

$$S_{x,\alpha(y)} = \{w : |w - x| < \delta_{x,\alpha(y)}\}.$$

Evidently, if $w \in S_{x,\alpha(y)} \cap D$, then $z = f(w) \in O_{\alpha(y)}$. Further, the collection of $S_{x,\alpha(y)}$'s forms an open cover of D. Thus, since D is compact, by the Heine-Borel Theorem (Theorem 3.2.5), there is a finite subcollection that covers D. We can identify this finite subcollection as $S_{x_1,\alpha(y)}, \ldots, S_{x_n,\alpha(y)}$. These, in turn, identify a finite subcollection of the $O_{\alpha(y)}$'s, $O_{\alpha(y_1)}, \ldots, O_{\alpha(y_n)}$, where $y_i = f(x_i)$. The reader can easily show (see Exercise 2) this finite subcollection covers $f(D)$. Since the $O_{\alpha(y)}$'s were an arbitrary open cover, it follows that $f(D)$ is compact. By the Heine-Borel Theorem, $f(D)$ is closed and bounded. $\square$

Discussion. The proof above illustrates the real power and utility of the compactness concept. With one punch, we are able to show $f(D)$ was both closed and bounded. An alternative approach would first directly establish $f(D)$ was closed, and then that it was bounded. Since D does not have to be an interval, the proof can be very complicated (see Exercise 3).

In concept, the proof above is easily described. Take an open cover of the image. Use the fact that the inverse image of an open set under a continuous function is open to generate an open cover of the domain. (In going from the range to the domain, we have to be aware of the comments following Theorem 3.1.8, which was why this theorem was not applied directly. Only the ideas in the proof of Theorem 3.1.8 were used.) Once an open cover of the domain was obtained, we applied the Heine-Borel Theorem to the domain to obtain a finite cover of the domain. But the images of the sets in this finite cover must comprise all of $f(D)$ and thus enable us to find a finite subcover of the original cover. $\square$

We have seen from the above theorems that continuous functions applied to closed bounded sets generate closed bounded sets. Closed bounded sets are very 'nice' subsets of **R**. It seems plausible then that the class of functions that can be continuous on a closed bounded set should be somewhat restrictive. Thus, one is led to ask whether a function that is continuous on a closed bounded set has any special properties. The answer turns out to be yes, and so we see that closed bounded sets actually make demands on the functions that are continuous on them.

Theorem 3.6.2. *If f is continuous on a closed bounded subset $D \subseteq \mathbf{R}$, then f is uniformly continuous on D.*

Proof. Let $\epsilon > 0$ be given. By continuity, for each $x \in D$, there exists $\delta_x > 0$ such that $|x - y| < \delta_x$ implies $|f(x) - f(y)| < \frac{\epsilon}{3}$. Note that the intervals $S_x = \left(x - \frac{\delta_x}{4}, x + \frac{\delta_x}{4}\right), x \in D$, form an open covering of D. By the Heine-Borel Theorem there is a finite collection of the x's, say, $x_1, \ldots, x_n$, such that the intervals $S_{x_i}, 1 \leq i \leq n$, still cover D. Now set

$$\delta = \frac{1}{2} \min\{\delta_{x_1}, \dots, \delta_{x_n}\},$$

and consider $u, v \in D$ such that $|u - v| < \delta$. Since the finite collection of intervals S_{x_i} covers D, we can find integers j, k such that $1 \leq j, k \leq n$, $u \in S_{x_j}$ and $v \in S_{x_k}$, whence

$$|u - x_j| < \frac{\delta_{x_j}}{4} \quad \text{and} \quad |v - x_k| < \frac{\delta_{x_k}}{4}.$$

By the definition of δ, we have $|x_j - x_k| < \max\{\delta_{x_j}, \delta_{x_k}\}$. It follows that

$$
\begin{aligned}
|f(u) - f(v)| &= |f(u) - f(x_j) + f(x_j) - f(x_k) + f(x_k) - f(v)| \\
&\leq |f(u) - f(x_j)| + |f(x_j) - f(x_k)| + |f(x_k) - f(v)| \\
&< \frac{\epsilon}{3} + \frac{\epsilon}{3} + \frac{\epsilon}{3} = \epsilon.
\end{aligned}
$$

Since ϵ was arbitrary, f is uniformly continuous on D. $\square$

Discussion. As with the previous theorem, the use of the compactness concept simplifies the proof. Conceptually, the argument runs as follows.

Fix ϵ. For this ϵ, about each $x \in D$, find a δ_x-interval that witnesses continuity at x. We use these subintervals to form a cover of D that is composed of the S_x's. The critical fact about the S_x's is that each has a total length that is half that of δ_x. By compactness, we generate a finite subcover of D. Since the list of covering intervals is finite, we know we can find a positive δ that yields an interval less than half the length of any of the intervals in the finite list. We now select two points, u, v, that are within a distance δ of one another. Each one of these points must be close to one of the special points that define the finite subcover. Thus, the points x_k, x_j are determined. We next observe x_k, and x_j cannot be very far apart. This is guaranteed by the definition of S_x. We now use the fact that u is close to x_k and f is known to be continuous at x_k to obtain that $|f(u) - f(x_k)|$ is small. Similarly, we get $|f(x_j) - f(v)|$ and $|f(x_j) - f(x_k)|$ are small, whence the sum of the three terms taken together must also be small. This completes the argument. In Exercise 8, the reader is asked to supply all the details related to the inequalities employed in the proof. $\square$

Exercises

1. Consider the functions in Exercise 3.5.1. Use the theoretical facts developed in this and the previous section to simplify the arguments establishing uniform continuity in the various cases.

2. Show the subcover generated in the proof of Theorem 3.6.1 actually covers $f(D)$.

3. Give a proof of Theorem 3.6.1 that does not resort to compactness.

4. Let $f(x) = \frac{1}{x - \pi}, x \in \mathbf{Q}$. Show f is continuous on $[0, 6]$, but has an unbounded range.

5. Let f be defined and continuous on a closed interval of $\mathbf{Q}$. Show the range of f is not necessarily bounded, and even if it is bounded, it does not have to be closed.

6. Show there are uniformly continuous functions defined on closed bounded subsets of $\mathbf{Q}$ and having range in $\mathbf{R}$ whose range is not closed in $\mathbf{R}$. Can you do the same with $\mathbf{R}$ replaced by $\mathbf{Q}$?

7. Let f be continuous on (a, b) and suppose the right-hand limit at a exists and likewise the left-hand limit at b. Show f is uniformly continuous on (a, b).

8. A function $f : \mathbf{R} \to \mathbf{R}$ is said to be **periodic** on $\mathbf{R}$ if there exists a number $p > 0$ such that $f(x + p) = f(x)$ for all $x \in \mathbf{R}$. Prove that a continuous periodic function on $\mathbf{R}$ is uniformly continuous and bounded on $\mathbf{R}$.

9. Recall the Lipshitz condition explored in Exercise 3.5.21. Give an example of a uniformly continuous function defined on a closed interval, $[a, b]$, that is not Lipshitz.

10. Write out all the details concerning the inequalities used to complete the proof of Theorem 3.6.2.

11. Give an alternative proof of Theorem 3.6.2 for functions defined on closed intervals that runs along the following lines. Suppose f is not uniformly continuous. Then there is an exceptional ϵ, say, ϵ_0. Consider $\delta_n = \frac{1}{n}$. For each choice δ_n, there must be an x_n such that for some x, $|x - x_n| < \delta_n$, but $|f(x) - f(x_n)| \geq \epsilon_0$. By pushing this reasoning to the limit, f can be shown to be discontinuous at a particular $x \in I$, resulting in a contradiction. Complete the details.

12. Let A be a noncompact subset of $\mathbf{R}$. Show

    (a) there exists a continuous function on A that is unbounded;

    (b) there exists a continuous, bounded function on A that has no maximum;

    (c) if, in addition, A is bounded, then there exists a continuous function on A that is not uniformly continuous.

13. Let D be a bounded set and suppose f is continuous but not uniformly continuous on D. Show D has a limit point a, such that $a \notin D$ and no extension of f to $D \cup \{a\}$ is continuous.

14. Let D be closed, and suppose f is continuous on D. Show f has a continuous extension to all of $\mathbf{R}$. Note, this extension may not be unique. Show if D is not closed, this result is not generally true.

15. Let f be uniformly continuous on a bounded set D. Then for every limit point a of D such that $a \notin D$, there is a continuous extension of f to $D \cup \{a\}$.

16. Let f be continuous on a bounded set D. Then f is uniformly continuous on D if and only if f has a continuous extension to all of $\overline{D}$.

17. Let f and g be uniformly continuous on D.

    (a) Show for any constant a, af is uniformly continuous on D.

    (b) Show if D is compact, fg is uniformly continuous on D.

    (c) Suppose D is merely bounded. Will the product fg be uniformly continuous?

    (d) What can be said about the composition of two uniformly continuous functions? Explore the possibilities.

18. Let $f(x) = \sqrt{1 - x^2}$, $|x| \leq 1$. Discuss the uniform continuity properties of f (see Exercise 3.5.12).

19. Let f be continuous on $\mathbf{R}$. What can be concluded if both limits at infinity exist? Can anything be said if f is bounded?

20. Let f be upper semicontinuous (Exercise 3.5.26) on a compact set, D. Show Rng f is bounded above.

21. Let f be upper semicontinuous (Exercise 3.5.26) on a compact set, D. Show f assumes its maximum.

Chapter 4

Differentiation

In this chapter, we consider a somewhat more complicated type of limit. One might think our order reflected the historical development. However, this is not the case, as the differential calculus was laid down long before the real numbers had been put on the firm foundation we have so carefully sketched. Moreover, the concept of limit, as we have framed it, did not exist. The reason for our development is it is only by utilizing the precision of expression of the nineteenth and twentieth centuries that today's mathematicians can work with any form of limit. But if one is prepared to work on a completely intuitive level, then certain problems lead on naturally to the development of the derivative of a function. Such problems as

(i) given a function describing the position of a moving object with respect to time, find its velocity and

(ii) given a curved surface of a lens, find the tangent plane to the surface at a given point

were at the center of the scientific thrust of the seventeenth century, and their solution not only demanded the differential calculus, but also led naturally to its development on an intuitive level. It is not surprising in view of this that the two problems always treated in elementary calculus books are

(i) using the derivative to obtain the velocity and acceleration functions and

(ii) finding the tangent to a curve at a given point.

Our approach will employ the problem of finding a tangent to a curve as the basis for our discussion of the derivative because so many of the important insights about the nature of the derivative are present in this example.

As is usually the case with creations, once conceived, they have a life of their own. And so is it with the derivative. Its achievements have far surpassed the expectations of Newton and Leibnitz, and it is these achievements that are the subject of this chapter.

4.1   Definition and Basic Facts

The notion of tangency arises in the geometry of the ancient Greeks long before the existence of analysis. In this context, one considers the problem of finding a straight line that passes through 'exactly one point' on a curve. The earliest recognition of tangents appears to be with respect to circles where a tangent line was a straight line passing through one and only one point on the circle. Another form of this definition includes the assertion that the tangent line must lie on only one side of the curve (Kline, 1972[1]). In any event, for circles, there is a unique straight line that passes through a given point P on the circle and only through that given point. This straight line is referred to as the **tangent line to the circle** at P. It was known to the Greeks that the tangent line to a circle had many other properties. For example, it was known the tangent at P was perpendicular to the diameter of the circle through P.

The fact that tangents to circles existed suggested tangents to other curves might exist as well, and the Greeks searched, and indeed found, tangents to other curves such as various spirals and conic sections.

As soon as one has developed a theory of functions and coordinate geometry, one is practically forced to wonder about which curves will have tangents and. at a more basic level, just what would be a suitable analytic definition for the tangent concept. More precisely, let us suppose we have a function f which is continuous on an open interval I and a is a fixed member of I. We want to know under what condition there will be a straight line passing through the point $P = (a, f(a))$, which can reasonably be said to be the tangent line to the graph of f at P.

To answer this question, consider Figure 4.1.1. In this figure, we have pictured a curve defined by the graph of f and a tangent line passing through the point $(a, f(a))$. Let us ignore for a moment the problem of whether, or under what conditions, the tangent to the curve at $(a, f(a))$ will exist and suppose that it does. We can then ask:

- How can we find the tangent to the curve at $(a, f(a))$?

To find the tangent line, consider that it is a straight line and, thus, is completely determined by any point on the line and the angle γ formed between the line and the x-axis (see Figure 4.1.1). Because the line we require is to be tangent to the curve at the point $(a, f(a))$, a point on the tangent line is specified as a condition of the problem. Thus, we need only find the angle γ. The key fact to be noticed here is that γ is completely determined by the value of its tangent, and this value can be found from the ratio

$$\tan \gamma = \frac{y_2 - y_1}{x_2 - x_1}$$

where (x_1, y_1) and (x_2, y_2) are any two points on the tangent line. The fact that the ratio

$$\frac{y_2 - y_1}{x_2 - x_1}$$

---

[1]M. Kline, *Mathematical Thought from Ancient to Modern Times* (New York: Oxford University Press, 1972).

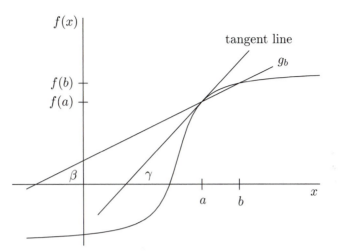

Figure 4.1.1 The tangent to the graph of f at $(a, f(a))$; g_b is the approximation to the tangent line determined by $(b, f(b))$. See text for additional discussion.

is a constant that can be calculated from any pair of points on the straight line led to the realization that the ratio could be used as part of the description of a straight line. For this reason, the ratio was given a name, **slope**, a quantity that should be well familiar to any student of calculus. It was also realized that the slope could be used as part of the analytic description of a line. Thus, in our usual functional notation, we have the point-slope description of a straight line,

$$y = g(x) = m(x - x_1) + y_1,$$

where (x_1, y_1) is any point on the line and m is its slope. Again, these are concepts that are dealt with in the early chapters of almost any calculus book.

With this characterization of a straight line in mind, assume the tangent to the curve exists. We might then think of finding an approximation to it. Given we already have one point on the line, namely, $(a, f(a))$, it seems obvious that any approximation to the tangent line should include this point. If we had another point, we would be done, since any second point will completely determine the line. The only possible source for a second point is the curve itself. Thus, we can pick any point $b \in I$ and take the straight line passing through $(a, f(a))$ and $(b, f(b))$ as an approximation to the tangent line. By employing the point-slope form of the straight line, we see the approximation to the tangent line is

$$g_b(x) = \frac{f(b) - f(a)}{b - a}(x - a) + f(a),$$

where we employ the notation g_b to denote the approximation to the tangent line determined by $(b, f(b))$, as shown in Figure 4.1.1. This line is referred to as a **secant** to the curve. Evidently, a better approximation to the tangent line can be obtained by employing a point $(c, f(c))$ where $a < c < b$. If we compare the analytic form of g_c,

$$g_c(x) = \frac{f(c) - f(a)}{c - a}(x - a) + f(a),$$

with that of g_b, we see the only thing that has changed is the slope of the line. Thus, all the approximation process is doing is finding estimates of the slope of the tangent line. This leads naturally to looking at the quantity

$$\frac{f(b) - f(a)}{b - a}$$

with a fixed, and asking what happens to the ratio if we let b approach a. Evidently, this is a question about limits and, although somewhat different in form, should be no different in substance from questions previously considered.

Definition. Let $f : D \to \mathbf{R}$ be defined in a neighborhood about a; that is, $(a - c, a + c) \subseteq D$ for some $c > 0$. We will say f is **differentiable** at a, provided there is an $L \in \mathbf{R}$ such that for every $\epsilon > 0$, there exists $\delta > 0$ such that if $0 < |h| < \delta$, then

$$\left| \frac{f(a + h) - f(a)}{h} - L \right| < \epsilon.$$

In the case f is differentiable at a, we will write

$$f'(a) = \lim_{h \to 0} \frac{f(a + h) - f(a)}{h} = L$$

and say that the **derivative of f at a is L**.

Discussion. The quantity

$$\frac{f(a + h) - f(a)}{h}$$

is referred to as the **difference quotient** and looks different from something of the form

$$\frac{f(b) - f(a)}{b - a}.$$

To see they are really the same, let $b = a + h$ to obtain that

$$\frac{f(b) - f(a)}{b - a} = \frac{f(a + h) - f(a)}{h}.$$

Since the numerator of the difference quotient is the change, or **increment**, in the function values calculated at the points a and $a + h$, while the denominator, h, is the distance between a and $a + h$, the value of the difference quotient can always be interpreted as the slope of the straight line joining the two points: $(a, f(a))$ and $(a + h, f(a + h))$.

A detail that should be noted is the use of $|h|$ in the definition. This ensures h is permitted to be both positive and negative or, equivalently, that h is permitted to approach 0 from both directions. Thus, in Figure 4.1.1, we could as easily have selected a point b that lay to the left of a on the x-axis as one that lay to the right.

In the event the limit L exists as h tends to 0, in line with the discussion above, we would interpret the number L to be the slope of the tangent line to the curve at the point $(a, f(a))$. Analytically, this means the tangent to the curve at $(a, f(a))$ is given by

$$g(x) = f'(a)(x - a) + f(a).$$

The reader should understand this is essentially a definition. The notion of tangent is a geometric notion that provides motivation for the definition of the derivative. However, the limit as defined above, which is analytic in nature, does not necessarily have to produce a tangent to the curve at the point $(a, f(a))$ in the geometric sense. For one thing, prior to defining the derivative, we don't have tangents to most types of curves. In such a case, where no geometric tangent is available, how are we to decide that the straight line passing through $(a, f(a))$ and having slope $f'(a)$ really is the correct tangent line? We cannot decide, so when we assert that it is, we are essentially making a definition. What we can do is check that the analytic specification for a tangent line agrees with, or has the same properties as, the geometric tangent line in cases for which the latter exists. This type of check will be illustrated in Example 5. □

We have stressed that whenever a new limit definition arises, a number of natural questions arise. For example, is the limit unique, how does it behave with respect to numerical operations, and so on. Since these results are essential to the remaining development, we state two summary theorems.

Theorem 4.1.1. *Let f be defined in a neighborhood about a and differentiable at a. Then $f'(a)$ is unique.*

Proof. We offer a sketch; details are left to Exercise 20.

Assume for the sake of argument that $f'(a)$ is not unique and both L and M satisfy the definition. Then set $\epsilon = \frac{|L-M|}{3}$. A contradiction is now easily obtained. □

Discussion. Note the similarities in the reasonings here and those of Theorems 0.4.1, 1.2.1, 2.1.1, and 2.3.1(a). □

Theorem 4.1.2. *Let f and g be defined in a neighborhood about a and differentiable at a. Then*

(i) *for any $c \in \mathbf{R}$, $(cf)'$ exists and $(cf)' = cf'$;*

(ii) *$(f + g)'(a)$ exists and $(f + g)'(a) = f'(a) + g'(a)$;*

(iii) *$(fg)'(a)$ exists and $(fg)'(a) = f'(a)g(a) + f(a)g'(a)$;*

(iv) *if $g(a) \neq 0$, $\left(\dfrac{f}{g}\right)'(a)$ exists and $\left(\dfrac{f}{g}\right)'(a) = \dfrac{f'(a)g(a) - f(a)g'(a)}{[g(a)]^2}$.*

Proof. We prove (iii) and leave the rest to Exercises 21–24. Thus, assume f

and g are defined on (c, b) and differentiable at $a \in (c, b)$. Then

$$
\begin{aligned}
\left| \frac{(fg)(a+h) - (fg)(a)}{h} \right| &= \left| \frac{f(a+h)g(a+h) - f(a)g(a)}{h} \right| \\
&= \left| \frac{[f(a+h)g(a+h) - f(a)g(a+h)] + [f(a)g(a+h) - f(a)g(a)]}{h} \right| \quad (1) \\
&= \left| \frac{f(a+h)g(a+h) - f(a)g(a+h)}{h} + \frac{f(a)g(a+h) - f(a)g(a)}{h} \right|.
\end{aligned}
$$

It follows that

$$
\begin{aligned}
\left| \frac{(fg)(a+h) - (fg)(a)}{h} - [f'(a)g(a) + f(a)g'(a)] \right| &\leq \\
\left| \frac{f(a+h)g(a+h) - f(a)g(a+h)}{h} - f'(a)g(a) \right| & \quad (2) \\
+ \left| \frac{f(a)g(a+h) - f(a)g(a)}{h} - f(a)g'(a) \right|.
\end{aligned}
$$

To complete the proof, we must show each term on the right-hand side of Eq. 2 can be made arbitrarily small merely by taking h small. Consider the second term first;

$$
\left| \frac{f(a)g(a+h) - f(a)g(a)}{h} - f(a)g'(a) \right| = \left| \frac{g(a+h) - g(a)}{h} - g'(a) \right| |f(a)|.
$$

Since $g'(a)$ exists, the quantity

$$
\left| \frac{g(a+h) - g(a)}{h} - g'(a) \right|
$$

can be made arbitrarily small by taking h sufficiently small. Since $f(a)$ is a constant, the product of this quantity with $f(a)$ can also be made arbitrarily small by taking h small. Now consider the first term on the left-hand side of Eq. 2. Set $g(a+h) = g(a) + \psi$, then

$$
\begin{aligned}
\left| \frac{f(a+h)g(a+h) - f(a)g(a+h)}{h} - f'(a)g(a) \right| & \\
= \left| \frac{f(a+h)(g(a) + \psi) - f(a)(g(a) + \psi)}{h} - f'(a)g(a) \right| & \\
\leq \left| \frac{f(a+h)g(a) - f(a)g(a)}{h} - f'(a)g(a) \right| + \left| \frac{f(a+h) - f(a)}{h} \psi \right| & \quad (3) \\
= \left| \frac{f(a+h) - f(a)}{h} - f'(a) \right| |g(a)| + \left| \frac{f(a+h) - f(a)}{h} \right| |\psi|.
\end{aligned}
$$

We must show both terms in the last line of Eq. 3 can be made arbitrarily small by taking h to be small. The argument for the first term is essentially the same as the argument for the second term in Eq. 2 and is left to the reader. This leaves us to consider the product, $\left| \dfrac{f(a+h) - f(a)}{h} \right| |\psi|$. Since $f'(a)$ exists, we may assume the first term satisfies

$$\left| \frac{f(a+h) - f(a)}{h} \right| \leq |f'(a)| + 1,$$

which may not be small, but is bounded. If we assume assume g is continuous at a, then, given $\epsilon > 0$, we can find a $\delta = \delta(\epsilon) > 0$ such that

$$|\psi| = |g(a+h) - g(a)| < \epsilon$$

whenever $|h| < \delta(\epsilon)$. Thus, the continuity of g at a would be enough to guarantee $|\psi|$ can be made arbitrarily small by taking h small, and hence, the product will be small. Thus, the required conclusion is achieved, provided g is continuous at a. We will prove this fact in the next section (Theorem 4.2.1). $\square$

Discussion. As an aid to understanding the proof, let us review the proof of Theorem 2.3.2. in which the product of the limits is shown to be the limit of the product. In that proof, the essential inequality is

$$|(fg)(x) - AB| \leq |f(x)||g(x) - B| + |f(x) - A||B|. \tag{4}$$

The object is to show both terms on the right-hand side of (Eq. 4) become small as $x \to a$. The second term, like the second term in Eq. 2, is easy to deal with because it is the product of a constant and something that can be made arbitrarily small. On the other hand, the first term on the left of Eq. 4 is difficult to handle because it the product of two variable terms. However, it is tractable because the term $|f(x)|$ is bounded, whence it can be replaced by a constant, for example any upper bound on $|f(x)|$.

If one attempts to repeat this argument with the two terms on the left-hand side of Eq. 2, one succeeds with the second term. The first term is particularly difficult to deal with because there is no common factor in

$$\left| \frac{f(a+h)g(a+h) - f(a)g(a+h)}{h} - f'(a)g(a) \right|.$$

The source of the problem is that two of the terms involve $g(a+h)$ while the third involves $g(a)$. The way around this difficulty is to force the issue by rewriting $g(a+h)$ as $g(a) + \psi$ and performing the manipulations leading to Eq. 3. Both terms on the left-hand side of Eq. 3 are products. The first is manageable because it is the product of a constant and something that can be made to be arbitrarily small. The second does not obviously satisfy either of these conditions, that is, both terms are variable, and it is not clear why either term can be made small. If one examines $\left| \frac{f(a+h) - f(a)}{h} \right|$, one concludes in general it cannot be made small, since its limiting value is $f'(a)$. Thus, the only hope lies with ψ. Since $\psi = g(a+h) - g(a)$, we observe ψ will become small as $h \to 0$ exactly if $\lim_{h \to 0} g(a+h) = g(a)$, which is another way of saying g is continuous at a. $\square$

Example 1. Let $f(x) = x^2, x \in \mathbf{R}$. Show $f'(a)$ exists for each $a \in \mathbf{R}$.

Solution. Let $a \in \mathbf{R}$ be fixed, and let us examine the difference quotient

$$\frac{f(a+h) - f(a)}{h} = \frac{(a+h)^2 - a^2}{h}$$

$$= \frac{a^2 + 2ah + h^2 - a^2}{h}$$

$$= \frac{2ah + h^2}{h} = 2a + h.$$

Evidently, we should set $L = 2a$. Now let $\epsilon > 0$ and $\delta = \epsilon$. Then if $0 < |h| < \delta$, we have

$$\left| \frac{f(a+h) - f(a)}{h} - 2a \right| = |2a + h - 2a| = |h| < \delta = \epsilon.$$

Thus, $f'(a) = 2a$ is the limit. Since a was arbitrary, we are done. $\square$

Discussion. Algebraic computations of the type illustrated are done in every calculus course and do not require further review. $\square$

The definition of the derivative is based on a computation at a particular point. However, as the example illustrates, a function may have a derivative at each point of its domain. This is similar to the situation with respect to continuity, where we began by discussing continuity at a point and ended up with the concept of a continuous function on a domain. This leads to the following definition.

Definition. Let f be defined on an open interval I. We say f is **differentiable on** I provided that for each $a \in I$, f is differentiable at a. In the event f is differentiable on I, we set

$$f' = \{(a, f'(a)) : a \in I\}$$

and refer to f' as the **derivative** of f on I.

Discussion. By Theorem 4.1.1, $f'(a)$ is unique for each $a \in I$. Thus, f' is a function, which by assumption has domain I. We have used a to denote the independent variable in the construction of f', however, we will generally use x in the sequel, as in $f'(x)$. $\square$

Example 2. Discuss the differentiability of $f(x) = x^{\frac{1}{3}}$.

Solution. By definition, f is differentiable at $x \in \mathbf{R}$ exactly if $f'(x)$ exists. Now,

$$f'(x) = \lim_{h \to 0} \frac{(x+h)^{\frac{1}{3}} - x^{\frac{1}{3}}}{h} = \lim_{h \to 0} \frac{(x+h)^{\frac{1}{3}} - x^{\frac{1}{3}}}{(x+h) - x}$$

$$= \lim_{h \to 0} \frac{1}{(x+h)^{\frac{2}{3}} + (x+h)^{\frac{1}{3}} x^{\frac{1}{3}} + x^{\frac{2}{3}}} \quad \textbf{(WHY?)}$$

$$= \frac{1}{3x^{\frac{2}{3}}}.$$

The sequence of computations above is meaningful only if $x \neq 0$. Hence f is differentiable at all points except $x = 0$, and the derivative is given by

$$f'(x) = \frac{1}{3} x^{-\frac{2}{3}}, \ x \neq 0. \qquad\qquad \square$$

Discussion. In the computations above, we have made full use of the limit theorems. Again the reader will perceive the bulk of the calculations are algebraic.

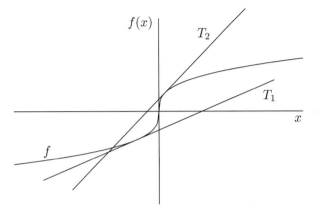

Figure 4.1.2 The curve is defined by $f(x) = x^{1/3}$. The tangent, T_1, is calculated at $x = -0.85$ and T_2 at $x = 0.2$. Both tangents intersect (or will intersect) the graph of f at more than one point. The y-axis is also tangent to the graph (at $x = 0$) even though the derivative does not exist there.

The graph of the function $f(x) = x^{\frac{1}{3}}$ is presented in Figure 4.1.2. The graph is smooth, and it would appear that a tangent to the curve at each point should exist in the geometric sense of the definition (a few tangents have been drawn in to illustrate this). However, the computations above fail to produce a tangent to the curve at $(0,0)$. The reason for this failure is not that the tangent doesn't exist, since the y-axis satisfies the requirement of intersecting the curve only at the point $(0,0)$. Rather, it is because the analytic definition cannot capture this situation that involves a vertical tangent line for which the slope is not defined.

In a case where the derivative fails to exist because of a vertical tangent line, the problem can be eliminated by changing the coordinate system. Thus, in the present case, if we were to rotate the axis slightly in a counterclockwise direction, the derivative would then exist everywhere.

Examination of Figure 4.1.2 also illustrates that a tangent line, in the analytic sense, may cut the graph at more than one place, contrary to the geometric definition. But the original definition was developed based on figures such as circles. Such figures 'curve' in only one direction, while the graph for which we are finding tangent lines curves in two directions. To see what we are getting at, consider a point P that moves from left to right along the graph of f starting at $(-8, -2)$. At first, the path that P follows turns from right to left. As soon as P passes through $(0,0)$, the path reverses and begins turning from left to right. Circles do not behave in this manner. As long as a point moves in the same direction along a circle, the direction of turning is the same. In conclusion, the problem is not that a tangent to f should not exist, but rather the geometric requirement of cutting the curve in only one point must be replaced by an alternative property. This is what the analytic definition supplies and thus permits us to find tangents to a wider variety of curves. □

The example above illustrates a situation in which we have a curve for which the geometric tangent exists everywhere, but for which the analytic tangent fails to exist at certain points. This brings up the obvious problem of identifying those situations in which the derivative fails to exist. If the reader again considers Figure 4.1.1, and the related comments, it should seem plausible that if a function failed to be continuous at a, it would be unreasonable to expect it to be differentiable at a. In other words, differentiability at a should imply continuity at a. This fact is the first theorem of the next section, so we will not pursue it, although the ambitious reader may wish to generate the proof at this stage. But it suggests that what we want to find are examples of functions that are continuous but for which the derivatives do not exist at some point of their domain. As well, we want to understand why the derivative fails to exist.

Example 3. Discuss the differentiability of the function $f(x) = |x|$.

Solution. From the definition of absolute value, if $x > 0$, then $f(x) = x$, whereas if $x < 0$, then $f(x) = -x$. In both cases, a simple calculation shows the derivative f' exists and that $f'(x) = 1$ for $x > 0$ and $f'(x) = -1$ for $x < 0$. Thus, we only need worry about $x = 0$.

For the case $x = 0$,

$$\frac{f(x+h) - f(x)}{h} = \frac{|0+h| - |0|}{h} = \frac{|h|}{h}.$$

As $h \to 0$, $\lim_{h \to 0} \frac{|h|}{h}$ does not exist, since the expression is 1 for $h > 0$ and -1 for $h < 0$. Hence, the function is not differentiable at $x = 0$. □

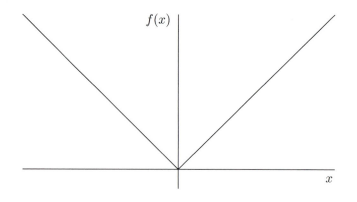

Figure 4.1.3 The function $f(x) = |x|$ has a corner at $x = 0$, and so is not differentiable there.

Discussion. Figure 4.1.3 presents the graph of $|x|$ that is composed of two straight half-lines emanating from $(0,0)$, with slopes of -1 for the half-line on the left and 1 for the half-line on the right. Thus, for $x \neq 0$ tangents to the curve exist. (In this context we are considering that a straight line is tangent to itself, although the Greeks would never have done this.)

The difficulty exists at $x = 0$. The problem is not that there does not exist a straight line that intersects the graph of $f(x) = |x|$ only at the point (0,0). Rather, the problem is that there are too many straight lines that have this property. Thus, the geometric definition has broken down because there is no unique straight line intersecting the curve in one and only one point.

One solution to the problem posed by this example is to restrict the type of curve to which one seeks tangents. It seems plausible that this would have been the tack taken by the Greeks. They would have considered $f(x) = |x|$ to be an inappropriate example for two reasons. First, the graph consists of straight lines, and a straight line that cuts another straight line at one and only one point is obviously not a tangent. Second, to try and find a tangent to a corner of a geometric figure such as a triangle or a square would have seemed nonsensical for the obvious reason that no unique tangent line could exist. The difficulty with an approach based on examination of individual curves is it leaves too much to the eye of the beholder.

The analytic approach deals with the problems posed by the example as follows. First, it makes straight lines tangent to themselves. Second, it provides an analytic test that restricts the situations in which we can discuss tangency. Both are accomplished with a single definition by appealing to the idea of limit.

With respect to the limit, the current situation is similar to that of a jump discontinuity. Namely, the derivative on each side of 0 exists, and is constant. Since the constants on the two sides are different, where the sides join together, a problem is bound to arise. $\square$

We have generated situations where the derivative is infinite, and where the derivative fails to exist because the computation yields one value as h approaches 0^+ and another as h approaches 0^-. When we considered limits of functions, another case existed that involved an infinity of oscillations. This leads us to wonder whether such an example exists in the context of the derivative.

Example 4. Show

$$f(x) = \begin{cases} x \sin \frac{1}{x}, & \text{if } x \in \mathbf{R} \sim \{0\} \\ 0, & \text{if } x = 0 \end{cases}$$

is not differentiable at $x = 0$.

Solution. As shown in Example 2.2.5, the function $g(x) = \sin \frac{1}{x}$ assumes the values 1 and -1 in any interval of the form $(-\delta, \delta)$. It follows the difference quotient

$$\frac{h \sin \left(\frac{1}{h}\right) - 0}{h}$$

must also assume the value 1 and the value -1 (infinitely often) in any interval containing 0. Hence, the limit of the difference quotient cannot exist as $|h|$ tends to 0. $\square$

Discussion. A graph of this function is presented in Figure 4.1.4. The figure also includes the two straight lines $y = x$ and $y = -x$. The graph of f oscillates between these two straight lines with an infinite number of oscillations occurring in any neighborhood of 0. Thus, the approximating secant lines will have slopes that oscillate between plus and minus 1 as the approximating point slides toward 0, whence the derivative at 0 cannot exist.

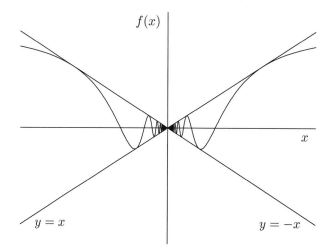

Figure 4.1.4 The graph of $f(x) = \begin{cases} x\sin\frac{1}{x}, & x \neq 0 \\ 0, & x = 0. \end{cases}$
The infinity of oscillations near 0 prevents the existence of a derivative at $x = 0$.

Another way of thinking about the problem is to consider a point P on the graph of f; then the tangent to f at P exists provided P is not $(0,0)$. As the point P slides in toward $(0,0)$, we obtain a series of tangent lines. The slopes of these tangent lines also oscillate, although in this case, they oscillate without bound. (To see this is the case, the reader has only to apply the usual rules for differentiating the trigonometric functions to these functions.) Since the tangent lines near $(0,0)$ are also oscillating infinitely, it is apparent that a limiting line could not exist at $(0,0)$. $\square$

The discussion thus far has presented examples of functions that are continuous on an interval but that fail to be differentiable at some point of the interval. Obviously, it is possible to have an example of a continuous function with a finite number of points in an interval for which it is not differentiable. But a really ugly function would be continuous on an interval but not differentiable anywhere on the interval. Such functions do indeed exist and an example will be outlined in Chapter 8. The reader may wish to try and construct such a beast in the interim.

The reader will also have noticed that in all cases where a derivative failed to exist at a point a, the derivative f' did not have a continuous extension that included the point a. For example, the derivative of $f(x) = |x|$ is $f'(x) = \text{sgn}\,(x)$ for $x \in \mathbf{R} \sim \{0\}$, and there is no way to include 0 in the domain of f' in such a way that a continuous function results. This leads to the question: If f is differentiable on I, must f' be continuous on I? The answer is no (Exercise 38).

We conclude this section by illustrating that a tangent constructed via the analytic methodology satisfies the geometric definition.

Example 5. Show the analytic tangent to the parabola given by $f(x) = x^2, x \in \mathbf{R}$ satisfies the geometric properties required of a tangent; namely, it includes only one point on the curve.

Solution. In Example 1 we showed for any $a \in \mathbf{R}$, $f'(a) = 2a$. Thus, the analytic tangent line to the parabola at $(a, f(a))$ is given by

$$g_a(x) = 2a(x - a) + a^2.$$

Now, the point P having first coordinate x is common to both the parabola and the tangent exactly if $g_a(x) = f(x)$. This yields

$$2a(x - a) + a^2 = x^2 \quad \text{if and only if} \quad 2a(x - a) + a^2 - x^2 = 0,$$
$$\text{if and only if} \quad 2a(x - a) + (a + x)(a - x) = 0,$$
$$\text{if and only if} \quad (x - a)(2a - (a + x)) = 0,$$
$$\text{if and only if} \quad (x - a)(a - x) = 0.$$

The last equality is satisfied exactly if $x = a$, whence the only point common to both the parabola and the tangent line is $(a, f(a))$. $\square$

Discussion. It would be nice if the tangent line was the unique straight line cutting the parabola at $(a, f(a))$. There is one other, namely, the vertical straight line determined by the condition $x = a$. Nevertheless, the analytic tangent line, which is obtained as a limit of secants, has all the properties one would wish of a tangent line. More specifically, in all cases where a geometric tangent line is known to exist and is not vertical, the analytic tangent line also exists and has the same properties. For this reason, the analytic tangent line can reasonably be referred to as the tangent line. $\square$

The reader may well feel we have placed undue emphasis on the notion of tangency in the development of the derivative. We would argue the essence of the derivative notion is present in Figure 4.1.1, and it is this picture that the student should fully understand. Indeed, it is our feeling the intuition required to correctly develop applications of the derivative is inherent in the tangency concept. For this reason we urge the reader to fully come to terms with the ideas contained in Figure 4.1.1 and the material related to it.

Exercises

1. Show the derivative of f at the point a is given by the formula

$$f'(a) = \lim_{x \to a} \frac{f(x) - f(a)}{x - a}$$

provided that the limit exists.

2. Use the definition of derivative to find derivative for the following functions on the indicated domain:

    (a) $f(x) = c,\ x \in \mathbf{R},\ c$ a constant;   (b) $f(x) = x^3,\ x \in \mathbf{R}$;

    (c) $f(x) = \frac{1}{x},\ x \in \mathbf{R} \sim \{0\}$;   (d) $f(x) = \sqrt{x},\ x \in \mathbf{R}^+ \cup \{0\}$;

    (e) $f(x) = \begin{cases} \frac{x^3}{3}, & x \leq 0 \\ \frac{5}{2}x^2 - 4x, & x > 0, \end{cases} x \in \mathbf{R}$;   (f) $f(x) = x^{\frac{2}{3}},\ x \in \mathbf{R}$;

    (g) $f(x) = \begin{cases} \frac{x^3}{3}, & x \leq 0 \\ \frac{5}{2}x^2 - 4x + 1, & x > 0, \end{cases} x \in \mathbf{R}$;   (h) $f(x) = \frac{1}{x^2},\ x \in \mathbf{R} \sim \{0\}$;

    (i) $f(x) = \begin{cases} 0, & -2 \leq x \leq 1 \\ 1 - x^2, & -1 < x \leq 1, \end{cases} x \in \mathbf{R}$;   (j) $f(x) = \frac{1}{\sqrt{x}},\ x \in \mathbf{R}^+$.

3. Let
$$f(x) = \begin{cases} x^2, & \text{if } x \leq 1 \\ x^3, & \text{if } x > 1. \end{cases}$$

Show f is not differentiable at $x = 1$.

4. Let
$$f(x) = \begin{cases} x^2, & \text{if } x \leq 1 \\ \sqrt{x}, & \text{if } x > 1. \end{cases}$$

Is f differentiable at $x = 1$?

5. Let
$$f(x) = \begin{cases} x^2, & \text{if } x \leq 1 \\ 2x - 1, & \text{if } x > 1. \end{cases}$$

Is f differentiable at $x = 1$? What if in the definition of f, $2x - 1$ is replaced by $2x$?

6. Let
$$f(x) = \begin{cases} x^2, & \text{if } x \in \mathbf{Q} \\ 0, & \text{if } x \notin \mathbf{Q}. \end{cases}$$

Is f differentiable at $x = 0$?

7. Let
$$f(x) = \begin{cases} 3x + 2, & \text{if } x \in \mathbf{Q} \\ x^2 - 3x + 5, & \text{if } x \notin \mathbf{Q}. \end{cases}$$

Is f differentiable at $x = 3$?

8. Let f be defined on $[-1, 1]$ by
$$f(x) = \begin{cases} x^2, & \text{if } x \text{ is rational} \\ x^4, & \text{if } x \text{ is irrational}. \end{cases}$$

Show f is differentiable at exactly one point of its domain and find its derivative there.

9. Given $f(x) = \sin \frac{1}{x}$, $g(x) = x \sin \frac{1}{x}$, and $h(x) = x^2 \sin \frac{1}{x}$, with $f(0) = g(0) = h(0) = 0$, prove the following:

    (a) f is neither continuous nor differentiable at 0;

    (b) g is continuous, but not differentiable at 0;

    (c) h is differentiable at 0, but h' is not continuous at 0;

    (d) f' is unbounded near 0.

10. Let $f : \mathbf{R} \to \mathbf{R}$ be such that $f'(c)$ exists for some $c \in \mathbf{R}$. Prove the sequence $\left\{ \dfrac{f(b_n) - f(a_n)}{b_n - a_n} \right\}$ converges to $f'(c)$, where $\{a_n\}$ and $\{b_n\}$ are two sequences converging to c, where $a_n \neq b_n$.

11. Let f be defined on an interval I, and let $a \in I$. Prove f is differentiable at a if and only if for each sequence $\{x_n\}$ in $I \sim \{a\}$ that converges to a, $\left\{ \dfrac{f(x_n) - f(a)}{x_n - a} \right\}$ converges. Show the latter sequence converges to $f'(a)$ in case f is differentiable at a.

12. Show f is differentiable at a if and only if there exists a function $\alpha(h)$ defined in an open interval $(-\delta, \delta)$ and a constant k such that

$$f(a + h) = f(a) + kh + \alpha(h),$$

where $\frac{\alpha(h)}{h}$ tends to 0 as h tends to 0. If f is differentiable at a, prove $k = f'(a)$.

13. Give examples to illustrate that not every function is the derivative of some other function.

14. Consider the graphs of $f(x) = x^2$ and $f(x) = x^3$. Can pieces of the two graphs be 'patched together' in such a way that a function that is differentiable on $\mathbf{R}$ results?

15. Let $f(x) = x^2$, $x \in \mathbf{R}$. Show there are exactly two straight lines that intersect the curve at (a, a^2) and only at that point.

16. **Left-side** and **right-side derivatives** of f are defined by letting $h \to 0^-$, and $h \to 0^+$, respectively, in the definition of the derivative. Show f is differentiable at x if and only if both the right- and left-hand derivatives exist and are equal. Further, find a function for which both the one-sided derivatives exist at x but which is not differentiable there.

17. Let f be defined at all points in an open interval I, except possibly at the point $a \in I$. We define the **generalized derivative** of f **from the left** at the point a by

$$\lim_{x \to a^-} \frac{f(x) - \lim\limits_{x \to a^-} f}{x - a}$$

if it exists. Similarly, define the generalized derivative from the right at the point a. Prove if f has a one-sided derivative at a (Exercise 16), the corresponding generalized derivative also exists and the two are equal. Give an example to show the converse need not be true.

18. Obtain the derivatives of $f(x) = \sin x, x \in \mathbf{R}$ and $f(x) = \cos x, x \in \mathbf{R}$.
    [Hint: Use the limits: $\lim\limits_{h \to 0} \dfrac{\sin h}{h} = 1$ and $\lim\limits_{h \to 0} \dfrac{1 - \cos h}{h} = 0$, in addition to familiar properties of trigonometric functions.]

19. Discuss the differentiability of the following functions:

    (a) $f(x) = |x| + |x + 1|, x \in \mathbf{R}$;
    (b) $f(x) = x|x|, x \in \mathbf{R}$.

20. Complete the proof of Theorem 4.1.1.

21. Prove Theorem 4.1.2 (i).

22. Prove Theorem 4.1.2 (ii).

23. Complete the missing details in the prof of Theorem 4.1.2 (iii).

24. Prove Theorem 4.1.2 (iv).

25. Establish a formula for differentiating each of the following functions:

    (a) $f(x) = x^n$, $n \in \mathbf{N}$;   (b) $f(x) = \sqrt{x}$;   (c) $f(x) = \frac{1}{x^2}$;

    (d) $f(x) = \dfrac{1}{\sqrt{x}}$;   (e) $f(x) = \dfrac{1}{x\sqrt{x}}$.

26. Let $f(x) = x^{\frac{1}{n}}$, $x > 0$ and n a fixed member of $\mathbf{N}$. Show f is differentiable for each $x > 0$, and find a formula for f'.

27. Left- and right-side derivatives are defined in Exercise 16. Suppose a specification for $f'(x)$ is known over its domain, except possibly at $x = a \in D$, and $\lim_{x \to a-} f'(x) = \lim_{x \to a+} f'(x)$; that is, the left- and right-hand limits exist and are equal at a using the specification. Does it follow that the derivative of f exists at $x = a$?

28. If $f : \mathbf{R} \to \mathbf{R}$ is differentiable at a, show

$$f'(a) = \lim_{n \to \infty} \left(n \left\{ f \left(a + \frac{1}{n} \right) - f(a) \right\} \right).$$

Show by an example, the existence of the limit on the right-hand side does not imply the existence of the derivative.

29. Let f be differentiable at x. Show

$$\lim_{h \to 0} \frac{f(x + h) - f(x - h)}{2h} = f'(x).$$

30. Let $\alpha(h)$ be a nonzero function of h having limit 0 as $h \to 0$. Show if f is differentiable at x, then

$$\lim_{h \to 0} \frac{f(x + \alpha(h)) - f(x)}{\alpha(h)} = f'(x).$$

31. Let f be differentiable at x and $\alpha, \beta \in \mathbf{R}$. Show

$$\lim_{h \to 0} \frac{f(x + \alpha h) - f(x - \beta h)}{h} = (\alpha + \beta)f'(x).$$

32. Let $x(t)$ and $y(t)$, $t \in [0, 1]$ be two functions from $[0, 1] \to \mathbf{R}$. The set

$$C = \{(x(t), y(t)) : t \in [0, 1]\}$$

will generate a graph in the plane which we will refer to as the **plane curve** C with **parameterization** $x(t), y(t)$ whenever x and y are continuous on $[0, 1]$. Define the concept of derivative at a point on a plane curve C in such a way that when this derivative exists, it will be the slope of the tangent to the curve at that point.

33. Let C be a plane curve and suppose $x(t)$ and $y(t)$ are differentiable functions of t with $x'(t) \neq 0$ for $t \in [0, 1]$. Show the tangent to C exists at each point. Moreover , show if $\frac{y'(t)}{x'(t)}$ is a constant, then C is a straight line.

34. Consider $f(x) = \sqrt{1 - x^2}$, $x \in [0, 1]$. Show $f'(x)$, $x \in [0, 1]$ is the slope of a line perpendicular to the radius joining $(0, 0)$ and $(x, f(x))$. What can you conclude from this computation?

35. The derivative of f' is called the **second-order derivative** of f (denoted by f''). Similarly, all the **higher-order derivatives**

$$f''', f'''', f^{(v)}, \ldots, f^{(n)}$$

are defined. If f and g have second-order derivatives at a point a, prove

$$(fg)''(a) = f''(a)g(a) + 2f'(a)g'(a) + f(a)g''(a).$$

More generally (stating the conditions needed), prove **Leibnitz's Formula** for the nth-order derivative of the product fg:

$$(fg)^{(n)}(a) = \sum_{k=0}^{k=n} \binom{n}{k} f^{(k)}(a) g^{(n-k)}(a),$$

where $f^{(0)} = f$, $g^{(0)} = g$.

36. Obtain the nth-order derivative $f^{(n)}$ for the following functions [You may use the fact that $(\ln x)' = \frac{1}{x}$ and $(e^x)' = e^x$.]:

(a) $f(x) = x^m$, $m > n$;

(b) $f(x) = \frac{1}{x}$;

(c) $f(x) = \ln x$;

(d) $f(x) = x^m \ln x$, $m > n$;

(e) $e^x \cos x$.

37. Sketch the graph of the following function and discuss its differentiability at the point $x = 0$:

$$f(x) = \begin{cases} 0, & x = 0 \\ \frac{1}{4^{n-1}} - x, & \frac{1}{24^{n-1}} \le x \le \frac{1}{4^{n-1}} \\ 2\left(x - \frac{1}{4^n}\right), & \frac{1}{4^n} \le x \le \frac{1}{24^{n-1}} \\ f(-x), & -1 \le x < 0, \end{cases} \qquad n \in \mathbf{N}.$$

38. If f is differentiable on an interval I, must f' be continuous on I?

39. If f' exists and is bounded on A, prove f is uniformly continuous on A.

40. Prove if f is differentiable on an interval I and f' is bounded on I, then f satisfies Lipschitz's condition (see Exercise 3.5.21) on I.

41. We say f satisfies **Lipschitz's condition of order** α at $c \in$ Dmn f if there exists a constant $K > 0$ such that

$$|f(x) - f(c)| \le K|x - c|^\alpha$$

for all x in a neighborhood of c, $x \ne c$. If $\alpha > 0$ and f satisfies Lipschitz's condition of order α at c, show f is continuous at c. If $\alpha > 1$, show f is also differentiable at c. Illustrate these concepts by considering $f(x) = x^n$ for $n \in \mathbf{Q}$ and $c = 0$. Give an example where f satisfies Lipschitz's condition of order 1, but f' does not exist.

42. Let f and g be thrice differentiable in $\mathbf{R}$ and satisfy the identity $f(x)g(x) = 1$ throughout $\mathbf{R}$. Prove the following (whenever the denominator is nonzero):

(a) $\dfrac{f'(x)}{f(x)} + \dfrac{g'(x)}{g(x)} = 0;$

(b) $\dfrac{f''(x)}{f'(x)} - \dfrac{g''(x)}{g'(x)} = 2\dfrac{f'(x)}{f(x)};$

(c) $\dfrac{f'''(x)}{f'(x)} - \dfrac{g'''(x)}{g'(x)} = 3\dfrac{f''(x)}{f(x)} - \dfrac{g''(x)}{g'(x)} + 3\dfrac{f'(x)g''(x)}{f(x)g'(x)}.$

4.2   Continuity, the Differential, and the Chain Rule

Once we have settled on the derivative as a useful tool, the problem becomes one of finding the derivative of an arbitrary function. Merely applying the definition is not particularly satisfactory. Certain general methods were developed in Exercises 20 through 26 of the last section, but these do not generate derivatives for much more than the polynomials and rational functions. The most powerful single tool we have for obtaining derivatives rests in the chain rule.

Our first result in this section relates the two familiar notions of continuity and differentiability:

Theorem 4.2.1. *If f is differentiable at x, then f is continuous at x.*

Proof. Let f be differentiable at x. Then, given $\epsilon > 0$, there exists $\delta > 0$ such that $0 < |h| < \delta$ implies $\alpha(h) < \epsilon$, where

$$\alpha(h) = \frac{f(x+h) - f(x)}{h} - f'(x).$$

Let h be small enough to guarantee $\alpha(h) < 1$. For such an h, we have

$$
\begin{aligned}
|f(x+h) - f(x)| &= |f'(x)h + \alpha(h)h| \\
&< (|f'(x)| + 1)|h|.
\end{aligned}
$$

As $h \to 0$, the right-hand side must also approach the limit 0. Thus,

$$\lim_{h \to 0} f(x+h) = f(x),$$

which is another way of expressing that f is continuous at x. $\square$

Discussion. The theorem implies the condition for differentiability is much more stringent than that for continuity, and $f(x) = |x|$ at $x = 0$, considered in Section 4.1, illustrates this fact. Thus, differentiable functions form a proper subclass of the class of all continuous functions. Geometrically speaking, for differentiability, the graph of the function should not only be continuous, but it must also be 'smooth'. While the absolute value function possesses only one point of nondifferentiability, there exist real-valued functions that are continuous everywhere on the real line but not differentiable even at a single point. One such example, due to Weierstrass, is discussed in Chapter 8 (Example 8.1.8).

Two details contained in the proof are worthy of further consideration. First, the defining equation for $\alpha(h)$ can be rewritten in the form

$$\alpha(h) = \frac{f(x+h) - f(x)}{h} - \frac{f'(x)h}{h}.$$

In this form, the two denominators represent a change, or **increment**, in the independent variable; that is, $h = (x+h) - x = \Delta x$ (we shall denote the change in a quantity x by Δx). The numerator of the first term also represents an increment, $f(x+h) - f(x) = \Delta f$. It is not so obvious, but the second term, $f'(x)h$, is an increment as well. To see this, set

$$g_x(h) = f'(x)h + f(x) = f'(x)((x+h) - x) + f(x).$$

Then $f'(x)h$ is given by

$$f'(x)h = g_x(h) - g_x(0) = f'(x)((x+h) - x) + f(x) - f(x).$$

As shown in Figure 4.2.1, this latter increment, associated with the function

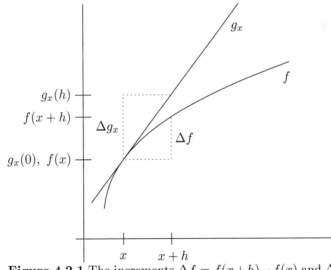

Figure 4.2.1 The increments $\Delta f = f(x+h) - f(x)$ and $\Delta g_x = g_x(h) - g_x(0)$ between x and $x+h$. As is apparent from the figure, g_x is the tangent to the graph of f at $(x, f(x))$.

g_x, amounts to an increment determined by the tangent to the graph of f at the point $(x, f(x))$. Since the graph of g_x is a straight line, the increment in the numerator, Δg_x, is linear. Thus, we see that $\alpha(h)$ is composed of two terms, both of which are ratios of increments. The first, $\frac{\Delta f}{\Delta x}$, is the ratio of the change in the function for a given change in the independent variable. The second, $\frac{\Delta g_x}{\Delta x}$, is the ratio of the same change in the independent variable, to a change determined by the linear function g_x. Since g_x is linear, $\frac{\Delta g_x}{\Delta x}$ is a constant, independent of h, and $\alpha(h)$ is simply the difference between these two ratios of increments.

The second point concerns the inequality

$$|f(x+h) - f(x)| < (|f'(x)| + 1)|h|.$$

The right-hand side is linear in h. (Recall x is fixed.) What this is saying is Δf is approximately linear. Another way of saying this is that near x, f is approximately a straight line. What straight line? Why the tangent line to the graph at $(x, f(x))$! These ideas are made precise in the corollary that follows. $\square$

Corollary 1. *Let f be differentiable at x. Then there is a function $\alpha(h)$, with limit 0 as $h \to 0$, defined in a neighborhood of 0 such that*

$$f(x+h) - f(x) = f'(x)h + \alpha(h)h.$$

Proof. The required function is defined in the proof of Theorem 4.2.1. $\square$

Discussion. Consider the following form of the equality above:

$$f(x+h) = f'(x)h + f(x) + \alpha(h)h.$$

If we drop the term $\alpha(h)h$, which is negligible when h is sufficiently small due to the fact that both h and $\alpha(h)$ are tending to 0, we get

$$f(x+h) \approx f'(x)h + f(x)$$

(the symbol $\approx$ means 'approximately equal to'), which expresses the fact that if h is small, then $f(x+h)$ is well approximated by a linear function of the increment h. $\square$

The idea of using the derivative at x to provide a linear approximation to f near x permits us to obtain approximate values for functions near known values for these same functions. These ideas form the basis for our next definition.

Definition. If f is differentiable at x, the **differential** of f at x is the function $d_x f : \mathbf{R} \to \mathbf{R}$ defined by $d_x f(h) = f'(x)h$ for $h \in \mathbf{R}$.

Discussion. In the above definition, if we take for f, the identity function defined by $f(x) = x$, then $f'(x) = 1$, so that $d_x f(h) = 1 \cdot h = h$, showing the differential of the identity function at the point x is the identity function. Since h is the increment in x, we can say the differential of x is just the increment Δx. By abbreviating $d_x x$ simply as dx, we are stating that $dx = h$. Then, for $y = f(x)$, what is $d_x y$ (or dy for short)? The definition tells us this quantity is precisely $f'(x)h = f'(x)dx$, since $h = dx$. In other words, we obtain the identity $dy = f'(x)dx$ or restated, $\frac{dy}{dx} = f'(x)$. In this formula, $\frac{dy}{dx}$ is a true fraction, which is just the quotient of two differentials. Hence, it is not surprising $\frac{dy}{dx}$ is alternate notation for the derivative $f'(x)$ of the function, $y = f(x)$, with respect to x. In the sequel, we will often use df instead of dy to denote the differential in the dependent variable.

While the increment h in x is the differential dx itself, the increment, $\Delta f = f(x+h) - f(x)$, in the function f is not the differential, df, but the quantity

$f'(x)h + \alpha(h)h = df + \alpha(h)h$. For small enough increments h, the quantity $\alpha(h)h$ is negligibly small, and $\Delta f \approx df$. This is point is most clearly seen in a picture, as in Figure 4.2.2. In summary, then, for any differentiable function, the differential gives approximate values for Δf by a linear computation in h. $\square$

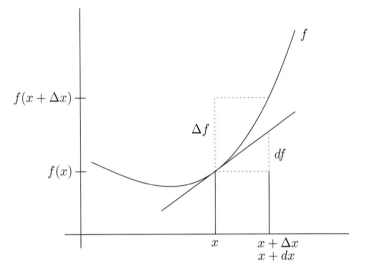

Figure 4.2.2 Graph illustrating how the differential $df = f'(x)dx$ provides a linear approximation to the increment Δf. Notice how the approximation will improve if Δx is small.

Before continuing toward our goal we want to give a concrete application of the use of the tangent line (differential) as a method of approximation.

Example 1. Develop a method for finding approximate solutions to an equation $f(x) = 0$ that can be applied to functions differentiable on $\mathbf{R}$ for which it is known that solutions to the equation exist.

Solution. Let us suppose we have an approximate solution and ask how we could find a better approximate solution. Consider the picture presented in Figure 4.2.3. If x_n is our present solution, and $x = a$ is an actual solution, it is apparent that if we merely slide down the tangent line to x_{n+1}, we will have an even better approximation to a. Recall the tangent line to f at $(x_n, f(x_n))$ is given by

$$g_{x_n}(x) = f'(x_n)(x - x_n) + f(x_n).$$

Since x_{n+1} is defined by $g_{x_n}(x_{n+1}) = 0$, our task is to solve the linear equation:

$$f'(x_n)(x_{n+1} - x_n) + f(x_n) = 0,$$

which leads to

$$x_{n+1} = x_n - \frac{f(x_n)}{f'(x_n)}.$$

Evidently, this method can be applied repetitively to find an even better solution to the problem. $\square$

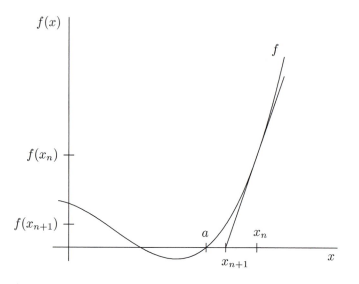

Figure 4.2.3 A graphical illustration of Newton's method for generating approximate solutions of $f(x) = 0$. In the figure, x_{n+1} is a better approximation than x_n.

Discussion. This method for solving equations is known as **Newton's method**, after its inventor, Sir Isaac Newton. While it is impossible to reconstruct the precise reasoning that led him to the method, it is likely it was based on consideration of a picture similar to that presented in Figure 4.2.2.

To illustrate how the method works, consider the polynomial function given by $f(x) = x^5 + 2x + 1$. It is easily checked that this has a root. If we take $x_0 = 0$ as the first approximate solution (0 is really easy to compute with), we see $f(0) = 1$ and $f'(0) = 2$. But $x_1 = -\frac{1}{2}$ and $f(x_1) = -\frac{1}{32}$, which is a considerably better solution. Additional iterations yield even better results as the reader will find from Exercise 1.

The reader may well wonder to what degree the method depends on the function of interest yielding a picture similar to that presented in Figure 4.2.2. An example of an unsuitable function is presented in Exercise 3, although no sensible mathematician would ever use Newton's method for this equation. In general, the effectiveness of the procedure depends of both the nature of the function and the given point at which an approximation is being computed. Much interesting and illustrative mathematics arises from Newton's method and we refer the interested reader to *Saari and Urenko*[2] for a survey of this area. □

We now turn to the task of proving the most important result of this section, namely the Chain Rule for differentiating composite functions.

Theorem 4.2.2 (Chain Rule). *Let g be differentiable at x and f differentiable*

---

[2]Donald G. Saari, and John B. Urenko, Newton's method, circle maps, and chaotic motion, *The American Math. Monthly*, **91**, No. 1 (1984): 3–17.

at $y = g(x)$. Then, $f \circ g$ is differentiable at x and

$$(f \circ g)'(x) = f'(g(x))g'(x).$$

Proof. Since g is differentiable, setting $k = g(x + h) - g(x)$, we observe

$$k = g'(x)h + \alpha(h)h,$$

where $\alpha(h)$ approaches 0 with h. Now, letting $y = g(x)$, consider the numerator of the difference quotient;

$$
\begin{aligned}
(f \circ g)(x + h) - (f \circ g)(x) &= f(g(x + h)) - f(g(x)) \\
&= f(y + k) - f(y) \\
&= f'(y)k + \alpha_1(k)k,
\end{aligned}
$$

where $\alpha_1(k)$ is found by the corollary and $\alpha_1(k) \to 0$ as $k \to 0$. Now, substituting the expression for k, we have

$$(f \circ g)(x + h) - (f \circ g)(x) = f'(y)(g(x + h) - g(x)) + \alpha_1(k)(g(x + h) - g(x)).$$

We have then that

$$\frac{(f \circ g)(x + h) - (f \circ g)(x)}{h} = f'(y)\frac{g(x + h) - g(x)}{h} + \alpha_1(k)\frac{g(x + h) - g(x)}{h}.$$

If we let $h \to 0$, then the first quantity on the right-hand side tends to $f'(y)g'(x) = f'(g(x))g'(x)$, and the second quantity tends to 0 since $\alpha_1(k) \to 0$ and $\frac{g(x+h)-g(x)}{h}$ tends to $g'(x)$. It is immediate $f \circ g$ is differentiable at x with the derivative shown. $\square$

Discussion. We have not formulated the argument above in terms of $\epsilon - \delta$. Rather we have adopted a looser approach that can be turned into an $\epsilon - \delta$ argument with little difficulty. The argument uses continuity in two ways: first, we use the fact that the composition of continuous functions is again continuous, and second, we use the corollary to Theorem 4.2.1. To write out an $\epsilon - \delta$ proof, one would have to spell out the application of these facts clearly. We ask the reader to do so in Exercises 5.

The main use of the Chain Rule is in finding derivatives for composite functions. Several applications appear in the exercises. $\square$

A theorem that is useful in connection with the Chain Rule is

Theorem 4.2.3. *Let f be strictly increasing on (a, b) with inverse g. If f is differentiable at $x \in (a, b)$, $f'(x) \neq 0$, and $y = f(x)$, then g is differentiable at y and $g'(y) = \frac{1}{f'(x)}$.*

Proof. For small h, consider the difference quotient

$$
\begin{aligned}
\frac{g(y + h) - g(y)}{h} &= \frac{g(y + h) - g(y)}{(y + h) - y} \\
&= \frac{(x + k) - x}{f(x + k) - f(x)} \\
&= \frac{1}{\frac{f(x+k)-f(x)}{(x+k)-x}}
\end{aligned}
$$

for some k that depends on h. Now as $k \to 0, h \to 0$ also. Since $f'(x) \neq 0$, the result follows by allowing $k \to 0$. $\square$

Discussion. One would like to obtain this result by a direct application of the Chain Rule; however, not all the hypotheses for that theorem are met. Again, we shall ask the reader to rewrite the proof in terms of ϵ, δ in Exercise 6. Last, the theorem gives us the equation

$$g'(y) = \frac{1}{f'(x)},$$

which expresses the derivative of g as a function of x, the range variable, rather than as a function of y, the domain variable. To correct this, we can use $x = g(y)$, to obtain

$$g'(y) = \frac{1}{f'(g(y))}.$$

Loosely stated, this amounts to the equation $\frac{dx}{dy} = \frac{1}{\frac{dy}{dx}}$, where $x = g(y)$ and $y = f(x)$. $\square$

Exercises

1. For the function $f(x) = x^5 + 2x + 1$, continue the process begun in the discussion following Example 1 to find x_2 and x_3.

2. Find an approximate solution to $x^5 + x^3 + x + 1 = 0$.

3. Show Newton's method fails to produce better approximate solutions to the equation $x^{\frac{1}{3}} = 0$ for any value of $x_n \neq 0$.

4. Formalize the proof of Theorem 4.2.1 in terms of $\epsilon - \delta$.

5. Formalize the proof of the Chain Rule (Theorem 4.2.2) in terms of $\epsilon - \delta$.

6. Formalize the proof of Theorem 4.2.3 in terms of $\epsilon - \delta$.

7. Give an example of functions f and g such that $f \circ g$ is differentiable at x, g is differentiable at x, but f is not differentiable at $y = g(x)$.

8. Use the formula for inverses to find the derivative of $f(x) = x^{\frac{1}{4}}$, $x > 0$.

9. Use the formula for inverses to find the the derivative of $f(x) = x^{\frac{1}{n}}$, $n \neq 0$, and $n \in \mathbf{Z}$ on an appropriate domain.

10. Find the derivative of $f(x) = x^q$, $q \in \mathbf{Q}$ on $x \in \mathbf{R}^+$.

11. Show the function $f(x)$ defined by

$$f(x) = \begin{cases} x^2 \sin \frac{1}{x}, & x \in \mathbf{R} \sim \{0\} \\ 0, & x = 0, \end{cases}$$

   is differentiable everywhere but does not have a continuous derivative.

12. If $t > 0$ is a rational number, and $f : \mathbf{R} \to \mathbf{R}$ is defined by

$$f(x) = \begin{cases} x^t \sin \frac{1}{x}, & x \neq 0 \\ 0, & x = 0, \end{cases}$$

   determine the values of t for which $f'(0)$ exists.

13. Let f be defined on $\mathbf{R}$ by

$$f(x) = \begin{cases} x^3, & x < 2 \\ ax + b, & 2 \leq x, \end{cases}$$

for some choice of $a, b \in \mathbf{R}$. Show a and b can be chosen so that f has a continuous derivative on $\mathbf{R}$.

14. Let $f : D \to E, g : E \to F$, and $h : F \to \mathbf{R}$ be such that f is differentiable at $a \in D$, g is differentiable at $f(a) \in E$, and h is differentiable at $(g \circ f)(a) \in F$. Prove $h \circ (g \circ f)$ is differentiable at a. Find the derivative $(h \circ (g \circ f))'$.

15. Obtain a formula for the second derivative of $(f \circ g)$, stating the conditions needed.

16. If f is differentiable, obtain the derivatives of the following (stating the conditions required):
   (a) f^n; (b) $\frac{1}{f}$; (c) $\sin f$;
   (d) $\ln f$; (e) e^f (f) $f^m g^n$;
   (g) $\left(\frac{f}{g}\right)^{\frac{p}{q}}, \frac{p}{q} \in \mathbf{Q}$.

17. If f is differentiable, what can you say about the differentiability of $|f|$?

18. Use mathematical induction to extend the Chain Rule to differentiate $f_1 \circ f_2 \circ \cdots \circ f_n$.

19. Suppose f has a left-hand derivative at x. Show f must exhibit left-hand continuity at x.

20. Assuming there exists a function $L : (0, \infty) \to \mathbf{R}$ such that $L'(x) = \frac{1}{x}$, calculate the derivative of the following:
   (a) $L(3x + 5)$; (b) $[L(x^3)]^5$;
   (c) $L(\alpha x), \alpha > 0$; (d) $L(L(L(x)))$, where $L(x) > 0, L(L(x)) > 0$.

21. Let $t : \left(-\frac{\pi}{2}, \frac{\pi}{2}\right) \to \mathbf{R}$ be such that $t'(x) = \frac{1}{1+x^2}$. Find the derivatives of the following:
   (a) $t(1 + x^2)$; (b) $t(t(t(3x + 4)))$; (c) $\frac{1}{t\left(\frac{1}{x}\right)} + t\left(\frac{1}{x}\right)$.

22. Let $s : (-1, 1) \to \left(-\frac{\pi}{2}, \frac{\pi}{2}\right)$ be such that $s'(x) = \frac{1}{\sqrt{1 - x^2}}$. Find the derivatives of the following:
   (a) $\sin s(x^2)$; (b) $\frac{1}{s(s(s(x)))}$; (c) $\ln \tan s\left(\frac{1}{1 + x^2}\right)$.

23. Denoting $f''(x)$ by $\frac{d^2 y}{dx^2}$, show Theorem 4.2.3 does not hold for second derivatives; that is, $\dfrac{d^2 x}{dy^2} \neq \dfrac{1}{\dfrac{d^2 y}{dx^2}}$.

4.3   The Mean Value Theorem

One of the most useful results in the theory of differentiation is the Mean Value Theorem. Its consequences are many, as we shall see throughout the remainder of this book. Moreover, the result is essentially intuitive. The essentials to this theorem stem from the idea of local maximum and minimum points on a graph. We begin by studying these concepts.

Let f be defined on $[a, b]$ and $x_0 \in [a, b]$. The point x_0 is called a **maximum point** if $f(x_0) \geq f(x)$ for all $x \in [a, b]$. Such a maximum can be **strict** or **weak** depending on whether the inequality can be made strict or not. The value $f(x_0)$ is called a **maximum** for f on $[a, b]$ provided x_0 is a maximum point. Again, we can have strict or weak maxima. The point $x_0 \in [a, b]$ is called a **local maximum point** if there is a neighborhood $(x_0 - c, x_0 + c), c > 0$, around x_0 such that $f(x_0) \geq f(x)$ for each $x \in (x_0 - c, x_0 + c)$. Similarly, the concepts of **minimum** and **local minimum** can be defined. These ideas are illustrated in Figure 4.3.1.

Theorem 4.3.1. *Let f be defined and continuous on $[a, b]$. If f has a local maximum (minimum) at $x_0 \in (a, b)$ and if f is differentiable at x_0, then $f'(x_0) = 0$.*

Proof. Let us suppose f has a local maximum at x_0. Then, if $h < 0$,

$$\frac{f(x_0 + h) - f(x_0)}{h} \geq 0,$$

since the numerator and the denominator are both negative. On the other hand, for $h > 0$, the numerator is still negative while the denominator is now positive, whence the difference quotient is now negative. By the result on one-sided derivatives (Exercise 4.1.16), we conclude $f'(x_0) = 0$. The corresponding result for local minimum is left to the reader (Exercise 1). $\square$

Discussion. The result is truly simple in nature. All one has to do is to write down the difference quotients on each side of the point x_0 and look at the sign. Since they have opposite signs, the left- and the right-side derivatives of f at x_0 are of opposite sign. But f being differentiable, the two one-sided derivatives must coincide. This can happen only if $f'(x_0) = 0$. Note there is no requirement of continuity, except at x_0. The function in Exercise 4.1.8 shows x_0 may be the only point of continuity or differentiability in the domain of the function. $\square$

Our next theorem is the precursor to the so-called Mean Value Theorem and is known as **Rolle's Theorem**.

Theorem 4.3.2. *Let f be continuous on $[a, b]$, $a < b$ and differentiable on (a, b). If $f(a) = f(b)$, there is a point $x_0 \in (a, b)$ such that $f'(x_0) = 0$.*

Proof. Since f is continuous on $[a, b]$, by Theorem 3.6.1, f attains both its maximum and its minimum. If both of these values coincide with $f(a)$, then f reduces to the constant function $f(x) = f(a)$ throughout $[a, b]$, and the result is immediate. If one of them, say, the minimum, is different from $f(a)$, then there exists $x_0 \in (a, b)$ such that $f(x_0) \leq f(x)$ for all $x \in (a, b)$ (**WHY?**). Since x_0 is a local minimum point and since by hypothesis, f is differentiable at x_0, we must have $f'(x_0) = 0$ by our last result. $\square$

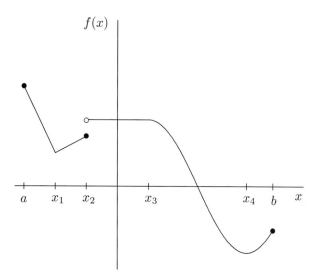

Figure 4.3.1 A graph illustrating the various max/min concepts. Note a is both a maximum and a local maximum point; x_1 is a local minimum point, although f is not differentiable there; every point in (x_2, x_3) is both a local minimum and a local maximum; x_4 is a minimum and a local minimum; and b is a local maximum.

Discussion. Rolle's Theorem, proved above, is simply an application of Theorem 4.3.1 and the fact that a function continuous on a closed bounded interval must attain both its maximum and its minimum. Of course, this latter result has at its heart the supremum principle. The reader should make note of the hypothesis of the theorem: continuity on the closed interval with differentiability on the interior of that interval, since this will be the hypothesis required in all our applications of the Mean Value Theorem. To understand the geometry of Rolle's Theorem, observe the difference quotient evaluated between a and b satisfies

$$\frac{f(b) - f(a)}{b - a} = 0,$$

whence the secant line joining $(a, f(a))$ with $(b, f(b))$ has slope 0. The theorem then asserts that there ought to be a point in between where this slope is actually achieved as a value for the derivative, and in this particular case, the tangent is horizontal (see Figure 4.3.2). The existence of this point is in fact guaranteed, but only if all the hypotheses of the theorem are satisfied, as our exercises make clear. □

In Rolle's Theorem, the point x_0 can be thought as providing a 'mean value' for the derivative. The idea extends naturally to the **Mean Value Theorem**, which is our next result.

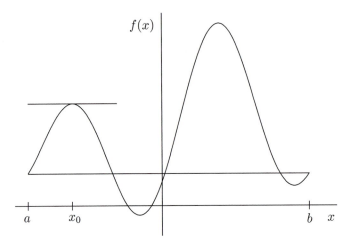

Figure 4.3.2 A graph illustrating Rolle's Theorem. Three other points, besides x_0, exist in (a, b) that are suitable choices to satisfy the theorem.

Theorem 4.3.3. *Let f be continuous on $[a, b]$ where $a < b$ and differentiable in (a, b). Then there exists $x_0 \in (a, b)$ such that*

$$f'(x_0) = \frac{f(b) - f(a)}{b - a}.$$

Proof. Define a function F on $[a, b]$ by

$$F(x) = f(x) - \frac{f(b) - f(a)}{b - a}(x - a) - f(a).$$

It is immediate F is continuous on $[a, b]$, differentiable on (a, b), and further, $F(a) = F(b) = 0$. By Rolle's Theorem, there exists an $x_0 \in (a, b)$ such that $F'(x_0) = 0$ so that x_0 satisfies the conclusion of the theorem. $\square$

Discussion. The proof given above is completely trivial, once we have the function F. In fact, the construction of F is not mere trickery. Consider Figure 4.3.3. We have f together with the straight line corresponding to the secant joining $(a, f(a))$ with $(b, f(b))$. Let g denote the function whose graph is this indicated straight line. From the picture, we see $(f - g)$ must satisfy $(f - g)(a) = (f - g)(b)$, since $f(a) = g(a)$ and $f(b) = g(b)$. Moreover, $f - g$ is continuous on $[a, b]$, differentiable on (a, b) whence we can apply Rolle's Theorem. Last, when $(f - g)'$ equals 0, the value of the derivative of f must equal the value of the derivative of g, which is the slope of the straight line and is given by evaluating the difference quotient at a and b! If we now write out g and $f - g$, we get, using the secant form of the equation of a line

$$\frac{y - f(a)}{x - a} = \frac{f(b) - f(a)}{b - a},$$

whence

$$g(x) = y = \frac{f(b) - f(a)}{b - a}(x - a) + f(a)$$

and

$$(f - g)(x) = f(x) - \frac{f(b) - f(a)}{b - a}(x - a) - f(a) = F(x).$$

From the figure, notice that at x_0, the slope of the tangent to the curve is the same as that of the secant line, whence the two lines are parallel. The tangent line to the curve is indicated by the dashed line. To conclude, if we set out to find a function that naturally incorporates the function, f, together with the linear function defined by the secant joining $(a, f(a))$ with $(b, f(b)$ and that in addition satisfies the hypotheses of Rolle's Theorem, we are led to F and the proof of the theorem follows.

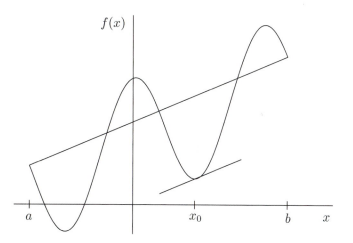

Figure 4.3.3 A graphical illustration of the Mean Value Theorem. Three points other than x_0 exist that satisfy the theorem.

The principle contained in this theorem can be adopted to manufacture many interesting and useful Mean Value Theorems. The strategy is very simple. If a 'nice' (that is, continuous in a closed interval, and differentiable in its interior) function vanishes at two points, then the derivative must vanish somewhere in between. Using this technique, we ask the reader to arrive at several Mean Value Theorems, a project we pursue in the exercises. $\square$

Example 1. Use the Mean Value Theorem to estimate the sixth root of 65.

Solution. If f is differentiable on (a, b), then for some x_0

$$f(b) = f(a) + f'(x_0)(b - a).$$

In the present instance we let $f(x) = x^{\frac{1}{6}}$. This function is differentiable on $(64, 65)$ and its derivative is given by $f'(x) = \frac{1}{6x^{5/6}}$. Since f' achieves an absolute maximum at $x = 64$, we have

$$2 < 65^{\frac{1}{6}} < 2 + \frac{1}{192}. \qquad \square$$

Discussion. The reader may well wonder about the relevance of this type of calculation in an age when a programmable calculator can be purchased for no more than \$50. The answer lies not in the fact that the reader will be likely to avail himself of this calculation, instead of employing the calculator on the desk. Rather, it lies in the fact that the use of this type of approximation is at the root of a great deal of numerical mathematics. We have already seen this in Newton's method. We will see it again in a more general form in Taylor's Theorem, which also goes by the name of the Generalized Mean Value Theorem. These approximation methods are at the heart of many numerical techniques occurring in differential equations and other branches of applied mathematics, as well as the methods used by calculators. Thus, we are dealing with a simple, but extremely important, idea whose time has not passed. $\square$

The following generalization of the Mean Value Theorem is known as **Cauchy's Mean Value Theorem**.

Theorem 4.3.4. *Let f and g be both continuous in $[a, b]$ with $a < b$, and differentiable in (a, b). If $g'(x) \neq 0$ for all $x \in (a, b)$, then $g(a) \neq g(b)$ and there exists $x_0 \in (a, b)$ such that*

$$\frac{f(b) - f(a)}{g(b) - g(a)} = \frac{f'(x_0)}{g'(x_0)}.$$

Proof. First, note $g(a) = g(b)$ contradicts Rolle's Theorem. Now, define F on $[a, b]$ by

$$F(x) = f(x) - f(a) - \frac{f(b) - f(a)}{g(b) - g(a)}[g(x) - g(a)].$$

It is immediate F satisfies the hypotheses of Rolle's Theorem. Thus, there exists $x_0 \in (a, b)$ such that $F'(x_0) = 0$. The reader may now check that x_0 has the required property. $\square$

Discussion. The proof given here is similar to that of the Mean Value Theorem 4.3.3. Once we specify the 'magic function', F, that satisfies the hypothesis of Rolle's Theorem, the proof follows immediately. The reader should understand clearly how to construct such a function. Exercises 29, 30, and 32 provide additional practice. $\square$

Exercises

1. Complete the proof of Theorem 4.3.1.

2. Let $f(x) = x^3 - 2x$, $x \in \mathbf{R}$. Find all local extrema for f.

3. Let $f(x) = x^3 - 3x^2 - x + 3$, $x \in [-1, 3]$. Find the value of x_0 that will satisfy Rolle's Theorem.

4. Let $f(x) = x^3 - 4x^2 + 3x + 1$, $x \in [0, 2]$. Find the value of x_0 that will satisfy Mean Value Theorem.

5. Let $f(x) = 4x^2$, $g(x) = x^3 + 1$, $x \in [0, 2]$. Find the value of x_0 that will satisfy Cauchy's Mean Value Theorem.

6. If f is continuous on $[a, b]$, differentiable on (a, b) and if $f'(x) = 0$ throughout (a, b), prove f is a constant on $[a, b]$.

7. Let f and g satisfy the hypothesis of the Mean Value Theorem on $[a, b], a < b$. Show if $f'(x) = g'(x)$ for all $x \in (a, b)$, then there is a constant C such that $f(x) = g(x) + C$ for all $x \in (a, b)$.

8. If f is twice differentiable on an interval I, and if $f''(x) = 0$ throughout the interval, show $f(x) = kx + l$ for suitable constants k and l. If f is thrice differentiable on I and if $f'''(x) = 0$ on I, what is the form of f in the interval I?

9. Let $f : \mathbf{R} \to \mathbf{R}$ be such that $|f(x) - f(y)| \leq (x - y)^2$ for all $x, y \in \mathbf{R}$. Must f reduce to a constant?

10. Let f be continuous on $[a, b]$ and let $y \in (a, b)$. If $f'(y)$ exists for each $y \in (a, b)$, $y \neq x$, and if $\lim_{y \to x} f'(y)$ exists, prove $f'(x)$ exists.

11. Give an example of a function f that is differentiable on all of $\mathbf{R}$ but having a point x_0 such that $\lim_{x \to x_0} f'(x) \neq f'(x_0)$.

12. Suppose f is differentiable on $[a, b]$. Show f' must assume every value between its extremes on this interval. Deduce that if f is differentiable on $[a, b]$, then f' cannot have discontinuities of the first kind.

13. Let f be defined by
$$f(x) = \begin{cases} x^2 \sin \frac{1}{x^2}, & x \neq 0 \\ 0, & x = 0. \end{cases}$$

Show f has an absolute minimum at $x = 0$, and f is differentiable at $x = 0$, but f' does not have a simple change of sign at $x = 0$.

14. Let f be continuous on $[0, a)$ with $f(0) = 0$. Show if f is differentiable on $(0, a)$ and f' is positive and increasing, then $\frac{f(x)}{x}$ is also increasing.

15. Give an example of a bounded, monotonically increasing, differentiable function having the property $\lim_{x \to \infty} f'(x) \neq 0$.

16. As shown in Example 1, the Mean Value Theorem can be used to form estimates for quantities like $28^{\frac{1}{3}}$, $17^{\frac{1}{4}}$, and $\sin 61°$. Find estimates for these quantities.

17. Let f be increasing on $[a, b]$. Show if f is differentiable at $x_0 \in [a, b]$, then $f'(x_0) \geq 0$. If f is strictly increasing, can we conclude $f'(x_0) > 0$?

18. Give an example of a function that is continuous on $[0, 2]$, differentiable on $(0, 1) \cup (1, 2)$, and takes the same value at 0 and 2, but that does not satisfy Rolle's Theorem.

19. Give an example of a function that is differentiable on $(0, 1)$ and such that $f(0) = f(1)$ but that does not satisfy Rolle's Theorem.

20. Give a careful definition of minimum and local minimum. Write out a rigorous proof of Theorem 4.3.1 assuming a local minimum.

21. Fill in the missing details in the proofs of Theorems 4.3.3 and 4.3.4.

22. Let f satisfy the hypothesis of the Mean Value Theorem on $[a, b]$ and suppose $f'(x) > 0$ for all $x \in (a, b)$. Prove f is strictly increasing.

23. Let f and g satisfy the hypothesis of the Mean Value Theorem on $[a, a+h]$, and let $g' \neq 0$ throughout $[a, a+h]$. Show if f' and g' are not simultaneously zero, then there exists $\theta \in (0, 1)$ such that

$$\frac{f'(a+\theta h)}{g'(a+\theta h)} = \frac{f(a+h) - f(a)}{g(a+h) - g(a)}.$$

Taking $f(x) = \sin x, g(x) = \cos x$, show $\theta = \frac{1}{2}$ for any h.

24. Explain the fallacy involved in the following proof of Theorem 4.3.4:

$$\frac{f(b) - f(a)}{g(b) - g(a)} = \frac{\frac{f(b)-f(a)}{b-a}}{\frac{g(b)-g(a)}{b-a}} = \frac{f'(x_0)}{g'(x_0)}$$

invoking Theorem 4.3.3.

25. If $a_i \in \mathbf{R}$, $0 \leq i \leq n$, and if

$$a_0 + \frac{a_1}{2} + \cdots + \frac{a_{n-1}}{n} + \frac{a_n}{n+1} = 0,$$

show the polynomial $a_n x^n + a_{n-1} x^{n-1} + \cdots + a_1 x + a_0$ has at least one root in $(0, 1)$.

26. Let f be defined in $(0, 1]$ and possess a bounded derivative in $[0, 1]$. Show the sequence $\{a_n\}$ converges, where $a_n = f(\frac{1}{n})$.

27. Let f be thrice differentiable in $[a, b]$. If $f(a) = f(b) = f'(a) = f'(b)$, prove there exists $\xi \in (a, b)$ such that $f'''(\xi) = 0$.

28. Let C be a curve in the plane with derivatives at all $t \in (0, 1)$. Show if $x(a) \neq x(b)$, then for some $t \in (a, b)$, the tangent to C at $(x(t), y(t))$ is parallel to the secant joining $((x(a), y(a))$ with $((x(b), y(b))$.

29. Let

$$F(x) = \begin{vmatrix} f(x) & f(a) & f(b) \\ g(x) & g(a) & g(b) \\ h(x) & h(a) & h(b) \end{vmatrix},$$

where f, g, h satisfy the hypothesis of Mean Value Theorem in $[a, b]$, $a < b$. Prove there exists $\xi \in (a, b)$ such that

$$F'(\xi) = \begin{vmatrix} f'(\xi) & f(a) & f(b) \\ g'(\xi) & g(a) & g(b) \\ h'(\xi) & h(a) & h(b) \end{vmatrix} = 0.$$

Choosing the functions f, g, and h appropriately, derive Rolle's, Mean Value, and Cauchy's Mean Value Theorems as particular cases of this example.

30. If f and g satisfy hypothesis of Cauchy's Mean Value Theorem in $[a, b]$, $a < b$, prove there exists $\xi \in (a, b)$ such that

$$\frac{f(\xi) - f(a)}{g(b) - g(\xi)} = \frac{f'(\xi)}{g'(\xi)}.$$

31. Suppose f and g satisfy the hypothesis of Cauchy's Mean Value Theorem on (a, ∞). Moreover, suppose $\lim_{x \to \infty} f(x) = \lim_{x \to \infty} g(x) = 0$. Show for any $x \in (a, \infty)$, there exists $x_0 \in (a, \infty)$ such that $\frac{f(x)}{g(x)} = \frac{f'(x_0)}{g'(x_0)}$.

32. If ϕ and ψ are twice differentiable functions in (a, b), prove there exists $\xi \in (a, b)$ satisfying

$$\frac{\begin{vmatrix} \phi(x) & x & 1 \\ \phi(a) & a & 1 \\ \phi(b) & b & 1 \end{vmatrix}}{\begin{vmatrix} \psi(x) & x & 1 \\ \psi(a) & a & 1 \\ \psi(b) & b & 1 \end{vmatrix}} = \frac{\phi''(\xi)}{\psi''(\xi)}.$$

33. Let f and g be n times differentiable in (a, b) and simultaneously vanish at n distinct points in (a, b). Show there exists $\xi \in (a, b)$ such that $\dfrac{f(x)}{g(x)} = \dfrac{f^{(n)}(\xi)}{g^{(n)}(\xi)}$.

34. Let f be as in Theorem 4.3.3, and $a, b > 0$. Use Cauchy's Mean Value Theorem to show exists $\xi \in (a, b)$, where $f(b) - f(a) = \xi f'(\xi) \ln \frac{b}{a}$. Deduce the sequence $a_n = n[a^{\frac{1}{n}} - 1]$ converges to $\ln a$.

35. Give a geometrical interpretation of Cauchy's Mean Value Theorem.

36. Let f be k times continuously differentiable in (a, b). Prove **Taylor Formula**: there exists $x_0 \in (a, b)$ such that

$$f(b) = f(a) + (b - a)f'(x) + \frac{(b - a)^2}{2}f''(a) + \cdots + \frac{(b - a)^{(k)}}{k!}f^{(k)}(x_0).$$

37. Prove (stating carefully the conditions needed) the following generalization of Cauchy's Mean Value Theorem that guarantees the existence of $\xi \in (a, b)$ such that

$$\frac{f(b) - f(a) - \sum_{k=1}^{n-1} \frac{(b-a)^k}{k!} f^{(k)}(a)}{g(b) - g(a) - \sum_{k=1}^{n-1} \frac{(b-a)^k}{k!} g^{(k)}(a)} = \frac{f^{(n)}(\xi)}{g^{(n)}(\xi)}.$$

38. Let f satisfy conditions of Rolle's Theorem in $[a - h, a + h]$, $(h > 0)$.

(a) Prove there exists θ, $(0 < \theta < 1)$ such that

$$\frac{f(a + h) - f(a - h)}{h} = f'(a + \theta h) + f'(a - \theta h).$$

(b) Prove there exists ξ, $(0 < \xi < 1)$ such that

$$\frac{f(a + h) - 2f(a) + f(a - h)}{h} = f'(a + \xi h) + f'(a - \xi h).$$

(c) If $f''(a)$ exists, prove

$$f''(a) = \lim_{h \to 0} \frac{f(a + h) - 2f(a) + f(a - h)}{h^2}.$$

39. Show if f satisfies the hypotheses of the Mean Value Theorem in the interval $[a, a + h]$, then there exists $\theta, 0 < \theta < 1$ such that $f(a + h) = f(a) + hf'(a + \theta h)$. Obtain a similar statement for Cauchy's Mean Value Theorem in $[a, a + h]$.

40. If f is twice differentiable in $[a, a + h]$, show there exists θ, $(0 < \theta < 1)$ such that

$$f(a + h) = f(a) + hf'(a) + \frac{h^2}{2}f''(a + \theta h).$$

41. If f'' is continuous in $[a, a + h]$ and differentiable in $[a, a + h)$, prove there exists θ $(0 < \theta < 1)$ satisfying

$$f(a + h) = f(a) + \frac{h}{2}[f'(a) + f'(a + h)] - \frac{h^3}{12}f'''(a + \theta h).$$

42. Assuming f'' is continuous on $[a, b]$, and $a < c < b$, prove there exists $\xi \in (a, b)$ satisfying

$$(b - a)f(c) - (c - a)f(b) - (b - c)f(a) = \frac{1}{2}(b - a)(c - a)(c - b)f''(\xi).$$

43. P, Q, and R are points on the curve $y = f(x)$, whose x-coordinates are x_P, x_Q, x_R, respectively. Assuming f' exists at all points on the curve, prove there exists $z \in (x_P, x_Q) \cup (x_Q, x_R)$ such that

$$\pm \frac{1}{4}f''(z) = \frac{\text{area of triangle } PQR}{(x_P - x_Q)(x_Q - x_R)(x_R - x_P)}.$$

44. The **difference operators** with spacing $h > 0$ are defined by

$$\Delta^0 f(x) = f(x);$$
$$\Delta^1 f(x) = \Delta f(x) = f(x + h) - f(x);$$
$$\Delta^{n+1} f(x) = \Delta(\Delta^n f(x)), n \geq 1.$$

   (a) Show $\Delta^n f(x) = \sum_{k=0}^{n}(-1)^{n-k}\binom{n}{k}f(x + kh)$;

   (b) If $P(x) = a_0 + a_1(x - t) + \cdots + a_n(x - t)^n$, show $\Delta^n P(x) = n!h^n a_n$;

   (c) If f is continuous on $[a, b]$ and n-times differentiable on (a, b) and $x + nh \in [a, b]$ for $x \in [a, b]$, $h \neq 0$, show there exists $\theta, (0 < \theta < 1)$ such that $\Delta^n f(x) = f^{(n)}(x + n\theta h)h^n$.

4.4   L'Hospital's Rule

We conclude this chapter with the first major application of the Mean Value Theorem. In practice, one often encounters limits of the form $\lim_{x \to c} \frac{f(x)}{g(x)}$, where the expression reduces to the meaningless form $\frac{0}{0}$, when $x = c$. In such situations, the following theorem (often called **L'Hospital's Rule**) enables us to compute the limit.

Theorem 4.4.1. *Let f and g be defined on (a, b) with $a < b$. Suppose f' and g' exist on (a, b) with $g'(x) \neq 0$ on (a, b). If $\lim_{x \to a^+} f(x) = \lim_{x \to a^+} g(x) = 0$, and*

$$\lim_{x \to a^+} \frac{f'(x)}{g'(x)} = A, \text{ then } \lim_{x \to a^+} \frac{f(x)}{g(x)} = A.$$

Proof. Consistent with the one-sided limits, define f and g at a by $f(a) = g(a) = 0$. By Cauchy's Mean Value Theorem applied to f and g on the interval (a, x) (where $x < b$), there exists $x_0 \in (a, x)$ such that

$$\frac{f(x) - f(a)}{g(x) - g(a)} = \frac{f'(x_0)}{g'(x_0)}.$$

By hypothesis, there is a $\delta > 0$ such that $x_0 \in (a, a + \delta)$ will imply

$$\left| \frac{f'(x_0)}{g'(x_0)} - A \right| < \epsilon,$$

where ϵ has been prescribed. It is immediate, if $x \in (a, a + \delta)$, consideration of the interval (a, x) will have the same effect, and the proof follows. $\square$

Discussion. The reader should carefully verify the hypothesis of Cauchy's Mean Value Theorem. As well, in the exercises, we ask the reader to formalize this argument into strict $\epsilon - \delta$ reasoning.

The force of the theorem is that we are allowed, in the special circumstance of the hypothesis, to replace the functional values with those of the derivative of the functions for the purpose of calculating the limit. This form of L'Hospital's Rule is often referred to as the $\frac{0}{0}$ form. Other forms, such as $\frac{\infty}{\infty}, \infty \cdot 0, \infty - \infty, 0^0$, and 1^∞, will be explored in the exercises. $\square$

Example 1. Evaluate $\lim\limits_{x \to 0} \dfrac{\sin 4x}{\sin 3x}$.

Solution. Based on facts from elementary calculus (details presented formally in Chapter 8), both $\sin 4x$ and $\sin 3x$ are differentiable in an open interval about 0. Further, $(\sin x)' = \cos x$, and with no loss in generality we may assume that the interval chosen about 0 does not include a zero of $\cos x$. Thus, the hypothesis of Theorem 4.4.1 is satisfied and

$$\lim_{x \to 0} \frac{\sin 4x}{\sin 3x} = \lim_{x \to 0} \frac{4 \cos 4x}{3 \cos 3x}.$$

Since $\lim\limits_{x \to 0} \cos x = 1$, it follows that the limit of the right-hand side is $\frac{4}{3}$. $\square$

Discussion. Theorem 4.4.1 is based on the right-hand limits at a. We have applied the theorem in a situation where we want to compute a limit at an interior point. What the theorem requires us to do is to first to calculate the right-hand limit, then the left-hand limit, and then to check that the left- and right-hand limits are the same. $\square$

Example 2. Let f be twice differentiable in a neighborhood of a. Evaluate

$$\lim_{h \to 0} \frac{f(a + h) - (f(a) + f'(a)h)}{h^2}.$$

Solution. Since the hypothesis of Theorem 4.4.1 is satisfied,

$$\lim_{h \to 0} \frac{f(a + h) - (f(a) + f'(a)h)}{h^2} = \lim_{h \to 0} \frac{f'(a + h) - f'(a)}{2h} = \frac{1}{2} f''(a). \qquad \square$$

Discussion. This example has a certain look of artificiality to it. However, to dismiss it as such would be a mistake.

The required limit is in the form of a ratio as h tends to 0. Since both the numerator and the denominator have the limit 0, the force of the computation is to compare the rate at which the numerator is tending to 0 with the rate at which the denominator is tending to 0. There are several possible outcomes. For example, if the numerator approaches 0 much, much faster than the denominator, the final limit will be 0. If the denominator goes to 0 much, much faster than the numerator, the result will be ∞, or $-\infty$. Other cases will yield a real number other than 0.

Now consider the numerator, $f(a+h) - (f(a) + f'(a)h)$. The first term is the value of the function near a. The second quantity, $f(a) + f'(a)h$, is the linear approximation to f obtained from the tangent line to the graph of f at $(a, f(a))$. Recall that the difference between these two quantities is given by $\alpha(h)h$, where $\alpha(h)$ has the property that it approaches 0 as h goes to 0. It is natural to ask about how rapidly $\alpha(h)$ tends to 0 as h tends to 0. This accounts for why one would consider a limit having $f(a+h) - (f(a) + f'(a)h)$ as its numerator.

What about the denominator? Obviously, the denominator should look like $\alpha(h)h$. One could, therefore, in an experimental way, simply postulate a possible form for $\alpha(h)$. The simplest possible form would have us set $\alpha(h) = h$, which then gives rise to a denominator consisting of h^2.

Having decided to perform this computation, the outcome is almost like finding *nirvana*, because it leads immediately to an approximation of f near a of the form

$$f(a+h) \approx f(a) + f'(a)h + \frac{1}{2}f''(a)h^2$$

that contains the second derivative of f evaluated at a multiplied by h^2. Indeed, it is starting to look like f might be approximately a polynomial function. Given that polynomials are so easy to deal with, this result is just fraught with possibilities and certainly must have created a great deal of excitement in its initial discoverer. $\square$

The example above suggests that a function that is twice differentiable in a neighborhood of a can be represented as a quadratic polynomial involving derivatives evaluated at a. There are a number of ways to pursue this result. First, if a third derivative of f exists, one could attempt to find a representation for f as a cubic polynomial involving three derivatives. This idea is explored in Exercise 2. Second, we could take the approach of the Mean Value Theorem and consider $f(b)$ for b near a as a linear function of the form

$$f(b) = f(a) + K(b-a).$$

Under these circumstances, it turned out that $K = f'(x_0)$ for some $x_0 \in (a, b)$.

We can ask exactly the same question about the quadratic approximation suggested by Example 2. Namely, for

$$f(b) = f(a) + f'(a)(b-a) + \frac{1}{2}K(b-a)^2,$$

what can be said about K? (We have separated out the factor $\frac{1}{2}$ precisely because it turned up in the computation performed in Example 2.)

Example 3. Let f be twice differentiable in the interval (a, b) and continuous on $[a, b]$. If we represent $f(b)$ by

$$f(b) = f(a) + f'(a)(b - a) + \frac{1}{2}K(b - a)^2,$$

what can be said about the value of K?

Solution. Consider the function of t defined by

$$F(t) = f(b) - f(t) - f'(t)(b - t) - \frac{1}{2}K(b - t)^2.$$

Observe that F is continuous on $[a, b]$ and differentiable on (a, b). Further, substitution shows that $F(b) = 0$, while the equation that defines K forces $F(a) = 0$. Thus, the hypothesis of Rolle's Theorem is satisfied. Differentiation yields

$$\begin{aligned} F'(t) &= -f'(t) - f''(t)(b - t) + f'(t) + K(b - t) \\ &= K(b - t) - f''(t)(b - t). \end{aligned}$$

By Rolle's Theorem, there is a $t_0 \in (a, b)$ such that $F'(t_0) = 0$, whence $K = f''(t_0)$. $\square$

Discussion. Consider the quantity

$$f(b) - \left[f(a) + f'(a)(b - a) + \frac{1}{2}K(b - a)^2 \right].$$

There are two possible substitutions into this expression to obtain a function $F(t)$ to which one could apply Rolle's Theorem. The most natural substitution is to replace each occurrence of b by t to obtain

$$G(t) = f(t) - \left[f(a) + f'(a)(t - a) + \frac{1}{2}K(t - a)^2 \right].$$

This looks promising, since $G(b) = 0 = G(a)$, and the remainder of the hypothesis of Rolle's Theorem holds as well. However, the approach falls apart when $G'(t)$ is computed as the reader will show in Exercise 3. Thus, one is forced to the other alternative, namely substituting $t = a$, that leads to our $F(t)$.

Once K has been determined, we have established the existence of $x_0 \in (a, b)$ such that

$$\frac{1}{2}f''(x_0)(b - a)^2 = f(b) - (f(a) + f'(a)(b - a)).$$

From this equation it is easily seen why $\frac{1}{2}f''(x_0)(b - a)^2$ is thought of as giving the error between the linear approximation to f obtained at the point a and evaluated at b, and the function value at b, namely, $f(b)$. What is especially important to notice is this error depends on $(b - a)^2$. Thus, if $f''(x)$ is bounded near a, the error term will be small when $b - a$ is small. An example of the use of this idea is given in Exercise 4. $\square$

Exercises

In the exercises below, you may use whatever facts are required concerning the derivatives of special functions.

1. Evaluate the following limits:

   (a) $\lim\limits_{x \to 0} \dfrac{6 \sin x - 6x + x^3}{2x^2 \ln(1 + x) - 2x^3 + x^4}$;

   (b) $\lim\limits_{x \to 0} \left(\ln \dfrac{1}{x} \right)^{\ln(1-x)}$;

   (c) $\lim\limits_{x \to 0} \dfrac{\sin \ln(1 + x)}{\ln(1 + \sin x)}$;

   (d) $\lim\limits_{x \to 0} \left(\dfrac{x}{e^x - 1} \right)^{\frac{1}{x}}$;

   (e) $\lim\limits_{x \to 0} \dfrac{\sin(x \sin x) - (x \cos x)^2}{x^2}$;

   (f) $\lim\limits_{x \to 1} \tan^2 \left(\dfrac{\pi x^2}{2} \right) (1 + \sec \pi x)$;

   (g) $\lim\limits_{x \to \infty} [(x + 1)^\alpha - x^\alpha], \alpha > 0$;

   (h) $\lim\limits_{x \to 1} \left[\dfrac{x}{x - 1} - \dfrac{1}{\ln x} \right]$;

   (i) $\lim\limits_{x \to 0+} \left(\dfrac{1}{x} \right)^{\sin x}$;

   (j) $\lim\limits_{x \to 1-} x^{\frac{1}{1-x}}$;

   (k) $\lim\limits_{x \to 0+} x^x$;

   (l) $\lim\limits_{x \to \left(\frac{\pi}{2} \right)^-} (\tan x)^{\cos x}$;

   (m) $\lim\limits_{t \to 1} (1 - t) \ln(1 - t^3)$;

   (n) $\lim\limits_{x \to 0} \dfrac{\sinh(\sin x) - \sin(\sinh x)}{x^7}$;

   (o) $\lim\limits_{x \to 0} (1 - \cos x) \cdot \cot(x^2)$;

   (p) $\lim\limits_{x \to 0} \left(\dfrac{1}{e^x - 1} - \dfrac{1}{\sin x} \right)$;

   (q) $\lim\limits_{x \to 0} \dfrac{1 - \cos x^2}{x^3 \sin x}$.

2. Suppose f is thrice differentiable in a neighborhood of a. Repeat the analysis of Example 2 to find an approximation to f that employs $f'''(a)$. In the process, evaluate
$$\lim_{h \to 0} \frac{f(a + h) - \left[f(a) + f'(a)h + \frac{1}{2} f''(a)h^2 \right]}{h^3}.$$

3. Show the remainder of the argument in Example 3 falls apart if one tries to use $G(t)$ as formulated in the discussion following the example.

4. Let $f(x) = \ln x$, $x > 0$. Find a linear estimate to $\ln 1.5$, and calculate the error involved in this estimate. Find an $x_0 \in (1, 1.5)$ that will determine the actual error in the estimate.

5. Let f be thrice differentiable on (c, d) and suppose $a, b \in (c, d)$ with $a < b$. Show there is an $x_0 \in (a, b)$ such that
$$f(b) = f(a) + f'(a)(b - a) + \frac{1}{2} f''(a)(b - a)^2 + \frac{1}{6} f'''(x_0)(b - a)^3.$$

6. Let $f(x) = \ln x$, $x > 0$. Find a quadratic estimate to $\ln 1.5$, and calculate the error involved in this estimate. Find an $x_0 \in (1, 1.5)$ that will determine the actual error in the estimate.

7. Suppose f is four times differentiable in a neighborhood about a. What sort of approximation to f near a can be developed?

8. State and prove a formulation of the result in Exercise 5 for a function having four derivatives.

9. Let f and g satisfy the hypothesis of Cauchy's Mean Value Theorem on (a, ∞). Show if $\lim\limits_{x \to \infty} f(x) = \lim\limits_{x \to \infty} g(x) = 0$, then
$$\lim_{x \to \infty} \frac{f(x)}{g(x)} = \lim_{x \to \infty} \frac{f'(x)}{g'(x)}.$$

10. What is the fallacy in the following computation?

$$\lim_{x \to 2} \frac{3x^2 - 4x - 4}{x^2 - 2x} = \lim_{x \to 2} \frac{6x - 4}{2x - 2} = \lim_{x \to 2} \frac{6}{2} = 3.$$

11. Show $\lim_{x \to 0} \dfrac{x^2 \sin \frac{1}{x}}{\sin x} = 0$. Explain why L'Hospital's Rule does not apply.

12. Fill in all the missing details in Theorem 4.4.1, so as to put it into proper $\epsilon - \delta$ form of proof.

13. State and prove a form of Theorem 4.4.1 that involves infinite limits, that is, the $\frac{\infty}{\infty}$ form of L'Hospital's Rule.

14. Give an example where $\frac{f(x)}{g(x)}$ has a limit as $x \to a$, but $\frac{f'(x)}{g'(x)}$ fails to have a limit as x tends to a.

15. Determine A and B such that $\lim_{x \to 0} \dfrac{A \sin x - x(1 + B \cos x)}{x^3} = 1$

16. Let

$$f(x) = \begin{cases} e^{-1/x^2}, & x \neq 0 \\ 0, & x = 0. \end{cases}$$

Prove f possesses derivatives of all orders, and show further for each $n \in \mathbf{N}, f^{(n)}(0) = 0$.

17. Find $\lim_{x \to \infty} \dfrac{x f'(x)}{f(x}$, given $\lim_{x \to \infty} f(x) = \lim_{x \to \infty} f'(x) = $

$\lim_{x \to \infty} f''(x) = \infty$, and $\lim_{x \to \infty} \dfrac{x f'''(x)}{f''(x)} = k.$

18. Given a circle, center O, radius r, and a tangent line AT; P any point on the circle; and M any point on the tangent at A such that $AM = AP$. Letting MP meet AO at B, find the limiting position of B as P approaches A.

19. Prove the following version of L'Hospital's Rule for sequences: If $\{a_n\}, \{b_n\}$ are sequences of real numbers such that $\{b_n\}$ increases and diverges to $+\infty$, then

$$\lim_{n \to \infty} \frac{a_{n+1} - a_n}{b_{n+1} - b_n} = L \text{ implies } \lim_{n \to \infty} \frac{a_n}{b_n} = L.$$

20. Let $f(x) = x + \sin x \cos x$ and $g(x) = e^{\sin x} f(x)$. Show:

(a) $\lim_{x \to \infty} f(x) = \lim_{x \to \infty} g(x) = +\infty$;   (b) $\lim_{x \to \infty} \dfrac{f'(x)}{g'(x)} = 0$;

(c) $\lim_{x \to \infty} \dfrac{f(x)}{g(x)}$ fails to exist.

Does this contradict L'Hospital's Rule?

21. Suppose f is defined on $[a, b]$, $c \in (a, b)$, and $f''(c)$ exists. Prove

$$\lim_{h \to 0} \frac{f(c+h) - 2f(c) + f(c-h)}{h^2} = f''(c).$$

Give an example where the limit exists, but $f''(c)$ fails to exist.

22. Let $\{r_n\}$ denote the sequence of all rational numbers in the interval $(0, 1)$. Show $\varliminf \{r_n{}^{r_n}\} = e^{-\frac{1}{e}}$ and $\varlimsup \{r_n{}^{r_n}\} = 1$.

23. Suppose f has two continuous derivatives and $f(0) = 0$. If g is defined by

$$g(x) = \begin{cases} f(x)/x, & \text{if } x \neq 0 \\ f'(0), & \text{if } x = 0, \end{cases}$$

prove g has a continuous derivative.

24. Let f be differentiable in (a, ∞). Prove the following:

(a) If $\lim\limits_{x \to \infty} f(x) = 1$ and $\lim\limits_{x \to \infty} f'(x) = k$, then $k = 0$;

(b) If $\lim\limits_{x \to \infty} f'(x) = 1$, then $\lim\limits_{x \to \infty} \dfrac{f(x)}{x} = 1$;

(c) If $\lim\limits_{x \to \infty} f'(x) = 0$, then $\lim\limits_{x \to \infty} \dfrac{f(x)}{x} = 0$.

25. Show there does not exist a polynomial $p(x)$ with coefficients from $\mathbf{R}$ such that for each $n \in \mathbf{N}$, $p(n) = n \ln n$.

Chapter 5

Integration

In this chapter, another fundamental notion called the **integral** of a function is introduced in a formal manner. After discussing the basic properties of integrable functions and various criteria for ensuring the integrability of a function, we will indicate two directions in which the notion can be fruitfully generalized. The integral is defined using the supremum principle, and is realized as the limit of a set of suitable sums, thus ruling out the common misconception that integration is always a reverse process of differentiation. The Fundamental Theorem of Integral Calculus is then established that brings forth the relationship between differential and integral calculus; namely, for a certain class of functions, it turns out that integration is indeed a reverse process of differentiation, in a sense to be made precise later.

The earliest evidences of the so-called integral calculus are to be found in the works of Greek geometers who employed the **Method of Exhaustion** to give a meaning to, and to calculate, areas of plane regions[1] having circular or parabolic boundaries. Centuries later, subsequent to the invention of calculus by Newton and Leibnitz, attention was focused on the inverse character of differentiation and techniques of evaluating both definite and indefinite integrals. But a rigorous and systematic mathematical formulation was first attempted by Riemann for the notion of the definite integral, and this, together with Cauchy's extension to unbounded functions, resulted in a complete and formal expression of the concept of the integral as the limit of a certain sum that, incidentally, justifies the literary meaning of the word. Toward the close of the last century, Stieltjes introduced a broader concept of integration replacing certain linear functions crucial to Riemann's definition by functions of a more general character. The beginning of this century saw the development of the notion of the measure of a set of real numbers that paved the way to the foundations of the modern theory of the Lebesgue Integral, now accepted as a beautiful and inevitable generalization of the Riemann Integral.

---

[1]As we use the word 'region' in this book, it refers to any connected set of points in the plane.

5.1   Motivation for Definition of the Riemann-Darboux Integral

Since the study of the integral began with the geometrical considerations of calculating areas of plane figures, we begin our deliberations with a discussion of area.

One of the prime reasons for developing a notion of area is to provide a means of comparing plane figures. Simply stated, by assigning a number to each plane figure, the notion of area provides an answer to the question: Which is bigger? Quite obviously, not just any number will do. The number chosen for area must have certain properties. For example, it should be well defined, which means that two competent mathematicians will assign the same number to the same figure in all cases. As well, if figure F_1 fits inside figure F_2, then the number assigned as the area of F_1 should be smaller than the number assigned as the area of F_2. (The industrious reader might want to make a list of other important properties that a notion of area should satisfy.) The point of this is that it is by no means a trivial task to come up with a means for assigning a number to each plane figure in such a way that a consistent answer to the question 'which is bigger?' is provided.

If one thinks about this problem, and we encourage the reader to do so, one sees that a simple first cut might consist of creating a small standard figure, perhaps a square, and then counting the number of these which one could fit inside a given figure of unknown area without overlap. Approaching the area in this manner would very quickly convince one that a good solution for the problem of assigning a number to plane figures existed for figures which were rectangles. Indeed, it seems likely that the well-known formula for computing the area of a rectangle would arise from this approach almost immediately. This formula,

$$\text{area of a rectangle} = \text{length} \times \text{width},$$

amounts to a definition. It is a definition in which we have great confidence, in the sense that if we apply it to calculate the number of square tiles required to lay a floor in a rectangular room, we know it will provide us with the correct number. The problem that follows from this is that of finding the correct generalization of this definition which we can apply to other plane figures.

Thus, our point of departure is the familiar concept of the area of a rectangular region, namely, the product of its length and breadth. This concept is abstracted by considering a function defined on the closed interval $[a, b]$ of the real line and that assumes a constant value $k \geq 0$ throughout the interval. In this situation, the graph of the function gives rise to a rectangular region bounded by the x-axis and the ordinates $x = a$ and $x = b$. Obviously, the area enclosed is $k(b - a)$. If further, $[a, b]$ is broken up into smaller intervals by inserting points of division between a and b, say,

$$a = x_0 \leq x_1 \leq \ldots \leq x_{n-1} \leq x_n = b,$$

and if the function f is defined so as to take a constant value at each of the resulting subintervals, say (as for example in Figure 5.1.1),

$$f(x) = k_i \geq 0 \text{ if } x \in [x_{i-1}, x_i), \ i = 1, \ 2, \ \ldots, n, \text{ and } f(b) = k_n,$$

and if d_i denotes the length $(x_i - x_{i-1})$ of the ith subinterval, then we obtain n

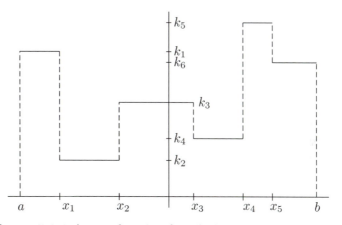

Figure 5.1.1 A step function for which we can exactly compute the area under the graph by decomposing the region into rectangles.

rectangular regions and the total area enclosed by them above the x-axis is the finite sum of the various rectangles, namely,

$$\text{area} = k_1 d_1 + k_2 d_2 + \cdots + k_n d_n.$$

Notice that in this last equation, we have generalized the notion of area. That is, we now are able to compute the area of a figure (see Figure 5.1.1) that is not a rectangle. How? By breaking up the figure into a series of nonoverlapping rectangles that make up the totality of the figure, and summing their respective areas. This is merely the natural abstraction of the same process used by the early geometers.

Since the graph of the function above consists of n different **steps**, such a function is usually called a **step function** and what we just saw was that the area of a region bounded by a nonnegative step function, the vertical lines defined by $x = a$, $x = b$, and the x-axis is just the sum of the areas of a finite number of rectangles resulting from the various steps of the graph.

The analytic task we have set for ourselves is to introduce the notion of an **integral** of a function. This integral will be a mapping from a subcollection of the class of all functions defined on an interval $[a, b]$ into **R**. As well, when the integral is applied to a nonnegative function on $[a, b]$, we will require that the real number obtained from the integral (mapping) will be the area of the region bounded by the graph of f, the vertical lines $x = a$ and $x = b$ and the x-axis. This task will be achieved by approximating the given function by suitable step functions. The area of the region will be then be approximated by the areas enclosed by these step functions, which in turn are obtained as a sum of the areas of nonoverlapping rectangles as described in the computations above. This, then, precisely summarizes the main ideas behind the formal treatment of the integral in the next section.

5.2   Definition of the Riemann-Darboux Integral

We begin by introducing some terminology and basic notions that will be standard throughout this section.

Definition. Let $a, b \in \mathbf{R}$ with $a \leq b$. By a **partition** of the interval $[a, b]$, we mean a finite collection of points $P = \{x_0, x_1, \ldots, x_n\}$, where

$$a = x_0 \leq x_1 \leq \cdots \leq x_n = b.$$

If P and Q are two partitions of $[a, b]$, we call Q a **refinement** of P provided $P \subseteq Q$.

Discussion. The basic idea of a partition is to divide the interval $[a, b]$ into a finite collection of subintervals. Specifically, we have $n + 1$ points of division, with the first point being $x_0 = a$ and the last point being $x_n = b$. The result is to divide $[a, b]$ into n subintervals $[x_{i-1}, x_i]$, where $i = 1, 2, \ldots, n$. In terms of our goal of approximating arbitrary functions with step functions, partitions will play an essential role by delineating the subintervals associated with each given step.

On the technical side, we have allowed a subinterval to consist of only one point, since it is quite possible that $x_{i-1} = x_i$. This amounts to permitting points to be repeated, although as we shall see, this makes no difference to the computations. For a given partition, P, we can produce a refinement Q by adding a finite number of additional points of $[a, b]$ to P. This will in general produce more divisions of the interval, whence the name. Last, given two partitions P and Q, it is not in general the case that one is a refinement of the other. The reader should produce an example. $\square$

Definition. For a given partition $P = \{x_0, x_1, \ldots, x_n\}$ of $[a, b]$, we let

$$d_i = x_i - x_{i-1}, \; i = l, \, 2, \, \cdots, \, n$$

be the **length** of the ith subinterval. Further, we set

$$\|P\| = \max\{d_1, \, d_2, \, \ldots, d_n\}$$

and call this quantity the **norm** of the partition P.

Discussion. It is immediate from the definition of length of a subinterval that (Exercise 1)
$$d_1 + d_2 + \ldots + d_n = b - a.$$

If we refine a partition, P, to obtain a partition, Q, then, clearly,

$$\|Q\| \leq \|P\|.$$

As a general rule, the intent of forming a refinement of P will be to obtain a Q with a strictly smaller norm. However, it is definitely not the case that refining a partition will of necessity result in a reduced norm. On the other hand, it is the case that by taking refinements we can produce a partition of any interval with a norm which is less than any previously assigned positive number δ (**WHY?**). These ideas are illustrated in the following example. $\square$

Example 1. Find a partition of $I = [-3, 4]$ having norm $\frac{1}{10}$.

Solution. The length of this interval is $4 - (-3) = 7$. Hence, a partition of I having norm $\frac{1}{10}$ must have at least 71 members. Fix δ such that $0 < \delta \leq \frac{1}{10}$. Set $x_0 = -3$ and $x_1 = -2.9$. For $i \geq 2$ set

$$x_i = \begin{cases} x_{i-1} + \delta, & \text{if } x_{i-1} + \delta < 4 \\ 4, & \text{otherwise.} \end{cases}$$

Let n be the least i such that $x_i = 4$, and set $P = \{x_0, x_1, \ldots, x_n\}$. It is easily checked that $\|P\| = \frac{1}{10}$. $\square$

Discussion. There are minor details that have been left to the reader. These should be completed. The reader should notice that the simplest possible partition satisfying the requirement is one in which the subintervals are of equal length. This partition would result from setting $\delta = \frac{1}{10}$. $\square$

The motivation that we have used as the foundation on which to develop the integral concept is the problem of finding the area of a region bounded by the graph of a nonnegative function, f, defined on $[a, b]$, the vertical lines given by $x = a$ and $x = b$, and the x-axis; formally, this region is the set of points given by

$$\{(x, y) : a \leq x \leq b \text{ and } 0 \leq y \leq f(x)\}$$

and is pictured in Figure 5.2.1. However, as stated in the outline of our program, it is intended that we be able to compute an integral for other functions as well. In these cases, we will simply not be able to interpret the numerical result as measuring an area. Hence, we proceed by making a definition that for nonnegative functions yields area, but that can be applied to functions taking negative values as well. Thus, we will assume that when area is mentioned in the discussion below, the function being considered is nonnegative.

Let f be a real-valued function defined on $[a, b]$, and let us further assume that f is bounded. By the supremum principle, then, the set of values of f admit a supremum and an infimum. Let

$$M = \sup \{f(x) : x \in [a, b]\}$$

and

$$m = \inf \{f(x) : x \in [a, b]\}.$$

If $P = \{x_0, x_1, \ldots, x_n\}$ is any partition of $[a, b]$, since f is bounded on $[a, b]$, f is also bounded on each of the subintervals $[x_{i-1}, x_i]$, $i = 1, 2, \ldots, n$. Let M_i, m_i, respectively, denote the supremum and the infimum of $f(x)$ in the ith subinterval $[x_{i-1}, x_i]$ (see Figure 5.2.1). These quantities exist by the completeness of $\mathbf{R}$, and they are unique by Theorem 0.4.1. This enables us to define two step functions, s^P and s_P, on $[a, b]$ as follows:

$$s^P(x) = M_i \text{ if } x \in [x_{i-1}, x_i), \ i = 1, 2, \ldots, n$$

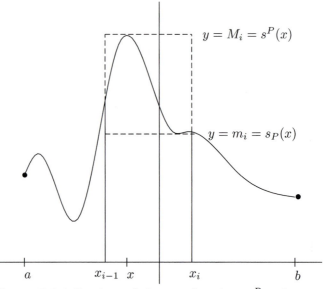

Figure 5.2.1 Portions of the step functions s^P and s_P on the subinterval $[x_{i-1}, x_i]$.

and

$$s_P(x) = m_i \text{ if } x \in [x_{i-1}, x_i), \ i = 1, \ 2, \ \ldots, \ n,$$

with $s^P(b) = M_n$ and $s_P(b) = m_n$. Evidently, the step functions, s^P and s_P, satisfy the fundamental inequality

$$s_P(x) \leq f(x) \leq s^P(x)$$

for all x in $[a, b]$ (see Exercise 3). Thus, the graph of the function, f, has now been 'sandwiched' between the graphs of two step functions, s_P and s^P, as can be seen from inspection of the ith subinterval in Figure 5.2.1. From this picture, it is intuitively obvious that any value we might want to assign as the area of the region under the function, f (see Figure 5.2.1) must lie between the areas obtained from the step functions. (Values for the areas under the step function are known, since we can calculate them as a finite sum of the areas of nonoverlapping rectangles.)

Figure 5.2.1 relates to a nonnegative function. In the general case, the function of interest may take both positive and negative values. In Figures 5.2.2 and 5.2.3, graphs are presented that illustrate how a more general f is sandwiched between the step functions, s_P and s^P. Specifically, in Figure 5.2.2 the function f takes only negative values. Thus, on the ith subinterval we have $s_P(x) = m_i \leq f(x) \leq M_i = s^P(x) \leq 0$. In Figure 5.2.3 we illustrate a subinterval in which the graph of f crosses the x-axis. For such a case, M_i will be positive and m_i will be negative.

Returning now to our main theme, the aim of our program is to associate a real number with each bounded function defined on $[a, b]$. In the discussion related to a nonnegative f, we suggested computing the area determined by

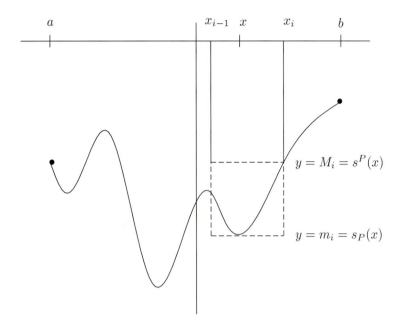

Figure 5.2.2 Portions of the step functions s^P and s_P for an interval over which the function is negative.

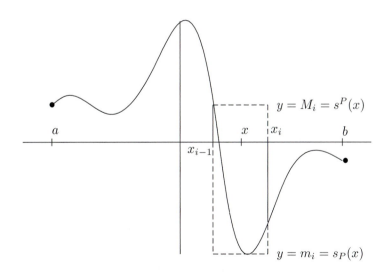

Figure 5.2.3 Portions of the step functions s^P and s_P for an interval over which the function takes both positive and negative values.

each step function. Let us make this notion precise. Thus, in what follows, $P = \{x_0, \ldots, x_n\}$ is a partition of $[a, b]$, f is a bounded function on $[a, b]$, and M_i and m_i are the supremum and infimum on the ith subinterval, respectively.

Definition. By the **upper (Darboux) sum**, of the function, f, corresponding to the partition, P, we mean the number, $\overline{S}(f, P)$, given by

$$\overline{S}(f, P) = M_1 d_1 + M_2 d_2 + \ldots + M_n d_n.$$

Similarly, the **lower (Darboux) sum** of the function, f, corresponding to the partition, P, is given by

$$\underline{S}(f, P) = m_1 d_1 + m_2 d_2 + \ldots + m_n d_n.$$

Discussion. First and foremost, the reader should observe that a Darboux sum is exactly that, a *sum*. We emphasize this point, even though it is so patently obvious. Second, since every partition generates a finite collection of subintervals, every Darboux sum is *finite*. For this reason, a Darboux sum should be thought of as a *finite approximating sum*.

In the case of a general function, there is no universal interpretation that motivates this approximation. However, if f is nonnegative, the numbers that arise as Darboux sums corresponding to a partition, P, are nothing but the areas enclosed by the step functions s^P and s_P, respectively, $x = a$ and $x = b$, and the x-axis. As such, they represent finite approximations to the area of the region determined by f (once again, see Figure 5.2.1).

The dependence of the Darboux sums on the particular function, f, and partition, P, is indicated by the notations, $\overline{S}(f, P)$ and $\underline{S}(f, P)$. Because of the uniqueness of the supremum and infimum of a bounded set of real numbers and the fact that we are dealing with a finite sum, for a particular partition, P, and a fixed bounded function, f, the two numbers $\overline{S}(f, P)$ and $\underline{S}(f, P)$ are unique. This means $\underline{S}$ and $\overline{S}$ are functions. The domain of these functions is the collection of pairs (f, P), where f is bounded on $[a, b]$, and P is a partition of $[a, b]$. While it is important to be able to think about $\underline{S}$ and $\overline{S}$ as functions, and to ask questions about them as functions, all of the intuition about them arises from one's understanding of what these functions do. In this sense, what is important is not the function aspect, but rather that each value returned by the function is an approximating sum over a finite partition.

As a matter of notation, when we are dealing with only one bounded function f, we can dispense with the dependence on f, and simply write $\overline{S}(P)$ and $\underline{S}(P)$ for the upper and lower sums given by P. $\square$

Upper and lower Darboux sums serve as the cornerstone of our development. As such, it will be useful to establish some pertinent facts about them. This we do in the following lemmas that will assume that f and g are bounded functions on $[a, b]$ and that P is a partition of that interval.

Lemma 5.2.1. *If $P = \{a, b\}$, then $\overline{S}(P) = M(b - a)$ and $\underline{S}(P) = m(b - a)$. Further, if P is an arbitrary partition of $[a, b]$, then*

$$m(b - a) \leq \underline{S}(P) \leq \overline{S}(P) \leq M(b - a).$$

Proof. The first statement is left to Exercise 4. Thus, let P be an arbitrary partition of $[a, b]$. Then

$$m \leq m_i \leq M_i \leq M,$$

for $i = 1, \ldots, n$. It follows that

$$m d_i \leq m_i d_i \leq M_i d_i \leq M d_i,$$

for each i. If we now sum the n inequalities for $i = 1, \ldots, n$, we obtain

$$m(b - a) \leq \underline{S}(P) \leq \overline{S}(P) \leq M(b - a)$$

as desired. $\square$

Discussion. This lemma establishes that for a fixed bounded function, f, the collection of all upper sums as well as the collection of all lower sums over f is bounded below by $m(b-a)$ and bounded above by $M(b-a)$. This fact is essential to the development, and it is the reason why f is required to be bounded on $[a, b]$. $\square$

Lemma 5.2.2. *Let f and g be defined on $[a, b]$ and P be a partition of $[a, b]$. Then*

(i) $\overline{S}(af, P) = a\overline{S}(f, P)$ *for any real constant $a \geq 0$;*

(ii) $\overline{S}(af, P) = a\underline{S}(f, P)$ *for any real constant $a < 0$;*

(iii) $|\overline{S}(f, P)| \leq \overline{S}(|f|, P)$;

(iv) $\overline{S}(f, P) + \overline{S}(g, P) \geq \overline{S}(f + g, P)$.

Proof. We prove (iii) and leave the rest to Exercise 5. Consider the ith subinterval. For all $x \in [x_{i-1}, x_i]$, $-|f|(x) = -|f(x)| \leq f(x) \leq |f(x)| = |f|(x)$, by definition of $|f|$. Thus,

$$-M_{i,|f|} \leq M_i = \sup\{f(x) : x \in [x_{i-1}, x_i]\}$$
$$\leq \sup\{|f(x)| : x \in [x_{i-1}, x_i]\} = M_{i,|f|},$$

where $M_{i,|f|}$ is the supremum of $|f|$ over the ith subinterval, whence

$$-M_{i,|f|} \times d_i \leq M_i \times d_i \leq M_{i,|f|} \times d_i.$$

But this means that

$$|\overline{S}(f, P)| = \left| \sum_{i=1}^{n} M_i \times d_i \right| \leq \sum_{i=1}^{n} |M_i \times d_i|$$
$$\leq \sum_{i=1}^{n} M_{i,|f|} \times d_i = \overline{S}(|f|, P),$$

which is the desired inequality. $\square$

Discussion. This proof works because we are dealing with finite sums. Thus, the basic manipulations all reduce to versions of theorems proved in Chapter 0. For example, the final inequality, whose primary content is

$$|\overline{S}(f,P)| = \left|\sum_{i=1}^{n} M_i \times d_i\right| \leq \sum_{i=1}^{n} |M_i \times d_i|,$$

is merely a version of the Triangle inequality.

The only infinite process involved in the computation is the use of the Supremum Principle to obtain the various M_i's. However, this does not affect the computations, since it acts merely to produce the finite list of numbers that are to be manipulated.

It is critical that the reader understand that it is only by keeping the approximating sums finite that we are able to perform these computations. This requirement is the reason why we insist that a partition be a finite list of numbers from $[a, b]$. $\square$

For various partitions P of $[a, b]$, and a fixed bounded function, f, on $[a, b]$, we set

$$\{\overline{S}(P)\} = \{\overline{S}(f,P)\} = \{\overline{S}(f,P) : P \text{ is a partition of } [a,b]\}$$

and

$$\{\underline{S}(P)\} = \{\underline{S}(f,P)\} = \{\underline{S}(f,P) : P \text{ is a partition of } [a,b]\}$$

and note that these sets of real numbers are both bounded above by $M(b-a)$ and below by $m(b-a)$. Hence, by the Supremum Principle, we may make the following definition.

Definition. Let f be bounded on $[a, b]$. Then the number given by $\inf\{\overline{S}(f,P)\}$ is called the **Upper Darboux Integral** of f on $[a, b]$ and is denoted by $\overline{\int}_a^b f$. Similarly, the number given by $\sup\{\underline{S}(f,P)\}$ is called the **Lower Darboux Integral** of f on $[a, b]$ and is denoted by $\underline{\int}_a^b f$. Further, if

$$\overline{\int}_a^b f = \underline{\int}_a^b f,$$

then the common value is called the **Riemann-Darboux Integral** of the function f on the interval $[a, b]$ and is denoted by $\int_a^b f$. Moreover, when the Riemann-Darboux Integral of f exists, we will say that f is **Riemann-Darboux integrable** (or **R-D integrable**) on $[a, b]$.

Discussion. Let us review the development leading to this definition. First, the function f is assumed to be bounded, so that we are able to invoke the Supremum Principle over any subset of the range of f. Second, we produce two step functions by using the bounds on f over various subintervals; these step functions sandwich the function, f, between them. Third, we associate with each step function a number that can be thought of as the value of an approximating sum. For the step function above f, this number is referred to as the upper sum. For the step function below f, this number is referred to as the lower sum. Fourth, we find the infimum of all the upper sums and the supremum of all the lower sums. This process defines the upper and lower Darboux integrals. Finally, we observe that in some cases the upper and lower Darboux integrals generate the same number. In this last case, we say that the function is Riemann-Darboux integrable on $[a, b]$.

The reader will observe that this achieves our intent of developing a process that associates a real number with some subclass of the functions defined on $[a, b]$. We have generated such a process, but a number of questions are outstanding. The most obvious of these is related to characterizing, in a simple manner, those functions that are Riemann-Darboux integrable.

The other portion of our intent related to the concept of area. Thus, consider a nonnegative function, f, on $[a, b]$ such as that shown in Figure 5.2.1. As we have already noted, for each partition, P, any number that we might choose to assign as the area under f must satisfy

$$\underline{S}(P) \leq \text{ the area under } f \leq \overline{S}(P).$$

The argument for this assertion is *geometric*, not analytic. It is based on our belief that any sensible notion of area must satisfy the condition that if a plane figure F_1 is enclosed in a plane figure F_2, then the areal measure of F_1 should be no more than the measure of F_2. In this case, the region defined by s^P on $[a, b]$ encloses the region determined by f, as shown in Figure 5.2.1. Moreover, we know from sound geometric principles how to obtain the number corresponding to the area under s^P, namely, calculate $\overline{S}(P)$. Similar reasoning establishes the left-hand side of inequality. Since this argument holds for every partition, we can assert that the area under f must satisfy

$$\sup \underline{S}(P) \leq \text{ the area under } f \leq \inf \overline{S}(P).$$

For this reason, under the assumption that the supremum and infimum are the same, we would be confident in *defining* the value produced by the integral to be the area under f. In making this definition, we would certainly want to perform a variety of checks to ensure that the definition was consistent with what we had been trying to achieve. Thus, we would want to verify that all the important features of our intrinsic notion of area had been captured. Suffice it to say that all such checks validate this approach to the problem of defining area.

Most important, the reader should come to grips with the way the notion of integral takes our knowledge of how to solve a problem in a very simple case, namely, finding the area of a rectangle, and extends it to more general situations by forming finite approximating sums followed by applying a limiting process to the collection of approximations. This idea is the essence of all applications of the integral to real situations. □

Note. The notations introduced above will be standard for the remainder of this chapter.

We now generate some of the basic facts about the integration process.

Theorem 5.2.1. *If f is a bounded function defined on $[a, b]$ and Q is a refinement of a partition P of $[a, b]$, then $\underline{S}(P) \leq \underline{S}(Q) \leq \overline{S}(Q) \leq \overline{S}(P)$. Moreover, if Q was obtained by adjoining at most k more points to the partition P, then,*

$$|\overline{S}(P) - \overline{S}(Q)| \leq 2kM\|P\| \quad and \quad |\underline{S}(Q) - \underline{S}(P)| \leq 2kM\|P\|,$$

where $M = \sup\{|f(x)| : x \in [a,b]\}$.

Proof. Let $P = \{x_0, x_1, \dots, x_n\}$ and let us first assume that Q is obtained by adjoining one more point y to P, say, between x_{r-1} and x_r. Let d' and d'' denote the lengths of the new subintervals $[x_{r-1}, y]$ and $[y, x_r]$ so generated, and let M' and M'' be the suprema of f in these subintervals, respectively. Then M_r is never less than M' or M''. Now,

$$
\begin{aligned}
\overline{S}(Q) - \overline{S}(P) &= M'd + M''d'' - M_r d_r \text{ (since all other terms cancel)} \\
&= (M' - M_r)d' + (M'' - M_r)d'', \text{ (since } d_r = d' + d''), \\
&\le 0,
\end{aligned}
$$

proving $\overline{S}(Q) \le \overline{S}(P)$. Similarly, considering the infima of f in these subintervals, the reader can prove that $\underline{S}(P) \le \underline{S}(Q)$.

Next, if Q is *any* refinement of P, obtained by adding k more points, we construct a succession of k partitions $Q_1, Q_2, \dots, Q_k = Q$, where each refinement Q_i contains just one more point than the preceding. It now follows that

$$\overline{S}(Q) \le \overline{S}(Q_{k-1}) \le \dots \le \overline{S}(Q_1) \le \overline{S}(P),$$

and a similar result for lower sums.

The second assertion follows by an easy induction on k, the number of points adjoined to obtain Q from the partition P. For $k = 1$,

$$
\begin{aligned}
|\overline{S}(P) - \overline{S}(Q)| &= |(M' - M_r)d' + (M'' - M_r)d''| \\
&\le (M + M)d' + (M + M)d'' \\
&= 2M(d' + d'') \le 2M\|P\|.
\end{aligned}
$$

We leave the reader to complete this induction argument (Exercise 6). $\square$

Discussion. Only two facts are used in this proof. First, if an interval is dissected into two pieces by an intermediate point, the sum of the lengths of the two subintervals generated is exactly equal to the length of the original interval. Second, we have used the fact that the supremum of a superset is greater than or equal to that of the subset. $\square$

In the next theorem, it is shown that regardless of the partition involved, no lower sum can exceed any upper sum of any partition whatsoever.

Theorem 5.2.2. *If f is a function bounded in $[a, b]$ and if P and Q are any two partitions of $[a, b]$, then $\underline{S}(Q) \le \overline{S}(P)$.*

Proof. We consider the partition $R = P \cup Q$ that is clearly a refinement of P, as well as Q. By Theorem 5.2.1, we have

$$\underline{S}(Q) \le \underline{S}(R) \le \overline{S}(R) \le \overline{S}(P). \qquad \square$$

Theorem 5.2.3. *Let f be a bounded function defined on $[a, b]$. Then*

$$\int_{\underline{a}}^{b} f \le \overline{\int_a^b} f.$$

Proof. Let P be an arbitrary partition of $[a, b]$. Then for all partitions Q, $\underline{S}(Q) \leq \overline{S}(P)$, whence $\underline{\int}_a^b f \leq \overline{S}(P)$. Since P was an arbitrary partition of $[a, b]$, we have $\underline{\int}_a^b f \leq \overline{\int}_a^b f$ as desired. $\square$

Discussion. The content of these two results merely spells out in analytic form what is obvious from Figures 5.2.1 through 5.2.3, namely, that every lower sum is less than or equal to every upper sum, regardless of the partitions involved. After all, f sits between the two types of step functions, s_Q and s^P.

The simplicity of the arguments above is another illustration of the essential power of the supremum concept. $\square$

Let us now turn to some examples to illustrate the nature of integrability.

Example 2. Discuss the Riemann-Darboux integrability of the function f defined on $[a, b]$ by $f(x) = k$ (a constant) for all x.

Solution. For any partition, P, $M_r = m_r = k$ on any subinterval; hence

$$\overline{S}(P) = kd_1 + kd_2 + \cdots + kd_n = k(b - a).$$

Similarly, $\underline{S}(P) = k(b - a)$. Since P was arbitrary, it follows that $\underline{\int}_a^b f = \overline{\int}_a^b f = k(b - a)$, whence f is R-D integrable on $[a, b]$ and $\int_a^b f = k(b - a)$. $\square$

Discussion. The computations are straightforward. If k is positive, the region bounded by the graph of f, the vertical lines $x = a$ and $x = b$, and the x-axis is a rectangle. We have asserted that for nonnegative functions, integration produces area. For this function, we already know how to calculate the area, that is, length times width, which analytically is $k(b - a)$. $\square$

Example 3. Discuss the Riemann-Darboux integrability of f defined on $[a, b]$ by

$$f(x) = \begin{cases} k > 0, & \text{if } x \in [a, b] \text{ and } x \neq c \in [a, b] \\ 0, & \text{if } x = c. \end{cases}$$

Solution. Let us show that f is integrable. For any partition P, we will have that the components of the upper sum agree with the components of the lower sum except on the interval(s) that contain c (there can be at most two of these with nonzero lengths). Suppose that $c \in [x_{i-1}, x_i]$. Then the contribution arising from this subinterval to $\underline{S}(P)$ will be 0, whereas the contribution to $\overline{S}(P)$ will be kd_i, since $k > 0$. It follows that

$$\overline{S}(P) \leq \underline{S}(P) + 2k\|P\|.$$

Since $\|P\|$ can be made as small as we want, we conclude that for any positive ϵ we can choose a partition, P, such that

$$|\overline{S}(P) - \underline{S}(P)| \leq \epsilon.$$

Since

$$\left| \underline{\int}_a^b f - \overline{\int}_a^b f \right| \leq |\overline{S}(P) - \underline{S}(P)|,$$

this forces $\underline{\int}_a^b f = \overline{\int}_a^b f$, whence f is R-D integrable. $\square$

Discussion. The purpose of these examples is to develop some intuition about how and why functions are integrable. Consider the present example in relation to the approximation process. This process works by sandwiching f between the two step functions, s_P and s^P. Integrability is achieved when we can sandwich so closely that

$$\overline{S}(P) - \underline{S}(P) \text{ can be made as small as we please.}$$

In the present example, the sandwiching is perfect on all subintervals that do not have c as a member. For any partition, there are at most two of these, which we identify by the subscripts $i, i+1$. On any other subinterval, say, $[x_{j-1}, x_j]$, we have that $m_j = M_j = k$. It follows that the contribution to the jth subinterval is the same for both the upper and lower sum. Equivalently, we have,

$$(M_j - m_j) \times d_j = 0.$$

Now,

$$\overline{S}(P) - \underline{S}(P) = \sum_{k=0}^{n} (M_k - m_k) \times d_k.$$

Thus, the contribution to $\overline{S}(P) - \underline{S}(P)$ from the jth subinterval is small precisely because $M_j - m_j$ is small, in fact 0.

On the other hand, if we turn to either of the remaining two subintervals, the contribution to $\overline{S}(P) - \underline{S}(P)$ from either will be small exactly if d_i and/or d_{i+1} is small. These can be made arbitrarily small by choosing the norm of P to be small, and this is the method used to conclude the argument. The essential features of this discussion are pictured in Figure 5.2.4.

We did not compute the value of the integral. However, it is straightforward to see that $\int_a^b f = k(b-a)$, which is the same as that for the function in Example 2, even though the functions differ by changing the value of the first function at a single point. This suggests that by altering the value of the function at a point in $[a, b]$ (or even at a finite number of points), the value of the integral and the integrability of the function are unchanged. These facts are dealt with later in this section.

Implicit in the suggestions above is the idea that to be integrable, a function does not necessarily have to be continuous. This example demonstrates that fact, and the reader should study carefully the methods used to deal with the lack of continuity. At a deeper level, the reader should try to come to grips with what this suggests. Specifically, what functions are integrable? $\square$

Example 4. Discuss the integrability of the function f defined on $[a.b]$ by

$$f(x) = \begin{cases} 0, & \text{if } x \text{ is irrational} \\ 1, & \text{if } x \text{ is rational.} \end{cases}$$

Solution. Let P be an arbitrary partition of $[a, b]$. Evidently, the ith subinterval of this partition must contain an infinite number of rational and irrational points.

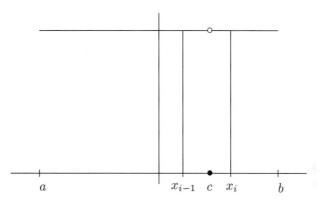

Figure 5.2.4 Only the area above $[x_{i-1}, x_i]$ contributes to the quantity $\overline{S}(P) - \underline{S}(P)$.

So $M_i = 1$ and $m_i = 0$. But this holds for every i. Thus, for any partition P of $[a, b]$, $\overline{S}(P) = b - a$ and $\underline{S}(P) = 0$. In consequence,

$$\underline{\int_a^b} f = 0, \ \text{while} \ \overline{\int_a^b} f = b - a,$$

whence f is not R-D integrable in $[a, b]$. □

Discussion. First and foremost, this example establishes that not all functions that are bounded on $[a, b]$ are integrable there. Further, it is apparent that continuity, or its lack, should play a role in the characterization of R-D integrable functions.

Given this comment, it is essential that the reader understand the difference in the situation of Example 3 versus Example 4. As suggested in the discussion following Example 3, a function will be integrable exactly if we can make the quantity $\overline{S}(f, P) - \underline{S}(f, P)$ as small as we please by a careful selection of the partition. The basic tool that we have available for manipulation in arguments concerning the integral is the norm of P. This is the quantity that we adjust, and it plays the same role as the δ in arguments concerning limit and continuity. In Example 3, we were able to establish the integrability of f precisely because we were able to use the norm of P to bound the difference between the upper and the lower integrals. This was accomplished by using the fact that there was only one point at which we had a problem, namely, c, and that this resulted in a difference between the sums of at most $2k\|P\|$. Further, this quantity "approached" 0 as the norm of P "approached" 0. The question that arises is why we cannot do the same thing in the present case.

Consider now the situation of Example 4. We would like to construct an argument of the type given in Example 3 to show that f (of Example 4) is integrable. But this time, no matter how small the norm of P, each subinterval will still contain points of discontinuity of the function. Thus, no matter how small we choose the norm of the partition, on the ith subinterval, the difference between M_i and m_i will not be small. Indeed, it will be constant and will contribute

$$(M_i - m_i) \times d_i = d_i$$

to the quantity $\overline{S}(f, P) - \underline{S}(f, P)$. Thus, the resulting difference between an upper sum and a lower sum will also be constant; indeed, it will be $b - a$. As a result, the upper and lower integrals differ, and f is not integrable.

The essential point here is that since the d_i's always sum to $b - a$, we cannot use the length as the factor that makes $(M_i - m_i) \times d_i$ small in all instances. Geometrically, the reader should observe that no matter what the choice of P, s^P and s_P do not sandwich the function of this example closely. $\square$

Our next theorem formalizes some of the points made in the discussion above.

Theorem 5.2.4. *A bounded function f defined on $[a, b]$ is Riemann-Darboux integrable on $[a, b]$ if and only if for every $\epsilon > 0$, there is a partition P of $[a, b]$ such that $\overline{S}(P) - \underline{S}(P) < \epsilon$.*

Proof. Let f be integrable. Then by Theorem 0.4.2, given $\epsilon > 0$, there is a partition, P, of $[a, b]$ such that

$$\int_{\underline{a}}^{b} f - \frac{\epsilon}{2} < \underline{S}(P) < \overline{S}(P) < \overline{\int}_{a}^{b} f + \frac{\epsilon}{2}.$$

The desired inequality now follows. Conversely, suppose for each $\epsilon > 0$ there is a partition P such that $\overline{S}(P) - \underline{S}(P) < \epsilon$. Since $\underline{S}(P) \leq \int_{\underline{a}}^{b} f \leq \overline{\int}_{a}^{b} f \leq \overline{S}(P)$, the result follows. $\square$

Discussion. This is the theorem we will often apply to check the integrability of a function. The tool for obtaining the desired partition will be the clever manipulation of the norm; specifically, we will make the norm small, as in Example 3. In general, if ϵ is small, the norm of P will have to be small as well to guarantee that the difference $\overline{S}(P) - \underline{S}(P) < \epsilon$. The reader can check that the function of Example 4 cannot be made to satisfy the theorem for $\epsilon = 1$. $\square$

As we have suggested, it is of real importance to identify the class of integrable functions on a given interval $[a, b]$. By our basic assumption, they must all be bounded, but as Example 4 shows, this is by no means sufficient. Our next theorem shows that a continuous function will be integrable, but again as Example 3 shows, this is not a necessary condition.

Theorem 5.2.5. *If f is continuous on $[a, b]$, then f is Riemann-Darboux integrable on $[a, b]$.*

Proof. By Theorem 3.6.2, f is uniformly continuous on $[a, b]$. Thus, given $\epsilon > 0$, we can find a $\delta > 0$ such that for all $x, y \in [a, b]$, with $|x - y| < \delta$, it is true that $f(x) - f(y) < \frac{\epsilon}{(b-a)}$. Now, choose a partition, P, with norm less that δ. The continuity of f guarantees that on each subinterval, $[x_{i-1}, x_i]$, of P, the infimum, m_i, and the supremum, M_i, of f are actually attained at points, say, y_i, z_i, in that subinterval. Since $y_i - z_i < \delta$, we have $M_i - m_i < \frac{\epsilon}{(b-a)}$. This being the case for $i = 1, 2, \ldots, n$, we get

$$
\begin{aligned}
\overline{S}(P) - \underline{S}(P) &= \sum_{i=1}^{n} M_i d_i - \sum_{i=1}^{n} m_i d_i \\
&= \sum_{i=1}^{n} (M_i - m_i) d_i \\
&\leq \frac{\epsilon}{(b-a)}(b-a) \qquad \textbf{(WHY?)} \\
&= \epsilon
\end{aligned}
$$

whence Theorem 5.2.4 guarantees that f is R-D integrable. $\square$

Discussion. We urge the reader to observe the way continuity and Theorem 5.2.4 have been melded in this argument. The key is the fact that a continuous function on a closed interval, $[a, b]$, is uniformly continuous there. For this reason, we obtain a uniform bound, on $|f(y_i) - f(z_i)|$. This bound has the extraordinary property that it can be made as small as we please provided y_i and z_i are sufficiently close. The required closeness can be ensured by the single step of making the norm of P sufficiently small. It is the existence of this uniform, arbitrarily small, bound that permits us to replace $M_i - m_i$ by a fixed, arbitrarily small, quantity that may then be factored out of the sum, as in the following line from the proof,

$$
\sum_{i=1}^{n} (M_i - m_i) d_i \leq \frac{\epsilon}{(b-a)}(b-a).
$$

Notice that if the uniform bound could not be made arbitrarily small, we would be forced back into the situation of Example 4, a situation where R-D integrability failed. $\square$

Example 5. Discuss the integrability of the function f on the interval $[0, 1]$ defined by

$$
f(x) = \begin{cases} 0, & \text{if } x \in \mathbf{R} \sim \mathbf{Q} \\ \frac{1}{q}, & \text{if } x = \frac{p}{q} \in \mathbf{Q} \text{ in lowest terms.} \end{cases}
$$

Solution. We will show that f is integrable. Let $\epsilon > 0$ be given and let N be the least integer such that $\frac{1}{N} < \frac{\epsilon}{2}$. Let k denote the number of rationals $\frac{p}{q} \in [0, 1]$ with denominator q less than N. Observing that k is finite, we can enclose these finite number of points in subintervals of total length less than $\frac{\epsilon}{2}$ by choosing a partition, P, such that $\|P\| < \frac{\epsilon}{2k}$. Since f is bounded by 1, for such a partition, the total contribution to the difference $\overline{S}(P) - \underline{S}(P)$ from the subintervals containing these k points cannot exceed $\frac{\epsilon}{2}$. Further, on the intervals not containing any of these k points, $m_i = 0$, while $M_i \leq \frac{1}{N} < \frac{\epsilon}{2}$, whence the total contribution to $\overline{S}(P) - \underline{S}(P)$ from these subintervals is less than $\frac{\epsilon}{2}$. It is now immediate that

$$
\overline{S}(P) - \underline{S}(P) < \frac{\epsilon}{2} + \frac{\epsilon}{2} = \epsilon
$$

as desired. $\square$

Discussion. In Exercise 2.5.1(h), the reader was asked to discuss the continuity properties of f. The conclusion was that f was continuous on the irrationals and discontinuous on the rationals. Thus, this function possesses a countable dense set of discontinuities, in the sense that the closure of the set containing the points of discontinuity consists of the entire domain, $[0, 1]$. With this in mind, and recalling our earlier remarks after Example 4, it seems remarkable that f should be integrable. We should find out why.

There is a fundamental intuitive difference between the functions of Examples 4 and 5 that can be seen from Figures 5.2.5(a) and 5.2.5(b). If we draw a horizontal line given by $1 > y = c > 0$ across the graph of f (dashed line in Figure 5.2.5(a)), we see that an infinite number of points of the graph lie above this line. On the other hand, this is not the case for the function of Example 5 as Figure 5.2.5(b) shows; in this case only a finite number of points of the graph lie above any line defined by $y = c > 0$ (Exercise 10). This difference is the key. Here's why.

Our problem is to make $\overline{S}(P) - \underline{S}(P)$ small, in either case. To achieve this goal, we must make the contribution from each subinterval, $(M_i - m_i) \times d_i$ small. For the function of Example 4, we saw that $M_i - m_i$ had a fixed value of 1. Thus the only way to make the product small, was to have d_i small. However, this could not suffice to solve the problem because the sum of the d_i's was always the length of the interval $[a, b]$.

In Example 5, every subinterval contains points of discontinuity. Now suppose we choose a partition where $\|P\|$ is very small. The only way this can happen is if n is very large, where $P = \{x_0, \dots, x_n\}$. Relatively speaking, there will only be a few points, and hence a few subintervals, for which M_i, and hence $M_i - m_i$, is large. Thus, the contribution of most subintervals to the sum $\overline{S}(P) - \underline{S}(P)$ will be small, not because $\|P\|$ is small, but because $M_i - m_i$ is small! Since we can make the norm arbitrarily small, the contribution to the sum from the few subintervals where M_i is large can be made arbitrarily small.

The argument can also be approached geometrically, as follows. First, observe that for any partition, P, $s_P = 0$ on $[0, 1]$, whence, $\underline{S}(P) = 0$. Next we observe that the area under a line of constant height, $\frac{1}{N}$, in the unit square is exactly $\frac{1}{N}$. Thus, if we have all the functional values (or all but a finite number) trapped in the strip of height $\frac{1}{N}$, then $\overline{S}(P)$ should be near $\frac{1}{N}$ provided that P is suitably chosen. This is the intuition that the argument implements and similar intuitions are the heart of many arguments concerning integration. □

The following theorem yields another rich class of integrable functions.

Theorem 5.2.6. *A monotonic function defined on an interval $[a, b]$ is R-D Integrable.*

Proof. Let us assume that f is monotonically increasing and fix $\epsilon > 0$. Choose a partition P such that $\|P\| < \frac{\epsilon}{f(b) - f(a)}$ (the case where $f(b) - f(a) = 0$ is trivial). By monotonicity, we have that $M_{i-1} = m_i$ so that $\sum (M_i - m_i) = f(b) - f(a)$.

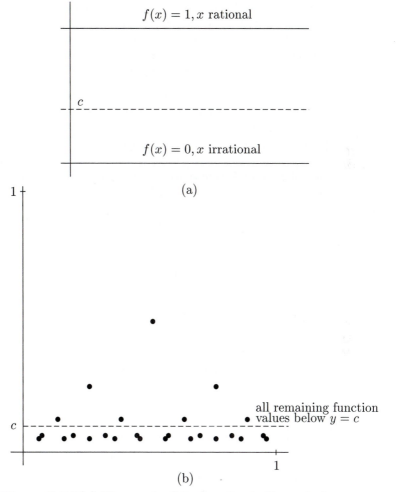

Figure 5.2.5(a) The graph of the function in Example 4.
(b) The graph of the 'ruler' function in Example 5. Only a few points are plotted. The points get really crowded near the x-axis.

Now,

$$\overline{S}(P) - \underline{S}(P) = \sum_{i=1}^{n}(M_i - m_i)d_i$$

$$< \left(\sum_{i=1}^{n}(M_i - m_i)\right) \cdot \frac{\epsilon}{f(b) - f(a)}$$

$$= (f(b) - f(a)) \cdot \frac{\epsilon}{f(b) - f(a)} = \epsilon \qquad (\textbf{WHY?}),$$

whence f is integrable. $\square$

Discussion. As in Theorem 5.2.5 and Example 5, the strategy here is to come up with a partition P that satisfies the condition of Theorem 5.2.4. The monotonicity of the function f guarantees that the maximum and the minimum values occur at the two endpoints of each subinterval. So, if we choose a partition with all subintervals having an equal length, say d, then since $\sum(M_i - m_i) = f(b) - f(a)$ (see Exercise 11), it turns out that $\overline{S}(P) - \underline{S}(P) = d \times [f(b) - f(a)]$. This justifies our choice of the quantity d to be less than $\frac{\epsilon}{f(b)-f(a)}$. $\square$

Theorems 5.2.5 and 5.2.6 together generate a large class of integrable functions. They do not, however, characterize the class of all integrable functions. It might seem that there is no connection between the two theorems, but this is not the case. It is a fact that a monotone function can have at most a countable set of discontinuities; thus, a theorem that showed that functions having at most a countable set of discontinuities were integrable would subsume Theorem 5.2.6. Such a theorem is plausible and is in fact the case. However, at this moment, we do not have techniques of sufficient delicacy to deal with this situation. To substantially improve on the results in this section requires consideration of the nature of the set of discontinuities of the function in question. Such considerations are the subject of Lebesgue integration.

We close this section with a few more improvements on the class of integrable functions, the proofs of which are left as exercises.

Theorem 5.2.7. *If f is R-D integrable on $[a, b]$ and g is obtained by altering the values of f at a finite number of points, then g is R-D integrable in $[a, b]$; furthermore, $\int_a^b f = \int_a^b g$.*

Corollary. *A bounded function on a closed interval $[a, b]$ whose set of discontinuities is finite is R-D integrable on $[a, b]$.*

Exercises

1. Verify for an arbitrary partition P of $[a, b]$, $\displaystyle\sum_{i=1}^{n} d_i = b - a$.

2. Find a partition of $[-e, \pi]$ such that $\|P\|$ is less than
   (a) $\frac{1}{20000}$;   (b) $\frac{1}{\pi}$;   (c) $\frac{1}{e^5}$;
   where e and π have their usual meanings.

3. Let f be bounded on $[a, b]$ and P be an arbitrary partition of $[a, b]$. Verify that $s_P(x) \leq f(x) \leq s^P(x)$ for all $x \in [a, b]$.

4. Supply a proof for the first part of Lemma 5.2.1.

5. Establish the relations stated in parts (i), (ii), and (iv) of Lemma 5.2.2.

6. Supply the missing proof for $\underline{S}(P) \leq \underline{S}(Q)$ in Theorem 5.2.1. Also, complete the induction argument in the latter half of the proof, for $\overline{S}(P)$ as well as $\underline{S}(P)$.

7. State and prove results corresponding to those in Lemma 5.2.2 (a), (c), and (d) for $\underline{S}(P)$.

8. Give an example of a function defined on $[0, 1]$ and a partition, P, such that $|\overline{S}(f, P)| < \overline{S}(|f|, P)$.

9. Give an example of functions, f and g, defined on $[0,1]$ and a partition, P, such that $\overline{S}(f+g, P) < \overline{S}(f, P) + \overline{S}(g, P)$.

10. Consider the function of Example 5, and let k denote the number of rationals in lowest terms in $[0,1]$ having denominator less than N. Show $k \leq \frac{N^2 - 3N + 6}{2}$.

11. Let f be monotone increasing on $[a, b]$. Show

$$\sum_{n=1}^{n} (M_i - m_i) = f(b) - f(a).$$

How will this result change if f is monotone decreasing?

12. Show $\int_a^a f = 0$, where a is in the domain of f.

13. Evaluate $\int_0^a [x]$.

14. Evaluate $\int_0^a \lceil x \rceil$, where $\lceil x \rceil$ is the least integer greater than or equal to x.

15. Evaluate $\int_0^n f$, where $f(x) = \frac{(-1)^n}{n}$, if $x \in [n-1, n)$.

16. Evaluate $\int_{-1}^{1} \text{sgn}$.

17. Evaluate $\int_0^1 \frac{1}{n} [nx^2]$.

18. Evaluate $\int_0^1 \frac{1}{n^2} [nx^2]$.

19. Prove that a function that is constant except at a finite number of points of a closed interval is R-D integrable in that interval. What is the value of the integral?

20. Discuss the integrability of the following functions in the intervals indicated, using Theorem 5.2.4. Also evaluate the integral of those functions that are R-D integrable.

(a) for $x \in [-M, M]$, $f(x) = \begin{cases} a, & \text{if } x \in \mathbf{Z} \\ b, & \text{otherwise.} \end{cases}$

(b) for $x \in [0,1]$, $f(x) = \begin{cases} 2^{-n}, & \text{if } x \in (2^{-n-1}, 2^{-n}], n \in \mathbf{N} \\ 0, & x = 0 \end{cases}$

(c) for $x \in [0,1]$ and $a \in \mathbf{N}$, $a \geq 2$, $f(x) = \begin{cases} a^{1-n}, & \text{if } x \in (a^{-n}, a^{1-n}], n \in \mathbf{N}, \\ 0, & x = 0; \end{cases}$

(d) for $x \in [0,1]$, $f(x) = \begin{cases} x, & \text{if } x \in \mathbf{Q} \\ 1 - x, & x \in (\mathbf{R} \sim \mathbf{Q}); \end{cases}$

(e) for $x \in [0,1]$, $f(x) = \begin{cases} (-1)^{n-1}, & \text{if } x \in \frac{1}{n+1} < x \leq \frac{1}{n}\, n \in \mathbf{N} \\ 0, & x = 0; \end{cases}$

(f) for $x \in [0,1]$, $f(x) = \begin{cases} \frac{1}{n}, & \text{if } x \in \frac{1}{n+1} < x \leq \frac{1}{n}\, n \in \mathbf{N} \\ 0, & x = 0; \end{cases}$

(g) for $x \in [0,1]$, $f(x) = \begin{cases} 2nx, & \text{if } x \in \frac{1}{n+1} < x \leq \frac{1}{n}\, n \in \mathbf{N} \\ 0, & x = 0; \end{cases}$

(h) for $x \in [0,2]$, $f(x) = \begin{cases} 1, & \text{if } x \in (\mathbf{R} \sim \mathbf{Q}) \cap [0,1] \\ \frac{q-2}{q}, & x = \frac{p}{q} \in \mathbf{Q} \text{ in lowest terms;} \end{cases}$

(i) for $x \in [0,4]$, $f(x) = \begin{cases} x, & \text{if } x \in \mathbf{Q} \\ -x, & \text{if } x \notin \mathbf{Q}; \end{cases}$

(j) for $x \in [0, 1]$, $f(x) = \begin{cases} 0, & \text{if } x = \frac{n}{n+1} \text{ or } x = \frac{n+1}{n} \\ 1, & \text{otherwise.} \end{cases}$

21. Discuss the integrability of the function f in $[0, 1]$, where

$$f(x) = \begin{cases} \sin \frac{1}{x}, & \text{if } x \in (0, 1] \\ 0, & x = 0. \end{cases}$$

22. Prove Theorem 5.2.7 and its corollary.

23. If the word 'finite' is replaced by 'countable' in the statement of Theorem 5.2.7, do we get a theorem? Prove or give a counterexample.

24. Show a bounded function in $[a, b]$, whose set of discontinuities has a single limit point, is integrable.

25. Show f is R-D integrable in $[a, b]$ if and only if there exists $I \in \mathbf{R}$ with the property that given $\epsilon > 0$, there exists a partition P of $[a, b]$ such that $|\overline{S}(P) - I| < \epsilon$, $|I - \underline{S}(P)| < \epsilon$.

26. In the proof of the necessity of Theorem 5.2.4, we claimed that there is a single partition P satisfying

$$\int_{\underline{a}}^{b} f - \frac{\epsilon}{2} < \underline{S}(P) \leq \overline{S}(P) < \overline{\int_{a}^{b}} f + \frac{\epsilon}{2}.$$

Show such a single partition must exist.

27. Let f be continuous and bounded on (a, b). Show f is integrable in $[a, b]$.

28. Let f be defined on $[0, 1]$ by: $f(x) = \begin{cases} \frac{1}{\sqrt{x}}, & \text{if } x \neq 0 \\ 0, & \text{if } x = 0. \end{cases}$ Let P be any partition of $[0, 1]$ and M be an arbitrary positive real number. Show there is a set $\{c_i : i \in \mathbf{N}\}$ such that $x_i \leq c_i \leq x_{i+1}$ and $\sum f(c_i)d_i \geq M$. Note that f is "improperly" Riemann-Darboux integrable over this interval.

29. Let f be unbounded on $[a, b]$, but defined there. Show for any partition P of $[a, b]$ there is a set $\{c_i : i \in \mathbf{N}\}$ as in Exercise 28, such that $\sum f(c_i)d_i$ exceeds any arbitrary positive real number M.

30. Let f be integrable on $[a, b]$ and $[c, d] \subseteq [a, b]$. Show that f is integrable on $[c, d]$.

31. Let f be continuous and nonnegative on $[a, b]$. If $f(x) > 0$ for some $x \in [a, b]$, can the same be said for $\int_a^b f$? What about if continuity is replaced by mere integrability?

32. Let f and g be defined and integrable on $[a, b]$ and satisfy $f \leq g$, that is, $f(x) \leq g(x)$ for each $x \in [a, b]$. Show for any partition P, $\overline{S}(f, P) \leq \overline{S}(g, P)$ and that $\underline{S}(f, P) \leq \underline{S}(g, P)$. Conclude that $\int_a^b f \leq \int_a^b g$.

33. Let f be continuous on the range of g and g be continuous on $[a, b]$. Show $f \circ g$ is integrable on $[a, b]$.

34. Let f be continuous on the range of g, and g be monotone on $[a, b]$. Show $f \circ g$ is integrable on $[a, b]$.

5.3   The Problem of Computing an Integral

At this stage in our development, we have proven several theorems for testing whether a given function is integrable on an interval $[a, b]$. For example, we can see from the fact that $f(x) = x^2, x \in [0, 2]$ is continuous (or monotone) on the given interval, that it is integrable there. Unfortunately, this fact does not give us a method for finding the value of the integral. In practice, this is a difficult procedure, even given the so-called Fundamental Theorem of Calculus. One reason is that there are unpleasant functions that have no elementary function for an antiderivative (see Section 4). The elementary functions (see Chapter 9) are those that are commonly treated in calculus books. An example of a function that is not elementary is e^{x^2}, and we leave it to the reader to try to find a simple antiderivative for this function! In such situations, if one wants to evaluate the integral, one is left with applying the definition. Indeed, it is the definition of the integral as a limit of an approximating sum that underlies the sophisticated techniques of numerical integration explored at the end of section 5.4.

Example 1. Find the value of $\int_0^2 x^2$.

Solution. We shall proceed numerically. First, we partition $[0, 2]$ into n equal subdivisions of length $\frac{2}{n}$ each, to obtain a partition

$$P_n = \left\{ 0, \frac{2}{n}, \frac{4}{n}, \dots, \frac{2i}{n}, \dots, \frac{2n}{n} = 2 \right\}.$$

For this partition,

$$
\begin{aligned}
\overline{S}(P_n) &= \sum_{i=1}^{n} \left(\frac{2i}{n} \right)^2 \left(\frac{2}{n} \right) \\
&= \frac{8}{n^3} \sum_{i=1}^{n} i^2 \\
&= \frac{8}{n^3} \cdot \frac{n(n+1)(2n+1)}{6},
\end{aligned}
$$

where the last inequality is obtained by applying the formula for the sum of squares up to n. The sequence $\{\overline{S}(P_n)\}$ is monotonically decreasing and bounded below. By taking the limit as n tends to infinity, we get

$$\inf \left\{ \overline{S}(P_n) \right\} = \lim_{n \to \infty} \overline{S}(P_n) = \frac{16}{6} = \frac{8}{3}.$$

On the other hand, evaluating $\underline{S}(P_n)$ leads to

$$
\begin{aligned}
\underline{S}(P_n) &= \sum_{i=0}^{n-1} \left(\frac{2i}{n} \right)^2 \left(\frac{2}{n} \right) \\
&= \frac{8}{n^3} \sum_{i=1}^{n-1} i^2 \\
&= \frac{8}{n^3} \cdot \frac{(n-1)(n)(2n-1)}{6}.
\end{aligned}
$$

On taking the limit of $\underline{S}(P_n)$ as n tends to infinity we again get quantity $\frac{8}{3} =$ sup $\underline{S}(P_n)$, a calculation that the reader should verify. Now,

$$\frac{8}{3} = \inf\{\overline{S}(P_n)\} \ge \overline{\int_0^2} x^2 \quad \text{and} \quad \frac{8}{3} = \sup\{\underline{S}(P_n)\} \le \underline{\int_0^2} x^2.$$

Hence, $\underline{\int_0^2} x^2 = \frac{8}{3} = \overline{\int_0^2} x^2$, and we conclude that $\int_0^2 x^2 = \frac{8}{3}$. $\square$

Discussion. In the calculation above, we used a rather special collection of partitions, namely, partitions where each subdivision was of equal length. Partitions of this type are exceedingly useful for calculation purposes, but as of this juncture, there is no guarantee that calculations involving only this type of partition will always generate either the upper or lower integral. Thus, we did not write

$$\lim_{n \to \infty} \overline{S}(P_n) = \overline{\int_0^2} x^2.$$

Technically speaking, we should have computed the supremum of $\underline{S}(P)$ and the infimum of $\overline{S}(P)$ for *all possible* partitions, P, of $[0, 2]$, and identified both these quantities with the number $\frac{8}{3}$. Instead, we only used the special subcollection $\{P_n\}$ consisting of all partitions that generated subintervals of equal length, $d = \frac{2}{n}$. For these special partitions, we were able to show that

$$\inf\{\overline{S}(P_n)\} = \sup\{\underline{S}(P_n)\} = \frac{8}{3}.$$

This fact, together with Theorem 5.2.2, is sufficient to guarantee that x^2 is not only integrable, but to establish the value of the integral on $[0, 2]$.

While it is the case that if f is integrable on $[a, b]$, that calculations using equally spaced intervals will yield the value of the integral, it is not obviously the case that the equivalent statement can be made if f is not integrable. Even so, the calculations above are worth studying because of their ease. They exhibit two features of importance:

(i) The partitions used consist of $(n + 1)$ equally spaced points.

(ii) The components of the upper and lower sums associated with suprema and infima over subintervals have been replaced by values obtained by evaluating the function at the endpoints of the subintervals.

The second item is a result of the fact that the function being integrated is monotone on the intervals in question. This may not hold for other functions. Indeed, the supremum and/or infimum may not be attained on a given subinterval (see Exercise 1).

In conclusion it is clear that a computation that involves equal length subintervals and evaluation of the function only at the end points of these subintervals, or other convenient places, is exceptionally nice. Thus it would behoove us to find out under what conditions such a calculation gives a complete answer to the question of whether a function is integrable, and if so, what the value of the integral is. $\square$

To answer the question raised above, we will return to the approach originally taken by Riemann to the development of the integral that bears his name.

Definition. Let $P = \{x_0, \dots, x_n\}$ be a partition of $[a, b]$. A finite sequence, c_i, such that $1 \le i \le n$ and $c_i \in [x_{i-i}, x_i]$ is called an **intermediate partition** for P.

Discussion. The intent of this definition is to generate a collection of points, with one point being a member of each of the subintervals generated by the partition, P. Some special choices of the c_i's are the left endpoint x_{i-1}, or the right endpoint x_i, or even the midpoint $\frac{x_{i-1}+x_i}{2}$.

In our previous calculations of the upper and lower Riemann-Darboux sums, we have used some special choices. Specifically, if f is continuous on $[a, b]$, then both M_i, m_i are attained on the ith interval. Thus, we can find $c_{i,M}$ and $c_{i,m}$ such that $f(c_{i,M}) = M_i$ and $f(c_{i,m}) = m_i$, respectively. These choices enter the calculations via $\overline{S}(P) = \sum f(c_{i,M})d_i$ and $\underline{S}(P) = \sum f(c_{i,m})d_i$, where $d_i = x_i - x_{i-1}$.

Thus, an intermediate partition should be thought of as a set of choices c_i that will enter the calculation $\sum_{i=1}^{n} f(c_i)d_i$. This sum resembles an upper or lower sum over a partition except that we are no longer requiring that $f(c_i)$ be either M_i or m_i. Calculations of this general type, that is, $\sum f(c_i)d_i$ where $\{c_i\}$ is an intermediate parion, will be used in the next definition. This more general approach leads to the concept of a Riemann Integral. $\square$

Definition. Let f be defined on $[a, b]$, P be a partition of $[a, b]$ and $Q = \{c_i\}$ an intermediate partition for P. The **Riemann sum**, $R(f, P, Q)$ of the function f over the partition P, and the intermediate partition Q is defined by

$$R(f, P, Q) = \sum_P f(c_i)d_i,$$

where the P under the summation sign indicates summation extends over all the subintervals generated by P and the c_i's are the members of Q.

Discussion. It should be emphasized that $R(f, P, Q)$ is a real number, and for a given partition P there will be many different numbers $R(f, P, Q)$, in fact, one for each distinct intermediate partition Q. Two obvious candidates for Riemann sums are the upper and the lower Riemann-Darboux sums $\overline{S}(P), \underline{S}(P)$ defined in Section 5.2, although these may not exist, since f is not required to be bounded on $[a, b]$; rather, it is merely required to be defined on the total interval. For this reason, the Riemann sum is a generalization of the Riemann-Darboux sum. In the event that f is bounded, for any partition, P, and any intermediate partition, Q, of P, we have

$$\underline{S}(P) \le R(f, P, Q) \le \overline{S}(P).$$

As shown in Figure 5.3.1, the Riemann sum is also an approximating sum associated with a step function. However, for a given intermediate partition, Q, the associated step function neither lies above the function, f, as is the case with s^P, nor does it lie completely below the function as is the case with s_P. Instead, it lies somewhere in the middle and, as a result should more closely conform to f than either s^P or s_P. For the case of a nonnegative function, as show in Figure

5.3.1, the Riemann sum also generates area. By virtue of the inequality above, this area must always lie between the area given by s_P and that given by s^P. $\square$

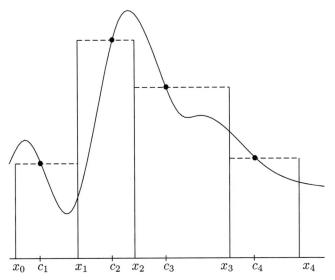

$$x_0 \quad c_1 \qquad x_1 \qquad c_2 \ x_2 \quad c_3 \qquad\qquad x_3 \quad c_4 \qquad\qquad x_4$$

Figure 5.3.1 The step function generated by a typical intermediate partition for a Riemann sum, $R(f, P, Q)$, is shown.

Definition. Let f be defined on $[a, b]$. We say that f is **Riemann integrable (R-integrable)** on $[a, b]$ provided there is an $L \in \mathbf{R}$ such that for every $\epsilon > 0$ there exists $\delta > 0$ such that for every partition P, and intermediate partition Q of P,

$$\|P\| < \delta \text{ implies } |R(f, P, Q) - L| < \epsilon.$$

The real number L is called the **Riemann Integral** of the function f over the interval $[a, b]$, and we write $R\int_a^b f = L$.

Discussion. The definition must be viewed as a limit definition in the sense of our previous limit definitions. The number L is clearly the limit of the quantity $R(f, P, Q)$, a functional entity that "approaches" L. The distinction between this and our previous definitions is in the variable quantity P (or more appropriately $\|P\|$), which is being made less than δ. In the earlier versions, the variable quantity, x, to be manipulated was constrained to move along the real line, and associated with each x was a unique real number, $f(x)$. In this case, the norm of P is also constrained to be a real number, but there are an infinity of partitions associated with this real number. Moreover, given a particular partition, P, there are infinitely many choices for Q and in consequence, there are many numbers of the form $R(f, P, Q)$ for this particular P. The result is that we are dealing with a much more complex form of limit.

 The thrust of this definition is to isolate out the quantities $\|P\|$ and L as being those of interest. Thus, we must find a single L such that it is enough to have knowledge only about $\|P\|$ to ensure that $R(f, P, Q)$ is close to L. Note that for the definition to be satisfied, this must happen in spite of the fact that there are infinitely many partitions having a fixed small norm, and infinitely many intermediate partitions associated with any given one of these partitions.

While this definition contains all we need to know about integration, it suffers from the fact that it is difficult to work with. One may ask why then introduce it. Suppose for a minute that we were able to show that Riemann Integrability and R-D Integrability were equivalent concepts and further that the L of the definition above was $\int_a^b f$ whenever f was integrable on $[a, b]$. It would solve the problem suggested at the end of Example 1, since the L arrived at is independent of the choice of P and Q. Thus, if f is known to be integrable, the limits obtained using the methods of Example 1 must be the value of the integral! In other words, for integrable functions, we could compute our sums by evaluating the function at any point of convenience within each subinterval and use subintervals of any type that was convenient, so long as the norm goes to 0.

Finally, the reader will realize that since we are dealing with a limit, there are some rather standard questions that must be asked, such as, Is this limit unique?, and so forth. The problem of uniqueness is dealt with below. Answers to many of the other standard questions are presented in Section 5.4. $\square$

Theorem 5.3.1. *If f is Riemann integrable on $[a, b]$, then $R \int_a^b f$ is unique.*

Proof. The proof is left to Exercise 2. $\square$

The next theorem constitutes the first step in establishing an equivalence between the two types of integrals.

Theorem 5.3.2. *If f is Riemann integrable on $[a, b]$, then f is bounded.*

Proof. Let $\epsilon > 0$ be given, and let $\delta > 0$ be chosen satisfying the definition, with L the limit. For any P such that $\|P\| < \delta$ and any two intermediate partitions Q and Q', we have

$$|R(f, P, Q) - R(f, P, Q')| \leq |R(f, P, Q) - L| + |R(f, P, Q') - L| < 2\epsilon.$$

Thus, let P be a fixed partition having $\|P\| < \delta$, and let $Q = \{c_i\}$ be a fixed intermediate partition for P.

Now, if f is not bounded on $[a, b]$ then f is unbounded over a subinterval, say, the subinterval $[x_{j-1}, x_j]$. Hence, we can choose a point $x \in [x_{j-1}, x_j]$ such that $|f(x) \times d_j|$ is as large as we please. In particular, we can find c_j' such that

$$|[f(c_j) - f(c_j')] \times d_j| \geq 2\epsilon.$$

Now, if Q' is obtained from Q by replacing c_j by c_j', then

$$|R(f, P, Q) - R(f, P, Q')| \geq 2\epsilon,$$

contrary to our initial inequality. It follows that f is bounded. $\square$

Discussion. The proof hinges on the fact that we can tie down the unboundedness of f to a single subinterval. This, of course, is due to the fact that there are only finitely many subintervals to deal with. Had there been an infinite number, the situation would have been different. Once we have found an interval on which to focus, we use the fact that the function must assume arbitrarily large vales to produce a contradiction.

The essential fact contained in this theorem is that there is no loss of generality in restricting our attention to bounded functions only, as we were forced to do in Section 5.2. Without this result, we could not hope to develop a general equivalence between the two results.

Last, the reader should compare this result with those in Exercises 5.2.27–29. □

The following result is known as Darboux's Theorem. It links the two definitions of integral by establishing that as the norm of the partition, $\|P\|$, approaches 0, the upper and lower Riemann-Darboux sums, $\overline{S}(P)$ and $\underline{S}(P)$, of a partition P indeed "approach" the upper and lower R-D integrals $\overline{\int}_a^b f$ and $\underline{\int}_a^b f$, respectively.

Theorem 5.3.3. (Darboux's Theorem) *If f is bounded on $[a, b]$, then given $\epsilon > 0$, there exists a $\delta > 0$ such that if P is any partition of $[a, b]$ with $\|P\| < \delta$, then*

$$\overline{S}(P) < \overline{\int_a^b} f + \epsilon \quad and \quad \underline{S}(P) > \underline{\int_a^b} f - \epsilon$$

Proof. Let $\epsilon > 0$ be given. By the property of the infimum, there exists a partition P_1 of $[a, b]$ satisfying

$$\overline{S}(P_1) < \overline{\int_a^b} f + \frac{\epsilon}{2}.$$

Suppose P_1 consists of k points other than a and b and set $\delta = \frac{\epsilon}{4kM}$, where M is the supremum of f in $[a, b]$. If we let P be any partition such that $\|P\| < \delta$ and $P_2 = P_1 \cup P$, then by Theorem 5.2.1 we have

$$|\overline{S}(P) - \overline{S}(P_2)| \le 2kM\|P\| \le 2kM \cdot \frac{\epsilon}{4kM} = \frac{\epsilon}{2}.$$

Hence,

$$\begin{aligned} \overline{S}(P) \quad &\le \quad \overline{S}(P_2) + \frac{\epsilon}{2} \le \overline{S}(P_1) + \frac{\epsilon}{2} \\ &< \quad \overline{\int_a^b} f + \frac{\epsilon}{2} + \frac{\epsilon}{2} \\ &= \quad \overline{\int_a^b} f + \epsilon. \end{aligned}$$

The assertion for the lower sums is left to Exercise 6. □

Discussion. This theorem establishes the connection between two apparently disparate notions. Specifically, it takes the fact that the infimum of a set must have points of that set close to it, and replaces it by the fact that we can find partitions of small norm. The proof is simple, but clever, because it takes Theorem 5.2.1 and applies it in an unexpected way. Let's see how.

Theorem 5.2.1 tells us that if we obtain Q from P by adding n points to P, then

$$|\overline{S}(P) - \overline{S}(Q)| \leq 2nM \cdot \|P\|.$$

In the present case, we find the partition P_1 that has the property that $\overline{S}(P_1)$ is close to the upper integral. The obvious way to proceed is start with P_1, and add points. This approach fails because the number of points that we must add to P_1 to obtain an arbitrary partition, P, having norm less than δ, is unbounded. Thus, the n in $2nM\|P_1\|$ can be arbitrarily large, making this a meaningless bound. The clever application of Theorem 5.2.1 comes in turning this situation around by thinking of adding P_1 to P. The key to this idea is that we notice that a common refinement of P_1 and P differs from P by at most k points. On the other hand, a common refinement of P and P_1 must yield an upper sum that is even closer to the upper integral than the upper sum over P_1. Moreover, the application of Theorem 5.2.1 is meaningful because k is fixed, while $\|P\|$ can be as small as we like.

The vital force behind this theorem is the properties of the supremum and infimum that enable us to write the first inequality in the proof.

Roughly speaking, this theorem asserts that as the norm of P decreases, the upper sums "approach" the value of the upper integral and the lower sums "approach" the lower integral. Since the collection $\{\overline{S}(P) : P \text{ is a partition of } [a,b]\}$ do not form a sequence (the domain is not $\mathbf{N}$, and there are uncountably many partitions of $[a,b]$), we can not speak of $\lim_{\|P\|\to 0}\{\overline{S}(P)\}$ in our usual sense. But if we agree to define this quantity as a real number L, provided given $\epsilon > 0$, there exists a $\delta > 0$ such that if P is any partition with $\|P\| < \delta$, we have $|\overline{S}(P) - L| < \epsilon$, then the above theorem states

$$\lim_{\|P\|\to 0} \overline{S}(P) = \overline{\int_a^b} f$$

and

$$\lim_{\|P\|\to 0} \underline{S}(P) = \underline{\int_a^b} f.$$

We ask the reader to calculate the upper and lower R-D integrals in Example 1 above, using this theorem. $\square$

Theorem 5.3.4. *Let f be Riemann integrable on $[a,b]$ with L as the Riemann integral. Then f is Riemann-Darboux integrable on $[a,b]$ and $\int_a^b f = L$.*

Proof. By Theorem 5.3.2, f is bounded on $[a,b]$ whence both the upper and the lower integrals exist. Let $\epsilon > 0$ be given. Then there is a partition, P_1 of $[a,b]$, such that

$$\underline{\int_a^b} f - \frac{\epsilon}{2} \leq \underline{S}(P_1) \leq \overline{S}(P_1) \leq \overline{\int_a^b} f + \frac{\epsilon}{2}.$$

Let $\delta > 0$ be chosen that satisfies the definition of the Riemann Integral for ϵ and L. Choose P to be any refinement of P_1 that satisfies $\|P\| < \delta$. For this partition, we have

$$\overline{S}(P) - \frac{\epsilon}{2} < \overline{\int_a^b} f \leq \overline{S}(P) + \frac{\epsilon}{2} \qquad \textbf{(WHY?)}$$

and
$$\overline{S}(P) - \frac{\epsilon}{2} < L < \overline{S}(P) + \frac{\epsilon}{2},$$
whence
$$\left| \overline{\int_a^b} f - L \right| < \epsilon.$$

Since ϵ was arbitrary, we have $\overline{\int_a^b} f = L$. Similarly, we can show that $\underline{\int_a^b} f = L$, whence the conclusion follows. $\square$

Discussion. The basic intuition on which the proof is based is simple. The definition of Riemann Integrability allows us to choose partitions such that $R(f, P, Q)$ will be close to L, in particular, $\overline{S}(P)$ will be close to L. Further, we can choose partitions so that $\overline{S}(P)$ is close to $\overline{\int_a^b} f$. But taking common refinements allows us to come up with a single partition P such that $\overline{S}(P)$ is simultaneously close to L and also to $\overline{\int_a^b} f$. Thus, L must be close to $\overline{\int_a^b} f$, in fact as close as we please! $\square$

Next, we prove the converse of Theorem 5.3.4, establishing the equivalence of the two approaches to integration.

Theorem 5.3.5. *If f is Riemann-Darboux integrable on $[a, b]$, then f is Riemann integrable on $[a, b]$, and $\int_a^b f = R \int_a^b f$.*

Proof. Clearly, f is bounded. Let $\epsilon > 0$ be given. By Theorem 5.3.3, there exists $\delta > 0$ such that if $\|P\| < \delta$, we have
$$\underline{\int_a^b} f - \frac{\epsilon}{2} < \underline{S}(P) \le \overline{S}(P) < \overline{\int_a^b} f + \frac{\epsilon}{2}.$$

Since f is Riemann-Darboux integrable, we have
$$\underline{\int_a^b} f = \overline{\int_a^b} f = L,$$

where L denotes the R-D integral of f. Thus,
$$L - \frac{\epsilon}{2} < \underline{S}(P) \le R(f, P, Q) \le \overline{S}(P) < L + \frac{\epsilon}{2}$$

for any Riemann sum $R(f, P, Q)$ obtained from the partition P. Thus,
$$|R(f, P, Q) - L| < \epsilon$$

proving that f is Riemann integrable, with L as the Riemann Integral. $\square$

Discussion. The bulk of this proof is already contained in the argument for Darboux's Theorem. The only additional fact needed is that the Riemann sum $R(f, P, Q)$ lies between the numbers $\underline{S}(P)$ and $\overline{S}(P)$. $\square$

Note. In view of Theorems 5.3.4 and 5.3.5, we shall use the phrase 'integrable' to denote either of Riemann or Riemann-Darboux integrability.

We began this section with a problem of finding the value of a particular integral, $\int_0^2 x^2$. Our method was to obtain a formula for $R(f, P, Q)$, where

(i) P was a partition of $[a, b]$ into n equal subintervals;

(ii) Q consists of the endpoints of the subintervals, either left hand or right hand (or any other convenient point).

Having found such a formula, we then proceeded to find the limit as n tended to infinity. Based on the results above, we know that for any function that is integrable (in either sense) on the interval $[a, b]$, this procedure will yield the correct value of the integral.

Example 2. Evaluate $\int_{-1}^{3} 3x + 4$.

Solution. The function defined by $f(x) = 3x + 4$ is continuous on $[-1, 3]$, whence it is integrable there. If we subdivide $[-1, 3]$ into n subintervals of equal length, each subinterval has length $\frac{3-(-1)}{n} = \frac{4}{n}$. The points of division for this partition are given by $c_i = -1 + \frac{4i}{n}$, where $i = 0, 1, \ldots, n$. If we use the right-hand endpoint of each subinterval to define the intermediate partition, then the Riemann sum is given by

$$
\begin{aligned}
R(3x + 4, P, Q) &= \sum_{i=1}^{n} (3c_i + 4) \times \frac{4}{n} = \frac{4}{n} \times \sum_{i=1}^{n} (3c_i + 4) \\
&= \frac{4}{n} \times \sum_{i=1}^{n} (3(-1 + \frac{4i}{n}) + 4) \\
&= \frac{4}{n} \times \sum_{i=1}^{n} \left(1 + \frac{12i}{n}\right) = \frac{4}{n} \times \left(\sum_{i=1}^{n} 1 + \sum_{i=1}^{n} \frac{12i}{n}\right) \\
&= \frac{4}{n} \times \left(n + \frac{12}{n} \times \sum_{i=1}^{n} i\right) \\
&= 4 + \left(\frac{48}{n^2} \times \frac{n(n+1)}{2}\right).
\end{aligned}
$$

The limit as $n \to \infty$, of the last expression is 28, whence we see that the value of the integral is also 28. $\square$

Discussion. The computation above illustrates the utility of the formula for evaluating $\sum_{i=1}^{n} i$. Other examples require other formulae. This may have been the motivation behind the search for polynomials in n that would evaluate $\sum_{i=1}^{n} i^k$. Such formulae exist and can be found by solving a series of simultaneous equations that are generated under the assumption that such a formula does exist. The degree of the polynomial in n is always $k + 1$. $\square$

The method of defining the integral has been based on the use of step functions to approximate the function. Approximating a function with a step function amounts to suggesting that a given function is locally constant, that is, around every x there is an interval on which the function is constant. This assumption is not likely to be true for most functions. Even so, as we have seen, the procedure still yields the correct value for the integral. The effect of this 'bad' assumption is felt in the rate of convergence. That is, we must generally work with partitions having very small norms if we want the approximate value to be close to the true value. Evidently, if one is going to use approximating sums to evaluate integrals, one would like to have a procedure that converged more quickly. As well, for a given approximation, one would like to be able to estimate the error. Further exploration of these questions must await development of some of the basic theory in the next two sections.

Exercises

1. Give an example of a function on $[0, 1]$ for which there are infinitely many distinct subintervals such that f neither achieves its maximum nor its minimum on the subinterval.

2. Let f be Riemann integrable on $[a, b]$. Show the value of the integral is unique.

3. Use the methods developed to evaluate:

   (a) $\int_0^a x^2$;   (b) $\int_2^3 x$;   (c) $\int_{-3}^{-1} x$;

   (d) $\int_1^4 x^3$;   (e) $\int_{-1}^1 (1 - x^2)$;   (f) $\int_0^2 (2 - 3x + x^3)$.

4. Give an example of a function that is not Riemann integrable.

5. Fill in the missing details in Example 1.

6. Complete the missing proof for lower sums in Theorem 5.3.3.

7. Complete the proof of the fact that $\int_{\underline{a}}^b f = L$ in Theorem 5.3.4.

8. Let f be a function that is defined on $[0, 1]$. If f takes one fixed value on the rationals and another on the irrationals, will f be integrable?

9. Let f be defined on $[a, b]$ but discontinuous at every point of the interval. Show f is not integrable there.

10. In earlier chapters we have discussed why limits fail to exist. Let f be a function defined on an interval $[a, b]$ but that is not integrable there. What, if any, conclusions can be drawn about f?

11. Let f and g be defined and bounded on $[a, b]$. For $c, d \in \mathbf{R}$ show

$$c \cdot R(f, P, Q) + d \cdot R(g, P, Q) = R(cf + dg, P, Q),$$

   where P is an arbitrary partition of $[a, b]$ and Q is an intermediate partition of P. Use this result to show

$$c \cdot \int_a^b f + d \cdot \int_a^b g = \int_a^b (cf + dg).$$

12. Let $a > 0$. Use the partition $\{a, ar, ar^2, \dots, ar^n = b\}$ of $[a, b]$ to compute $\int_a^b \frac{1}{x}$.

13. Show $\int_0^a \sin x = 1 - \cos a$, where $0 < a \leq 1$. [Hint: Use

$$\sum_{k=1}^n \sin kx = \frac{\sin \frac{(n+1)x}{2} \sin \frac{nx}{2}}{\sin \frac{x}{2}}, \quad x \neq 2m\pi,\ m \in \mathbf{Z}.]$$

14. Given f is integrable on $[a, b]$, show for every $\epsilon > 0$, there exist step functions g, h defined on $[a, b]$ such that $g \leq f \leq h$ and

$$\int_a^b (f - g) < \epsilon, \text{ and } \int_a^b (h - f) < \epsilon.$$

15. Show f is integrable on $[a, b]$ if for every positive ϵ there exist step functions g, h such that $g \leq f \leq h$ and $\int_a^b (h - g) < \epsilon$.

16. Let f be bounded on $[a, b]$. If there exists a sequence of partitions, P_n, such that

$$\lim_{n \to \infty} \left(\overline{S}(f, P_n) - \underline{S}(f, P_n) \right) = 0,$$

prove f is integrable. Conversely, show if f is integrable, there exists a sequence of partitions, P_n, such that

$$\int_a^b f = \lim_{n \to \infty} \overline{S}(f, P_n) = \lim_{n \to \infty} \underline{S}(f, P_n).$$

[Note: This justifies our approach to Example 1.]

17. If f is continuous on $[a, b]$, prove

$$\int_a^b f \quad = \quad \lim_{n \to \infty} \frac{b - a}{n + 1} \sum_{k=1}^{n} f\left(a + \frac{k(b - a)}{n + 1} \right)$$

$$= \quad \lim_{n \to \infty} \frac{b - a}{n + 1} \sum_{k=1}^{n} f\left(a + \frac{(k + 1)(b - a)}{n + 1} \right).$$

18. Prove the following Cauchy's criterion for Riemann integrability: f is integrable in $[a, b]$ if and only if given $\epsilon > 0$, there exists $\delta > 0$ such that for all partitions P_1, P_2 such that $\|P_i\| < \delta (i = 1, 2)$, and for all intermediate partitions Q_1, Q_2 of P_1, P_2,

$$|R(f, P_1, Q_1) - R(f, P_2, Q_2)| < \epsilon.$$

19. Let $g : [0, 1] \to [0, 1]$ be a one-to-one, onto, continuous function. Show geometrically $\int_0^1 g(x) + \int_0^1 g^{-1}(x) = 1$.

20. If f is Riemann integrable in $[0, 1]$, and if $a_n = \frac{1}{n} \sum_{k=1}^{n} f(\frac{k}{n})$, prove a_n converges to $\int_0^1 f$. Show also if f is not Riemann integrable, $\{a_n\}$ may fail to converge. Give an example where $\{a_n\}$ converges, but f is not integrable.

21. If f satisfies Lipschitz's condition of order 1 in $[0, 1]$, with Lipschitz's constant M (see Exercise 3.5.21), prove

$$\left| \int_0^1 f - \frac{1}{n} \sum_{k=1}^{n} f\left(\frac{k}{n} \right) \right| < \frac{M}{2n}.$$

5.4   Properties of the Integral

In the previous sections, we considered methods that enable us to associate with each integrable function f defined on $[a, b]$ a unique real number called the integral (in the sense of Riemann-Darboux, as well as Riemann) and denoted by the symbol $\int_a^b f$. This passage from integrable functions to the value of the integral is a limiting process and as such should exhibit certain nice properties. We have already established uniqueness. Additional properties are presented in the following set of theorems. Our first result proves that the integral is additive.

Theorem 5.4.1. *If f and g are both integrable on $[a, b]$, then $f + g$ is integrable on $[a, b]$, and moreover,*

$$\int_a^b (f + g) = \int_a^b f + \int_a^b g.$$

Proof. Let $\epsilon > 0$ be given. Since f and g are both integrable on $[a, b]$, we can choose a partition P such that

$$\int_a^b f - \frac{\epsilon}{2} < \underline{S}(f, P) \leq \overline{S}(f, P) < \int_a^b f + \frac{\epsilon}{2}$$

and

$$\int_a^b g - \frac{\epsilon}{2} < \underline{S}(g, P) \leq \overline{S}(g, P) < \int_a^b g + \frac{\epsilon}{2}.$$

Adding these two inequalities yields

$$\int_a^b f + \int_a^b g - \epsilon \quad < \quad \underline{S}(f, P) + \underline{S}(g, P)$$

$$\leq \quad \overline{S}(f, P) + \overline{S}(g, P) < \int_a^b f + \int_a^b g + \epsilon.$$

By Lemma 5.2.2(iv), $\overline{S}(f + g, P) \leq \overline{S}(f, P) + \overline{S}(f, P)$. Since a similar statement is valid for lower sums, we have

$$\int_a^b f + \int_a^b g - \epsilon < \underline{S}(f + g, P) \leq \overline{S}(f + g, P) < \int_a^b f + \int_a^b g + \epsilon.$$

It is immediate from Theorem 5.2.4 that $f + g$ is integrable and further that

$$\int_a^b (f + g) = \int_a^b f + \int_a^b g. \qquad \square$$

Discussion. Observe that the heart of this argument is the calculation implicit in Lemma 5.2.2(iv). The proof of this lemma depends only on our ability to manipulate finite sums.

This theorem is analogous to our previous theorems that relate limiting processes to the operation of addition. In essence, it tells us that the integral (limit of a sum process) is well behaved with respect to addition. $\square$

Theorem 5.4.2. *If f is integrable on $[a, b]$ and c is any constant, then cf is integrable on $[a, b]$ and*

$$\int_a^b cf = c \int_a^b f.$$

Proof. There are two cases, namely, $c \geq 0$ and $c < 0$. We treat the latter and leave the former as an exercise. Let $\epsilon > 0$ be fixed, and let $\epsilon_1 = \min\{\epsilon, \frac{|\epsilon|}{|c|}\}$. Then there is a partition P of $[a, b]$ such that

$$\int_a^b f - \epsilon_1 < \underline{S}(f, P) \leq \overline{S}(f, P) < \int_a^b f + \epsilon_1.$$

Since $c < 0$, by Lemma 5.2.2(ii), we obtain

$$\left(c \int_a^b f \right) + c\epsilon_1 < \overline{S}(cf, P) \leq \underline{S}(cf, P) < \left(c \int_a^b f \right) - c\epsilon_1.$$

Since $0 < |c\epsilon_1| \leq \epsilon$, it is immediate that

$$\int_a^b cf = c \int_a^b f. \qquad \square$$

Discussion. The reader should note the implicit use of Theorem 5.2.4 in the arguments we have given. $\square$

Theorem 5.4.3. *If f is integrable on $[a, b]$, then so is f^2.*

Proof. Assume $f \neq 0$. Let $\epsilon > 0$ be fixed and M be a positive bound for $|f|$. Set $\epsilon_1 = \frac{\epsilon}{2M}$. Let P be a partition of $[a, b]$ satisfying $\overline{S}(P) - \underline{S}(P) < \epsilon_1$. For this partition, let M_i, m_i, and d_i have their usual meanings (as in Section 5.2). We want to calculate $\overline{S}(f^2, P) - \underline{S}(f^2, P)$. We claim that the contribution to the difference arising from the interval $[x_{i-1}, x_i]$ is $|M_i^2 - m_i^2|d_i$. To see this, note that one of M_i^2 and m_i^2 will be the supremum for f^2, while the other will be the infimum for f^2 on this subinterval, unless m_i is negative, M_i is positive, and $f(x) = 0$ for some x in the interval. Thus, there are two cases to be considered. In the former,

$$\begin{aligned} |M_i^2 - m_i^2| \cdot d_i &= |M_i + m_i| \cdot (M_i - m_i)d_i \\ &\leq 2M(M_i - m_i)d_i. \end{aligned}$$

In the latter, one of the quantities M_i^2 and m_i^2 will be replaced by 0 to obtain either $M_i^2 d_i$ or $m_i^2 d_i$. Using the former quantity, we get

$$\begin{aligned} M_i^2 d_i &= M_i(M_i - 0)d_i \\ &< M_i(M_i - m_i)d_i \\ &< 2M(M_i - m_i)d_i. \end{aligned}$$

The latter quantity yields a similar inequality. Thus,

$$\begin{aligned} \overline{S}(f^2, P) - \underline{S}(f^2, P) &< 2M(\overline{S}(f, P) - \underline{S}(f, P)) \\ &< 2M\epsilon_1 = \epsilon. \qquad \square \end{aligned}$$

Discussion. The proof is simple. The only algebra used there is the identity $M_i^2 - m_i^2 = (M_i + m_i)(M_i - m_i)$. The quantity $M_i + m_i$ is less than or equal to twice the maximum value of the function in $[a, b]$. We are not readily concluding that M_i^2 and m_i^2 are the maxima and minima of f^2 in the ith subinterval. We have to take into consideration the points where the function $f(x) = 0$, and simultaneously m_i is negative. In such cases, the minimum value of f^2 in the subinterval is 0 and not m_i^2. $\square$

Theorem 5.4.4. *If f and g are integrable on $[a, b]$, so is their product fg.*

Proof. By our previous results, $(f + g)^2$ and $(f - g)^2$ are both integrable on $[a, b]$. The observation that

$$fg = \frac{(f + g)^2 - (f - g)^2}{4}$$

completes the proof. $\square$

The results of this section have dealt with the relationship of the integral to the various algebraic operations that have been defined on functions. It is reasonable to ask about the interaction of the integral with the various other properties defined on functions, and we now turn to this direction.

Theorem 5.4.5. *If f is nonnegative and integrable on $[a, b]$, then $\int_a^b f$ is non-negative.*

Proof. Observe that every lower sum is nonnegative. $\square$

Corollary. *If f and g are integrable on $[a, b]$, and $f \geq g$ throughout $[a, b]$, then $\int_a^b f \geq \int_a^b g$.*

Theorem 5.4.6. *If f is integrable on $[a, b]$, then $|f|$ is integrable on $[a, b]$, and further,*

$$\left| \int_a^b f \right| \leq \int_a^b |f|.$$

Proof. Let $\epsilon > 0$ be fixed, and choose a partition, P, such that $\overline{S}(P) - \underline{S}(P) < \epsilon$. For a given subinterval of this partition, $[x_{i-1}, x_i]$, let M_i' and m_i' denote the supremum and infimum of $|f|$, respectively. The reader can check that $M_i - m_i \geq M_i' - m_i'$, whence we have

$$\overline{S}(|f|, P) - \underline{S}(|f|, P) \leq \overline{S}(P) - \underline{S}(P) < \epsilon.$$

Thus, $|f|$ is integrable on $[a, b]$. The last inequality follows from Lemma 5.2.2(iii) that establishes the analogous result for upper sums. $\square$

Discussion. The inequality established in Theorem 5.4.6 may be thought of as the ultimate generalization of the triangle inequality, $|a + b| \leq |a| + |b|$; that is, the absolute value of the limit of a sum never exceeds the limit of the sum of the absolute values. $\square$

Let us summarize our work in this section so far. Consider a fixed interval, say, $[a, b]$. Let $\Re[a, b]$ denote the class of all Riemann integrable functions on this interval. We have shown that if $f, g \in \Re[a, b]$, then $f + g$, fg, λf ($\lambda \in \mathbf{R}$) and $|f|$ belong to $\Re[a, b]$. In other words, we say that this class is 'closed' under addition and multiplication, scalar multiplication and the formation of the absolute value. Moreover, if we think of the integral as a function

$$\text{Int} : \Re[a, b] \to \mathbf{R}$$

defined by Int $(f) = \int_a^b f$, with domain $\Re[a, b]$ and range contained in $\mathbf{R}$, then this function has the properties Int $(f + g) = $ Int $(f) + $ Int (g) and Int $(\lambda f) = \lambda$Int (f). In other words, the function Int preserves 'vector sums' and 'scalar products'. In the language of linear algebra, the function, Int, is a linear mapping in the usual sense of vector spaces. This function also enjoys some additional properties such as Int $(f) \geq $ Int (g) whenever $f \geq g$.

The last theorem in this section shows that the integral is additive on an interval.

Theorem 5.4.7. *If f is integrable on $[a, b]$ and $c \in [a, b]$, then f is integrable on $[a, c]$ and $[c, b]$ and further*

$$\int_a^b f = \int_a^c f + \int_c^b f.$$

Proof. Let P be any partition of $[a, b]$ that contains c. Let P' be the points of P that lie to the left of c, including c, and P'' be the points of P to the right of c, including c. Thus, $P = P'' \cup P'$. It is immediate that

$$\overline{S}(P') - \underline{S}(P') \leq \overline{S}(P) - \underline{S}(P) \quad \text{and} \quad \overline{S}(P'') - \underline{S}(P'') \leq \overline{S}(P) - \underline{S}(P).$$

Since all differences are nonnegative, and may be made less that any prescribed positive number, we obtain integrability in $[a, c]$ and $[c, b]$. The rest of the proof left to the reader (Exercise 3). $\square$

Discussion. This particular theorem has an important interpretation for non-negative functions. Namely, it tells us that if we split the interval over which we are integrating into two parts, the value of the integral over the whole will be the sum of the two integrals over the subintervals. This amounts to dividing the region whose area must be found into two separate parts and observing that the total area is the sum of the areas of the separate portions. $\square$

We conclude this section by applying some of the theory just developed. As we have already noted, when one approximates an integral via the limit of a sum process, one would like information on two aspects of the approximation. First, one would like to know how good a given approximation is. Second, one would like information on how rapidly one is approaching the limit.

Various techniques for approximating integrals have been developed. These techniques vary according to the entry corresponding to $f(c_i)$ in the Riemann sum. (Recall that we have already used M_i, m_i and $f(x_i)$ as choices for this entry in an approximating sum.) The point is that by shrewdly choosing these values, one can force a more rapid convergence. Our next example illustrates this for the case of $f(x_i)$, which amounts to assuming the function is locally constant.

In the following example let f be defined on $[a, b]$ and have a continuous derivative there, with M being a bound for f' on $[a, b]$. Let $[a, b]$ be divided into n subintervals of equal width, $h = \frac{b-a}{n}$, with right-hand endpoints $x_0, x_1, \ldots, x_{n-1}$ and let

$$y_i = f(x_i), \quad 0 \le i \le n - 1.$$

Example 1. Establish the **rectangular rule**:

$$\int_a^b f = \lim_{n \to \infty} (y_0 + y_1 + \ldots + y_{n-1})h.$$

Moreover, show

$$\left| \int_a^b f - (y_0 + y_1 + \ldots + y_{n-1})h \right| \le Mh(b - a),$$

where $M > 0$ is any bound on $|f'|$.

Solution. The first statement follows from Theorems 5.2.5 and 5.3.3. We prove the second statement. To this end, consider the ith subinterval, $[x_{i-1}, x_i]$. By Theorem 5.4.7,

$$\int_a^b f = \sum_{i=0}^{n-1} \int_{x_i}^{x_{i+1}} f.$$

Thus the problem reduces to estimating

$$\left| \int_{x_i}^{x_{i+1}} (f - y_i) \right| = \left| \int_{x_i}^{x_{i+1}} (f - f(x_i)) \right|.$$

The function $f - f(x_i)$ satisfies the hypothesis of the Mean Value Theorem on $[x_i, x_{i+1}]$, whence for each $x \in [x_i, x_{i+1}]$, we can choose c_x such that

$$|f(x) - f(x_i)| = |f'(c_x)|(x - x_i) \le Mh,$$

where the last inequality follows from the fact that $x - x_i \le h$. It is immediate that

$$\int_{x_i}^{x_{i+1}} |(f - f(x_i))| \le Mh^2.$$

If we now sum over i, we get

$$\left| \int_a^b f - \sum_{i=0}^{n-1} y_i \right| \le \sum_{i=0}^{n-1} Mh^2$$

$$= Mh \times \sum_{i=0}^{n-1} h = Mh(b - a). \quad \square$$

Discussion. The important feature of the argument in this example is the application of the Mean Value Theorem. It derives from the fact that we are approximating $f(x)$ on $[x_i, x_{i+1}]$ by $f(x_i)$, which is the function evaluated at the left-hand endpoint of the subinterval. This sets up the direct application of the theorem as in Example 4.3.1. The trick is to realize that we can apply it in each interval from x_i to x, which yields

$$|f(x) - f(x_i)| = |f'(c_x)|(x - x_i).$$

Completing the proof is then merely a matter of completing the calculation. This application will undoubtedly give the reader further insight into why the Mean Value Theorem and its generalization are such powerful tools in analysis and its applications.

The reader should also look at the bound, $Mh(b - a)$. Evidently, rapid convergence will be achieved if M is small, which means that the function is flat. If the reader thinks about it, it should seem reasonable that $(b - a)$ appears in the bound. Clearly, the longer the interval over which we are integrating, the greater the error for a given number of subdivisions, n. These are the two unchanging parts of the bound. What is used to make them small is decreasing h, which is accomplished by increasing n. Quite obviously, we would like a term that gets small more rapidly than h in the bound.

It would seem obvious that the assumption that f is locally a constant is not a good one. A more tenable assumption would be that f is piecewise linear, that is, a finite number of straight lines not necessarily having slope 0. If one thinks about this idea, one quickly realizes that the best piecewise linear approximation to f on $[x_i, x_{i+1}]$ is the one that connects the points $(x_i, f(x_i))$ and $(x_{i+1}, f(x_{i+1}))$. This leads to the trapezoidal rule (Exercise 32). Evidently, one can also use more complicated approximations, say by requiring the approximating function to go through more points on each subinterval. These topics are treated in the discipline of Numerical Analysis. We suggest that the interested reader consult Conte and de Boor [1972].[2] □

Exercises

1. Fill in the missing details in the proof of Theorem 5.4.1.

2. Give a proof for the case not treated in Theorem 5.4.2.

3. Fill in the missing details of Theorem 5.4.6, and complete the proof for the equality of the two integrals in Theorem 5.4.7.

4. Give an example of a function f that is not integrable on $[0, 1]$ but such that f^2 is integrable there.

5. If f and g are integrable in $[a, b]$, prove that $f - g$ is integrable in $[a, b]$ and $\int_a^b (f - g) = \int_a^b f - \int_a^b g$.

6. Generalize Theorem 5.4.1 to the case of the sum of n integrable functions.

7. Let f be integrable on $[a, b]$, and let $P = \{a = x_0 < x_1 < \ldots < x_n = b\}$ be any partition of $[a, b]$. Show f is integrable in each subinterval, $[x_{i-1}, x_i]$,

---

[2]S. D. Conte, and Carl de Boor, *Elementary Numerical Analysis: An Algorithmic Approach* (New York: McGraw-Hill Book Company. 1972).

($i = 1, 2, \ldots, n$), and further,

$$\int_a^b f = \sum_{k=1}^n \int_{x_{k-1}}^{x_k} f.$$

8. If f is integrable on $[a, b]$ and $g : [a + c, b + c] \to \mathbf{R}$ is defined by $g(x) = f(x - c)$, for $x \in [a + c, b + c]$, prove g is integrable on $[a + c, b + c]$ and

$$\int_{a+c}^{b+c} g = \int_a^b f.$$

9. If $\int_a^b f > 0$, must f be nonnegative on $[a, b]$? If $\int_a^b f = 0$, must $f = 0$ on $[a, b]$?

10. Is there any relationship between $\int_a^b f$ and $\int_a^b f^2$?

11. Let f be defined and integrable on $[a, b]$. Show f^+ and f^- are integrable on $[a, b]$.

12. Use Exercise 11 above to show that the integrability of f on $[a, b]$ implies the integrability of $|f|$. Give an example to show that the converse need not hold.

13. If f and g are both integrable in $[a, b]$, what can you say about the integrability of $\max(f, g)$ and $\min(f, g)$?

14. If f is integrable, and if there exists m, M satisfying $0 < m \le f \le M$ in $[a, b]$, prove that $\frac{1}{f}$ is integrable in $[a, b]$.

15. Prove or disprove: If f is continuous on $[a, b]$ and $\int_a^b fg = 0$ for every integrable function g on $[a, b]$, then $f(x) = 0$ on $[a, b]$.

16. Let $g \le f \le h$ on $[a, b]$, where g and h are integrable, and $\int_a^b g = \int_a^b h = A$, prove that f is integrable and $\int_a^b f = A$.

17. Is there any relationship between the value of $\int_a^b fg$ and $(\int_a^b f)(\int_a^b g)$, assuming that all integrals exist.

18. Show $\int_a^b fg$ can exist while neither of the separate integrals in Exercise 17 exist.

19. Let $\int_a^b fg$ exist along with $\int_a^b f$. Can we draw any conclusions about $\int_a^b g$? Suppose we assume f is continuous, or monotone, or strictly monotone, are your conclusions going to be changed?

20. If f is continuous and nonnegative on $[a, b]$ and if $M = \max f(x)$ in $[a, b]$, show that

$$\lim_{n \to \infty} \left[\int_a^b (f(x))^n \right]^{\frac{1}{n}} = M.$$

21. If f is periodic on $\mathbf{R}$ with period $\alpha > 0$ and integrable on $[0, \alpha]$, show f is integrable on $[t, t + \alpha]$ for any $t \in \mathbf{R}$, and $\int_t^{t+\alpha} f = \int_0^\alpha f$,

22. Let f be strictly monotone on $[a, b]$. Show if f is nonnegative on $[a, b]$ then $\int_a^b f$ is positive.

23. Let f be positive on $[a, b]$ and integrable there. Must the value of the integral be positive (assuming that $a < b$) ?

24. Let f be nonnegative on $[a, b]$. Show

$$\int_a^b f \ge \int_c^d f$$

for any subinterval $[c, d]$ of $[a, b]$.

25. Let f be integrable on $[a, b]$ and define $F(x) = \int_a^x f$. Find a condition on f that will ensure that F is nondecreasing; strictly increasing; monotone.

26. Let g be the function of Example 5.2.5 and f be defined by

$$f(x) = \begin{cases} 1, & \text{if } x > 0 \\ 0, & \text{if } x = 0. \end{cases}$$

Show $f \circ g$ is not integrable on $[0, 1]$ even though both f and g are integrable.

27. A function f is **even** on the interval $[-a, a]$, $a > 0$, provided $f(-x) = f(x)$ for each $x \in [-a, a]$. It is **odd** provided $f(-x) = -f(x)$ in that interval. Let F be integrable on $-a, a]$. Show if f is even on $[-a, a]$, then $\int_{-a}^a f = 2 \int_0^a f$. Show if f is odd, then $\int_{-a}^a f = 0$.

28. Let f and g be integrable on $[a, b]$. Establish the **Cauchy-Schwartz inequality:**

$$\left[\int_a^b fg \right]^2 \le \left[\int_a^b f^2 \right] \left[\int_a^b g^2 \right].$$

29. Give a suitable definition of $\int_a^b f$ where $b < a$. For this definition, establish the formula $\int_a^b f = -\int_a^b f$.

30. Let $f \le M$ (a constant) throughout $[a, b]$ and integrable there. Show $\int_a^b f \le M(b - a)$.

In the remaining exercises, let f be defined on $[a, b]$ and have an appropriate number of derivatives. Let $[a, b]$ be divided into n subintervals of equal width, $h = \frac{b-a}{n}$, with right-hand endpoints $x_0, x_1, \ldots, x_n$, and let

$$y_i = f(x_i), 0 \le i \le n - 1; \quad y_n = f(b).$$

31. Establish the following **forward rectangular rule**: there exists $\psi \in (a, b)$ such that

$$\int_a^b f = (b - a)f(a) + \frac{(b - a)^2}{2} f'(\psi).$$

More generally, show if $x_0 \in [a, b]$ then there exists $\psi \in (a, b)$ such that

$$\int_a^b f = (b - a)f(x_0) + \frac{(b + a - 2x_0)^2}{2} f'(\psi),$$

which is the general **rectangular rule**.

32. Establish the **trapezoidal rule:**

$$\int_a^b f = \lim_{n \to \infty} \left(\frac{y_0}{2} + y_1 + \cdots + y_{n-1} + \frac{y_n}{2} \right) h.$$

Moreover, show that for any n there is a $c \in [a, b]$ such that

$$\int_a^b f - \left(\frac{y_0}{2} + y_1 + \cdots + y_{n-1} + + \frac{y_n}{2} \right) = -(b - a)h^2 \frac{f''(c)}{12}.$$

Why is this referred to as the trapezoidal rule?

33. Establish **Simpson's rule**

$$\int_a^b f = \lim_{n \to \infty} S_n,$$

where $S_n = \frac{1}{3}(y_0 + 4y_1 + 2y_2 + 4y_3 + 2y_4 + \cdots + 4y_{n-1} + y_n)h$, where n is even. Moreover, show that for some $c \in [a, b]$

$$\int_a^b f - S_n = -\frac{(b-a)h^4 f''''(c)}{180}.$$

34. Establish **midpoint rule** $\int_a^b f = \lim_{n \to \infty} M_n$, where

$$M_n = h \sum_{k=1}^{n} f\left(a + \left(k - \frac{1}{2}\right)h\right)$$

35. Why is Simpson's rule a generally more effective method for numerical integration than the rectangular rule?

36. For a polynomial $P(x)$ of degree at most 3, show that Simpson's rule yields the exact value.

37. Use the Simpson and the trapezoidal rules with $n = 4$ and 8, respectively, to obtain approximations for $\int_0^1 f$ for the functions

   (a) $f(x) = \dfrac{1}{x}$;   (b) $f(x) = \dfrac{1}{1 + x^2}$; $f(x) = e^{-x^2}$.

   Compare the two results.

38. If $|f''(x)| < M$ in $[0, 1]$, show that the error in calculating $\int_0^1 f$ by Simpson's rule is less than $\dfrac{M}{2880n^4}$, where the interval $[0, 1]$ is divided into $2n$ equal parts.

5.5   The Relationship Between Integration and Differentiation

The preceding sections saw the development of the notion of integral as a limit of approximating sums that have no obvious relationship to the process of differentiation. In this section, we bring forth the intimate connection between the notions of differentiation and integration for a certain class of functions and show that in the case of continuous functions, the so-called Fundamental Theorem of Integral Calculus is true; namely, integration is the reverse process of differentiation. To begin our discussion, we associate with a function f defined on the interval $[a, b]$, a new function that has the role of an 'antiderivative'. More precisely,

Definition. If f is a function defined on $[a, b]$, a function F is called a **primitive** (or an**antiderivative**) of f on $[a, b]$, provided F is differentiable and $F'(x) = f(x)$ for all $x \in [a, b]$.

Discussion. The essential point about antiderivatives is that they are not unique. However, the key additional fact regarding primitives is that two functions that are primitives for the same function must differ by a constant. □

Theorem 5.5.1. *Let F and G be primitives for f on $[a, b]$. Then $F - G$ is a constant on $[a, b]$.*

Proof. Let $H = F - G$ on $[a, b]$. The reader can check that H satisfies the hypothesis of the Mean Value Theorem on $[a, b]$. Suppose, for the sake of argument, that H is not constant of $[a, b]$. Then there exist $c, d \in [a, b]$ such that $c < d$ and $H(c) \neq H(d)$. By the Mean Value Theorem, we can find $x \in (a, b)$ such that

$$H'(x) = \frac{H(d) - H(c)}{d - c} \neq 0.$$

However, $H'(x) = 0$ for all $x \in [c, d]$, whence we have a contradiction. □

Discussion. The proof of this theorem is another of the many applications we have seen for the Mean Value Theorem. The reader should come to an understanding of why the Mean Value Theorem is the natural tool to apply in this context. An attempt to construct a more direct proof from the definition of derivative might aid in developing this appreciation.

The converse of this theorem is left as Exercise 1. The theorem and its converse imply that if a function has a primitive, then it has an infinite number of primitives. □

The next result that brings forth the relationship between integrals and primitives enables us to evaluate a Riemann Integral easily, provided we have the knowledge and access to a primitive of that function. However, not all functions possess primitives, and the theorem is applicable only if the primitive is already available in advance.

Theorem 5.5.2 (Fundamental Theorem of Calculus). *If f is integrable on $[a, b]$ and F is any primitive for f on $[a, b]$, then*

$$\int_a^b f = F(b) - F(a).$$

Proof. Let $P = \{a = x_0, x_1, \ldots, x_n = b\}$ be any partition of $[a, b]$. Consider the interval $[x_{i-1}, x_i]$. By the Mean Value Theorem (Theorem 4.3.3), we have that for some $c_i \in (x_{i-1}, x_i)$

$$F(x_i) - F(x_{i-1}) = F'(c_i)(x_i - x_{i-1}) = f(c_i)d_i,$$

where $d_i = x_i - x_{i-1}$. Let Q be the intermediate partition of P consisting of the c_i 's. It is immediate that

$$F(b) - F(a) = \sum [F(x_i) - F(x_{i-1})]$$
$$= \sum f(c_i)d_i = R(f, P, Q),$$

a Riemann sum. Thus, for every partition, P, regardless of norm, there is an intermediate partition, Q, such that $R(f, P, Q) = F(b) - F(a)$. Hence, we conclude

$$\int_a^b f = F(b) - F(a). \qquad \square$$

Corollary. *If the derivative, f', of f is integrable on $[a, b]$, then*

$$\int_a^b f' = f(b) - f(a).$$

Discussion. Theorem 5.5.2 is known as the Fundamental Theorem of Calculus. Its purpose is to make the process of integration easier. Specifically, to obtain the integral of f on $[a, b]$, find *any* primitive, F, of f; evaluate this primitive at a and b, and finally, compute $F(b) - F(a)$. We emphasize that any primitive will do. It is clear that when a primitive is known, the process described above is much simpler than trying to use the methods suggested in the previous section. However, as the reader well knows from previous studies, the work involved in finding a primitive for a given function can be substantial, and as we have already remarked, a simple primitive may not exist. It is clear then that we should try to elucidate the conditions that will guarantee the existence of a primitive, F, although mere existence will not guarantee that we can produce a primitive in any kind of a usable form. Moreover, simply because a function has a primitive on a given closed interval does not ensure that it is integrable there (see Exercise 8).

Let us turn now to the proof of the theorem. As with the previous theorem, it is an application of the Mean Value Theorem, and a beautifully simple one at that. The key line is the proof is

$$F(x_i) - F(x_{i-1}) = F'(c_i)(x_i - x_{i-1}) = f(c_i)(x_i - x_{i-1}).$$

The essential fact that must be understood is that if we sum over i, the left-hand side sums to $F(b) - F(a)$, while the right-hand side is a Riemann sum. Once this is realized, letting the norm of the partition go to 0 completes the proof.

The Mean Value Theorem also provides possible insight about the original thinking that led to the Fundamental Theorem. Consider the problem of finding the area under a curve, as shown in Figure 5.5.1. Suppose the function, F, evaluated at c yields the area bounded by the graph of f, the x-axis, the lines $x = a$ and $x = c$ for any $c \in [a, b]$. The portion of the figure bounded by $x = c$ and $x = c + \Delta x$ has an area is given by

$$F(c + \Delta x) - F(c) = f(c_0) \cdot \Delta x,$$

where $c_0 \in [c, c + \Delta x]$. One could supply a variety of arguments for this position, but the initial one was probably that it was 'intuitively obvious'. In any case, if one supposes that F is differentiable, then division by Δx yields

$$\frac{F(c + \Delta x) - f(c)}{\Delta x} = f(c_0).$$

Since the left-hand side contains a quantity that in the limit becomes the derivative of F, one concludes that f must be the derivative of F.

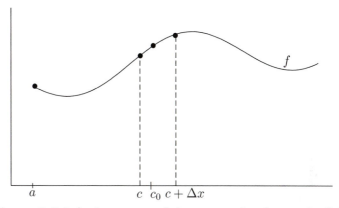

Figure 5.5.1 An increment of the area under the graph of f; the numerical value is given by $F(c + \Delta x) - F(c)$.

The Fundamental Theorem also gives rise to the more standard notation for integrals. If we set $\Delta x_i = x_i - x_{i-1}$, then

$$\int_a^b f = \lim_{\|P\| \to 0} R(f, P, Q) = \lim_{\|P\| \to 0} \sum_{i=1}^n f(c_i) \Delta x_i.$$

In the limit, $\Delta x \to dx$, which may be thought of as an infinitesmally small number. This forces $f(c_i) \to f(x)$, where $x \in [c_i, c_i + dx]$. Thus we have the notational form

$$\int_a^b f(x) dx = \int_a^b f. \qquad \square$$

Our aim now is to find a condition on an **integrand** (function to be integrated) that will guarantee the existence of a primitive.

Theorem 5.5.3. *Let f be integrable on $[a, b]$, and define*

$$F(x) = \int_a^x f, \; x \in [a, b],$$

then F is a continuous function with domain $[a, b]$. Further, if $c \in (a, b)$ and f is continuous at c, then F is differentiable at c and $F'(c) = f(c)$.

Proof. The fact that $F(x)$ is defined for each $x \in [a, b]$ follows from Theorem 5.4.7. That F is a function follows from the fact that the limit of a Riemann sum is unique. The continuity of F is left as Exercise 2.

To establish differentiability, consider the quantity $\frac{F(c+h)-F(c)}{h}$, where h is small enough to satisfy $c + h \in (a, b)$. From the definition of F, we have

$$\frac{F(c + h) - F(c)}{h} = \frac{1}{h} \int_c^{c+h} f.$$

Let M_h and m_h denote the maximum and the minimum of f respectively, on the interval $[c, c+h]$ (the specification for this interval would be reversed if $h < 0$). It is immediate that

$$m_h h \leq \int_c^{c+h} f \leq M_h h.$$

Thus,

$$m_h \leq \frac{F(c+h) - F(c)}{h} < M_h.$$

Now, since f is continuous at c, we can find a $\delta > 0$ such that if $|h| < \delta$, we have

$$f(c) - \epsilon < m_h \leq \frac{F(c+h) - F(c)}{h} \leq M_h < f(c) + \epsilon.$$

But this means that

$$\left| \frac{F(c+h) - F(c)}{h} - f(c) \right| < \epsilon$$

whenever $|h| < \delta$, whence $F'(c) = f(c)$, as claimed. $\square$

Discussion. This theorem makes four important points. First, it tells us that if f is integrable on $[a, b]$, then there is a function, F, that is naturally associated with f via the integration process. The domain of F is the same as interval over which f, is integrated, namely, $[a, b]$. Second, F is continuous. Thus, the process of integration generates continuous functions. Third, if the function, f, is continuous on $[a, b]$, then F is differentiable on $[a, b]$. Thus, the process of integration applied to continuous functions generates differentiable functions. Fourth, the theorem tells us that at any point of continuity of f, we will have $F'(c) = f(c)$. This means that if f is continuous on the whole of $[a, b]$, then F will be a member of the family of primitives of f on $[a, b]$.

For the case of continuous functions, this leads to the notation

$$\int f(x)\, dx$$

for the family of primitives of f. This collection of symbols is often given the name **indefinite integral** of f. It does not denote a single function, but a family of functions. Thus, a member of the indefinite integral of f will always be an antiderivative for f. More generally, we will refer to any function, F, that is obtained by integration of f on some interval $[a, x]$ as an **integral** of f.

Naturally, the discussion above leaves us to wonder about cases where f is not continuous. Evidently, the integral of f is not of necessity differentiable at a particular point of $[a, b]$. An example illustrating this fact is contained in Exercise 4. Moreover, there are functions that when integrated generate functions that are not differentiable at any point of their domain. Worse yet, even when the integral of f is differentiable, the derivative may not coincide with f; that is, $F'(c) \neq f(c)$ for various values of c in that interval. Thus, the question of just what can happen when we first integrate f and then try to differentiate its integral, F, is exceedingly complex. Of course, there is also the reverse process: differentiate f and then try to integrate the derivative f'. Suffice it to say that the corollary to Theorem 5.5.1 notwithstanding, the situation is equally complex, and we shall explore it further in the exercises. □

The theorems proved so far lead naturally to

Theorem 5.5.4. *Let f be continuous on $[a, b]$. Then there is a point $c \in (a, b)$ such that*

$$f(c) = \frac{1}{b-a} \int_a^b f.$$

Discussion. The straightforward proof of this theorem is left to the reader. For obvious reasons, it is usually referred to as the **Mean Value Theorem for Integrals**. Notice that one conclusion that can be drawn from this theorem is that for a nonnegative continuous function, f, the area between f, the lines $x = a$ and $x = b$ and the x-axis can be realized as the area of a rectangle having one side of length $(b - a)$ and the other $f(c)$ for some $c \in (a, b)$. With a little work, the result can be tracked back to its ultimate home, namely, the Completeness Axiom. □

Since we have developed the Fundamental Theorem, we would not want to leave this section without looking at the two most important techniques for finding primitives. The first is the well-known formula for **integration by parts**.

Theorem 5.5.5. *If f and g are differentiable functions on $[a, b]$ such that the derivatives f' and g' are both integrable on $[a, b]$, then*

$$\int_a^b fg' = f(b)g(b) - f(a)g(a) - \int_a^b f'g.$$

Proof. By hypothesis, f and g are both continuous and, hence, Riemann integrable. Therefore, fg' as well as $f'g$ are integrable. Consequently, $fg' + f'g = (fg)'$ is also integrable and by Theorem 5.4.1,

$$\int_a^b (fg)' = \int_a^b fg' + \int_a^b f'g.$$

By the Fundamental Theorem, we may then write

$$\int_a^b (fg)' = f(b)g(b) - f(a)g(a).$$

The result is now obvious. □

Discussion. The above theorem is a clever device by which we can write down the integral of the product of two functions. What we need to know is that the primitive of one of the two should be expressible in a simple form and that the derivative of the other should also be simple so that the product of these two is easily integrable. In many instances, the fact that the identity function $f(x) = x$ is the primitive of the constant function 1 is exploited to compute $\int f$, by invoking the integration by parts rule to the product $f \cdot 1$.

The source of this theorem is the product rule for differentiation. The reader should ascertain this relationship, as the proof will then become transparent. □

The next theorem is another device by which we can perform integration by composing the given function with another function g so that the new function $f \circ g$, admits an easy integral. This procedure is known as **the change of variable formula**, or **simple substitution**.

Theorem 5.5.6. *Let f be defined and continuous on the range of the function g. If g' is integrable on $[c, d]$, then*

$$\int_a^b f = \int_c^d (f \circ g)g',$$

where $a = g(c)$ and $b = g(d)$.

Proof. Let $F(x) = \int_a^x f$ be a primitive of f. Note that F is defined on the range of g. Since f is continuous, F is differentiable by Theorem 5.5.3, and $F' = f$. Thus, $G = F \circ g$ is defined on $[c, d]$. By the chain rule, G is differentiable, and $G' = (F' \circ g)g' = (f \circ g)g'$. Also $f \circ g$ is continuous since f and g are, and hence integrable. By hypothesis, g' is also Riemann integrable; thus, $(f \circ g)g'$ is integrable. Hence,

$$
\begin{aligned}
\int_c^d (f \circ g)g' &= \int_c^d G' = G(d) - G(c) \\
&= F(g(d)) - F(g(c)) \\
&= F(b) - F(a) = \int_a^b f. \qquad \square
\end{aligned}
$$

Discussion. The reader should note that there is no requirement that the quantity a be less than b. In fact, the opposite may be true with no effect on the correctness of the theorem. The source of the proof is in the Chain Rule for differentiation, and it is in this connection that the reader should seek insight into the proof.

In fact, the theorem might be thought of as a chain rule for integration, except that it is used exactly the opposite way from the Chain Rule; that is, the Chain Rule tells us how to differentiate a composite function while the Change of Variable Theorem tells us how to simplify an integral by rewriting it as a composite function. Thus, we are using the equalities in the opposite directions. Suffice it to say that the value of this theorem depends on the user's skill at shrewdly rewriting the original function so as to come up with a simple integral. For a detailed treatment of the many uses of this theorem, see the section on techniques of integration in any calculus book. $\square$

We conclude this section with a theorem known as the **Second Mean Value Theorem for Integrals** or **Bonnet's Mean Value Theorem**.

Theorem 5.5.7. *If f is continuous and increasing on $[a, b]$ and g is nonnegative and integrable there, then there exists a $z \in [a, b]$ such that*

$$\int_a^b fg = f(a) \int_a^z g + f(b) \int_z^b g.$$

Proof. By an easily proved generalization of the First Mean Value Theorem (see Exercise 12), there exists $y \in [a, b]$ such that

$$\int_a^b fg = f(y) \int_a^b g.$$

Also, $f(a) \le f(y) \le f(b)$, since f is increasing. Now, consider the function G defined on $[a, b]$ by

$$G(x) = [f(b) - f(a)] \int_x^b g.$$

Since g is nonnegative, G is a decreasing function. Further,

$$
\begin{aligned}
G(a) & = [f(b) - f(a)] \int_a^b g \\
& \ge [f(y) - f(a)] \int_a^b g \\
& \ge 0 = G(b).
\end{aligned}
$$

Because G is continuous, we can apply the Intermediate Value Theorem to obtain $z \in [a, b]$ such that

$$G(z) = [f(y) - f(a)] \int_a^b g.$$

Now,

$$G(z) = [f(b) - f(a)] \int_z^b g = [f(y) - f(a)] \int_a^b g.$$

For this particular z, we have by expanding,

$$
\begin{aligned}
f(b) \int_z^b g - f(a) \int_z^b g & = f(y) \int_a^b g - f(a) \int_a^b g, \\
& = \int_a^b fg - f(a) \int_a^b g.
\end{aligned}
$$

Rearranging yields

$$\int_a^b fg = f(a) \left[\int_a^b g - \int_z^b g \right] + f(b) \int_z^b g,$$

which simplifies to the desired result. $\square$

Discussion. This theorem can best be understood in two steps. First, f is continuous and increasing, while g is nonnegative and integrable. It should seem obvious that there is a $y \in [a, b]$ such that

$$\int_a^b fg = f(y) \int_a^b g.$$

This is the first step. Since f is increasing, $f(a) \le f(y) \le f(b)$. Thus,

$$f(a) \int_a^b g \le f(y) \int_a^b g \le f(b) \int_a^b g.$$

It should therefore seem reasonable that a z should exist that will permit us to integrate $f(a) \int_a^z g$ and $f(b) \int_z^b g$ and sum to get the value in the middle. However, the reader will note that to achieve this 'reasonableness' requires the full power of the Intermediate Value Theorem. □

There are many applications of the Mean Value Theorem as we shall see when we develop the trigonometric functions and their inverses in Chapter 9.

Exercises

1. Prove the converse of Theorem 5.5.1.

2. Complete the proof of Theorem 5.5.3. Specifically, show that the function F is continuous on $[a, b]$.

3. Find the primitives for the following functions. Use whatever formulae are necessary for elementary functions.

  (a) $\int \dfrac{x}{\sqrt{x^2 + 5}}$;

  (b) $\int (2x^3 + 1)^7 x^2$;

  (c) $\int \dfrac{x^2}{(x^3 - 2)^2}$;

  (d) $\int \dfrac{x + 1}{x^2 + 2x + 3}$;

  (e) $\int \sin^4 x \cos x$;

  (f) $\int \dfrac{e^x - e^{-x}}{e^x + e^{-x}}$;

  (g) $\int \dfrac{\sec x \tan x}{\sqrt{1 + \sec x}}$;

  (h) $\int \dfrac{x \ln(1 + x^2)}{(1 + x^2)}$;

  (i) $\int x \ln x$;

  (j) $\int x \arctan x$;

  (k) $\int x^2 \sin x$;

  (l) $\int x e^x$;

  (m) $\int \sec^5 x$;

  (n) $\int \dfrac{x \ln x}{(x^2 - 1)^{3/2}}$;

  (o) $\int \ln(x^2 + 1)$;

  (p) $\int \dfrac{x^3}{e^{x^2}}$;

  (q) $\int e^{ax} \cos(bx + c)$;

  (r) $\int \dfrac{x^5}{x + 1}$.

4. Let f be defined on $[-1, 1]$ by:

$$f(x) = \begin{cases} 1, & \text{if } x > 0 \\ 0, & \text{if } x = 0 \\ -1, & \text{if } x < 0. \end{cases}$$

Show the integral of f is not differentiable at 0.

5. Let f be monotone increasing on $[a, b]$ and suppose that f is discontinuous at $z \in (a, b)$. Show the integral of f cannot be differentiable at z.

6. Let $g : \mathbf{N} \to \mathbf{Q} \cap [0, 1]$ be one-to-one and onto. Set

$$M_x = \{i : g(i) \le x, \ i \in \mathbf{N}\}.$$

We define f on $[0, 1]$ by

$$f(x) = \sum \{2^{-i} : i \in M_x\}.$$

Using whatever fact from the theory of infinite series you need, show that f is monotone increasing and discontinuous at every rational, but is continuous at every irrational. Conclude that the integral of f is not differentiable at any rational.

7. Let f be differentiable on $[a, b]$ and suppose that f' is integrable there. Show $\int_a^b f' = f(b) - f(a)$.

8. Define f on $[-1, 1]$ by

$$f(x) = \begin{cases} x^2 \sin \frac{1}{x^2}, & \text{if } x \neq 0 \\ 0, & \text{if } x = 0. \end{cases}$$

Show

$$f'(x) = \begin{cases} 2x \sin \frac{1}{x^2} - \frac{2}{x} \cos \frac{1}{x^2}, & \text{if } x \neq 0 \\ 0, & \text{if } x = 0. \end{cases}$$

Conclude that f' is not integrable on $[-1, 1]$.

9. Consider the function f discussed in Example 5.2.5. Show that $F(x) = \int_0^x f$ is differentiable on $[0, 1]$ but that $F'(x) \neq f(x)$ at any rational other than 0.

10. For each of the integrals, find the mean value specified by Theorem 5.5.4:

    (a) $\int_0^2 x^2$;   (b) $\int_0^1 \sin x$;   (c) $\int_0^3 e^x$.

11. Suppose that g is nonnegative on $[a, b]$ and that both g and fg are integrable there. Show if $m < f(x) < M$ are bounds for f on this interval, then there exists $z \in (m, M)$ such that

$$z \int_a^b g = \int_a^b fg.$$

12. Show if f is continuous and g is integrable on $[a, b]$, then there exists $z \in [a, b]$ such that

$$\int_a^b fg = f(z) \int_a^b g.$$

13. Suppose f has a continuous derivative on $[a, b]$. Show f can be represented as the difference of two nondecreasing functions.

14. Let f be continuous on $[a, b]$ and suppose that for every function g that is integrable on $[a, b]$, $\int_a^b fg = 0$. What conclusions can be drawn about f?

15. Let f be defined on $[0, 1]$ by

$$f(x) = \begin{cases} 1 - x, & \text{if } x \text{ is irrational} \\ \sqrt{1 - x^2}, & \text{if } x \text{ is rational}. \end{cases}$$

Find the values of the upper and the lower integrals of f.

16. Let $a, b > 0$. Use the Change of Variable Theorem to show that

$$\int_a^{ab} \frac{1}{t} = \int_1^b \frac{1}{t}.$$

17. If f is continuous on $[a, b]$ and g and h are differentiable with range contained in $[a, b]$, show that

$$\left(\int_g^h f \right)' = (f \circ h)h' - (f \circ g)g'.$$

18. Let f be integrable in $[a, b]$ and $F(x) = \int_x^b f$ for $x \in [a, b]$. Prove that at each point of continuity of f, $g'(x) = -f(x)$.

19. For $x > 0$, set $L(x) = \int_1^x \frac{1}{t}$. Establish the following formulae:

    (a) $L(ab) = L(a) + L(b)$;   (b) $L(\frac{a}{b}) = L(a) - L(b)$;

    (c) $L(1) = 0$;   (d) $L(a^q) = qL(a)$, $q \in \mathbf{Q}$;

    (e) $L'(x) = \frac{1}{x}$;   (f) $L(x)$ is increasing and continuous on $\mathbf{R}^+$.

    [Note: The function $L(x)$ is the familiar logarithmic function.]

20. Define $E(x)$ to be the inverse of the function $L(x)$ considered in Exercise 19 for $x \in \mathbf{R}$. Find the derivative of $E(x)$ and use this to evaluate $\int_0^a E(x)$ for $a \in \mathbf{R}$.

21. Let $f(x) = x^2$. Show every primitive of f is realizable as $\int_a^x t^2$ for some choice of a. Is the same true for $f(x) = \cos x$?

22. Use your knowledge of primitives to compute the following limits:

(a) $\displaystyle \lim_{n \to \infty} \left(\frac{1}{n+1} + \frac{1}{n+2} + \ldots + \frac{1}{2n} \right);$

(b) $\displaystyle \lim_{n \to \infty} \sum_{k=0}^{n-1} \frac{1}{\sqrt{n^2 - k^2}};$

(c) $\displaystyle \lim_{n \to \infty} \sum_{i=1}^{n} \frac{i\sqrt{i^2 - n^2}}{n^3};$

(d) $\displaystyle \lim_{n \to \infty} \sum_{k=1}^{n} \frac{k^p}{n^{p+1}};$

(e) $\displaystyle \lim_{n \to \infty} \sum_{r=1}^{n} \frac{n^3}{n^2 + r^2}.$

23. Let

$$f(x) = \begin{cases} \frac{x^2 - 4}{x - 2}, & \text{if } x \neq 2 \\ 0, & \text{if } x = 2. \end{cases}$$

The primitive $F(x) = \frac{x^2}{2} + 2x$ is differentiable at $x = 2$, but $F'(2) = 4 \neq f(2)$. Does this contradict the Fundamental Theorem?

24. Let

$$f(t) = \begin{cases} t, & \text{if } t < 0 \\ t^2 + 1, & \text{if } 0 \leq t \leq 2 \\ 0, & \text{if } t > 2. \end{cases}$$

Find a primitive $F(x)$. Is F differentiable?

25. Let $F(x) = \int_{x-1}^{x+1} f$, where f is continuous on $\mathbf{R}$. Show F is differentiable, and compute F'.

26. Show $f(x) = \begin{cases} 1, & \text{if } 0 \leq x < 1 \\ 2, & \text{if } \leq x \leq 2. \end{cases}$ is discontinuous at $x = 1$, but a primitive of f; namely, $F(x) = \begin{cases} x, & \text{if } 0 \leq x < 1 \\ 2x - 1, & \text{if } 1 \leq x \leq 2. \end{cases}$ is continuous on $[0, 2]$. Is there any contradiction?

27. Let f be continuous in $[0, a]$. Define $f_0(x) = f(x)$, and for $n > 0$, $f_{n+1}(x) = \frac{1}{n!} \int_0^x f(t)(x - t)^n dt$. Show the nth derivative of $f_n(x)$ exists and coincides with $f(x)$.

5.6   Improper Riemann Integration

Once we have developed the Riemann integral, a number of outstanding problems remain. The first of these is to find a simple characterization of the class of Riemann integrable functions on $[a, b]$. The second is to look for ways to extend the integral concept so that it can be applied to functions that are not Riemann integrable. The first of these problems is beyond the scope of this book, although, as we have remarked, the answer is directly related to the nature of the set of discontinuities of the function. One approach to the problem of extending the integral concept is the subject of this section.

All functions that are Riemann integrable have bounded domains and bounded ranges. That functions have bounded domains arises from the fact that each of the integral definitions required the function to be defined on a closed interval. That functions have bounded ranges arose as a prerequisite in the case of the Riemann-Darboux integral due to the fact that we required the supremum and infimum of the function to exist as a finite number on each subinterval. For the case of the Riemann integral, even though we dropped boundedness as a prerequisite, the fact that the function was bounded turned out to be a consequence of Riemann integrability (see Theorem 5.3.2). Thus, an obvious question that arises directly from the theory is whether the integral concept can be extended to include functions having unbounded domains or unbounded ranges. In this section, we relax either or both of the bounded requirements and consider integrals of functions that are either defined on unbounded intervals, or the interval is bounded, but the function itself is unbounded there. This leads to the idea of **improper integrals**, and we discuss the convergence and divergence of such integrals. Two classical examples, namely, the beta and the gamma integrals, will serve as appropriate illustrations.

Definition. Let f be a function defined on an interval $[a, b)$, $a, b \in \mathbf{R}$. (We also allow b to be $+\infty$.) If f is Riemann integrable on $[a, c]$ for each c such that $a \leq c < b$, but not Riemann integrable on $[a, b]$, then the integral, $\int_a^b f$, is said to be an **improper integral**. In particular, we say that the integral is **improper** at the point b, and b is a **singularity** for the integrand f. We then define the **improper integral**, $\int_a^b f$, by

$$\int_a^b f = \lim_{c \to b^-} \int_a^c f.$$

If this limit exists as a real number, A, then we say that the improper integral **converges** to A. If the limit is infinite, or fails to exist, the improper integral is said to **diverge**.

Discussion. We want to address the question of why this definition is a reasonable way to extend the integral concept. Thus, consider the fact that f is Riemann integrable on $[a, c]$ for each $a \leq c < b$. This fact forces f to be bounded on $[a, c]$, whence it means that f can fail to be integrable on $[a, b]$ only if one of the following two statements is true:

(i) f is unbounded in a neighborhood of $b < \infty$;

(ii) $b = +\infty$.

Following the notation of Section 5.5, let us define a function $F : [a, b) \to \mathbf{R}$ by

$$F(x) = \int_a^x f, \ x \in [a, b).$$

For the former case, since f is Riemann integrable on $[a, c]$, by Theorem 5.5.3, F is continuous on $[a, c]$ for each $c \in [a, b)$. Indeed, if f is continuous on $[a, b)$, then F will be a primitive of f. Even so, we cannot guarantee that F is continuous at the point b. (The reader should be able to give many examples of functions that are continuous on an open interval but for which there is no continuous extension to a closed interval.) However, if it happens to be the case that F has a continuous extension at b, then it is clear that this extension must be given by

$$F(b) = \lim_{x \to b^-} F(x) = \lim_{x \to b^-} \int_a^x f.$$

Now, and this is the essential point, it may be the case that F has a continuous extension at b, *even though* f is unbounded at b. (Think of $f(x) = \frac{1}{x^{1/3}}$, $x \in [-1, 0)$.) For such a case, the fact that F is continuous, or can be defined so as to be continuous, at b argues strongly that the integral of f should exist over the entire closed interval, $[a, b]$. Having said that the integral over the closed interval should exist, what should its value be? Evidently, the only possible sensible value would be $\lim_{x \to b^-} F(x)$, which is the value given by the definition.

In the case where $b = +\infty$, the definition reduces to

$$\int_a^\infty f = \lim_{N \to \infty} \int_a^N f,$$

provided the latter limit exists. Expressing this in the language adopted above, we see that convergence of the improper integral implies $\lim_{x \to \infty} F(x)$ exists as a finite real number. Thus, it would appear reasonable to take this value as the value for the integral over the interval $[a, \infty)$. Again, the point is that there are functions for which this limit exists: $f(x) = \frac{1}{x^2}$ on $[1, \infty)$ is a good example.

Evidently, a dual situation exists for intervals of the form $(a, b]$. Thus, if f is bounded in $(a, b]$ and integrable in $[c, b]$ for each $c > a$ (we allow c to be $-\infty$ also), then the improper integral $\int_a^b f$ is defined to be $\lim_{c \to a^+} \int_c^b f$. If the limit exists as a real number, then the improper integral converges. If the limit is either infinite, or does not exist, we then have a divergent improper integral. In case f is bounded and integrable in $(-\infty, b]$, then $\lim_{N \to \infty} \int_{-N}^b f$ is defined to be the improper integral $\int_{-\infty}^b f$.

As in the case of Riemann integral, if $b < a$, we set $\int_a^b f = -\int_b^a f$. □

Our next definition is a straightforward generalization of the initial definition.

Definition. Let $a < c < b$ (we allow $a = -\infty, b = +\infty$). If the integrals $\int_a^c f$ and $\int_c^b f$ are both improper, where either the first is improper at a and the second is improper at b, or both are improper at c, and if each is convergent, then we define $\int_a^b f = \int_a^c f + \int_c^b f$. We also say that the integral $\int_a^b f$ is convergent.

Discussion. The purpose of this definition is to extend the initial definition to cover cases where f is unbounded at a point c, $c \in (a, b)$, or where $a = -\infty$ and $b = +\infty$. These cases are handled by splitting the integral into two integrals at the point $c \in (a, b)$. If at least one of the integrals is divergent, the improper integral $\int_a^b f$ is divergent. We take this opportunity to warn the reader that if

f is unbounded at a point c, and if $a = -\infty$ and $b = +\infty$, then the improper integral $\int_{-\infty}^{+\infty} f$ *should not be* evaluated as the limit: $\lim_{N\to\infty} \int_{-N}^{N} f$. These two quantities are not the same. (The reader is asked in Exercise 1 to provide an example to illustrate this situation.) What we are demanding in this definition is that for an arbitrary c, the integral $\int_{-\infty}^{c} f$ as well as the integral $\int_{-c}^{\infty} f$ should separately converge. $\square$

Example 1. Discuss the convergence or divergence of the integrals

(a) $\int_0^1 \frac{1}{x^p}$; (b) $\int_1^\infty \frac{1}{x^p}$; (c) $\int_0^\infty \frac{1}{x^p}$.

Solution. For (a) observe the function $f(x) = \frac{1}{x^p}$ is continuous on $(0, 1]$, but is undefined at $x = 0$ irrespective of the value of p. If $p < 0$, f is bounded in $(0, 1]$, so we can extend the definition to $x = 0$ by setting the value to be 0 when $x = 0$. If $p = 0$, the function is identically 1 throughout $[0, 1]$, so for $p \leq 0$, f, itself has a continuous extension to the whole of $[0, 1]$, whence it is Riemann integrable there. For $p > 0$, $\frac{1}{x^p}$ is unbounded at 0, and an improper integral results and three cases must be treated depending on the value of p.

If $0 < p < 1$, then $1 - p > 0$, and we then have

$$
\begin{aligned}
\int_0^1 \frac{1}{x^p} &= \lim_{c\to 0^+} \int_c^1 \frac{1}{x^p} = \lim_{c\to 0^+} \int_c^1 x^{-p} \\
&= \lim_{c\to 0^+} [F(1) - F(c)], \text{ where } F(x) = \frac{x^{1-p}}{1-p} \\
&= \lim_{c\to 0^+} \left(\frac{1}{1-p} - \frac{c^{1-p}}{1-p} \right) \\
&= \frac{1}{1-p},
\end{aligned}
$$

whence the integral converges to $\frac{1}{1-p}$.

For $p > 1$, $1 - p < 0$. If we repeat the computations above, we find that $F(c)$ is unbounded as c approaches 0^+. It follows that the above limit is $+\infty$; hence, the improper integral diverges.

Finally, when $p = 1$, using facts (to be) developed in Chapter 9, we have

$$
\int_0^1 \frac{1}{x} = \lim_{c\to 0^+} \int_c^1 \frac{1}{x} = \lim_{c\to 0^+} (-\ln c) = +\infty,
$$

whence the integral diverges, concluding the third case.

For (b), if $p > 1$, since there is no discontinuity in $(1, \infty)$, we have

$$
\begin{aligned}
\int_1^\infty \frac{1}{x^p} &= \lim_{N\to\infty} \int_1^N \frac{1}{x^p} \\
&= \lim_{N\to\infty} \left(\frac{N^{1-p}}{1-p} - \frac{1}{1-p} \right) \\
&= \frac{1}{p-1}.
\end{aligned}
$$

Hence, the integral converges for $p > 1$. We leave it to Exercise 2, to show that the integral diverges to $+\infty$ when $p \leq 1$.

Finally, for (c), consider separately the intervals $(0,1)$ and $(1,\infty)$. By the computations for the two previous cases, for any arbitrary p, one of the integral converges and the other diverges. Hence, the integral diverges to $+\infty$ for all p. □

Discussion. This is a straightforward application of the definition of convergence, with the knowledge of a primitive of the function considered. The only fact that we have not yet developed is that $\ln x$ is a primitive of $\frac{1}{x}$. □

Example 2. Discuss the convergence of the improper integral $\int_0^5 \frac{1}{(x-2)^3}$.

Solution. Observe that the function is unbounded at $x = 2$. So we must separately consider the intervals $(0,2)$ and $(2,5)$ and the two associated improper integrals given by

$$\lim_{h \to 0^+} \int_0^{2-h} \frac{1}{(x-2)^3} + \lim_{k \to 0^+} \int_{2+k}^5 \frac{1}{(x-2)^3}.$$

A simple computation shows that both the integrals diverge, and hence we conclude that this integral is divergent. □

Discussion. The reader might be tempted to perform the above calculations as

$$\lim_{h \to 0^+} \left\{ \int_0^{2-h} \frac{1}{(x-2)^3} + \int_{2+h}^5 \frac{1}{(x-2)^3} \right\}$$

without taking care to evaluate the two separate limits, one for h and the other for k which are taken *independently*. For $F(x) = \left[\frac{-1}{2(x-2)^2}\right]$, the result, then, would be

$$\lim_{h \to 0^+} (F(2-h) - F(0)) \quad + \quad (F(5) - F(2+h))$$

$$= \lim_{h \to 0} \left(\left[\frac{-1}{2h^2} + \frac{1}{8}\right] + \left[\frac{-1}{18} + \frac{1}{2h^2}\right] \right)$$

$$= \frac{5}{72},$$

showing that the integral converges to the value $\frac{5}{72}$. This answer makes no apparent sense, in view of the fact that both the separate limit calculations are divergent. However, on reflection, it suggests the following definition. □

Definition. If c is a point of discontinuity of the integral $\int_a^b f$ and

$$\lim_{h \to 0^+} \left(\int_a^{c-h} f + \int_{c+h}^b f \right) = A$$

exists, while the improper integral $\int_a^b f$ is divergent, then the value A is called **Cauchy's Principal Value** of the divergent integral.

Discussion. The Cauchy Principal Value for an integral acts to extend the integral concept even further. To understand the rationale for this extension, consider Figure 5.6.1, which presents part of the graph of $f(x) = \frac{1}{(x-2)^3}$. From Example 2, we know that $\int_2^{2+h} f$ does not exist, even as an improper integral. This argues that the area of the region bounded by $x = 2$, $x = 2+h$, $y = f(x)$, and $y = 0$ is infinite. However, what is apparent is that the area of this region should be identical to the area of the region bounded by $x = 2-h$, $x = 2$, $y = f(x)$, and $y = 0$. The reason underlying this assertion is that the two regions are congruent in the usual geometrical sense. Moreover, in any evaluation of an integral, $\int_a^b f$, where $a < 2 - h$ and $2 + h < b$, these two infinite areas should exactly cancel one another due to having opposite signs in the computation. If one accepts that these infinities can, and should, cancel, then one immediately gets the Cauchy Principal Value.

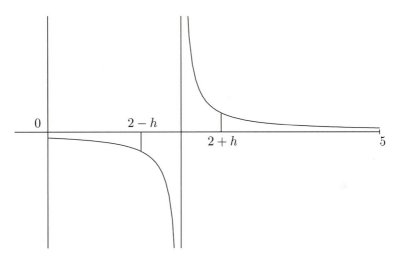

Figure 5.6.1 Cauchy's Principal Value for $\int_0^5 \frac{1}{(x-2)^3}$ is the limit as $h \to 0$ of the sums of the two areas from 0 to $2 - h$ and from $2 + h$ to 5.

Even if one accepts that reasoning above, it is clear that the existence of a Cauchy Principal Value in cases where the improper integrals do not exist is special and should be recognized as such. For this reason we avoid writing $\int_0^5 \frac{1}{(x-2)^3} = \frac{5}{72}$ and instead write

$$CPV \int_0^5 \frac{1}{(x-2)^3} = \frac{5}{72}$$

to indicate Cauchy's Principal Value of the divergent integral.

Two other forms of the Cauchy Principal Value are discussed in Exercises 4 and 5. □

Next, we consider functions that are non-negative throughout the interval of integration and obtain a test for deciding their convergence or divergence, by comparing them with an integral that is known to be convergent or divergent. Such a test is usually known as a **comparison test**. As a preliminary, we prove the following theorem, which resembles similar results on sequences and functions.

Theorem 5.6.1. *If f is nonnegative and Riemann integrable in $[a, K]$, $K > a$, then the improper integral $\int_a^\infty f$ converges if and only if the function $F : [a, \infty) \to \mathbf{R}$ defined by $F(x) = \int_a^x f$ is bounded. If the hypothesis is satisfied, then*

$$\int_a^\infty f = \sup \{F(x) : x > a\}.$$

Proof.   Note that F is nonnegative and monotonically increasing. Hence, $\lim_{x \to \infty} F(x)$ exists in $\mathbf{R} \cup \{\infty\}$, and further,

$$\sup \{F(x) : x > a\} = \lim_{x \to \infty} F(x) = \int_a^\infty f.$$

If F is bounded on $[a, \infty)$, say, by M, then the above limit exists and is less than or equal to M, whence the integral converges. On the other hand, if F is unbounded, by monotonicity, we conclude that $\lim_{x \to \infty} F(x) = \infty$, whence the integral diverges. $\square$

The following theorem is the so-called comparison test for convergence or divergence of the improper integral.

Theorem 5.6.2. *Let $0 \le f(x) \le g(x)$ for $x \in [a, \infty)$ and let f and g be integrable in $[a, K]$ for every $K > a$. If the improper integral $\int_a^\infty g$ converges, then the improper integral $\int_a^\infty f$ also converges. If $\int_a^\infty f$ diverges, then the integral $\int_a^\infty g$ diverges.*

Proof. If $\int_a^\infty g$ converges, then by Theorem 5.6.1, the function $G(x) = \int_a^x g$ is bounded. Also $0 \le F(x) = \int_a^x f \le G(x)$, proving the integral $\int_a^\infty f$ converges. The divergence case is left to the exercises. $\square$

Discussion. The proof given is straightforward. Nevertheless, because comparison tests are so common, it is worth examining the intuition. Consider then Figure 5.6.2. Because g dominates f and both are nonnegative, it is evident that the region defined by the graph of f, $x = 0$ and the x-axis is included in the region defined by the graph of g, $x = 0$ and the x-axis. It seems evident that if the latter region has a finite area, then the former must also. Similarly, if the former has an infinite area, then the latter must also.

The utility of comparison tests rests on having a large collection of functions available with two properties. First, the convergence properties of the individual functions must be known. Second, the collection must contain enough functions so that an appropriate comparison can be made between some member of the collection and any given function for which the convergence properties are unknown. The class of functions whose convergence properties are discussed in Example 1(b) is one useful collection of functions that can be used in comparison tests. $\square$

A practical way to use the above test is by a limit form of the comparison test. This is enunciated in the next theorem.

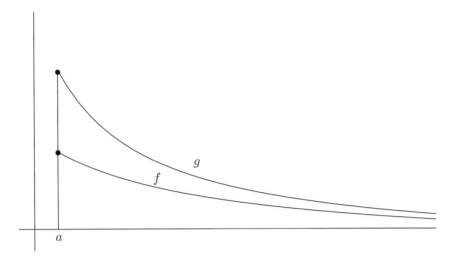

Figure 5.6.2 An illustration of the comparison test for improper integrals; if area under g exists on $[a, \infty)$, surely the area under f will also exist.

Theorem 5.6.3. *Let f be nonnegative and bounded on $[a, \infty)$, g be positive on $[a, \infty)$, and $\lim_{x \to \infty} \frac{f(x)}{g(x)} = L$. Further, suppose that f and g are integrable on $[a, K]$, for each $K \in [a, \infty)$.*

   (i) *If $L \neq 0$, then both the integrals $\int_a^\infty f$, and $\int_a^\infty g$ either converge or both of them diverge;*

   (ii) *If $L = 0$, and the integral $\int_a^\infty g$ is convergent, then $\int_a^\infty f$ is convergent;*

   (iii) *If $L = \infty$, and the integral $\int_a^\infty g$ is divergent, then $\int_a^\infty f$ is divergent.*

Proof. The proof is left to the reader as Exercise 7. $\square$

Discussion. Theorem 5.6.3 is a typical reformulation of the comparison test. It results from the recognition that to compare two functions, for convergence or divergence purposes over the interval $[a, \infty)$, one real interest is that the behavior of the one closely reflects the behavior of the other only on the tail of the interval, that is, on $[b, \infty)$, where b can be very large. This can be completely captured by looking at the limit of the ratio of the two functions. For this reason, this type of test is referred to as a **ratio comparison test**.

    The two tests given have been stated so as to apply to the problem of testing whether the improper integral over an interval of the form $[a, \infty)$ exists. In Exercise 8 we ask the reader to formulate an appropriate version of the ratio comparison test to apply in other situations. In the next example, such a test is applied.

    The comments regarding the utility of the comparison test also apply to the ratio comparison test. $\square$

Example 3. Discuss the convergence or divergence of the following improper integrals:

(a) $\int_0^1 \frac{1}{\sqrt{x}(1+x^2)}$; (b) $\int_0^1 \frac{1}{x^2(1+x)^2}$; (c) $\int_0^1 \frac{1}{\sqrt{x}\,\sqrt{1-x}}$.

Solution. For (a), 0 is the only singularity of the integrand f. Compare the integral f with the integral of g on $(0,1]$, where $g(x) = \dfrac{1}{\sqrt{x}}$. Since

$$\lim_{x\to 0+} \frac{\dfrac{1}{\sqrt{x}(1+x^2)}}{\dfrac{1}{\sqrt{x}}} = 1$$

and the integral $\int_0^1 \dfrac{1}{\sqrt{x}}$ converges, we conclude that the given integral is convergent by Theorem 5.6.3.

For (b), again we form a ratio to make the comparison. This time, set $g(x) = \frac{1}{x^2}$ on $(0,1]$. Since $\int_0^1 g$ is divergent, the integral of f is also divergent.

For (c), both 0 and 1 are discontinuities of the integrand. So, we separately consider the integrals $\int_0^{\frac{1}{2}} f$ and $\int_{\frac{1}{2}}^1 f$. The former converges, by comparing it with $\int_0^{\frac{1}{2}} \frac{1}{\sqrt{x}}$. For the latter, we obtain convergence again, if we compare the integral with $\int_{\frac{1}{2}}^1 \dfrac{1}{\sqrt{1-x}}$. Since both the integrals are convergent, we conclude that the given integral also converges. □

Discussion. The above examples illustrate our earlier remark that knowledge of the behavior of the improper integrals of the functions $\frac{1}{x^p}$ for various values of p is very useful. The reader should fill in all the missing details in parts (b) and (c), including checking that suitable hypothesis hold. □

Two important improper integrals, which have many practical applications are the **gamma** and the **beta** integrals. We study them in Examples 4 and 5. Some basic properties of the exponential function are assumed in what follows. These are formally developed in Chapter 9.

Example 4. Discuss the convergence or divergence of the integral $\int_0^\infty f$, where $f(x) = x^{a-1}e^{-x}$, $a > 0$.

Solution. Consider the intervals $(0,1]$ and $[1,\infty)$ separately and let $I_1 = \int_0^1 f$ and $I_2 = \int_1^\infty f$. Note that the improper integral $\int_1^\infty \frac{1}{x^2}$ converges by Example 1. Also, since $\frac{x^{a-1}e^{-x}}{x^{-2}}$ approaches 0 as $x \to \infty$, the integral I_2 converges for all a.

On the other hand, if $a < 1$, there is a discontinuity at $x = 0$. Comparing the integrand with $g(x) = x^{a-1}$, one readily sees that $\lim_{x\to 0+} \frac{x^{a-1}e^{-x}}{x^{a-1}} = 1$. Since $\int_0^1 x^{a-1}$ converges if $a-1 > -1$, and diverges otherwise, we see that I_1 converges only if $a > 0$. Thus, the given integral converges for all $a > 0$. □

Discussion. This integral is usually known as **Euler's Second Integral** or the **gamma integral**. The function $\Gamma : (0,\infty) \to \mathbf{R}$ defined by

$$\Gamma(a) = \int_0^\infty x^{a-1}\, e^{-x}$$

is called the **gamma function**. The gamma function first appeared in 1729 in a correspondence between L. Euler and Goldbach. Euler was interested in finding a function f with the property that $f(n) = n!$ for each natural number n. If such a function could be found, it then makes sense to talk about factorials of real numbers such as 2.973. Euler discovered the following 'infinite product' possesses this remarkable property:

$$n! = \left[\left(\frac{2}{1}\right)^n \frac{1}{n+1}\right]\left[\left(\frac{3}{2}\right)^n \frac{2}{n+2}\right]\left[\left(\frac{4}{3}\right)^n \frac{3}{n+3}\right]\cdots,$$

and this expression can be obtained as the 'limit' of a finite product,

$$n! = \lim_{m\to\infty} \frac{m!(m+1)^n}{(n+1)(n+2)\ldots(n+m)}.$$

Because this last expression is valid when n is replaced by x, it leads to following definition of the **gamma function**:

$$\Gamma(x) = \lim_{m\to\infty} \frac{m!(m+1)^x}{(x+1)(x+2)\ldots(x+m)},$$

where x is any real number, except negative integers. Euler also knew $n! = \int_0^1 (-\ln x)^n dx$, and he subsequently modified this expression to generate the gamma integral we have discussed above.[3]

One of the important properties of gamma integral is that $\Gamma(x+1) = x\Gamma(x)$, which the reader can easily verify. Motivated by this identity, the gamma function can be extended to negative real numbers also. If $x \in (-1,0)$, one defines $\Gamma(x) = \frac{1}{x}\Gamma(x+1)$, and more generally, if $x \in (n, n+1)$, we can define

$$\Gamma(x) = \frac{\Gamma(x+n)}{x(x+1)(x+2)\ldots(x+n-1)}.$$

The graph of $f(x) = \Gamma(x)$ is given in Figure 5.6.3. Some properties of this function are explored in the exercises. $\square$

Example 5. Discuss the convergence of the integral $\int_0^1 x^{p-1}(1-x)^{q-1}$.

Solution. If p and q are both greater that 1, this reduces to an ordinary Riemann Integral, and hence trivially converges. If $p, q < 1$, there are discontinuities at $x = 0$ and $x = 1$. It is easy to see that the integral converges if both $1 - p < 1$ and $1 - q < 1$, that is, when $p > 0$ and $q > 0$. The reader is asked to complete the details of this example in Exercise 9. $\square$

Discussion. This integral is called **Euler's First Integral** or the **Beta integral** and finds many interesting applications in statistics and applied mathematics. We define the beta function on the first quadrant of the Cartesian plane by $\beta(p,q) = \int_0^1 x^{p-1}(1-x)^{q-1}$. There are interesting relationships between the gamma and the beta functions, which we explore in the exercises. $\square$

---

[3]See M. Kline, *Mathematical Thought from Ancient to Modern Times* (New York: Oxford University Press, 1972), pp. 423–424.

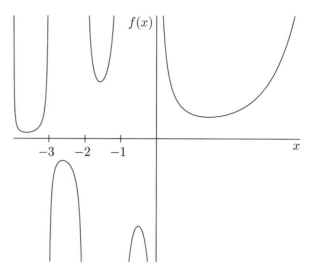

Figure 5.6.3 The graph of the gamma function $\Gamma(x)$. The function is discontinuous at $0, -1, -2, \ldots$, and has a vertical asymptote at each of these points.

Now, we consider improper integrals of functions that are not necessarily positive. The following theorem is a 'Cauchy Criterion' for the convergence of such types of improper integrals.

Theorem 5.6.4. *If f is integrable on $[a, K]$ for each K such that $a < K < \infty$, then the improper integral $\int_a^\infty f$ converges if and only if for each $\epsilon > 0$, there exists $X \geq a$ such that for all x_1, x_2, if $X < x_1 < x_2$, then*

$$\left| \int_{x_1}^{x_2} f \right| < \epsilon.$$

Proof. We prove the limit exists, that is, sufficiency. Necessity is left to the reader as Exercise 11. Following our usual tack, define $F(x) = \int_a^x f$, where $a \leq x$. The hypothesis guarantees that the sequence $\{F(n)\}$, $n \geq a$ is a Cauchy sequence (**WHY?**), whence it has a limit, L. Now fix $\epsilon > 0$ and choose X such that, if $X \leq n$ and $X < x_1 < x_2$, then $|F(n) - L| < \epsilon$ and $|F(x_1) - F(x_2)| < \epsilon$. For $X < K$, choose $n > K$ and apply the triangle inequality to obtain then $|F(K) - L| < 2\epsilon$ (**WHY?**), whence

$$\int_a^\infty f = \lim_{K \to \infty} \int_a^K f = \lim_{K \to \infty} F(K) = L. \qquad \square$$

Discussion. The reader should compare this theorem with the Cauchy Criterion for sequences in Section 3.3. Reflection should convince the reader that there is a natural generalization of the Cauchy Criterion that deals with limits of functions at infinity. The reader is asked to consider this problem in Exercise 12. $\square$

Definition. The improper integral $\int_a^b f$ is said to be **absolutely convergent**

provided $\int_a^b |f|$ is convergent. The integral is said to be **conditionally convergent** provided the improper integral $\int_a^b f$ converges and $\int_a^b |f|$ diverges.

Discussion. The requirement of absolute convergence is a strong one. It is easy to show that every absolutely convergent integral is convergent, but not conversely. We give an example of a convergent integral that fails to be absolutely convergent. $\square$

Example 6. Discuss the absolute convergence of the improper integral $\int_0^\infty \frac{\sin x}{x}$.

Solution. First, observe that, as will be seen from Chapter 9, $\lim_{x \to 0} \frac{\sin x}{x} = 1$. Thus, there is no singularity at 0, but only at the upper limit of integration. If $K > 0$, there exists $n \in \mathbf{N}$ and $t_n \in [0, \pi)$ such that $K = n\pi + t_n$. Hence,

$$\int_0^K \frac{\sin x}{x} = \int_0^{n\pi} \frac{\sin x}{x} + \int_{n\pi}^{n\pi + t_n} \frac{\sin x}{x}.$$

Let us call the first term A_n and the second B_n. Now,

$$|B_n| = \left| \int_{n\pi}^{n\pi + t_n} \frac{\sin x}{x} \right| \leq \int_{n\pi}^{n\pi + t_n} \left| \frac{\sin x}{x} \right| \leq \int_{n\pi}^{n\pi + t_n} \frac{1}{n\pi} \leq \frac{\pi}{n\pi} = \frac{1}{n},$$

proving that the $\lim_{n \to \infty} B_n = 0$. Further,

$$A_n = \sum_{i=1}^n \left[\int_{(i-1)\pi}^{i\pi} \frac{\sin x}{x} \right] = \sum_{i=1}^n a_i,$$

where the term $a_i = \int_{(i-1)\pi}^{i\pi} \frac{\sin x}{x}$ has an alternating sign, is decreasing and converges to 0. (This is easily checked based on computations from the calculus.) Hence, by Leibnitz's alternating series test (Theorem 7.4.1), $\sum a_i$ converges. Thus, we conclude that the improper integral also converges.

To see that this integral is not absolutely convergent, we consider $\int_0^\infty \frac{|\sin x|}{x}$. We have

$$\int_0^{n\pi} \frac{|\sin x|}{x} = \sum_{k=1}^n \int_{(k-1)\pi}^{k\pi} \frac{|\sin x|}{x}$$

$$= \sum_{k=1}^n \int_0^\pi \frac{\sin t}{(k-1)\pi + t} \qquad \text{(using } x = (k-1)\pi + t\text{)}$$

$$\geq \sum_{k=1}^n \frac{1}{k\pi} \int_0^\pi \sin t = \sum_{k=1}^n \frac{2}{k\pi}.$$

Since, as will be shown in Chapter 6, $\sum_{k=1^n} \frac{1}{k}$ approaches ∞ as n approaches ∞, we conclude that the integral diverges. Thus, the given integral is conditionally convergent. $\square$

Discussion. In this example, we have employed several results from the theory of infinite series, which are proved later in Chapter 6. But, this is not a disadvantage, since the student will already be familiar with the results, if not the proofs, from an elementary course in calculus. Moreover, the use of these results serves to establish the connection between certain types of improper integrals and infinite series. This connection is striking, and we explore further aspects in Chapter 6. $\square$

Exercises

1. Give an example of a function that illustrates the problem described following the definition of $\int_{-\infty}^{+\infty} f$.

2. Complete the proof of part (b) in Example 1.

3. Supply the missing details in Example 2.

4. Consider the problem of evaluating $\int_{-\infty}^{+\infty} f$. Define a Cauchy Principal Value for this type of integral and give an example of a function that is not improperly integrable on $[-\infty, \infty]$, but for which the Cauchy Principal Value exists.

5. Consider the problem of evaluating $\int_a^b f$ where for each u, v such that $a < u \le v < b$, $\int_u^v f$ exists. Define a Cauchy Principal Value for this type of integral and give an example of a function that is not improperly integrable on $[a, b]$, but for which the Cauchy Principal Value exists.

6. Prove the divergence case in Theorem 5.6.2.

7. Supply a proof of Theorem 5.6.3.

8. Formulate and prove a version of Theorem 5.6.3 that can be used to test whether the improper integral over $[a, b]$ exists for functions unbounded at b.

9. Fill in the missing details in Example 5.

10. Supply an example to show that in general,

$$\int_a^\infty (f + g) \ne \int_a^\infty f + \int_a^\infty g.$$

11. Complete the proof of Theorem 5.6.4.

12. Develop a Cauchy Criterion for the existence of limits of functions at infinity. Show this criterion is equivalent to the original limit definition.

13. Discuss the convergence or divergence of the following integrals:

(a) $\int_0^\infty x e^{-x}$;

(b) $\int_2^\infty \dfrac{1}{x\sqrt{x^3 - 1}}$;

(c) $\int_0^\infty \dfrac{1}{\sqrt{x}(1 + x)}$;

(d) $\int_0^\infty \dfrac{\arctan x}{1 + x^2}$;

(e) $\int_0^1 \dfrac{1}{x\sqrt{x}}$;

(f) $\int_0^{\frac{1}{e}} \dfrac{1}{x(\ln x)^2}$;

(g) $\int_3^5 \dfrac{x^2}{\sqrt{(x - 3)(x - 5)}}$;

(h) $\int_a^b \dfrac{1}{(b - x)^n}$;

(i) $\int_{-\infty}^\infty \dfrac{1}{x^2 + 9x + 10}$;

(j) $\int_0^{\frac{1}{2}} \dfrac{1}{x(\ln x)^2}$;

(k) $\int_0^\infty e^{\alpha x} \sin \beta x$;

(l) $\int_{-1}^1 \dfrac{1}{\sqrt{1 - x^2}}$;

(m) $\int_{-3}^{-1} \dfrac{x^2 - 1}{x + 1}$;

(n) $\int_{-\infty}^{+\infty} \dfrac{x}{\sqrt{2x^2 + 5}}$;

(o) $\displaystyle\int_{-3}^2 f$, where $f(x) = \begin{cases} \dfrac{1}{x^4}, & \text{if } -3 \le x < 0 \\ \dfrac{1}{\sqrt{x}}, & \text{if } 0 < x \le 2. \end{cases}$

14. Find Cauchy's Principal Value of the following improper integrals:

(a) $\int_{-1}^{1} \frac{1}{x}$;

(b) $\int_{0}^{4} \frac{2}{(x-2)^3}$;

(c) $\int_{-\infty}^{\infty} \sin x$;

(d) $\int_{-\infty}^{\infty} \cos x$;

(e) $\int_{-\infty}^{\infty} \frac{3}{x^3+2}$.

15. Show $CPV \int_{-1}^{1} \frac{1}{|x|}$ does not exist.

16. If f is continuous on $\mathbf{R}$ and the integral $\int_{-\infty}^{\infty} f$ converges to A, show $CPV \int_{-\infty}^{\infty} f = A$.

17. Prove an absolutely convergent integral is convergent.

18. Discuss the convergence and absolute convergence of the following improper integrals:

(a) $\int_{0}^{\infty} \frac{\sin \alpha x}{x}$;

(b) $\int_{0}^{\infty} \frac{x^{\alpha-1}}{1+x}$;

(c) $\int_{0}^{\infty} \frac{\cos x}{\sqrt{x+x^2}}$;

(d) $\int_{0}^{\infty} \frac{x^{2m}}{1+x^{2n}}$;

(e) $\int_{0}^{\infty} \frac{\cos ax - \cos bx}{x}$.

19. Give an example to show the improper integral $\int_{a}^{b} f$ exists, while $\int_{a}^{b} f^2$ does not exist.

20. Prove $\Gamma(x+1) = x\Gamma(x)$.

21. Prove $\Gamma(z) = \int_{0}^{1} \left(\ln \frac{1}{x}\right)^{z-1}$.

22. Prove the following properties of the β integral:

(a) $\beta(x,1) = \frac{1}{x}$;

(b) $\beta(x+1,y) = \frac{x}{x+y}\beta(x,y)$;

(c) $\beta(x,y) = \frac{\Gamma(x)\Gamma(y)}{\Gamma(x+y)}$;

(d) $\beta(p,q) = \int_{0}^{\frac{\pi}{2}} \sin^{2p-1} x \cos^{2q-1} x$;

(e) $\beta\left(\frac{a+1}{2}, \frac{1}{2}\right) = 2 \int_{0}^{\frac{\pi}{2}} \sin^a x$;

(f) $\beta(a, 1-a) = \int_{0}^{\infty} \frac{x^{a-1}}{1+x} = \frac{\pi}{\sin \pi a}$;

(g) $\beta\left(\frac{19}{2}, \frac{1}{2}\right) = \int_{0}^{1} \frac{x^9}{\sqrt{x}\sqrt{(1-x)}}$.

23. Show $\int_{0}^{1} \frac{1}{\sqrt{1-x^n}} = \frac{\sqrt{\pi}}{n} \frac{\Gamma\left(\frac{1}{n}\right)}{\Gamma\left(\frac{1}{n} + \frac{1}{2}\right)}$.

24. Use the following expression for π (**Wallis's Product**)

$$\pi = 2 \cdot \left(\frac{2 \cdot 2}{1 \cdot 3}\right)\left(\frac{4 \cdot 4}{3 \cdot 5}\right)\left(\frac{6 \cdot 6}{5 \cdot 7}\right)\left(\frac{8 \cdot 8}{7 \cdot 9}\right)\cdots$$

to show $\Gamma\left(\frac{1}{2}\right) = \sqrt{\pi}$.

25. Find $\Gamma\left(\frac{-5}{2}\right)$.

26. Show $\int_{-\infty}^{\infty} e^{-x^2} = \sqrt{\pi}$ [Hint: Use the fact that $\Gamma\left(\frac{1}{2}\right) = \sqrt{\pi}$.]

27. Let f be defined on $[a, b]$, $b \in \mathbf{R}$ and suppose f is integrable on $[a, c]$, for each c such that $a \le c < b$. Show if f is not integrable on $[a, b]$, then f is unbounded in any neighborhood of b.

28. Is it possible there could be a single nonnegative function f defined on $[0, \infty)$ such that for any bounded nonnegative function g, $\int_{a}^{\infty} g$ converges exactly if there is a c such that $g \le f$ on $[c, \infty]$?

29. Could there be a single test function for the limit form of the comparison test?

30. Could there be a single positive function, g, defined on $[0, \infty)$ that would serve to determine the convergence or divergence of the improper integrals of all other bounded functions on $[a, \infty)$?

31. Let f be bounded in $[a, \infty)$ and integrable in $[a, t]$ for $t > a$, and let $\int_a^\infty g$ be absolutely convergent. Prove $\int_a^\infty fg$ is absolutely convergent.

32. Prove **Abel's Test**: Let f be bounded and monotonic in $[a, \infty)$, and let $\int_a^\infty g$ is convergent. Then, $\int_a^\infty fg$ is convergent.

33. Prove **Dirichlet's Test**: Let f be bounded and monotonic in $[a, \infty)$, $\lim\limits_{x \to \infty} f(x) = 0$, let $\int_a^t g$ be bounded for $t \geq a$. Then $\int_a^\infty fg$ is convergent.

34. Test for convergence:

(a) $\int_0^\infty \sin x^2$;

(b) $\int_0^\infty e^{-ax} \cos x$;

(c) $\int_0^\infty e^{-ax} \frac{\sin x}{x}$;

(d) $\int_0^\infty \frac{\sin x}{x^p}$;

(e) $\int_e^\infty \frac{\ln x \sin x}{x}$.

5.7   Riemann-Stieltjes Integration

The theory of Riemann Integration is based on limits of sums that take the form of

$$\lim_{\|P\| \to 0} \sum_{i=1}^n f(c_i) \times \Delta x_i.$$

In the special case that the partition, P, defines subintervals of equal length, then $\Delta x_i = \Delta x_j$, for all $i, j \leq n$. This remark is stated for Riemann sums, but the same holds true for Darboux sums. The truth of this remark is of course due to the fact that the quantity Δx_i is itself the length of the subinterval.

Having made these observations, one might look at sums of the form

$$\lim_{\|P\| \to 0} \sum_{i=1}^n f(c_i) \times \Delta g_i,$$

where P is a partition of $[a, b]$, f is a function defined on $[a, b]$, $c_i \in [x_{i-1}, x_i]$, g is a function defined on $[a, b]$, and $\Delta g_i = g(x_i) - g(x_{i-1})$. While the theory of such limits may seem artificial at first glance, in fact, these limits arise in a natural way.

To motivate these ideas consider an urn filled with a thousand otherwise identical balls of ten different colors. If there are one hundred balls of each color, then the probability of picking a ball of a given color at random is simply $P(c_i) = 1/10$, where $P(c_i)$ stands for the probability of selecting a ball of color c_i. In probability theory, there may be values, V_i, attached to the selection of a ball of color, c_i, whence sums of the form

$$\sum_{i=1}^n V_i \times P(c_i)$$

arise. This sum is analogous to the familiar Riemann sum, since $P(c_i) = P(c_j)$.

Now suppose for contrast that the balls are not otherwise identical, so the value of $P(c_i)$ depends on c_i, even though the number of balls of each color remains unchanged. As an example the balls of one color might be larger than the others so that these balls are more likely to be chosen. The sum of interest is still

$$\sum_{i=1}^{n} V_i \times P(c_i),$$

but now $P(c_i) \neq P(c_j)$ for $i \neq j$, even though we still have 100 of each type of ball. This situation corresponds to the more general type of sum discussed above. Although our example is finite in nature, it is apparent similar sums could arise in situations involving continuous probability distributions. It is not our intent to discuss such situations here, but to mention them to illustrate the natural motivation underlying the considerations of this section.

In this section, then, we discuss a generalization of the concept of Riemann Integral. The integrals considered are of the form $\int f\, dg$ rather than the conventional form $\int f$. Darboux-Stieltjes (Riemann-Stieltjes) integrals are generalizations of the concept of Darboux (Riemann) Integral, in the sense that the integration is performed with respect to a function g. If we set $g(x) = x$, then the computation reduces to the ordinary Riemann Integral. With respect to the nature of the function, g, a new class of functions called **functions of bounded variation** forms a satisfactory setting for this development. We conclude this section by studying some basic properties of such functions.

We now define the concept of Stieltjes Integrals (both in the sense of Darboux and Riemann). First, we consider Darboux-Stieltjes integrals, which form a straightforward generalization of the concept of Riemann-Darboux integrals.

Definition. Let $f : [a,b] \to \mathbf{R}$ be bounded and $g : [a,b] \to \mathbf{R}$ be monotone increasing. Let $P = \{a = x_0 \leq x_1 \leq \cdots \leq x_n = b\}$ be a partition of $[a,b]$. We define the **upper** and **lower Darboux-Stieltjes sums** by

$$\overline{S}(f,g,P) = \sum_{i=1}^{n} M_i(g(x_i) - g(x_{i-1})),$$

$$\underline{S}(f,g,P) = \sum_{i=1}^{n} m_i(g(x_i) - g(x_{i-1})),$$

where M_i, m_i denote, respectively, the supremum and the infimum of f in the ith-subinterval $[x_{i-1}, x_i]$. We also define the **Upper Darboux-Stieltjes Integral**

$$\overline{\int}_a^b f\, dg = \inf \overline{S}(f,g,P)$$

and the **Lower Darboux-Stieltjes Integral**

$$\underline{\int}_a^b f\, dg = \sup \underline{S}(f,g,P),$$

where the infima, suprema are taken over all possible partitions P of $[a, b]$. If

$$\overline{\int}_a^b {}_a^b f \, dg = \underline{\int}_a^b {}_a^b f \, dg,$$

we say that f is **Darboux-Stieltjes integrable** with respect to g in $[a, b]$, and the common value is the **Darboux-Stieltjes Integral** D-S $\int_a^b f \, dg$.

Discussion. To help the reader understand the nature of the generalization being discussed, we compare the present definition with the original definition of the upper and lower Darboux sums presented in Section 5.2. Specifically, both definitions begin with a bounded function defined on a closed interval $[a, b]$. Both definitions take an arbitrary partition, P, of this interval. Both definitions employ the supremum and infimum of f, M_i, m_i on the ith subinterval. The distinction between the two is that the Stieltjes definition has a second function g defined and monotonically increasing on $[a, b]$. The original definition makes no mention of such a function. However, such a function is present in the original definition. This can be seen by considering how the function g is used in the extended definition. It appears as the multiplier in

$$M_i \times (g(x_i) - g(x_{i-1})) = M_i \times \Delta g_i.$$

If we ask whether an analogous function exists for the original definition, the answer is easily seen to be yes, since in that definition M_i is multiplied by Δx_i, the length of the ith subinterval. If one sets $g(x) = x$, which is a monotone increasing function on $[a, b]$, then the basic definition of Darboux integration is seen to be a special case of the more general situation. Moreover, this discussion also establishes the essential connection between the more usual notation for Riemann integrals and notation for Stieltjes integrals:

$$\int_a^b f dx \quad \text{and} \quad \int_a^b f dg.$$

The close similarity between the two definitions suggests the same techniques of proof used in the case of Darboux sums should also work in this generalized situation. This is indeed the case, and the reader can easily prove, for example, if Q is a refinement of the partition, P, $\overline{S}(f, g, Q) \leq \overline{S}(f, g, P)$ and $\underline{S}(f, g, Q) \geq \underline{S}(f, g, P)$. In fact *any* upper sum exceeds *any* lower sum (Exercise 1), and therefore $\underline{\int}_a^b f \, dg \leq \overline{\int}_a^b f \, dg$.

Quite obviously, theorems about Darboux-Stieltjes integrals with respect to a monotonic increasing function g apply to the Riemann-Darboux integrals as particular cases. In many instances, theorems and proofs for Riemann-Darboux integrals apply almost word by word for the generalized situation of the Darboux-Stieltjes Integral with respect to a monotonic increasing function. Three such instances are furnished in the following theorems. $\square$

Theorem 5.7.1. *Let f be bounded and g be monotonic increasing in $[a, b]$. A necessary and sufficient condition for D-S $\int_a^b f \, dg$ to exist is that given $\epsilon > 0$, there exists a partition P of $[a, b]$ such that $\overline{S}(f, g, P) - \underline{S}(f, g, P) < \epsilon$.*

Proof. Exercise 2. □

Discussion. This theorem provides the essential tool for working with the Darboux-Stieltjes sums. □

Theorem 5.7.2. *If f is continuous, and g is monotonic increasing in $[a, b]$, then the Darboux-Stieltjes Integral D-S $\int_a^b f\, dg$ exists.*

Proof. By Theorem 3.6.2 f is uniformly continuous on $[a, b]$. Thus, given $\epsilon > 0$, we can find a $\delta > 0$ such that for all $x, y \in [a, b]$, with $|x - y| < \delta$, it is true that $|f(x) - f(y)| < \frac{\epsilon}{g(b)-g(a)}$. (The case $g(a) = g(b)$ is trivial (**WHY?**)). Now, choose a partition, P, with norm less that δ. The continuity of f guarantees that on each subinterval, $[x_{i-1}, x_i]$, of P, the infimum, m_i, and the supremum, M_i, of f are actually attained at points, say, y_i, z_i, in that subinterval. Since $|y_i - z_i| < \delta$, we have $M_i - m_i < \frac{\epsilon}{g(b)-g(a)}$. This being the case for $i = 1, 2, \ldots, n$, we get

$$\overline{S}(P) - \underline{S}(P) = \sum_{i=1}^{n} M_i \Delta g_i - \sum_{i=1}^{n} m_i \Delta g_i = \sum_{i=1}^{n}(M_i - m_i)\Delta g_i$$
$$< \frac{\epsilon}{g(b) - g(a)} \left(g(b) - g(a)\right) = \epsilon,$$

whence Theorem 5.7.1 guarantees that f is D-S integrable. □

Discussion. This proof amply illustrates the contention that many of the proofs of statements for Riemann-Stieltjes integration can be proven by adapting, with little or no change, the proof of the analogous statement for Darboux integrals. To see how the new proofs reflect the old, we urge the reader to inspect the proof of Theorem 5.2.4.

The reader should realize that the function g does not have to be continuous. In reality, g could have an infinite number of discontinuities. The beauty of this proof is that it completely avoids any discussion of the continuity, or lack thereof, of g. It is essential that the reader understand how this is accomplished. □

Theorem 5.7.3. *If f is monotonic, and g is monotonic increasing and continuous in $[a, b]$, the Darboux-Stieltjes Integral D-S $\int_a^b f\, dg$ exists.*

Proof. Exercise 3. □

Example 1. Discuss the Darboux-Stieltjes integrability of the function, $f(x) = x$, $x \in [0, 2]$, with respect to the function, $g : [0, 2] \to \mathbf{R}$ defined by

$$g(x) = \begin{cases} 0, & \text{if } x \in [0, 1] \\ 1, & \text{if } x \in (1, 2]. \end{cases}$$

Solution. Given $\epsilon > 0$, choose a partition, P, of $[0, 2]$ such that $\|P\| < \epsilon$. Notice that if $\Delta g_k = g(x_k) - g(x_{k-1}) \neq 0$, then $[1, \delta) \subseteq [x_{k-1}, x_k]$, for some $\delta > 1$. It follows that there is exactly one k such that $\Delta g_k \neq 0$. For this k, we have

$$
\begin{aligned}
|\overline{S}(f,g,P) - \underline{S}(f,g,P)| &= |\sum_{i=1}^{n} M_i \Delta g_i - \sum_{i=1}^{n} m_i \Delta g_i| \\
&= M_k \Delta g_k - m_k \Delta g_k \\
&= (M_k - m_k)(1 - 0) = x_{k+1} - x_k \\
&\leq \|P\| < \epsilon.
\end{aligned}
$$

Since ϵ was arbitrary, the existence of the integral follows, and its value is easily seen to be 1. $\square$

Discussion. This example illustrates one of the useful properties of Stieltjes integrals, namely, their ability to assign different 'weights' to different values of f. In this case, the only point of f that is important for purposes of computing this integral is the value at 1. This is due to the fact that g is constant on any interval which does not include 1, while at 1, g undergoes rapid change. $\square$

Example 2. Discuss the Darboux-Stieltjes integrability of the function $f : [0,2] \rightarrow \mathbf{R}$ defined by

$$
f(x) = \begin{cases} 0, & \text{if } x \in [0,1] \\ 1, & \text{if } x \in (1,2], \end{cases}
$$

with respect to the function $g(x) = f(x)$.

Solution. For any partition, Q, there exists a refinement $P = \{0 = x_0 \leq x_1 \leq \ldots \leq x_n = 2\}$ of $[0,2]$, for which there exists a subinterval $[x_{k-1}, x_k]$ that contains the point 1 as well as a point greater than 1. As in the example above, we have $\Delta g_k = 1$ while $\Delta g_j = 0$ for $j \neq k$, whence

$$
\overline{S}(f,g,P) = \sum_{i=1}^{n} M_i \Delta g_i = M_k \Delta g_k = 1(1 - 0) = 1.
$$

On the other hand,

$$
\underline{S}(f,g,P) = \sum_{i=1}^{n} m_i \Delta g_i = m_k \Delta g_k = 0(1 - 0) = 0.
$$

Consequently, $\overline{\int}_0^1 f dg = 1$ and $\underline{\int}_0^1 f dg = 0$. Hence, $\int_0^1 f dg$ does not exist. $\square$

Discussion. As pointed out above, g does not have to be continuous. For this reason, Δg_i does not have to shrink to 0 as $\|P\| \rightarrow 0$. Now Theorem 5.7.2 uses the fact that $|f(x_i) - f(x_{i-1})|$ must shrink to 0 as the norm of P goes to 0 to avoid the problems with discontinuities in g. Theorem 5.7.3 uses the continuity of g, which forces Δg_i to shrink to 0 as the norm of P goes to 0 to avoid the difficulties with discontinuities in f. In this example, both f and g are discontinuous and the points of discontinuity match in such a way that the situation cannot be retrieved. $\square$

Next, we consider the approach of Riemann and discuss the Riemann-Stieltjes sums and integrals. The original definition of the Riemann Integral is presented in Section 5.3. Translation of that definition into the present context leads to:

Definition. Let f and g be defined on $[a, b]$ with g monotone increasing there. Let P be a partition of $[a, b]$ and $Q = \{c_i\}$ an intermediate partition for P. The **Riemann-Stieltjes sum** of f with respect to g over the partition, P, and intermediate partition, Q, is defined by the real number

$$R(f, g, P, Q) = \sum_P f(c_i)(g(x_i) - g(x_{i-1})) = \sum_P f(c_i)\Delta g_i,$$

where the P under the summation sign indicates summation extends over all the subintervals generated by P and the c_i's are the members of Q. We say that the function f is **Riemann-Stieltjes integrable** with respect to the function g in $[a, b]$, provided there exists a real number L with the property that given $\epsilon > 0$, there exists a $\delta > 0$ such that if $\|P\| < \delta$ and Q is any intermediate partition for P, we have

$$|R(f, g, P, Q) - L| < \epsilon.$$

In this case, we say that the **Riemann-Stieltjes Integral** of f with respect to g is L, and we write

$$\text{R-S} \int_a^b f dg = L.$$

Discussion. The reader can check that this definition is obtained from our earlier definition of the Riemann Integral by replacing Δx_i by Δg_i. Indeed, it is easily seen that this reduces to the Riemann Integral if we set $g(x) = x$. As in the case of Riemann integrals, the upper and lower Darboux sums are special cases of Riemann sums for special choices of the points c_i.

A principal feature of the two treatments of the ordinary integral is their equivalence. One might suspect that like the standard Riemann-Darboux and Riemann integrals, the concepts of Darboux-Stieltjes and Riemann-Stieltjes integrals coincide. Unfortunately, this is not the case as the following example shows. □

Example 3. Discuss the Darboux-Stieltjes and Riemann-Stieltjes integrability of the function f with respect to the function g in $[0, 2]$, where

$$f(x) = \begin{cases} 0, & \text{if } x \in [0, 1) \\ 1, & \text{if } x \in [1, 2] \end{cases}$$

and

$$g(x) = \begin{cases} 0, & \text{if } x \in [0, 1] \\ 1, & \text{if } x \in (1, 2]. \end{cases}$$

Solution. Consider separately the intervals $[0, 1]$ and $[1, 2]$. Since g is constant, hence continuous, and f is monotone in $[0, 1]$, it follows that f is Darboux-Stieltjes integrable with respect to g in $[0, 1]$ and D-S $\int_0^1 f dg = 0$. On the other hand, f is constant, hence continuous, in $[1, 2]$ and g is monotonic increasing, whence D-S $\int_1^2 f dg$ exists and equals 1. To establish 1 as the value for the integral, one merely notices that all contributions to any approximating sum will be 0 except the contribution associated with the unique subinterval that contains $[1, \nu)$ for some $\nu > 1$. The contribution of this unique subinterval is 1. Thus, we conclude that D-S $\int_0^2 f dg = 1$. (This computation is explored in detail in Exercise 6.)

We now show f is not Riemann-Stieltjes integrable with respect to g in $[0, 2]$. For each n, consider the partition P_n of $[0, 2]$ defined by

$$P_n = \left\{ 0, \frac{1}{n}, \frac{2}{n}, \dots, 1 - \frac{1}{n}, 1 - \frac{1}{2n}, 1 + \frac{1}{2n}, 1 + \frac{1}{n}, \dots, 2 - \frac{1}{n}, 2 \right\}$$

and choose b_i, c_i, respectively, to be the left and right endpoints of the subintervals. Then $R(f, g, P_n, b_i) = 0$, whereas $R(f, g, P_n, c_i) = 1$. Since this holds for each n, we conclude that $\lim_{\|P\| \to 0} R(f, g, P, W)$ does not exist. In other words, f is not Riemann-Stieltjes integrable with respect to g in $[0, 2]$. $\square$

Discussion. In this example, as in Example 2, the functions f and g have a common point of discontinuity, namely, $x = 1$. Nevertheless, f is D-S integrable with respect to g. While the reader may easily verify that we have correctly applied the relevant theorems, this verification will not lead to an essential understanding of difference between the two examples. Understanding requires going through the details of the calculation with the upper and lower sums. Specifically, the reader can check, by direct computation with the sums, that $\int_0^1 f dg = 0$. On the other hand, the reader should notice that for every subinterval of the form $[x_{i-1}, x_i] \subseteq [1, 2]$, $m_i = M_i = 1$. This ensures that the upper and lower sums generate the same value on the subinterval $[1, 2]$.

The key to the above argument establishing the integrability of f with respect to g is the fact that the computation is split at 1. This amounts to ensuring that $1 \in P$. We are permitted to use only partitions that include 1 because the process of taking the 'limit' in the Darboux case is based on finding the supremum for which there is a known upper bound or an infimum for which there is a known lower bound. If some select subset of the numbers whose supremum is required will already generate the supremum (infimum), then including the remaining numbers not change its value. That we have the supremum can be known, because we have a series of upper bounds available in the form of upper sums (lower sums in the case of infimum).

When considering Riemann-Stieltjes integration as defined above, we cannot ensure that any particular point belongs to every partition having a small norm. Indeed, the only points that are guaranteed to belong to every partition are a and b. However, the process of finding a limit requires us to produce a number, L, having the property that every partition having a small norm yields a Riemann-Stieltjes sum that is close to L. In the present example, every partition that fails to include 1 will have distinct intermediate partitions generating sums of 0 and 1, respectively. Quite obviously, this will destroy any possibility of having a limit.

It is easily checked that f is R-S integrable with respect to g on each of the two subintervals $[0, 1]$ and $[1, 2]$. This leads to the rather odd situation of a function that is not integrable on the interval $[0, 2]$ but that is integrable on $[0, 1]$ and $[1, 2]$. A slight change in the definition of R-S integrable permits us to rectify this situation. $\square$

Definition. We shall say that the function f is **revised Riemann-Stieltjes integrable** with respect to the function g in $[a, b]$, provided there exists a real number L with the property that given $\epsilon > 0$, there exists a partition P_ϵ of $[a, b]$ such that if P is any refinement of P_ϵ, we have

$$|R(f, g, P, Q) - L| < \epsilon,$$

for every intermediate partition Q of P. In this case, we say that the **Revised Riemann-Stieltjes Integral** of f with respect to g is L, and we write

$$\text{(r)R-S} \int_a^b f \, dg = L.$$

Theorem 5.7.4. *Let f be bounded and g be monotonic increasing in $[a, b]$. Then the Revised Riemann-Stieltjes Integral of f with respect to g exists if and only if the corresponding Darboux-Stieltjes Integral exists, and if they exist, they are equal.*

Proof. Exercise 4. $\square$

Theorem 5.7.5. *Let f be bounded, and g be monotonically increasing in $[a, b]$. Then $\text{(r)R-S} \int_a^b f \, dg$ exists if and only if the collection $\overline{S}(f, g, P, Q)$ is Cauchy, in the sense that given $\epsilon > 0$, there exists a partition P_ϵ such that if P and R are refinements of P_ϵ, and Q and S are any intermediate partitions for P and R, respectively,*

$$|\overline{S}(f, g, P, Q) - \overline{S}(f, g, R, S)| < \epsilon.$$

Proof. Exercise 7. $\square$

A new class of functions, called **functions of bounded variation**, forms a more general setting for the discussion of Darboux (Riemann)-Stieltjes integrals. In what follows, we define these functions, and we develop some of their elementary properties.

Definition. Let f be a function defined on $[a, b]$. For each partition, $P = \{a_0, a_1, \dots, a_n\}$, of $[a, b]$, let $V(f, P)$ denote the real number given by

$$V(f, P) = \sum_{i=1}^n |f(a_i) - f(a_{i-1})|.$$

The number $V(f, P)$ is called the **variation of f on $[a, b]$ for the partition P.**

Discussion. The reader should recall our early discussions about why functions fail to be continuous. One reason was that the function values were oscillating. This definition is an attempt to quantify the degree to which a function is oscillating. It does so by looking for changes in the values of the function.

Clearly, $V(f, P)$ is always bounded below by 0. We can sharpen this bound by noticing that

$$f(b) - f(a) = \sum_{i=1}^n f(a_i) - f(a_{i-1}),$$

whence by taking absolute values and using the triangle inequality, we see that $|f(b) - f(a)| \leq V(f, P)$. If the function f is monotone, then equality always holds (Exercise 19). However, for every function, there will be partitions for which $|f(b) - f(a)| = V(f, P)$. In general, the collection, $\{V(f, P)\}$, may not be bounded above.

If one is thinking of the notion of the variation of a function as being a measure of its oscillatory behavior, it is evident that what is important is not a lower bound on the collection of numbers $V(f, P)$, but an upper bound. Moreover, this measure should be independent of the choice of partition. $\square$

Definition. The function f is said to be of **bounded variation** on $[a, b]$ provided the set $\{V(f, P) : P \text{ is a partition of } [a, b]\}$ is bounded above. If f is a function of bounded variation, then we set

$$V(f, [a, b]) = \sup \{V(f, P) : P \text{ is a partition of } [a, b]\}.$$

We call this supremum, $V(f, [a, b])$ the **variation** of f on $[a, b]$.

Discussion. As indicated, the variation $V(f, P)$ corresponding to the partition, P, is an approximate measure of the 'vertical wiggle' of the graph of f that depends on the choice of the partition. The variation $V(f, [a, b])$ of f in $[a, b]$ is the total amount of vertical change in the graph of f throughout the interval, where the dependence on any particular partition has been eliminated by taking the supremum. We shall write $V(f)$ instead of $V(f, [a, b])$ if there is no confusion. $\square$

We briefly discuss some important properties of functions of bounded variation in the following set of theorems.

Theorem 5.7.6. *If f is a function of bounded variation in $[a, b]$ and $a < c < b$, then f is a function of bounded variation in $[a, c]$ and $[c, b]$, and furthermore,*

$$V(f, [a, c]) + V(f, [c, b]) = V(f, [a, b]).$$

Proof. Clearly, both $V(f, [a, c])$ and $V(f, [c, b])$ exist (**WHY?**). Let P be a partition of $[a, b]$ and P^* be obtained from P by adjoining the single point c. Let P' and P'' be the resulting partitions of $[a, c]$ and $[c, b]$, so that $P' \cup P'' = P^*$. Now,

$$V(f, P) \leq V(f, P^*) = V(f, P') + V(f, P''),$$

whence

$$V(f, P) \leq V(f, [a, c]) + V(f, [c, b]).$$

Taking supremum on the left-hand side, we obtain

$$V(f, [a, b]) \leq V(f, [a, c]) + V(f, [c, b]).$$

To prove the reverse inequality, if possible let $V(f, [a, b]) < V(f, [a, c]) + V(f, [c, b])$ and let $\epsilon = V(f, [a, c]) + V(f, [c, b]) - V(f, [a, b]) > 0$. There exist partitions P_1, P_2 of $[a, c]$ and $[c, b]$, respectively, such that $V(f, [a, c]) - \frac{\epsilon}{4} < V(f, P_1)$ and $V(f, [c, b]) - \frac{\epsilon}{4} < V(f, P_2)$. Therefore, adding the two, we get

$$V(f, [a, c]) + V(f, [c, b]) - \frac{\epsilon}{2} < V(f, P_1) + V(f, P_2) < V(f, [a, b]) \quad (\textbf{WHY?}).$$

Thus, since $V(f, [a, c]) + V(f, [c, b]) - V(f, [a, b]) < \frac{\epsilon}{2}$, we have a contradiction, proving the desired equality. $\square$

Theorem 5.7.7. *If f and g are functions of bounded variation in $[a, b]$, then $f + g$ is a function of bounded variation in $[a, b]$ and furthermore, $V(f + g) \leq V(f) + V(g)$.*

Proof. Exercise 20. $\square$

Theorem 5.7.8. *If f and g are functions of bounded variation in $[a, b]$, then fg is a function of bounded variation in $[a, b]$, and further, if M and N are the suprema of $|f|$ and $|g|$ in $[a, b]$, then $V(fg) \leq NV(g) + MV(f)$.*

Proof. Exercise 21. $\square$

Theorem 5.7.9. *If f is a function of bounded variation on $[a, b]$, then f is bounded in $[a, b]$.*

Proof. For every the partition, $P = \{a, x_1, b\}$, the variation satisfies

$$V(f, P) \leq V(f, [a, b]). \qquad \square$$

Theorem 5.7.10. *If f is a monotonic function in $[a, b]$, then f is a function of bounded variation in $[a, b]$ and $V(f) = |f(b) - f(a)|$.*

Proof. Exercise 22. $\square$

It will be shown in exercises that there exist continuous functions that are not functions of bounded variation; also there exist monotonic, countably discontinuous functions, that are clearly (by Theorem 5.7.8) functions of bounded variation. Hence, the connection between continuity and bounded variation is not a pleasant one. However, monotonicity is intimately tied with this notion. We have seen that every monotonic function is a function of bounded variation. In the converse direction, while a function of bounded variation need not be monotonic, it always admits a decomposition as the difference of two monotonic increasing functions. This is the aim of our next few results.

Let f be a function of bounded variation in $[a, b]$ and let $x \in [a, b]$. Then we can consider a function F on $[a, b]$ defined by $F(x) = V(f, [a, x]), x \in [a, b]$. This function will be called the **variation function** for the function f in $[a, b]$. Clearly, $F(a) = 0$ while $F(b) = V(f, [a, b])$. The following theorem relates the continuity of F to that of f.

Theorem 5.7.11. *The variation function F, defined above, is continuous at $x \in [a, b]$ if and only if f is continuous at x.*

Proof. Let F be continuous at x. Given $\epsilon > 0$ we can find $\delta > 0$ such that for all y in the domain, $|F(y) - F(x)| < \epsilon$ whenever $|y - x| < \delta$. Thus, $V(f, [x, y]) < \epsilon$. If $P = \{x, y\}$ is the trivial partition of $[x, y]$, then $V(f, P) = |f(y) - f(x)| \leq V(f, [x, y])$, and so $|f(y) - f(x)| \leq \epsilon$. Consequently, $\lim_{y \to x^+} f(y) = f(x)$. In a similar manner, we can prove left-continuity. The converse direction is left as an exercise to the reader. $\square$

The next theorem yields a nice characterization of functions of bounded variation as the difference of two monotonic functions, and the result is proved by considering the auxiliary function F described earlier.

Theorem 5.7.12. *If f is a function of bounded variation in $[a, b]$, then, f can be expressed as the difference $F - G$ of two nondecreasing functions.*

Proof. Let $F(x) = V(f, [a, x])$, $x \in [a, b]$, and $G = f - F$. If $x < y$, then $F(y) - F(x) = V(f, [x, y]) > 0$ proving F is nondecreasing. Now, set $G = F - f$. Note

$$
\begin{aligned}
G(y) - G(x) &= F(y) - F(x) - [f(y) - f(x)] \\
&= V(f, [x, y]) - [f(y) - f(x)] \\
&\geq V(f, [x, y]) - |[f(y) - f(x)]| \geq 0,
\end{aligned}
$$

since, for $P = \{x, y\}$, we have $V(f, P) \leq V(f, [x, y])$. $\square$

We conclude this section with a final remark on the use of functions of a bounded variation in the discussion of Riemann-Stieltjes integrals.

Let f be bounded and g be a function of bounded variation in $[a, b]$. By Theorem 5.7.12, there exist monotonic increasing functions g_1 and g_2 such that $g = g_1 - g_2$. If f is Riemann-Stieltjes integrable with respect to both g_1 and g_2, then we say that f is Riemann-Stieltjes integrable with respect to g in $[a, b]$ and we define

$$
\text{R-S} \int_a^b f \, dg = \text{R-S} \int_a^b f \, dg_1 - \text{R-S} \int_a^b f \, dg_2.
$$

Thus, we can extend all our earlier theorems where g is a monotonic increasing function to that of a function of bounded variation.

Exercises

1. Show every upper Darboux-Stieltjes sum exceeds every Darboux-Stieltjes lower sum and conclude $\underline{\int_a^b} f \, dg \leq \overline{\int_a^b} f \, dg$.

2. Prove Theorem 5.7.1.

3. Prove Theorem 5.7.3.

4. Prove Theorem 5.7.4.

5. Let f and g be as in Example 2. Let P be any partition of $[0, 2]$ that includes 1, that is, $x_i = 1$ for some i. Show every upper sum and every lower sum generate the same value for such a P.

6. Let f and g be as in Example 3. Write out all the details showing the Darboux-Stieltjes Integral exists. In particular, show that given $\epsilon > 0$ it is possible to choose a partition of $[0, 2]$ that will witness the truth of Theorem 5.7.1.

7. Prove Theorem 5.7.5.

8. If g is monotonic increasing and bounded in $[a, b]$ and $f(x) = k$ in $[a, b]$, where k is any constant, show D-S $\int_0^1 f dg$ exists and equals $k(g(b) - g(a))$.

9. Let f be bounded, and g be bounded and monotonic increasing in $[a, b]$. Prove the following:

   (a) If D-S $\int_a^b f \, dg$ exists, then there exists α lying between the bounds of f in $[a, b]$ such that D-S $\int_a^b f \, dg = \alpha[g(b) - g(a)]$.

(b) If f is continuous, there exists $\alpha[a, b]$ such that D-S $\int_a^b f\, dg = f(\alpha)[g(b) - g(a)]$.

(c) If $|f(x)| < K$ throughout $[a, b]$, and if D-S $\int_a^b f\, dg$ exists, then $\left| \text{D-S} \int_a^b f\, dg \right| \le K[g(b) - g(a)]$.

10. Give an example to show the value of the R-S integral may be affected, if we alter the value of the function at a single point.

11. Give an example where $|f|$ is R-S integrable with respect to g, but f is not R-S integrable with respect to g.

12. Evaluate the following Darboux-Stieltjes integrals:

(a) D-S $\int_0^1 x^2 d(x^3)$;

(b) D-S $\int_0^3 x\, d(x - [x])$;

(c) D-S $\int_{-1}^1 x\, d(e^x)$;

(d) D-S $\int_0^1 d([x^2])$;

(e) D-S $\int_0^1 x\, dg$, where $g(x) = \begin{cases} -1, & \text{if } x = 0 \\ x, & \text{if } 0 < x < 1 \\ 2, & \text{if } x = 1. \end{cases}$

13. Let f be bounded, g be differentiable in $[a, b]$, g' and f be both Riemann integrable in $[a, b]$. Prove f is Riemann-Stieltjes integrable with respect to g in $[a, b]$ and further

$$\text{R-S} \int_a^b f dg = \int_a^b f(x) g'(x)\, dx.$$

[This exercise shows that sometimes a Riemann-Stieltjes integral reduces to an ordinary Riemann integral.]

14. Evaluate the following Riemann-Stieltjes integrals. You may use Exercise 12.

(a) R-S $\int_\pi^{2\pi} \sin x\, d(\cos x)$; (b) R-S $\int_0^3 [x] d(e^x)$;

(c) R-S $\int_{-\frac{\pi}{2}}^0 e^x d(\cos x)$; (d) R-S $\int_{-10}^{10} (x^2 + e^x) d(\text{sgn } x)$;

(e) R-S $\int_0^1 x\, dg$ where

$$g(x) = \begin{cases} -1, & \text{if } x = 0 \\ x, & \text{if } 0 < x < 1 \\ 2, & \text{if } x = 1. \end{cases}$$

15. Let g_i, $(i = 1, 2, 3)$ be defined as follows:

(a) $g_1(x) = \begin{cases} 0, & \text{if } x \le 0 \\ 1, & \text{if } x > 0; \end{cases}$

(b) $g_2(x) = \begin{cases} 0, & \text{if } x < 0 \\ 1, & \text{if } x \ge 0; \end{cases}$

(c) $g_3(x) = \begin{cases} 0, & \text{if } x < 0 \\ \frac{1}{2}, & \text{if } x = 0 \\ 1, & \text{if } x > 0. \end{cases}$

Discuss the D-S and R-S integrability of $\int_{-1}^1 g_i dg_j$, $(1 \le i, j \le 3)$; also discuss $\int_{-1}^1 f dg_i$, where f is bounded in $[-1, 1]$ $(i = 1, 2, 3)$.

16. If f is Darboux-Stieltjes integrable with respect to every monotonic increasing function g, must f be continuous? Prove or give a counterexample.

17. Let f be Riemann-Stieltjes integrable with respect to an increasing function g in $[a, b]$ and let

$$F(x) = \text{R-S} \int_a^x f dg, \quad x \in [a, b].$$

Prove F is continuous on $[a, b]$.

18. In Exercise 17, let g be a function of bounded variation in $[a, b]$. Prove the following:

   (a) F is a function of bounded variation in $[a, b]$;

   (b) Every point of continuity of F is a point of continuity of g;

   (c) If g is increasing, g' exists, and f is continuous in $[a, b]$, then F is differentiable, and $F'(x) = f(x)g'(x)$.

19. Show if f is monotone on $[a, b]$, then for any partition, P, $V(f, P) = |f(b) - f(a)|$. Further, show even if f is not monotone on $[a, b]$ there are partitions for which $V(f, P) = |f(b) - f(a)|$.

20. Prove Theorem 5.7.7.

21. Prove Theorem 5.7.8.

22. Prove Theorem 5.7.10.

23. Let g be increasing continuous function on $[a, b]$, and f be Darboux-Stieltjes integrable with respect to g in $[a, b]$. Prove f is Riemann-Stieltjes integrable with respect to $[a, b]$ and the two integrals are equal.

24. Show if f possesses a bounded derivative in $[a, b]$, then f is a function of bounded variation in $[a, b]$.

25. Give an example of a function f that is of bounded variation, but f' is not bounded.

26. What is the relationship between Lipschitz's condition on f and the bounded variation of f on an interval $[a, b]$?

27. Find the variation of the following functions at the indicated intervals:

   (a) $f(x) = \frac{10}{3}$, $x \in [0, 2]$;   (b) $f(x) = |x|$, $x \in [0, 3]$;

   (c) $f(x) = 3x + 5$, $x \in [0, 1]$;   (d) $f(x) = x^2$, $x \in [0, 2]$;

   (e) $f(x) = x^3$, $x \in [0, 2]$;   (f) $f(x) = e^x$, $x \in [0, 1]$;

   (g) $f(x) = \cos x$, $x \in [0, 2\pi]$;   (h) $f(x) = 2\sin 3x$, $x \in [0, \pi]$;

   (i) $f(x) = x - [x]$, $x \in [0, 3]$;   (j) $f(x) = 2x^2 - 3x^3$, $x \in [-3, 4]$;

   (k) $f(x) = \begin{cases} x^2, & \text{if } 0 \leq x < 1 \\ x^3 + 1, & \text{if } 1 \leq x \leq 2 \end{cases}$ $x \in [0, 2]$;

   (l) $f(x) = \begin{cases} x, & \text{if } 0 \leq x < 1 \\ 2, & \text{if } x = 1 \\ 2 + x, & \text{if } 1 < x \leq 2 \end{cases}$ $x \in [0, 2]$;

   (m) $f(x) = \begin{cases} 0, & \text{if } x \notin \mathbf{Q} \\ \frac{1}{q}, & \text{if } x = \frac{p}{q} \in \mathbf{Q} \ (p, q \in \mathbf{N}, \text{ relatively prime }), \end{cases}$ $x \in [0, 1]$;

   (n) $f(x) = \begin{cases} x & \text{if } x \in \mathbf{Q} \\ 1 - x & \text{if } x \notin \mathbf{Q}. \end{cases}$ $x \in [0, 1]$.

28. Give an example of a function of bounded variation that fails to be continuous.

29. Consider the function $f(x) = \begin{cases} \sin \frac{1}{x}, & \text{if } x \in (0, \frac{2}{\pi}] \\ 0, & \text{if } x = 0. \end{cases}$

   Using the partition $P = \left\{ 0, \frac{2}{(2n-1)\pi}, \ldots, \frac{2}{3\pi}, \frac{2}{\pi} \right\}$, show f is continuous, but is not of bounded variation in $\left[0, \frac{2}{\pi} \right]$.

30. Let $f(x) = \begin{cases} x \cos \frac{\pi}{2x}, & \text{if } x \in (0,1] \\ 0, & \text{if } x = 0. \end{cases}$

Show f is continuous, but not a function of bounded variation in $[0,1]$.

31. If f is a function of bounded variation in $[a,b]$, show $|f|$ is also a function of bounded variation in $[a,b]$. What can you say about $V(f)$ as compared to $V(|f|)$?

32. Let f be defined on $[a,b]$ and let there exist a number k such that $|f(u) - f(v)| < k|u - v|$ for each u and v in $[a,b]$. Prove f is a function of bounded variation in $[a,b]$ and $|V(f)[a,v] - V(f)[a,u]| < k|u - v|$.

33. If f is a function of bounded variation in $[a,b]$ and if there exists $k > 0$ satisfying $|f(x)| > k$ for all $x[a,b]$, then show $\frac{1}{f}$ is a function of bounded variation in $[a,b]$. Is there any relationship between $V(f)$ and $V(1/f)$?

34. Show a polynomial $p(x)$ is a function of bounded variation in each closed interval $[a,b]$. Describe a method of finding the total variation of $p(x)$ on $[a,b]$ from the knowledge of the zeros of the derivative $p'(x)$.

35. If $V(x)$ denotes the variation function corresponding to a function f of bounded variation in $[a,b]$, set $P(x) = \frac{1}{2}(f(x) - f(a) + V(x))$, $N(x) = \frac{1}{2}(V(x) - f(x) + f(a))$. ($P(x)$ and $N(x)$ are called the **positive variation, negative variation** of f in $[a,b]$.) Compute the positive and negative variations of the functions in Exercise 27.

36. Prove the following form of **integration by parts formula**. Let f and g be functions of bounded variation over $[a,b]$. Then f is Riemann-Stieltjes integrable with respect to g if and only if g is Riemann-Stieltjes integrable with respect to f on $[a,b]$. In either case,

$$\text{R-S} \int_a^b f \, dg = f(b)g(b) - f(a)g(a) - \int_a^b g df.$$

37. Prove the **First Mean Value Theorem** for Riemann-Stieltjes integrals: If f is continuous and g is strictly increasing in $[a,b]$, then there exists $c \in (a,b)$ satisfying

$$\text{R-S} \int_a^b f \, dg = f(c)[g(b) - g(a)].$$

Give an example to show if g is not strictly increasing, the result need not be true.

38. Prove the **Second Mean Value Theorem** for Riemann-Stieltjes integrals: If f is strictly increasing and g is continuous on $[a,b]$, then there exists $c \in (a,b)$ such that

$$\text{R-S} \int_a^b f \, dg = f(a)[g(c) - g(a)] + f(b)[g(b) - g(c)].$$

Can we drop the requirement that f is strictly increasing?

39. Let f be continuous on $[0,n]$, $n \in \mathbf{N}$. Prove

$$\text{R-S} \int_0^n f d([x]) = \sum_{i=1}^{n} f(i).$$

[This exercise shows the Riemann-Stieltjes Integral with respect to the function $f(x) = [x]$ reduces to ordinary summation.]

40. Let h be continuous and increasing on $[a, b]$, let $h(a) = c, h(b) = d$. If $\int_c^d f \, dg$ exists, prove the integral $\int_a^b f(h(t)) \, dg(h(t))$ exists, and the two integrals are equal. Give an example where the second integral exists, but not the first.

Chapter 6

Infinite Series of Constants

In Chapter 1, we discussed at length the notion of a sequence and the limit of a given sequence of real numbers. We are already familiar with the summation symbol, $\sum$ (sigma notation), as applied to a finite number of terms of a sequence a_n of real numbers (see Exercise 0.4.20). In Chapter 5, finite sums played an essential role in the definition of the integral. However, as the reader can recall, the treatment of the integral carefully avoided the problem of an infinite sum. The purpose of this chapter is to assign meaning to sums of the form

$$\frac{1}{2} + \frac{1}{4} + \frac{1}{8} + \frac{1}{16} + \frac{1}{32} + \cdots,$$

where the '$\cdots$' is interpreted to indicate the remaining infinite number of additions that have to be performed. Our definition of this notion will lead to the conclusion that the 'addition' of an 'infinite number of real numbers', though it cannot be achieved physically, is made possible by applying the familiar limiting process to naturally associated sequences.

Infinite sums are not artificial. A simple example of an infinite sum is the infinite decimal expansion of the rational number $\frac{1}{3}$. Another example concerns the infinite sum presented above.

Consider a line segment of unit length, $\overline{AB}$, as shown in Figure 6.0.1. Imagine that we mark the midpoint of this segment, and then the midpoint of its right-hand segment, and then the midpoint of the next right-hand segment, and so forth. Now, let us construct an infinite sum as follows: Beginning at A, move to the first marked point to the right, that is, the midpoint of $\overline{AB}$, and take for the first entry to the sum, the length of the segment from A to the midpoint; each additional entry to the sum is obtained by moving to the midpoint adjacent and to the right of the point we are presently on, and the associated entry to the sum is the length of the line segment traversed.

Several features of this process are evident. First, each succeeding term of the sum is exactly half of its predecessor. Second, the process is discrete, in the sense that if one thinks of actually physically performing the process, it would require an infinite number of discrete operations. Third, a fixed real number clearly exists that should be identified as the sum, namely, the length of $\overline{AB}$, which is 1.

Figure 6.0.1 The first few terms in the summation of $\sum_{n=1}^{\infty} \frac{1}{2^n}$.

The three features mentioned capture the essential difficulty associated with infinite sums with which mathematicians have struggled since the time of Zeno; indeed, the example above in a somewhat different form is at the heart of Zeno's paradoxes. The difficulty revolves around assigning a suitable meaning to infinite sums. Clearly, such a meaning should exist, since the length of the line segment exists as a physical reality. Clearly, also, the physical process of performing an infinite summation can never be realized. Thus, the problem for mathematicians was to create a suitable framework that could accommodate an equation of the form

$$\frac{1}{2} + \frac{1}{4} + \frac{1}{8} + \frac{1}{16} + \frac{1}{32} + \cdots = 1.$$

In summary, then, it is the purpose of this chapter to give a satisfactory meaning to the summation of an infinite number of terms of a sequence of real numbers and to delineate those circumstances in which we can associate, in a meaningful way, a real number as a 'natural' **sum** of such an infinite series.

6.1   Infinite Series and Its Convergence

Definition. Let p be an integer, $\{a_n : n \geq p\}$ be a sequence of real numbers, and the sequence $\{s_n, n \geq p\}$ of numbers be defined by

$$s_n = a_p + a_{p+1} + \cdots + a_n = \sum_{i=p}^{n} a_i, \; n = p, \; p+1, \; \ldots .$$

The ordered pair $(\{a_n\}, \{s_n\})$ is formally called an **infinite series**. Elements of the sequence $\{a_n\}$ are referred to as **terms of the series**, and the variable subscript is called the **index**. The sequence $\{s_n\}$ is called the **sequence of partial sums** of the series.

Discussion. Let us see how this definition permits us to cope with the problem of assigning a meaning to expressions like

$$a_1 + a_2 + a_3 + a_4 + a_5 + a_6 + \cdots .$$

First, it requires us to have some method for specifying the individual terms of the series. Second, it replaces any notion involving an infinite summation process with that of a finite sum. This is exactly the technique that was employed in the definition of integral, and it is useful because it permits us to employ all our accumulated knowledge about the behavior of finite sums.

The reader will have noticed that we have described $\{a_n : n \geq p\}$ as a sequence. Strictly speaking, this is not true since the definition of sequence specifies functions having domain, the set $\mathbf{N}$ of positive integers, and mapping into $\mathbf{R}$. In this case, we have permitted our functions to have domains consisting of all integers greater than or equal to p. Nevertheless, it is easy to show (Exercise 2) that the above function is equivalent to a sequence. $\square$

Notation. The definition above is necessary but cumbersome. (This is again amplified in the subsection on notation at the end of this section.) We shall abuse our notation by saying that the collection of symbols

$$\sum_{i=p}^{\infty} a_i, \ p \text{ an arbitrary integer}$$

is an infinite series.

For the special case of $p = 1$, we have notations of the form

$$a_1 + a_2 + \cdots + a_n + \cdots,$$

or more simply $\sum a_n$ that represent infinite series. Attached to these convenient notations is the sequence $\{s_n\}$ of partial sums corresponding to the infinite series $\sum_{n=1}^{\infty} a_n$.

Discussion. The above definition tells us that an infinite series is much more than the mere sequence of symbols

$$a_1 + a_2 + \cdots + a_n + \cdots.$$

Rather, we start with a sequence $\{a_n\}$ of numbers and immediately construct the associated sequence, $\{s_n\}$, its sequence of partial sums. It is this sequence, $\{s_n\}$, that will provide most of the information about the series $\sum a_n$. Even though an infinite series can be simply written down from the sequence $\{a_n\}$, by replacing the commas by '+' signs, we emphasize that without the knowledge of the sequence of partial sums, $\{s_n\}$, we cannot further study the series. This is why we call the pair of these two sequences, together, the infinite series. In practice, we do not mention the sequence $\{s_n\}$, but merely say that $\sum a_n$ is an infinite series.

The definition of the series that we have given starts with the notion of sequence, which as we already know, is a function from $\mathbf{N}$ to $\mathbf{R}$. As discussed above, this is easily generalized to the situation of

$$\sum_{n=p}^{\infty} a_n,$$

where the initial value of the index is the arbitrary integer p. Moreover, in Exercise 0.4.20, we treated the problem of changing the index. Thus, we shall not be bound by the total formality of our definition of infinite series, but will feel free to consider the series of the form

$$\sum_{i=0}^{\infty} a_i, \ \sum_{n=-3}^{\infty} a_{2n}$$

or any other series that arises. As a further abuse of notation, some authors will use $\sum a_n$ as a generic notation for arbitrary infinite series, and even use it to denote specific series where the initial index value is other than 1. However, we agree that if there is no mention of the limits in the summation, $\sum a_n$ always means that the index n runs from 1 to ∞. $\square$

Next, we shall see how to realize a number, which can reasonably be called the **sum** of an infinite series. Naturally, this will be arrived at by using the sequence, $\{s_n\}$, which we constructed as an auxiliary sequence.

Definition. The series $\sum_{n=p}^{\infty} a_n$ is said to be **convergent** (or to **possess a sum**) if the associated sequence of partial sums, $\{s_n\}$, possesses a limit. Moreover, if $\lim_{n\to\infty} s_n = s$, then the number s is called the **sum** of the infinite series. We then write

$$\sum_{n=p}^{\infty} a_n = s.$$

The series $\sum_{n=p}^{\infty} a_n$ is said to be **divergent** provided it is not convergent, that is, if the sequence of partial sums fails to have a limit. Thus, we do not associate any 'sum' with a divergent series.

Discussion. It is quite clear that an infinite summation is a physical impossibility. Yet, by using the powerful tool of the limit concept, we are able to give an infinite sum a very concrete meaning. From the manner in which we constructed the sequence of partial sums, $\{s_n\}$, from the given series, $\sum_{n=p}^{\infty} a_n$, we observe that for the case $p = 1$,

$$\begin{aligned} s_1 &= a_1, \\ s_2 &= a_1 + a_2, \\ s_3 &= a_1 + a_2 + a_3, \\ \cdots \;\; &\cdots \quad \cdots \quad \cdots \, , \\ s_n &= a_1 + a_2 + \cdots + a_n, \end{aligned}$$

and so on. The natural 'termination' (if it were possible) would be to write

$$s_\infty = a_1 + a_2 + \cdots + a_n + \cdots .$$

However, to yield to this temptation would avoid any discussion of the conditions that make such an equality meaningful. For this reason, we are forced into exactly the type of considerations that led to the definition of the limit of a sequence. Such considerations, applied to the sequence of equalities given above, make it obvious that if a 'sum' of the series exists, it should be the limit of the sequence of its partial sums; this yields the definition that we have given above.

In addition to the benefits gained by looking at finite sums, for which a large body of knowledge has been developed, by specifying the existence of the infinite sum in terms of the limit of a sequence, we bring to bear on the summation problem the considerable body of theory that has already been developed for sequences. We would therefore encourage the ambitious reader to stop for a minute and consider the question: given my previous knowledge of sequences, how should the theory of series now be developed? What questions should be posed, and further, what are their answers? □

The remainder of this chapter is devoted to the study of the limit as it occurs in the context of infinite series. One might think that we have already developed a fair amount of information about the topic from our study of sequences and the fact that the definition of this limit boils down to the definition of the limit of a sequence. Certainly, all of what we have learned about the theory of sequences must have application to the theory of infinite series. For example, we know why sequences fail to converge; thus, we know why the sequence of partial sums $\{s_n\}$ of a series $\sum_{n=p}^{\infty} a_n$ will fail to converge. Our interest now becomes, what does this tell us about the sequence $\{a_n\}$? It is shown in Theorem 6.1.2 that for the sequence of partial sums to converge, the terms of the series must converge to 0. Of course, this observation is immediately followed by the question: Is the convergence of the sequence $\{a_n\}$ to 0 enough to ensure the convergence of $\{s_n\}$? As we shall see, the answer is no, and this fact generates many of the interesting subtleties that differentiate the theory of series from the theory of sequences. Let us first begin with several examples.

Example 1. Examine the convergence properties of the series

$$1 + \frac{1}{2} + \frac{1}{4} + \cdots + \frac{1}{2^{n-1}} + \cdots = \sum_{n=1}^{\infty} \frac{1}{2^{n-1}}.$$

Solution. For this series, we have

$$a_n = \frac{1}{2^{n-1}}.$$

Using the formula for the finite geometric series developed in the Exercise 0.4.22, we have

$$s_n = \sum_{k=1}^{n} \frac{1}{2^{k-1}} = \sum_{k=0}^{n-1} \left(\frac{1}{2}\right)^k = \frac{1 - \frac{1}{2^n}}{1 - \frac{1}{2}} = 2\left(1 - \frac{1}{2^n}\right).$$

It is immediate that $\{s_n\}$ converges to 2, whence the series converges with 2 as its sum. □

Discussion. The basic technique being employed here is to find an explicit formula for the nth term of the sequence of partial sums. Finding such an explicit formulation may not always be possible, and when it is, it may require the use of a variety of algebraic techniques for its accomplishment. However, if an explicit formula for s_n can be found, it is generally of real value, since it can be used as a tool for finding the sum as in the example above.

The formula developed above can be applied to the series generated from midpoints, which was discussed in the introduction to this section. That series is given by $\sum_{n=2}^{\infty} \left(\frac{1}{2}\right)^{n-1}$. This series is almost the series specified above. The difference is in the initial value, $n = 2$ in the midpoint case, $n = 1$ in the case of this example. It is plausible (Exercise 3) that

$$\sum_{n=2}^{\infty} \left(\frac{1}{2}\right)^{n-1} = (-1) + \sum_{n=1}^{\infty} \left(\frac{1}{2}\right)^{n-1} = (-1) + \frac{1}{1 - \frac{1}{2}}. \tag{1}$$

This leads to a sum of 1 for the midpoint case, which is the value we know we should get for the reasons discussed in the introduction. □

Example 2. Discuss the convergence or divergence of the series

$$1 - 1 + 1 - 1 + 1 - 1 + \cdots = \sum_{n=1}^{\infty} (-1)^{n-1}.$$

Solution. It can be shown by induction that $s_{2n} = 0$, while $s_{2n+1} = 1$. Thus, the sequence $\{s_n\}$ of partial sums oscillates between 0 and 1, and so does not have a limit. As a result, the series cannot converge, whence it is divergent. □

Discussion. This series is a canonical example of an 'alternating series'. It is so called because we alternate between adding a term followed by subtracting a term or, more precisely, adding a positive term followed by a negative term and so on. If we write down the string

$$1 - 1 + 1 - 1 + \cdots,$$

we might suspect that it is equal to

$$(1 - 1) + (1 - 1) + \cdots = 0 + 0 + \cdots = 0.$$

This calculation seems to be quite reasonable. However, the following is equally reasonable.

$$1 + (-1 + 1) + (-1 + 1) + \cdots = 1 + 0 + 0 + \cdots = 1.$$

The different values for the sum arise out of attempts to actually sum the entirety of the terms constituting the series and makes clear why such attempts have inherent difficulties. Thus, it is evident that permitting imprecise calculations, such as those above, would have many undesirable consequences. In this case, the problems arose as a result of the introduction of parenthesis into the summation process. For finite sums, we have seen that a generalized associative law holds, so that the introduction of parenthesis cannot change the sum. However in the calculation above, we are inserting an infinite number of parentheses simultaneously. Aside from the fact that the process would require very quick hands, it leads to nonsensical results, witness the above. (Still another introduction of parenthesis can be used to sum the above series to $\frac{1}{2}$, a fact pursued in the exercises.) This example shows that we must view a series as a collection of numbers that are added together sequentially according to a fixed order. To change this order in many instances will change the sum or produce a sum where none ought to exist. Last, the example shows why the notion of limit is essential to a proper definition of 'sum' for an infinite series. □

Example 3. Discuss the convergence properties of the **geometric series**

$$\sum_{n=1}^{\infty} x^{n-1} = 1 + x + x^2 + x^3 + \cdots,$$

where x is a fixed real number.

Solution. If $x \neq 1$, we have the formula

$$s_n = \frac{1 - x^n}{1 - x}.$$

If $|x| < 1$, then the limit exists and we have

$$\sum_{n=1}^{\infty} x^{n-1} = \lim_{n \to \infty} \frac{1 - x^n}{1 - x} = \frac{1}{1 - x}.$$

If $|x| > 1$, then $\{s_n\}$ will diverge to $+\infty$ or will oscillate in an unbounded manner. The case $x = -1$ was treated in Example 2. For $x = 1$, $s_n = n$, and so $\{s_n\}$ diverges to $+\infty$. □

Discussion. This example is the general case for the series considered in Example 1. Infinite geometric series occur frequently in analysis and elsewhere in mathematics and science. The above solution shows that once the common ratio x has its absolute value less than 1, the series converges to the sum $\frac{1}{1-x}$. This is one of the more useful formulae in mathematics. □

Example 4. Discuss the convergence of $\sum_{n=1}^{\infty} n^3$.

Solution. It can be shown by induction (Exercise 4; see also Exercise 0.4.48) that

$$s_n = \frac{n^2(n+1)^2}{4},$$

a fact that easily yields the divergence of the series. □

Discussion. We have established far more than is required to show that this series is divergent. As we have already remarked, if the series converges, then the a_n's must converge to 0. This is not true in this example. □

Example 5. Discuss the convergence of $\displaystyle\sum_{n=1}^{\infty} \frac{1}{n(n+1)}$.

Solution. Using the pleasant fact that

$$\frac{1}{n(n+1)} = \frac{1}{n} - \frac{1}{n+1},$$

we have (by induction, if necessary) that

$$
\begin{aligned}
s_n &= \frac{1}{1 \cdot 2} + \frac{1}{2 \cdot 3} + \cdots + \frac{1}{n(n+1)} \\
&= \left(1 - \frac{1}{2}\right) + \left(\frac{1}{2} - \frac{1}{3}\right) + \left(\frac{1}{3} - \frac{1}{4}\right) + \cdots + \left(\frac{1}{n} - \frac{1}{n+1}\right) \\
&= 1 - \frac{1}{n+1},
\end{aligned}
$$

whence it is immediate that this series converges and the sum is 1. □

Discussion. This example makes really clear how useful it is to establish a formula for s_n. The method applied to split up the quantity $\frac{1}{n(n+1)}$ into two fractions is a familiar technique often encountered in integration, namely, **the method of partial fractions.** (A description of this technique may be found in any elementary calculus book.) The formula for s_n is found by writing down several partial sums and noticing that cancellation takes place. A series that undergoes this type of cancellation is often referred to as a **telescoping series.**

We have not given a detailed proof of the formula for s_n, but a simple inductive argument will suffice, which is requested in Exercise 5. $\square$

From the examples above, the reader may get the impression that it is always possible to find a formula for s_n. Further, the reader may also believe that every convergent series has a 'sum' whose value we can know. Both of these ideas are wrong. It is always possible to *compute* a value for s_n for any n. However, it is not always possible to give a 'nice' formula for s_n of the type found in our previous examples and from which the sum can be found. Moreover, even if we have a series that is known to converge, we may not be able to 'know' what the limit is. To make this clear, consider the series of Example 5. Its sum is 1, and we 'know' very clearly what number has been identified as the sum. The series $\sum_{n=1}^{\infty} \frac{1}{n^2}$ is also known to converge, and the value of its sum is $\frac{\pi^2}{6}$ (in this case, s_n is not easy to calculate!). In this latter case, although the answer is irrational, we also feel that we 'know' the sum, since there is a well-recognized constant that can be proven to be the limit of the partial sums. On the other hand, at present, there is no identifiable constant that can be shown to be the sum of the series $\sum_{n=1}^{\infty} \frac{1}{n^3}$. Thus, in spite of the fact that we know that the series of inverse cubes has a sum, it cannot be said that we 'know' what that sum is.

The next example illustrates another situation where there is no known 'nice' formula for s_n, although this should not deter the reader from trying to find one!

Example 6 (Harmonic series). Discuss the convergence or divergence of the series $\sum_{n=1}^{\infty} \frac{1}{n}$.

Solution. We shall show that the series diverges. For this, we must prove that $\{s_n\}$ fails to have a limit. This is achieved by producing a suitable subsequence of $\{s_n\}$ that diverges. The desired candidate in this example is $\{s_{2^n}\}$. Now

$$
\begin{aligned}
s_{2^n} &= 1 + \frac{1}{2} + \cdots + \frac{1}{2^n} \\
&= 1 + \frac{1}{2} + \left(\frac{1}{3} + \frac{1}{4}\right) + \left(\frac{1}{5} + \cdots + \frac{1}{8}\right) + \cdots + \left(\frac{1}{2^{n-1}+1} + \cdots + \frac{1}{2^n}\right),
\end{aligned}
$$

where the introduction of this grouping is valid (**WHY?**). Further, note that the sum of the terms in each group is never less than $\frac{1}{2}$ (**WHY?**), whence,

$$
s_{2^n} \geq \frac{1}{2} + \frac{1}{2} + \frac{1}{2} + \cdots + \frac{1}{2} = \frac{(n+1)}{2}.
$$

Thus, $\{s_{2^n}\}$ can be made as large as we please by choosing n sufficiently large, so the subsequence $\{s_{2^n}\}$ diverges; hence, $\{s_n\}$ has no limit and consequently the harmonic series is divergent. $\square$

Discussion. As indicated above, there is no known formula for s_n that is simpler than the definition, that is, $\sum_{k=1}^{n} \frac{1}{k}$. Thus, we are forced to use a technique that does not depend on our finding an explicit formula for s_n. With no previous knowledge, it would require considerable effort to develop the convergence properties of this series. In the remainder of this chapter we will develop tests that enable the user to easily decide the basic issue of whether or not a given series converges.

The main insight is that if we can show that beyond any given n, the remaining terms will sum to more than a fixed constant, so the series cannot converge. In our particular case, we have shown that for every n, $\sum_{k=n}^{\infty} \frac{1}{k} \geq \frac{1}{2}$. In fact, we have shown more than this: we have proven that $\sum_{k=n}^{m} \frac{1}{k} \geq \frac{1}{2}$, where $n = 2^{i-1}+1$ and $m = 2^i$.

This series has a very important feature. It is the canonical example of a divergent series that satisfies the condition that its terms go to 0, that is, $a_n \to 0$. Initially, this is a startling result, yet not so startling when one recalls that not all improper Riemann Integrals converge, even though the integrand may go to 0 as $x \to \infty$ (this comparison will later lead us to the Integral Test).

The fact that this series does not converge illustrates that for series the real issue, which determines the existence or nonexistence of a sum, is not whether $a_n \to 0$, but rather 'how fast' $a_n \to 0$ as $n \to \infty$. This idea gives rise to the notion of searching for a test series, that is, one whose terms approach 0 at the slowest possible rate, while still forming a convergent series. Such a series would be immensely valuable, since we could compare other series to it via some sort of comparison test, and this comparison would then decide the issue of convergence.

The issues raised by the Harmonic series stimulated a good deal of mathematical effort, and it is worth looking at the historical method by which this series was shown to be divergent, a method used by Bernoulli.

Using Example 5,

$$1 = \frac{1}{1 \cdot 2} + \frac{1}{2 \cdot 3} + \cdots + \cdots,$$

$$1 - \frac{1}{1 \cdot 2} = \frac{1}{2} = \frac{1}{2 \cdot 3} + \frac{1}{3 \cdot 4} + \cdots,$$

$$\frac{1}{2} - \frac{1}{2 \cdot 3} = \frac{1}{3} = \frac{1}{3 \cdot 4} + \cdots,$$

and so on. Now adding (can we ??),

$$1 + \frac{1}{2} + \frac{1}{3} + \cdots = \frac{1}{1 \cdot 2} + \frac{2}{2 \cdot 3} + \frac{3}{3 \cdot 4} + \cdots,$$

$$= \frac{1}{2} + \frac{1}{3} + \frac{1}{4} + \cdots.$$

If the 'sum' on the right-hand side is A, then we are led to the paradoxical equation $A + 1 = A$, whence A must be infinite. In other words, we cannot associate a finite value to the series as a sum. The series must therefore diverge.

The argument of Bernoulli lends further substance to the difficulties involved in manipulating infinite sums. Just what does it mean to add an infinite number of infinite sums? These questions now have satisfactory answers, but they were not obtained without considerable mental struggle. $\square$

Example 7. Discuss the convergence properties of the series $\sum_{n=1}^{\infty} \frac{1}{n^2}$.

Solution. We shall not be concerned with determining the actual sum of this series, but merely deciding the issue of convergence. (It can be shown that the series sums to $\frac{\pi^2}{6}$.) Consider

$$A_n = \sum_{k=m}^{p} \frac{1}{k^2}, \text{ where } m = 2^{n-1} \text{ and } p = 2^n - 1, \ n \in \mathbf{N}.$$

We claim $A_n \leq \left(\frac{1}{2}\right)^{n-1}$. To see this, note that A_n is a sum of 2^{n-1} terms, since by subtraction,

$$p - (m-1) = [(2^n - 1) - 2^{n-1}] + 1 = 2^n - 2^{n-1} = 2^{n-1}.$$

Moreover, for each k such that $2^{n-1} \leq k \leq 2^n$,

$$\frac{1}{k^2} \leq \left(\frac{1}{2^{n-1}}\right)^2,$$

whence our claim about A_n follows. It seems plausible, and will be shown in the next section, that if we replace each A_n by $\frac{1}{2^{n-1}}$ and sum the series of replacements, the convergence of the replacement series should guarantee the convergence of $\sum_{n=1}^{\infty} A_n$. The sequence, $\{s_n\}$, of partial sums associated with the original series is easily seen to be monotonically increasing, and as such, its convergence properties are completely determined by any subsequence, in this case, by the subsequence $\{t_k\}$, where $t_k = \sum_{n=1}^{k} A_n$. It follows (**HOW**?) that $\sum_{n=1}^{\infty} \frac{1}{n^2}$ is convergent. $\square$

Discussion. In both the Examples 6 and 7, the arguments were based on examining a subsequence of the sequence of partial sums. For the case of using a subsequence to establish the divergence, there is never any problem, since if one subsequence does not converge to a limit, the sequence *in toto* cannot converge to a limit. However, this is not the case when we want to establish convergence. We well know that nonconvergent sequences may possess convergent subsequences. Thus, when we start with the series $\sum a_n$ and convert to the series $\sum A_n$, we must be very sure that the sequence of partial sums associated with the new series has the same convergence properties as the sequence of partial sums associated with the old series. In Example 7, all difficulties are taken care of by the fact that the sequence $\{s_n\}$ is monotonic increasing. However, this will not generally be true and the reader should take care when formulating this type of argument.

Looking at a subsequence of the sequence of partial sums, $\{s_n\}$, is equivalent to introducing parenthesis into the summation process. This can be seen in the definition of A_n, which is the sum of 2^{n-1} consecutive terms of the series. Thus, A_1 is (the sum of) the first term of the series, A_2 is the sum of the second and third terms of the series, A_3 is the sum of the next four terms of the series, and so forth. In the latter part of this chapter, we will consider the problem of identifying those conditions under which parenthesis can be introduced into the series without fear of altering the convergence properties of the original series. These considerations will lead to some of the more elegant results of the theory. □

We will complete this section by proving several simple, but basic, results. The first illustrates how results in the theory of finite sums carry over to the theory of infinite sums.

Theorem 6.1.1. *If $\sum_{n=p}^{\infty} a_n$ converges, then the sum is unique. Further, if the sum is a, then $\sum_{n=p}^{\infty} ca_n$ is a convergent series for any real constant, c, and*

$$\sum_{n=p}^{\infty} ca_n = ca.$$

In addition, if $\sum_{n=p}^{\infty} b_n = b$ is a convergent series, then $\sum_{n=p}^{\infty} (a_n \pm b_n)$ converges to $a \pm b$.

Proof. For uniqueness of the sum of a convergent series, recall that the limit of a sequence, when it exist, is unique. Thus, the sum, which is the limit of the sequence of partial sums, must be unique.

Let a be the limit of the sequence of partial sums of $\sum_{n=p}^{\infty} a_n$. The partial sum, s_k is given by $\sum_{n=p}^{k} a_n$, where $k \geq p$. If we let t_k denote the partial sum $\sum_{n=p}^{k} ca_n$, where $k \geq p$, of the series whose terms are ca_n, then by the generalized distributive law (Exercise 0.4.21), $cs_k = t_k$. By (a special case of) Theorem 1.2.5,

$$\lim_{k \to \infty} t_k = \lim_{k \to \infty} cs_k = ca.$$

Since this statement about the partial sums translates exactly into the required statement about the series, the proof is complete. The proof of the remainder of the theorem is left as Exercise 9. □

Discussion. The proof completely illustrates the use of the previously developed theory of finite sums and sequences. While there are many additional subtleties associated with the theory of series, nevertheless a substantial body of the theory can be obtained by direct analogy to the theory of sequences and finite series. □

Theorem 6.1.2. *If $\sum_{n=p}^{\infty} a_n$ is a convergent series, then $\lim_{n \to \infty} a_n = 0$.*

Proof. Let s_n denote the nth partial sum of the series so that

$$\sum_{n=p}^{\infty} a_n = \lim_{n \to \infty} s_n = s.$$

Note that $a_n = s_n - s_{n-1}$, whence

$$\lim_{n \to \infty} a_n = \lim_{n \to \infty} s_n - \lim_{n \to \infty} s_{n-1} = s - s = 0. \quad \square$$

Discussion. If the series is to have a sum, the theorem states that all the terms after some stage must become arbitrarily small. The contrapositive of the theorem states that if $\lim a_n \neq 0$, then the series cannot converge, and thus yields a simple first test for convergence, one that immediately takes care of the series in Examples 2 and 4. Unfortunately, or rather fortunately, as far as mathematicians are concerned, the result does not characterize convergent series, as the Harmonic series of Example 6 shows. From this, we see that the main use of the theorem will be in its contrapositive form, that is, for establishing nonconvergence of series. We suggest that the reader write out a complete justification for the equalities that form the last line of the proof. $\square$

So far, we have stressed the similarities between infinite series and sequences, namely, the series converges or diverges according as the associated sequence of partial sums converges or diverges. However, there is one important distinction. In studying sequences, it is essentially the behavior of the terms as n gets larger that is significant; thus, for sequences, for any fixed integer $p > 0$, we have the relation $\lim_{n \to \infty} a_n = \lim_{n \to \infty} a_{n+p}$ provided the limit in question exists. In other words, the convergence properties, *including the actual limit,* of a sequence are completely determined by any **tail** of the sequence.

But for a convergent series, every single term, and its position with respect to other terms in the sum, plays a vital role. If we omit a term a_i from the series $\sum_{n=p}^{\infty} a_n$, the sum will certainly diminish by that quantity. However, a moment's thought should convince the reader that the property of whether or not a series is convergent will be unaffected by the deletion of a 'finite' number of terms. Thus, we stress that the series $\sum a_n$ and the sequence $\{a_n\}$ are entirely different types of entities.

Notation. We have defined series so as to permit an arbitrary initial value for the index. In Exercise 2, the reader will be asked to establish a result that will have the effect of obviating the necessity for this more general approach. That is, the content of Exercise 2 tells us that all the theorems that can be established for series having initial index 1 and that do not depend in some essential way on the initial value can be translated into a form for series having any other integer as the initial value of the index. Since this is the case, throughout the remainder of this book we will discuss series of the form $\sum a_n$, where it is assumed that the initial index is 1. This not withstanding, we will take all theorems as being proven so that they may be applied in situations of other initial index values. In doing so, there is the tacit assumption that the reader could reformulate any proof so as to make it apply in the more general case. $\square$

We conclude this section with a theorem that provides a necessary and sufficient condition for the convergence of a given series, and which is very similar to the Cauchy Criterion for the convergence of a sequence (see Theorem 3.3.1).

Theorem 6.1.3 (Cauchy Criterion). *The series $\sum a_n$ is convergent if and only if given $\epsilon > 0$, there exists an $N \in \mathbf{N}$ such that*

$$(n > N) \text{ and } (p \in \mathbf{N}) \text{ .implies. } |a_{n+1} + a_{n+2} + \cdots + a_{n+p}| < \epsilon.$$

Proof. Assume the series is convergent and let s_n denote the nth partial sum of the series. This sequence is convergent and so must satisfy the Cauchy Criterion for sequences. Thus, for any $\epsilon > 0$, there exists an $N \in \mathbf{N}$ such that if $n > N$ and $p \in \mathbf{N}$, then $|s_{n+p} - s_n| < \epsilon$. Now

$$|s_{n+p} - s_n| = \left| \sum_{k=n+1}^{n+p} a_k \right|,$$

whence the desired conclusion follows. To establish the converse, assume the stated condition. Then, the sequence of partial sums is Cauchy, whence it has a limit. Call this limit a. We leave it to the reader (Exercise 10) to show a is the sum of the series. $\square$

Discussion. The above theorem asserts that if $\sum a_n$ is to have a sum, then 'blocks' of terms of arbitrary length must, after a certain stage, sum to a negligible quantity. For $p = 1$, this theorem is equivalent to the statement of Theorem 6.1.2. Moreover for each positive ϵ, there is an $N \in \mathbf{N}$ such that $|\sum_{n=N+1}^{\infty} a_n| < \epsilon$; that is, for each N, the Nth 'tail' of the series approaches 0 (see Exercise 19). Of course, if ϵ is very small, N will have to be correspondingly very large. $\square$

Exercises

1. Find the indicated term and partial sum of the given series:

    (a) $\sum_{n=1}^{\infty} n$, find a_1 and s_1;

    (b) $\sum_{n=1}^{\infty} n$, find a_2 and s_2;

    (c) $\sum_{n=1}^{\infty} n$, find a_{25} and s_{20};

    (d) $\sum_{n=1}^{\infty} -3$, find a_1 and s_1;

    (e) $\sum_{n=1}^{\infty} -3$, find a_8 and s_8;

    (f) $\sum_{n=1}^{\infty} -3$, find a_{80} and s_{80};

    (g) $\sum_{n=0}^{\infty} \left(\frac{1}{3}\right)^n$, find a_0 and s_0;

    (h) $\sum_{n=0}^{\infty} \left(\frac{1}{3}\right)^n$, find a_1 and s_1;

    (i) $\sum_{n=0}^{\infty} \left(\frac{1}{3}\right)^n$, find a_{25} and s_{25};

    (j) $\sum_{n=-3}^{\infty} (2n - 6)$, find a_{-3} and s_{-3};

    (k) $\sum_{n=-3}^{\infty} (2n - 6)$, find a_{-1} and s_{-1};

    (l) $\sum_{n=-3}^{\infty} (2n - 6)$, find a_1 and s_1;

    (m) $\sum_{n=-3}^{\infty} (2n - 6)$, find a_5 and s_5, and find the sum of the first five terms;

    (n) $\sum_{n>4} \frac{1}{n}$, find a_5 and the fifth partial sum;

    (o) $\sum_{2n>7} \frac{1}{2n}$, find the sixth term of the series and the fifth partial sum.

2. Let $\{a_n : n \geq p\}$, p any integer, be a collection of real numbers. Show there is a function $\sigma : \mathbf{N} \to \{n : n \geq p\}$ such that

    (a) $\sigma(1) = p$;

    (b) $\sigma(k + 1) = \sigma(k) + 1$ for each $k \in \mathbf{N}$, $k \geq 1$.

3. Let $\{a_n\}$ be any sequence and let $p \in \mathbf{N}$ be fixed. Show $\sum_{n=1}^{\infty} a_n$ exists, exactly if $\sum_{n=p}^{\infty} a_n$ exists. Further show if either sum exists, then

$$\sum_{n=1}^{\infty} a_n = \sum_{n=1}^{p-1} a_n + \sum_{n=p}^{\infty} a_n.$$

Use this result to establish the validity of Eq. 1 in the discussion following Example 1.

4. Give an inductive proof showing the formula for s_n given in Example 4 is correct.

5. Give an inductive proof showing the formula for s_n given in Example 5 is correct.

6. Show the addition of a finite number of terms to a series cannot alter its convergence properties. Do the same for the deletion of a finite number of terms of a series.

7. Show if we add or delete an infinite number of terms from a series, then it will not, in general, have the same convergence properties as the original series. In particular, if we delete terms from a convergent series, will it still stay convergent?

8. If $\sum_{n=1}^{\infty} a_n = s$, show for $k \in \mathbf{N}$, $\sum_{n=1}^{\infty} a_{n+k}$ converges to $s - \sum_{i=1}^{k} a_i$.

9. Let $\sum a_n = a$ and $\sum b_n = b$. Show $\sum (a_n \pm b_n) = a \pm b$.

10. Complete the proof of the Cauchy Criterion for series.

11. Let a_n and b_n be two sequences such that $a_n = b_{n+1} - b_n$. Show $\sum a_n$ converges if and only if $\lim_{n \to \infty} b_n$ exists. Moreover, in the event of convergence, what relationship exists between the sum and the limit?

12. Let $p : \mathbf{N} \to \mathbf{N}$ be strictly increasing with $p(1) = 1$, and let $\sum a_n$ be a infinite series. Set $m = p(n)$, $m' = p(n+1) - 1$ and

$$b_n = \sum_{m}^{m'} a_k.$$

Show $\sum b_n$ will converge whenever $\sum a_n$ converges. Show the converse is false. (Note this process amounts to introducing parentheses into the series $\sum a_n$.)

13. Let $\sum a_n$ and b_n be as in Exercise 12. Show if there is a constant $M > 0$ such that for all $n \in \mathbf{N}$, $p(n+1) - p(n) < M$ and $\lim_{n \to \infty} a_n = 0$, then the convergence of $\sum a_n$ is equivalent to the convergence of $\sum b_n$.

14. Does the series $\sum \frac{(-1)^{n+1}}{n}$ converge?

15. Let $a_n = \sqrt{n+1} - \sqrt{n}$. Does the series $\sum a_n$ converge?

16. Discuss the convergence, and where possible find the sum of the following series:

(a) $\frac{1}{1 \cdot 3} + \frac{1}{3 \cdot 5} + \frac{1}{5 \cdot 7} + \cdots$;

(b) $\frac{1}{1 \cdot 4} + \frac{1}{2 \cdot 5} + \frac{1}{3 \cdot 6} + \frac{1}{4 \cdot 7} + \cdots$;

(c) $\frac{1}{1 \cdot 3 \cdot 5} + \frac{1}{2 \cdot 4 \cdot 6} + \frac{1}{3 \cdot 5 \cdot 7} + \frac{1}{4 \cdot 6 \cdot 8} + \cdots$;

(d) $1 - \frac{1}{2 \cdot 3} - \frac{1}{4 \cdot 5} - \cdots$;

(e) $\sum_{n=1}^{\infty} (-1)^{n+1} \frac{(2^n + 3^n)}{5 \cdot 4^n}$;

(f) $\sum_{n=1}^{\infty} \frac{(-1)^{n+1}}{n} \left(\sum_{k=1}^{n} \frac{1}{k} \right)$;

(g) $\sum_{n=1}^{\infty} [nxe^{-nx^2} - (n-1)xe^{-(n-1)x^2}]$;

(h) $\sum_{n=1}^{\infty} \frac{(-1)^n (2n+1)}{(n+1)(n+2)}$;

(i) $1 - 1 + 0 + 1 - 1 + 0 + 1 - 1 + 0 + \cdots$;

(j) $\sum_{n=1}^{\infty} \frac{(n-2)}{(n^2-4)(n+1)}$;

(k) $(1 - 1 + 1) - 1 + (1 - 1 + 1) - 1 + \cdots$;

(l) $\sum_{n=1}^{\infty} \sin \frac{\pi}{n}$;

(m) $\sum_{n=1}^{\infty} [\arctan n - \arctan (n-1)]$;

(n) $\sum_{n=1}^{\infty} \frac{(-1)^{n+1}(n+3)}{(n+1)(n+3)}$;

(o) $1 - 1 + \frac{1}{2} - \frac{1}{2} + \frac{1}{3} - \frac{1}{3} + \cdots$;

(p) $\sum_{n=1}^{\infty} a_n$, where $a_n = \begin{cases} -1/n, & \text{if } n \text{ is divisible by 3} \\ 1/n, & \text{otherwise;} \end{cases}$

(q) $\sum_{n=1}^{\infty} a_n$, where $a_n = \begin{cases} 1/n, & \text{if } n \text{ is a perfect square} \\ 0, & \text{otherwise.} \end{cases}$

17. Consider the series of Example 2. Show the introduction of a finite number of parenthesis will lead to the conclusion that the series must sum to $\frac{1}{2}$.

18. Consider $\sum_{n=1}^{\infty} \frac{(-1)^n}{n}$. In Exercise 14 you showed this series had a sum. Call this sum a. Find a bound on the quantity $|s_n - a|$, where s_n is the nth partial sum.

19. If $\sum_{i=1}^{\infty} a_i$ converges, show $\lim_{n \to \infty} t_n = 0$, where $t_n = \sum_{n+1}^{\infty} a_i$ (t_n is usually referred to as the nth- **tail** of the series).

20. If $a_1 = 1, a_{n+1} = a_n + \frac{b_n}{a_n} (n > 1)$, where $b_n \geq 0$ is arbitrary, show $\lim a_n$ exists if $\sum b_n$ converges.

21. A sequence $\{b_n\}$ is said to be of **bounded variation** provided the series $\sum |b_n - b_{n+1}|$ converges. Prove every sequence of bounded variation is convergent, but not conversely.

22. Show every convergent monotone sequence is of bounded variation.

23. Which of the following sequences are of bounded variation?

(a) $\left\{ \frac{(-1)^n}{n} \right\}$;   (b) $\left\{ \frac{(-1)^n}{\sqrt{n}} \right\}$;   (c) $\left\{ \frac{(-1)^n}{n^2} \right\}$;   (d) $\left\{ \frac{\sin n}{n} \right\}$.

24. Show if the series $\sum a_n$ satisfies $\sum_{n=k}^{\infty} a_n \geq c$, for every $k \in \mathbf{N}$ where c is a fixed positive constant, then the series is divergent.

25. Let p, $\sum a_n$, and $\sum b_n$ be as in Exercise 12. Show if a_n is either nonnegative (or nonpositive) for all n, then the two series either both converge or both diverge.

26. Generalize the argument of Example 7 to show if $1 < p < 2$, then $\sum \frac{1}{n^p}$ is convergent. Explain why the argument will not extend to the case $p = 1$.

27. Let $\{a_n\}$ be a monotonic decreasing sequence of positive terms such that $\sum a_n$ is convergent. Prove $na_n \to 0$. Is this condition sufficient for the convergence of $\sum a_n$?

28. Prove or disprove:

   (a) If $\sum a_n$ converges, then $\sum a_n^2$ converges.

   (b) If $\sum a_n$ converges, then $\sum \sqrt{a_n}$ converges.

   (c) If $\sum a_n$ converges and $\{b_n\}$ is bounded, then $\sum a_n b_n$ converges.

   (d) If $\sum a_n$ converges, then $\sum |a_n - a_{n+1}|$ converges.

   (e) If $\sum a_n$ and $\sum b_n$ converge, then, $\sum a_n b_n$ converges.

   (f) If $\sum a_n$ converges, then $\sum \frac{1}{1+|a_n|}$ converges.

   (g) If $\sum a_n$ converges, then $\sum \frac{a_n}{n^2}$ converges.

   (h) If $\sum a_n$ converges, then the sequence $\left\{ \frac{a_1 + 2a_2 + \cdots + na_n}{n(n+1)} \right\}$ converges to 0.

29. Consider the series $\sum a_n$. What is the effect of 'diluting' this series by inserting 0's into the series at random?

30. Let $\{a_n\}$ be a sequence and consider the series $\sum (a_n - a_{n-1})$. Give any conditions on $\{a_n\}$ that will guarantee the existence of a sum for the series.

31. Rewrite the series in Example 5, as $\sum \left(\frac{n}{n+1} - \frac{n-1}{n} \right)$. Show if we remove the parenthesis in each term, the resulting series is not convergent.

32. Consider an arbitrarily large deck of playing cards. Show these cards may be stacked in such a way that the lead edge of the top card is arbitrarily far, horizontally, from the lead edge of the bottom card, that is, a badly leaning tower is formed.

6.2   Convergence of Series of Nonnegative Terms

The series $\sum a_n$ where all the terms a_n are nonnegative is called a **series of nonnegative terms.** Such a series is particularly well behaved, since the corresponding sequence $\{s_n\}$ of partial sums has the property that $s_n \leq s_{n+1}$ for all n. In other words, the sequence of partial sums is monotonically increasing, whence the sequence converges provided it is bounded above. If it is unbounded above, it diverges to $+\infty$. Thus, a series of nonnegative terms either converges or else diverges to $+\infty$. A series of nonnegative terms cannot 'wander' or 'oscillate' between two numbers. In the remainder of this section we develop some simple tests to determine whether a series of nonnegative terms is convergent or divergent, where the knowledge of the actual sum is unimportant.

Theorem 6.2.1 (Comparison Test). *Let $\sum a_n$, $\sum b_n$ be two series of nonnegative terms. If there exists a constant, $k > 0$, such that*

(i) *$a_n \leq k b_n$ for all n and $\sum b_n$ converges, then $\sum a_n$ converges;*

(ii) *$a_n \geq k b_n$ for all n and $\sum b_n$ diverges, then $\sum a_n$ diverges.*

Proof. Let s_n and t_n denote the nth partial sums of the series $\sum a_n$ and $\sum b_n$, respectively. If (i) holds, then since $\sum b_n$ converges, and there exists $M > 0$ such that $t_n < M$ for all n. Thus,

$$s_n = \sum_{i=1}^{n} a_i \leq \sum_{i=1}^{n} k b_i = k \sum_{i=1}^{n} b_i = k t_n < kM,$$

whence $\{s_n\}$ is also bounded. Being monotonic increasing, $\{s_n\}$ converges and, consequently, $\sum a_n$ converges as well.

If (ii) holds, observe that $s_n \geq k t_n$, and the latter is unbounded, since $\{t_n\}$ diverges. Hence, $\{s_n\}$ diverges. $\square$

Discussion. In loose expression, the above test tells us that a nonnegative series whose terms are less than the corresponding terms of a known convergent series is convergent, while one whose terms always exceed those of a divergent series must diverge. The proof is very simple, but invokes great power, the Completeness Axiom, in the form of Theorem 1.3.1.

To apply these tests successfully, we need to have advance knowledge of some standard convergent and divergent series that are used as test series. We must then be able to relate the corresponding terms with the right type of inequality.

Last, it is not necessary to insist that the inequalities in (i) and (ii) hold for all n. It suffices to require that they hold from some stage $n = p$ onward. $\square$

There is another form of the Comparison Test that involves limits and is somewhat more useful in actual practice that the previous test.

Corollary (Limit form of Comparison Test). *Let $\sum a_n$ and $\sum b_n$ be two series of positive terms $(a_n, b_n > 0)$ such that*

$$\lim_{n \to \infty} \frac{a_n}{b_n} = L.$$

(i) *If $L \neq 0$, then the both the series either converge or both of them diverge;*

(ii) *If $L = 0$, and $\sum b_n$ converges, then $\sum a_n$ converges;*

(iii) *If $L = \infty$, and if $\sum b_n$ diverges, then $\sum a_n$ also diverges.*

Proof. To prove (i), we observe $L > 0$, and given ϵ such that $L > \epsilon > 0$, we can find N satisfying

$$L - \epsilon < \frac{a_n}{b_n} < L + \epsilon$$

whenever $n > N$. Hence,

$$(L - \epsilon) b_n < a_n < (L + \epsilon) b_n.$$

The right-hand side of this inequality will establish convergence, while the left-hand side will establish divergence, simply by applying Theorem 6.2.1. The remainder of the proof is left as Exercise 2. □

Discussion. The limit form of the test is very useful. All we have to do is to compute the limit of the expression $\frac{a_n}{b_n}$. If the limit is nonzero, and one of the series is convergent (respectively, divergent), the other is also convergent (respectively, divergent), whence this is the ideal situation. If $\lim\limits_{n \to \infty} \dfrac{a_n}{b_n} = 0$, we can only conclude the convergence of $\sum a_n$ from that of $\sum b_n$. No conclusion can be obtained for divergence. Similarly, case (iii) provides information about divergence only. These tests make it clear why a universal test series, to which all others could be compared, would be very useful. □

The use of these tests is illustrated with two examples.

Example 1. Discuss the convergence properties of the series $\sum \frac{1}{n^p}$ for $p > 0$, $p \in \mathbf{R}$.

Solution. We claim that the series diverges for $0 < p \leq 1$ and converges for $p > 1$. For $p = 1$, the result in Example 6.1.6 gives divergence. Moreover, for $0 < p < 1$, we have $\frac{1}{n} \leq \frac{1}{n^p}$ for all $n \in \mathbf{N}$, whence $\sum \frac{1}{n^p}$ diverges by Theorem 6.2.1(ii). The case where $1 < p \leq 2$ was treated in Example 6.1.7 and Exercise 6.1.26. Finally, if $p > 2$, then $\frac{1}{n^p} < \frac{1}{n^2}$ and the series will converge by a direct application of Theorem 6.2.1(i). □

Discussion. The collection of series $\sum \frac{1}{n^p}$, $p > 0$, are referred to as p-**series** and provide a wealth of examples for use in the comparison and other tests. However, this is not what makes p-series fascinating to mathematicians. The fascination stems from the change from divergence to convergence as we go from $p = 1$ to $p > 1$. Surely, one thinks, there must be information contained in these results about just how fast the terms of the series must go to 0 in order to guarantee convergence. Another way of looking at this question is to ask: How many terms must we discard from the Harmonic series $\sum \frac{1}{n}$ in order to get a convergent series? We know, for example, that if we toss out all the terms that do not have perfect square denominators, we are left with a convergent series. Other series generated in this way are the series with 'prime' denominators, and the series obtained by discarding all terms whose denominator contains the digit nine. We note that the former is divergent, while the latter is convergent! □

Example 2. Discuss the convergence or divergence of the series $\sum \frac{2n+1}{n^3+3n+4}$.

Solution. We apply the limit form of the comparison test with a_n as above and $b_n = \frac{1}{n^2}$. Evidently,

$$\lim_{n\to\infty} \frac{a_n}{b_n} = \lim_{n\to\infty} \frac{(2n+1)n^2}{n^3+3n+4} = 2 \neq 0,$$

whence the series converges by the corollary and Example 1. □

Discussion. The strategy adopted here is very simple. We look for a series whose terms go to 0 at approximately the same rate as the series whose convergence we want to test. Recall that for ratios of integer polynomials of the form $\frac{P(n)}{Q(n)}$, as we have above in $\frac{2n+1}{n^3+3n+4}$, the limit as $n \to \infty$ is determined by the ratio of the term containing the highest power of n in the numerator and the term containing the highest power of n in the denominator. In the present case, this ratio is $\frac{2}{n^2}$, which converges to 0. However, if we introduce a factor of n^2 into the numerator, then a nonzero limit of 2 results. This is why we chose $\sum \frac{1}{n^2}$ as our test series. □

The comparison test and the series generated so far enable us to deal with a fairly large collection of infinite series, including all series whose terms are ratios of integer polynomials of the form $\sum \frac{P(n)}{Q(n)}$. However, the scope is really quite limited, and we are forced to look for tests of a substantially different character to handle other types of series. One such test is the Condensation Test due to Cauchy.

Theorem 6.2.2 (Cauchy's Condensation Test). *Let $\sum a_n$ be a series of nonnegative terms such that for some $N \in \mathbf{N}$, $n > N$ implies $a_n \geq a_{n+1}$. The series $\sum a_n$ converges (diverges) if and only if the series $\sum 2^n a_{2^n}$ converges (diverges).*

Proof. Let s_n, t_n, respectively, denote the nth partial sums of $\sum a_n$ and $\sum 2^n a_{2^n}$, respectively. Also, without loss of generality assume that $\{a_n\}$ is monotonically decreasing for $n \geq 1$. Given n, there is a k satisfying $n < 2^k$ **(WHY?)**, and hence

$$
\begin{aligned}
s_n &= a_1 + a_2 + \cdots + a_n \\
&\leq a_1 + (a_2 + a_3) + (a_4 + \cdots + a_7) + \cdots + (a_{2^k} + \cdots + a_{2^{k+1}-1}) \\
&\leq a_1 + 2a_2 + 4a_4 + \cdots + 2^k a_{2^k} = t_k.
\end{aligned}
$$

Thus, if the second series converges, then $\{t_k\}$ is bounded. Consequently, $\{s_n\}$ is bounded and $\sum a_n$ converges, since it is monotone increasing. Thus, the convergence of $\sum 2^n a_{2^n}$ implies the convergence of $\sum a_n$.

To see that the convergence of $\sum a_n$ implies the convergence of the condensed series, we treat the contrapositive. Thus, we assume that $\sum 2^n a_{2^n}$ diverges. Note that given any large k, there is an n such that n exceeds 2^k, and so

$$
\begin{aligned}
s_n &\geq a_1 + a_2 + (a_3 + a_4) + (a_5 + \cdots + a_8) + \cdots + (a_{2^{k-1}+1} + \cdots + a_{2^k}) \\
&\geq \frac{a_1}{2} + a_2 + 2a_4 + 4a_8 + \cdots + 2^{k-1} a_{2^k} = \frac{t_k}{2}.
\end{aligned}
$$

Hence if the condensed series diverges, $\{t_n\}$ is unbounded, whence $\{s_n\}$ is also unbounded and this forces the divergence of $\sum a_n$. $\square$

Discussion. The theorem captures and generalizes the techniques employed in Examples 6.1.6 and 6.1.7, and the reader should review these examples to see that this is so. The hypothesis is that the terms of the series are nonnegative and eventually become monotone decreasing. Without either of these conditions, this approach would be impossible. The usefulness of this theorem lies in the fact that when we apply it to certain series, they are converted to forms whose convergence properties are easily checked. Finally, the convergence properties of $\sum a_n$ are derived from those of a 'condensed' series $\sum 2^n a_{2^n}$, hence the name 'condensation test'. $\square$

Example 3. Discuss the convergence properties of $\sum \frac{1}{n^p}$, $p > 0$.

Solution. We apply the condensation test to obtain the condensed series, $\sum 2^n \frac{1}{(2^n)^p}$. Manipulation yields

$$\sum 2^n \frac{1}{(2^n)^p} = \sum \frac{1}{2^{-n}} \frac{1}{2^{np}} = \sum \left(\frac{1}{2^{p-1}} \right)^n,$$

which is a geometric series with ratio $\frac{1}{2^{p-1}}$ and, as such, converges exactly if $p - 1 > 0$, that is, $p > 1$. $\square$

Discussion. This example contains no new information about series. But it witnesses the ease with which the condensation test can be employed. $\square$

Example 4. Discuss the convergence of $\sum \frac{1}{n(\ln n)^p}$, $p > 0$.

Solution. Again we apply the condensation test to obtain

$$\sum 2^n \frac{1}{2^n (\ln 2^n)^p} = \sum \frac{1}{(n \ln 2)^p} = \frac{1}{(\ln 2)^p} \sum \frac{1}{n^p}.$$

It is clear from Example 1 (or Example 3) that the condensed series converges when $p > 1$ and diverges otherwise, whence the same is true of the original series. $\square$

Discussion. The reader should note that there is nothing sacred about using groupings of length 2^n in generating this result. The principle is to use the largest term in a group multiplied by the number of terms in the group to estimate the contribution of the whole group to the final sum. This is a simple idea (pursued in the exercises), and yet it has wide applications. $\square$

Another extremely useful test is Maclaurin's Integral Test, where we manufacture an improper integral from the terms of the series and then use the convergence or divergence of the integral to decide the issue of convergence for the series.

Theorem 6.2.3 (Integral Test). *Let* $f : \mathbf{R} \to \mathbf{R}$ *be a nonnegative function that is monotonically decreasing. If* $a_n = f(n)$ *for* $n \in \mathbf{N}$, *then* $\sum a_n$ *converges if and only if the improper Riemann Integral* $\int_1^\infty f$ *converges.*

Proof. Let f be as required. It is immediate that for any $k \in \mathbf{N}$,

$$a_k \geq \overline{\int}_a^b{}_k^{k+1} f \geq \int_k^{k+1} f \geq \underline{\int}_a^b{}_k^{k+1} f \geq a_{k+1}. \qquad \textbf{(WHY?)}$$

It follows that

$$\sum_{k=1}^n a_k \geq \overline{\int}_a^b{}_1^n f \geq \int_1^n f \geq \underline{\int}_a^b{}_1^n f \geq \sum_{k=2}^{n+1} a_k.$$

From the left-hand inequalities, it follows that if the improper integral diverges, then the series must also diverge. On the other hand, the right-hand set of inequalities establish that if the integral converges, then the series must also converge. $\square$

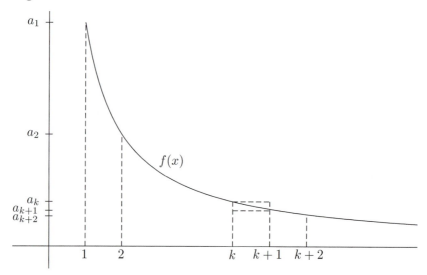

Figure 6.2.1 A graphical illustration of the integral test. Notice how the area under the curve provides an obvious bound for the sum.

Discussion. The intuition underlying this theorem and its proof is completely captured in Figure 6.2.1. Although we have not made their invocation explicit, each of the conditions on the function f in the statement of the theorem is crucial for the proof, and the reader should verify that there is no way to significantly weaken the hypothesis. The reader should convince himself that the stated inequalities are indeed true.

It should be pointed out that in the case where the series and the integral converge, the sum of the series need not coincide with the value of the improper integral. The reader can easily construct suitable examples.

The relationship between series and improper integrals leads to many interesting results. One of these is explored in Exercise 20. $\square$

Example 5. Discuss the convergence properties of $\sum n e^{-n}$.

Solution. Set $f(x) = xe^{-x}$, $x \in [0, \infty)$. Using properties of e^x established in calculus courses and proved in Chapter 9, f is decreasing and nonnegative and satisfies $f(n) = a_n$. Now, using the integration by parts formula, we see that

$$\int_1^n f = \frac{2}{e} - \frac{n+1}{e^n} \to \frac{2}{e}.$$

This integral converges to $\frac{2}{e}$, whence the series must also converge. $\square$

Discussion. The integral test employs the close relationship between integration and series of nonnegative terms. It is a simple test to apply. In a particular situation, we require only that the sequence of terms of the series be eventually monotonically decreasing and nonnegative. This is really all we need, although as the theorem is stated above, more is required.

As a general procedure for applying the theorem, we take the specification for a_n and use it to generate the specification for f. Thus, from ne^{-n}, we generated xe^{-x}. It must be checked that the function generated is also eventually monotone and nonnegative, since it is quite possible that the function generated from a_n in the natural way will not in general be either nonnegative or decreasing. Last, the basic value of the test lies in the fact that the function, f, generated from a_n has an easily calculated indefinite integral. Unless this is the case, this test will have no practical value for the series under consideration. Again, we point out that the sum of the series need not be $\frac{2}{e}$. $\square$

Exercises

1. Give complete details of the proof of Theorem 6.2.1(ii).

2. Supply proofs for parts (ii) and (iii) of the limit form of the comparison test (Corollary to Theorem 6.2.1).

3. Consider the series $\sum \frac{p(n)}{q(n)}$, where $p(n)$ and $q(n)$ are integer polynomials such that $q(n) \neq 0$ for all $n \in \mathbf{N}$. Find a necessary and sufficient condition for this series to converge.

4. State and prove a version of the condensation test that groups the terms of the series so that the nth group contains n terms.

5. Give an example of a series $\sum a_n$ whose terms are monotonically decreasing to 0 but for which the function f defined in the natural way; for example, $a_n = \frac{1}{n^2}$ and $f(x) = \frac{1}{x^2}$) does not share this property as $x \to \infty$.

6. Justify the string of inequalities that form the heart of the proof of the Integral Test (Theorem 6.2.3).

7. Test the following series for convergence or divergence. You may assume that the trigonometric, exponential, and logarithmic functions have their usual properties.

(a) $\sum \dfrac{\sqrt{n+1} - \sqrt{n}}{\sqrt{n}\sqrt{n+1}}$;   (b) $\sum n^2 e^{-n}$;   (c) $\sum n^k e^{-n}$, $k \in \mathbf{N}$;

(d) $\sum n^n e^{-n}$;   (e) $\sum \frac{1}{n!}$;   (f) $\sum n^{-n}$;

(g) $\sum a^n n^a$, where $a > 0$;   (h) $\sum (n^{\frac{1}{n}} - 1)$;   (i) $\sum (\sqrt{1 + n^2} - n)$;

(j) $\sum \frac{1+\sin n}{1+n^2}$;   (k) $\sum \ln \frac{n}{n^3+5}$;   (l) $\sum \ln \frac{n^3}{n^3+2}$;

(m) $\sum \frac{\ln n^3}{n^3+2}$;   (n) $\sum \frac{1}{n^{a+\frac{b}{n}}}$, $a, b \in \mathbf{R}$;

(o) $\sum_{n=2}^{\infty} \frac{1}{n \ln n \ln \ln n}$;   (p) $\sum n^a(\sqrt{n+1} - 2\sqrt{n} + \sqrt{n-1})$; $a \in \mathbf{R}$;

(q) $\sum_{n=0}^{\infty} \left(\frac{1}{3+(-1)^n}\right)^n$;   (r) $\sum_{n=2}^{\infty} \frac{1}{n \ln n (\ln \ln n)^p}$;

(s) $\sum_{n=2}^{\infty} \frac{1}{2^n \ln n \ln \ln \ln n \ln \ln \ln n}$;   (t) $\sum_{n=2}^{\infty} \frac{1}{n \ln n \ln \ln \ln \ln n (\ln \ln \ln n)^p}$, $p > 1$.

8. Generalize the results in (r) and (t) of Exercise 7 to the case where the denominator is
$$n \ln n \ln \ln \ln n \cdots (\ln \ln \ln \cdots \ln n)^p,$$
where $p > 1$ and the bracket consists of a string of k ln's.

9. Let $\sum a_n$ be a series with nonnegative terms and $\{t_n\}$ its sequence of partial sums. Show $\sum a_n$ converges if and only if $\{t_n\}$ is bounded. Moreover, if $\{t_n\}$ is bounded, then the supremum of $\{t_n\}$ is the sum of the series.

10. Let $a_n \geq 0$ for each n.

    (a) Let $\sum a_n$ converge and $b_n = \sum_{i=n}^{\infty} a_i$. Prove $\sum(\sqrt{b_n} - \sqrt{b_{n+1}})$ is convergent.

    (b) If $\sum a_n$ diverges and $s_n = \sum_{i=1}^{n} a_i$, prove $\sum(\sqrt{s_{n+1}} - \sqrt{s_n})$ is divergent.

11. Let $\sum a_n$ be a series of nonnegative terms. Show if for some $k > 1$, the sequence, $\{n^k a_n\}$ converges, then so does the series $\sum a_n$.

12. Is there a result similar to Exercise 11 that will establish divergence?

13. Let $\sum a_n$ be a convergent series of nonnegative terms. A nonnegative series $\sum b_n$ is said to **compare favorably** with $\sum a_n$ if there exists $M \in \mathbf{N}$ such that $n > M$ implies that $a_n \geq b_n$. Show that there does not exist a series $\sum a_n$ such that every convergent series of nonnegative terms compares favorably with $\sum a_n$.

14. Show that no finite list of nonnegative series will be such that every convergent series of nonnegative terms must compare favorably with at least one series in the list.

15. Show that even if we take a countably infinite list of convergent series, it will still not be adequate to test every convergent series of nonnegative terms.

16. If $\sum a_n$, $\sum b_n$ $(a_n, b_n > 0)$ are both convergent, and if $\lim_{n\to\infty} \frac{a_n}{b_n} = 0$, we say $\sum b_n$ **converges more slowly** than $\sum a_n$. Prove that given any convergent series $\sum a_n$, there always exists a series $\sum b_n$ that converges more slowly than $\sum a_n$.

17. If $\sum b_n$, $\sum a_n$ $(a_n \neq 0)$ are both divergent, and if $\lim_{n\to\infty} \frac{b_n}{a_n} = 0$, we say $\sum b_n$ **diverges more slowly** than $\sum a_n$. Prove that no series exists that diverges more slowly than all others.

18. If $d_n > 0$ and $\sum d_n$ diverges, discuss the behavior of the following:

    (a) $\sum \frac{d_n}{1+d_n}$;   (b) $\sum \frac{d_n}{1+nd_n}$;   (c) $\sum \frac{d_n}{1+n^2 d_n}$;   (d) $\sum \frac{d_n}{1+d_n^2}$.

19. Let $\sum a_n$ and $\sum b_n$ be two series of positive terms. Develop a comparison test based on the ratios, $\frac{a_{n+1}}{a_n}$ and $\frac{b_{n+1}}{b_n}$.

20. Using whatever results about $\ln x$ are required (see Chapter 9 for proofs), use the integral test to show that the Harmonic series, $\sum \frac{1}{n}$ is divergent. Further, set $h_n = \sum_{i=1}^n \frac{1}{i}$. Show the sequence $\{h_n - \ln n\}$ converges to a number in $(0, 1)$. The value of the limit is approximately 0.57722156649, (correct to 10 decimal places) but it is not known whether the limit is rational. The limit of this sequence is generally denoted by the symbol γ and known as **Euler's constant**.

21. Let $a \geq b \geq 0$. If $[x]$ has its usual meaning and h_n as in Exercise 20, show the sequence $\{h_{[na]} - h_{[nb]}\}$ converges and find an expression for its limit.

22. Prove the following generalization of the condensation test. Let $\{g_n\}$ be a sequence of positive integers such that for $k = 0, 1, \ldots$:

    (a) $g_k > g_{k-1} \geq 0$ and

    (b) $g_{k+1} - g_k \leq M(g_k - g_{k-1})$,

    where M is a positive constant. If $\sum a_n$ is a series of positive terms with the sequence $\{a_n\}$ monotonically decreasing, then $\sum_{n=0}^{\infty} a_n$ and $\sum_{n=0}^{\infty} (g_{n+1} - g_n) a_{g_n}$ both converge or both diverge.

23. Let $\sum a_n$ be a convergent series of nonnegative terms, and consider the sequence $\{n^p a_n\}$. For what values of p can it always be said that the sequence converges to 0?

24. Consider the collection of functions, f, which are defined and nonnegative on $[0, \infty)$. Show that in general there is no relationship between the convergence of $\int_0^{\infty} f$ and $\sum_{n=1}^{\infty} f(n)$; that is, give examples in which both converge, both diverge, and either one diverges while the other converges.

6.3   Absolute Convergence

In the last section, we developed tests for convergence that could be applied to series with nonnegative terms; moreover, this restriction was an essential hypothesis for the tests. In general, a series will have an infinite number of negative and an infinite number of positive terms. So it is important that we develop methods for dealing with such series.

Definition. A series, $\sum a_n$, is called **absolutely convergent** provided $\sum |a_n|$ is a convergent series. A series $\sum a_n$ is called **conditionally convergent** if $\sum a_n$ is convergent, but $\sum |a_n|$ is divergent.

Discussion. The importance of this definition is that it focuses our attention on the possibility that there may be series that are convergent but that are not absolutely convergent. It is immediate that a convergent but not absolutely convergent series must have terms that are both positive and negative, since otherwise (for series whose terms are all of the one sign) the two types of convergence are equivalent. An example of a conditionally convergent series is $\sum \frac{(-1)^n}{n}$, while the series $\sum \frac{(-1)^n}{n^2}$ is an absolutely convergent series.

For a series to be absolutely convergent, the terms must get small at a rapid rate. This requirement that the terms become rapidly small is effectively illustrated by the collection of series discussed in Example 6.2.1.

On the other hand, a series that is conditionally convergent can have its terms approach 0 as slowly as we please. Thus, the reason why a conditionally convergent series converges must be due to the interaction of the positive terms with the negative terms. This will provide one focus for our investigations in this section. □

Theorem 6.3.1. *If $\sum a_n$ is absolutely convergent, then it is convergent.*

Proof. Let s_n and t_n denote the nth partial sums of $\sum a_n$ and $\sum |a_n|$, respectively. Then

$$
\begin{aligned}
|s_{n+p} - s_n| &= |a_{n+1} + a_{n+2} + \cdots + a_{n+p}| \\
&\leq |a_{n+1}| + |a_{n+2}| + \cdots + |a_{n+p}| \\
&= |t_{n+p} - t_n|.
\end{aligned}
$$

We now apply the Cauchy Criterion and the hypothesis to complete the proof. □

Discussion. The proof illustrates the utility of two basic principles. The first is the generalized Triangle inequality, which is at the heart of the inequality string. The second is the Cauchy Criterion for convergence (where the completeness axiom is hidden).

This theorem provides us with the means to apply all the theory that we developed for series with nonnegative terms to series with terms of both signs. There is, however, one restriction. If the series is 'absolutely' divergent, we cannot in general conclude that the original series is also divergent. The canonical example in this regard is the series $\sum \frac{(-1)^{n+1}}{n}$, which was shown to be 'absolutely' divergent in Example 6.1.6 and convergent in Exercise 6.1.14. The reader who has not done this exercise may wish to do so or examine Theorem 6.4.1, which will effectively do the exercise for him. Thus, the application of our previous results will generally be in the form of testing for absolute convergence and concluding convergence of the original series when absolute convergence can be verified.

As noted before, a series will converge absolutely provided the rate at which its terms approach 0 is sufficiently rapid. The tests that we have developed to date check whether the terms of a series approach 0 with the necessary rapidity by comparing the terms of the series against those of a known series. Thus, it is obvious that if the terms go to 0 fast enough, changing the sign of some of them cannot affect the convergence properties, but if the terms go to 0 slowly, then changing the sign of some terms most definitely will affect the convergence. □

Definition. Let $\sum a_n$ be a series of arbitrary terms and set

$$
p_n = \frac{|a_n| + a_n}{2} \quad \text{and} \quad q_n = \frac{|a_n| - a_n}{2}.
$$

Then the series $\sum p_n$ is called the **positive part** of $\sum a_n$, and $\sum q_n$ is called the **negative part** of $\sum a_n$.

Discussion. The reader can verify that if $a_n \geq 0$, then $p_n = a_n$ and $q_n = 0$. On the other hand, if $a_n < 0$, then $p_n = 0$ and $q_n = |a_n|$. Thus, in essence, the series $\sum p_n$ collects together all the positive terms of the original series into a single sum. Similarly, the series $\sum q_n$ collects together all the negative terms, but with their signs changed so that the result is a series of positive terms. The motivation for doing this is that we think these two series may be of particular interest for conditionally convergent series. It is also important to notice that in the series $\sum p_n$, the index values associated with terms in $\sum a_n$ that were negative now are associated with terms that are 0. A similar comment holds regarding $\sum q_n$. For this reason, the original series can be reconstructed from

$$a_n = p_n - q_n$$

(see Exercise 2). Last, since both the positive and negative parts of $\sum a_n$ are series with nonnegative terms, the theory developed in Section 6.2 applies. $\square$

Theorem 6.3.2. *If $\sum a_n$ is conditionally convergent, then both $\sum p_n$ and $\sum q_n$ are divergent; if $\sum a_n$ is absolutely convergent, then both $\sum p_n$ and $\sum q_n$ converge, and furthermore,*

$$\sum a_n = \sum p_n - \sum q_n.$$

Proof. In the former case, $\sum a_n$ converges while $\sum |a_n|$ is divergent. We have

$$a_n = p_n - q_n \quad \text{and} \quad |a_n| = p_n + q_n.$$

Now, if $\sum q_n$ is convergent, since $p_n = a_n + q_n$, it follows that $\sum p_n$ is also convergent. But then, since $|a_n| = p_n + q_n$, it follows that $\sum |a_n|$ must converge as well (**WHY?**), which is a contradiction. So both the series must diverge. The second part is left to the reader as Exercise 3. $\square$

Discussion. In the exercises of Section 6.1, we explored the effect of adding or deleting terms from a series. The basic fact employed here is that addition of an infinite number of terms to a series will not affect the convergence properties of the series, provided the infinite collection of terms added form an absolutely convergent series. $\square$

Next we explore the value of the concept of absolute convergence. In the next set of results we shall see that it is a remarkable property to demand from a series. The first result shows that the absolute convergence of a series permits us to rearrange the terms in any manner while not destroying the convergence properties or changing the sum. Before we state the result, we must make precise what we mean by the rearrangement of a given infinite series. Recall that the rearrangement of a finite series is nothing but a permutation, which is a one-to-one, onto function from the set to itself. We shall extend this notion to infinite sets as well.

Definition. Let $f : \mathbf{N} \to \mathbf{N}$ be a one-to-one and onto function. Then $\sum_{n=1}^{\infty} a_{f(n)}$ is called a **rearrangement** of the series $\sum a_n$.

Discussion. If we think of what we ought to mean by 'rearrangement' of a series, then the motivation for this definition becomes very clear. Intuitively, a rearrangement of the terms of the series should take all the terms and mix them up and then add them in this different order. Thus, if we denote the rearranged series by $\sum b_n$, then each b_n should be identifiable as a particular a_k (and only one such a_k) from the original series. Further, each a_k should be found exactly once among the b_n's. These two properties create a one-to-one and onto function from **N** to **N** for us in the following way: given n as a subscript for the a_n, we know that $a_n = b_m$ for some term in the rearranged series; thus, we get $f(n) = m$. Conversely, given a term b_n in the rearranged series, it is there only because it was an a_m for some m in the original series. Thus, we see that the function f ought to be one-to-one and onto as required by the definition and that such functions are the tool required for creating a rearranged series. $\square$

Theorem 6.3.3. *Let $\sum a_n$ be an absolutely convergent series with sum, s. Then every rearrangement $\sum b_n$ of $\sum a_n$ converges (absolutely) to the sum, s.*

Proof. Our first step is to reduce the problem to series with nonnegative terms. Let $\sum p_n$ and $\sum q_n$ be the positive and negative parts of $\sum a_n$, while $\sum p_n^*$ and $\sum q_n^*$ are the positive and negative parts of $\sum b_n$. Now

$$a_n = p_n - q_n \quad \text{and} \quad b_n = p_n^* - q_n^*.$$

Thus, if $\sum p_n = \sum p_n^*$ and $\sum q_n = \sum q_n^*$, then $\sum a_n = \sum b_n$. Hence, it is enough to show that the positive and negative parts of the rearranged series preserve their original sum. To complete the reduction, we assert that the positive part of $\sum b_n$ is a rearrangement of the positive part of $\sum a_n$, and similarly for negative parts. (We leave the establishment of this claim to Exercise 4.)

Since the positive and negative parts of a series are series with nonnegative terms, we have left to show that if $\sum b_n$ is a rearrangement of $\sum a_n$, a series with no negative terms, then the convergence of $\sum a_n$ to s implies the convergence of $\sum b_n$ to s. Since we are dealing with series of nonnegative terms, s is the least upper bound of the sequence of partial sums of $\sum a_n$. Let $\{t_n\}$ denote the sequence of partial sums of $\sum b_n$. We first claim that s is an upper bound for the t_n's. To see this, let f be the function that defines the rearrangement. Then given t_n, there is an $N \in \mathbf{N}$ such that if $m \leq n$, then $f^{-1}(m) \leq N$. Evidently, $t_n \leq s_N \leq s$, whence s is an upper bound for the collection of t_n's. Next, let $\epsilon > 0$ be fixed. Since s is the supremum of the set of s_n's, there exists an s_N such that $s - \epsilon < s_N \leq s$, and there exists M such that $n \leq N$ implies $f(n) < M$. It follows that

$$s - \epsilon < s_N \leq t_M \leq s.$$

Thus, s is the supremum of the collection of partial sums of the rearranged series, and so is the sum, since the rearranged series is also a series of nonnegative terms. $\square$

Discussion. The argument given above is in two parts. The first part uses Theorem 6.3.2 to show that we need only prove the result for series that have nonnegative terms. The second half consists of proving the result for the reduced case. The reduction leads to considerable simplification; in fact, the result is almost trivial for this special case, due to the fact that the sequence of partial sums is monotone and consequently, so tractable. Without this reduction, we would have to trap the partial sums inside the interval centered at s, a task of much greater delicacy. We will ask the reader to carry out this type of proof in the exercises. $\square$

The rearrangement property is one of the most elegant results on absolutely convergent series. For this reason, an absolutely convergent series is sometimes referred to as an **unconditionally convergent** series, in the sense that whatever rearrangement we adopt, we still preserve the absolute convergence as well as the original sum. This is in contrast to nonabsolutely convergent series that we have earlier agreed to call conditionally convergent series. The hypothesis of absolute convergence is vital to the conclusion, as the absence of it may very well affect the convergence and the sum of the rearrangement. Our next theorem of this section, usually known as Riemann's theorem on the rearrangement of a conditionally convergent series, says that given any nonabsolutely convergent series, it can be suitably rearranged so as to yield any desired sum we want, or even divergence if we desire.

Theorem 6.3.4 (Riemann's Theorem). *Let $\sum a_n$ be a conditionally convergent series and let x and y be two given real numbers (including $\pm\infty$) with $x \leq y$. Then there exists a rearrangement $\sum b_n$ of $\sum a_n$ such that if t_n denotes the nth partial sum of $\sum b_n$, then*

$$\underline{\lim} \, t_n = x \quad and \quad \overline{\lim} \, t_n = y.$$

Proof. (Outline): We may safely assume that none of the a_n's is zero, since discarding those terms of a series that are zero does not affect the convergence or divergence (**WHY?**). As well, we assume that $x, y \in \mathbf{R}$, leaving the other cases to the reader as Exercise 7. Let $f : \mathbf{N} \to \mathbf{N}$ be a one-to-one function such that $a_{f(n)}$ is the nth positive term among the a_n's. Similarly, let $g : \mathbf{N} \to \mathbf{N}$ be a one-to-one function such that $a_{g(n)}$ is the nth negative term among the a_n's. It is a consequence of Theorem 6.3.2 both $\sum a_{f(n)}$ and $\sum a_{g(n)}$ are divergent. Next, we define a third function $h : \mathbf{N} \to \mathbf{N}$ inductively as follows:

(i) $h(1)$ is the least n such that $y < \sum_{i=1}^{h(1)} a_{f(i)}$;

(ii) $h(2)$ is the least n such that $\sum_{i=1}^{h(1)} a_{f(i)} + \sum_{i=1}^{h(2)} a_{g(i)} < x$;

(iii) $h(2n+1)$ is the least n such that $y < \sum_{i=1}^{h(2n+1)} a_{f(i)} + \sum_{i=1}^{h(2n)} a_{g(i)}$;

(iv) $h(2n+2)$ is the least n such that $\sum_{i=1}^{h(2n+1)} a_{f(i)} + \sum_{i=1}^{h(2n+2)} a_{g(i)} < x$.

It is straightforward to show

$$h(2n) < h(2n+2) \quad \text{and} \quad h(2n-1) < h(2n+1).$$

We now define the function $F : \mathbf{N} \to \mathbf{N}$ by

(i) $F(n) = f(n)$, if $1 \leq n \leq h(1)$;

(ii) $F(n) = g(n)$, if $h(1) < n \leq h(1) + h(2)$;

(iii) $F(n) = f(n)$, if $h(2j-1) + h(2j) < n \leq h(2j) + h(2j+1), 1 \leq j \in \mathbf{N}$;

(iv) $F(n) = g(n)$, if $h(2j) + h(2j+1) < n \leq h(2j+1) + h(2j+2), 1 \leq j \in \mathbf{N}$.

It is readily checked that F is one-to-one and

$$\{a_n : n \in \mathbf{N}\} = \{a_{F(n)} : n \in \mathbf{N}\}$$

(Exercise 7). We claim for any $n \in \mathbf{N}$, if $k = h(2n) + h(2n+1)$, then

$$\sum_{i=1}^{k-1} a_{F(i)} < y < \sum_{i=1}^{k} a_{F(i)},$$

and if $k = h(2n+1) + h(2n+2)$, then

$$\sum_{i=1}^{k} a_{F(i)} < x < \sum_{i=1}^{k-1} a_{F(i)}.$$

Since, $a_{F(i)} \to 0$ as $i \to \infty$ (Exercise 6), the result now follows (**HOW?**). $\square$

Discussion. The property of a conditionally convergent series to be rearranged to converge to any number whatever, or even to diverge to $\pm\infty$, is indeed remarkable; however, it is not really surprising once one has in hand the two significant facts that are employed in the proof of the theorem. These two facts are

(i) to be convergent, the terms, a_n, must go to 0;

(ii) to be conditionally convergent, both the positive and negative parts of a series must diverge.

To see how these two ideas are used, first consider the functions f and g. The positive integer $f(n)$ is the subscript in the original series of the nth positive term. Similarly, $g(n)$ is the subscript of the nth negative term. Since both $\sum a_{f(n)}$ and $\sum a_{g(n)}$ are divergent, it follows that for every $k \in \mathbf{N}$,

$$\sum_{n=k}^{\infty} a_{f(n)} \quad \text{and} \quad \sum_{n=k}^{\infty} a_{g(n)}$$

are both divergent. Intuitively, the proof should generate a new series for which the sequence of partial sums oscillates from just below x to just above y. Suppose that a given partial sum is just below x. The fact that $\sum_{n=k}^{\infty} a_{f(n)}$ diverges guarantees that we will always have enough positive terms left unused to obtain a new sum, which includes all previously used terms and which is greater than y. Similarly, if we have a partial sum that is above y, there will always be enough negative terms unused to obtain a new sum that is less than x. This fact is employed formally in the definition of the function, h, which counts the number of positive (negative) terms required to take a finite partial sum from just below x to just above y (from just above y to just below x). Of course, these terms have to be collected and added up in such a way as to accomplish this purpose. The definition of h also accounts for the order in which the terms must be added up, and the number of terms required to get from x to y, or vice versa. It does so by using the preestablished order on the positive (negative) terms, which is preserved by f (respectively, g) and the four defining properties. In summary, then, $h(2n-1) - h(2n-3)$, ($h(2n+2) - h(2n))$ gives the number of additional positive (negative) terms that are required to complete half of the nth oscillation.

Once we have defined h, it is possible to define the permutation F that defines the rearrangement. The definition of F yields a sequence of partial sums that will oscillate from 'just above' y to 'just below' x. In this case, 'just above' means: to exceed y by no more than the value of $a_{F(n)}$ for some n. It is at this point that we employ the fact that the terms of a convergent series must approach 0 to guarantee that x and y are the limit inferior and limit superior, respectively.

Many of the details of the proof have been left to the reader. These details can be proven by an appropriate induction. The reader who takes it upon himself to complete these details will find that a complete understanding of the proof results. $\square$

We give just one illustration of a rearrangement and show how the sum is altered.

Example 1. Show that

$$1 - \frac{1}{2} + \frac{1}{3} - \frac{1}{4} + \cdots = \ln 2$$

and further evaluate the sum of the rearranged series

$$1 + \frac{1}{3} - \frac{1}{2} + \frac{1}{5} + \frac{1}{7} - \frac{1}{4} + \frac{1}{9} + \frac{1}{11} - \frac{1}{6} + \cdots + + - \cdots .$$

Solution. We shall briefly outline how we compute the sum of the given series. Set

$$g_n = \sum_{k=1}^{n} \frac{1}{k} - \ln n.$$

It is known (see Exercise 6.2.20) that the sequence $\{g_n\}$ converges to a limit, to which we assign the symbol γ and which is known as **Euler's Constant**. If s_n denotes the partial sums of the original series, then

$$
\begin{aligned}
s_{2n} &= 1 - \frac{1}{2} + - \cdots + \frac{1}{2n-1} - \frac{1}{2n} \\
&= \left(1 + \frac{1}{2} + \cdots + \frac{1}{2n}\right) - 2\left(\frac{1}{2} + \frac{1}{4} + \cdots + \frac{1}{2n}\right) \\
&= \left(1 + \frac{1}{2} + \cdots + \frac{1}{2n}\right) - \left(1 + \frac{1}{2} + \cdots + \frac{1}{n}\right) \\
&= (g_{2n} + \ln 2n) - (g_n + \ln n) \\
&= g_{2n} - g_n + \ln 2.
\end{aligned}
$$

Since g_n is Cauchy, we have $\lim s_{2n} = \ln 2$. It is left to the reader to show that $\lim s_n = \ln 2$ for any partial sum.

Let t_n be the nth partial sum of the rearranged series. We then have

$$
\begin{aligned}
t_{3n} &= \left(1 + \frac{1}{3} - \frac{1}{2}\right) + \left(\frac{1}{5} + \frac{1}{7} - \frac{1}{4}\right) + \cdots + \left(\frac{1}{4n-3} + \frac{1}{4n-1} - \frac{1}{2n}\right) \\
&= \left(1 + \frac{1}{3} + \frac{1}{5} + \cdots + \frac{1}{4n-1}\right) - \frac{1}{2}\left(1 + \frac{1}{2} + \cdots + \frac{1}{n}\right) \\
&= \left(1 + \frac{1}{2} + \cdots + \frac{1}{4n}\right) - \frac{1}{2}\left(1 + \frac{1}{2} + \cdots + \frac{1}{2n}\right) - \frac{1}{2}\left(1 + \frac{1}{2} + \cdots + \frac{1}{n}\right)
\end{aligned}
$$

so that

$$
\begin{aligned}
t_{3n} &= (\ln 4n + g_{4n}) - \frac{1}{2}(\ln 2n + g_{2n}) - \frac{1}{2}(\ln n + g_n) \\
&= \left(g_{4n} - \frac{1}{2}g_{2n} - \frac{1}{2}g_n\right) + \frac{3}{2}\ln 2,
\end{aligned}
$$

whence, again since g_n is Cauchy,

$$
\lim t_{3n} = \frac{3}{2}\ln 2.
$$

Since

$$
t_{3n+1} = t_{3n} + \frac{1}{4n+1} \quad \text{and} \quad t_{3n+2} = t_{3n} + \frac{1}{4n+1} + \frac{1}{4n+3},
$$

it follows that $\lim t_n = \dfrac{3}{2}\ln 2$, as desired. $\square$

Discussion. The key to understanding this example is in coming to grips with the initial computations that establish the limit of $\{s_{2n}\}$. The easiest way to see what is going on in the string of equalities that begins $s_{2n} = \cdots$ is to replicate the string for some small value of n, say, $n = 3$. This will elucidate the changes involved in the first three steps. The clever part is the introduction of a form of 0, that is, $\ln n - \ln n$, which occurs in the fourth equality. The reader should recall the many previous occasions on which other forms of 0 have been introduced into equations. In any case, once this is done, the solution is apparent.

The rearrangement consisted of taking two positive terms followed by one negative term so that a grouping of three terms was obtained. This motivated us to compute the partial sum t_{3n}, and since t_{3n+1} and t_{3n+2} differ from t_n only in terms with limit 0, it suffices to study the behavior of t_{3n}. The computations that evaluate $\lim\limits_{n\to\infty} t_{3n}$ are similar to the computations for $\{s_{2n}\}$.

In the same vein, it is possible to show that if the above series for $\ln 2$ is rearranged so that p positive terms are followed by q negative terms each time, then the sum will be altered to $\ln 2 + \frac{1}{2}\ln \frac{p}{q}$ (Exercise 10). $\square$

Next, we introduce the notions of a subseries and subseries convergence and relate them to absolute convergence.

Definition. Let $f : \mathbf{N} \to \mathbf{N}$ be a strictly increasing, one-to-one (but not necessarily onto) function. If $b_n = a_{f(n)}$, then the series $\sum b_n$ is called a **subseries** of the series $\sum a_n$.

Discussion. The reader can check that this definition is nothing more than the definition of a subsequence phrased appropriately for series. $\square$

Theorem 6.3.5. *Let $\sum a_n$ be any series. Then $\sum a_n$ is absolutely convergent if and only if every subseries $\sum b_n$ of $\sum a_n$ is convergent. Moreover,*

$$|\sum b_n| \le \sum |b_n| \le \sum |a_n|.$$

Proof. We have already shown that a series that is not absolutely convergent has a divergent subseries. Thus, one half of the theorem is already proved. To complete the proof, let $\sum a_n$ be an absolutely convergent series, and let $\sum b_n$ a subseries, defined by $b_n = a_{f(n)}$. Now, given n, we can find M such that $k < n$ exactly if $f^{-1}(k) < M$. Then

$$|\sum_{k=1}^{n} b_k| \le \sum_{k=1}^{n} |b_k| \le \sum_{k=1}^{M} |a_k| \le \sum_{k=1}^{\infty} |a_k|.$$

All the required conclusions now follow. $\square$

Discussion. In this section we have studied three basic concepts related to an infinite series:

(i) absolute convergence;

(ii) rearrangement convergence, where every rearrangement of the series converges;

(iii) subseries convergence, where every subseries of the original series converges.

We can summarize our results by saying that if a series possesses any one of the above properties, it must have the other two as well. Finally, the alternating form of the Harmonic series, $\sum \frac{(-1)^{n+1}}{n}$, provides the canonical example of a conditionally convergent series where one and hence all the three conditions fail. $\square$

The reader may have noticed that Theorem 6.1.1, which discussed uniqueness and various arithmetic results for series, avoided any discussion of the product of two series. The reason for this is that multiplication of series requires one to assign a meaning to $(\sum a_n)(\sum b_n)$. If one reasons by analogy with sequences, the natural product is $\sum a_n b_n$, that is, term-by-term multiplication. There are three difficulties with this definition.

The first difficulty is illustrated by taking the series $\sum \frac{(-1)^n}{\sqrt{n}}$ and multiplying it by itself. While the initial series is convergent, its 'square' is not, as the reader can easily verify. The second difficulty relates to the Harmonic series, $\sum \frac{1}{n}$. It is divergent, but its 'square' is convergent. Neither of these results is particularly satisfying, and suggest that an appropriate concept has not been realized. Moreover, the definition does not appear to be the natural extension of a finite product concept. The third difficulty relates to the case where all three series converge, say, to a, b, and c, respectively. For this case, it is easily checked, using the alternating form of the Harmonic series, for example, that $a \cdot b \ne c$. Considerations such as these lead to

Definition. Let $\sum_{n=0}^{\infty} a_n$ and $\sum_{n=0}^{\infty} b_n$ be two infinite series. Set

$$c_n = a_0 b_n + a_1 b_{n-1} + \cdots + a_n b_0 = \sum_{k=0}^{n} a_k b_{n-k}$$

and define

$$\left(\sum_{n=0}^{\infty} a_n \right) \left(\sum_{n=0}^{\infty} b_n \right) = \sum_{n=0}^{\infty} c_n.$$

Discussion. This definition is due to Cauchy and carries the name **Cauchy product** or **convolution product**. We want to motivate the definition of c_n. Thus, consider the product of the two finite polynomials

$$\left(\sum_{k=0}^{n} a_k x^k \right) \times \left(\sum_{k=0}^{n} b_k x^k \right).$$

This product can be computed using the ordinary laws of arithmetic, since we are dealing with two finite sums. Let d_m denote the coefficient on x^m in the product and consider $m = i + j \leq n$. We claim (Exercise 20) that $d_m = c_m$. We stress that $m \leq n$ is a requirement for this equality. The important point to notice here is that the coefficient on x^m involves only coefficients associated with powers of x for which the degree is less than or equal to m. For this reason, if we consider the various finite products from the series $\sum a_k$ and $\sum b_k$ as n tends to infinity,

$$\left(\sum_{k=0}^{n} a_k x^k \right) \times \left(\sum_{k=0}^{n} b_k x^k \right),$$

the coefficient, c_m on x^m after multiplication, becomes fixed as soon as $n \geq m$. For this reason, we may be confident of the formal assertion

$$\sum_{k=0}^{\infty} c_k x^k = \left(\sum_{k=0}^{\infty} a_k x^k \right) \times \left(\sum_{k=0}^{\infty} b_k x^k \right),$$

which illustrates the utility of the Cauchy product. Finally, observe that in the definition of a Cauchy product, we started the index at 0, rather than at 1, since we want to account for the constant terms in the two polynomials and their product.

Given that $\sum a_n$, $\sum b_n$, and $\sum c_n$ converge to a, b, and c, respectively, one could hope that $ab = c$. We explore this and other questions below. $\square$

Example 2. Consider the series $\sum_{n=0}^{\infty} \dfrac{(-1)^n}{\sqrt{n+1}}$. What can be said about the convergence of the Cauchy product of this series with itself?

Solution. As noted above, the series is convergent. However,

$$c_n = (-1)^n \sum_{k=0}^{n} \frac{1}{\sqrt{(n-k+1)(k+1)}}.$$

Observe that

$$
\begin{aligned}
(n - k + 1)(k + 1) &= nk - k^2 + n + 1 \\
&\leq n^2 + n + 1 \leq (n+1)^2.
\end{aligned}
$$

Hence,

$$
|c_n| \geq \frac{n + 1}{\sqrt{(n + 1)^2}} = 1.
$$

Thus, c_n does not tend to zero, and the Cauchy product cannot converge. $\square$

Discussion. This example is due to Cauchy himself. It shows that even this formulation of products for series will not satisfy the natural requirement that if one starts with two convergent series, the product series will converge and have a sum that is the product of the sums of the factor series.

   If one thinks about this example, one sees that the difficulty is due to the fact that neither of the initial series is absolutely convergent. This suggests that a stronger hypothesis may be required to achieve the desired result. $\square$

Theorem 6.3.6 (Merten's).   *Suppose $\sum_{n=0}^{\infty} a_n$ converges absolutely to a, $\sum_{n=0}^{\infty} b_n$ converges to b.   Then the Cauchy product series $\sum_{n=0}^{\infty} c_n$ converges to ab.*

Proof. Let A_n, B_n, and C_n denote the partial sums up to n for the two series and the Cauchy product respectively. Also, set $\beta_n = B_n - b$. Now,

$$
\begin{aligned}
C_n = \sum_{k=0}^{n} c_k &= \sum_{k=0}^{n} \sum_{j=0}^{k} a_j b_{k-j} \\
&= \sum_{k=0}^{n} a_k B_{n-k} \\
&= \sum_{k=0}^{n} a_k (b + \beta_{n-k}) \\
&= A_n b + \sum_{k=0}^{n} a_k \beta_{n-k}.
\end{aligned}
$$

Evidently, $A_n b \to ab$ as $n \to \infty$. Thus, it suffices to show that

$$
\lim_{n \to \infty} \sum_{k=0}^{n} a_k \beta_{n-k} = 0.
$$

To establish this claim, let $\epsilon > 0$ be given. Now, by hypothesis, $\sum |a_n|$ converges, and we call its sum α. Further, β_n converges to 0, and there exists β such that $|\beta_n| \leq \beta$ for all n. Choose $N \in \mathbf{N}$ such that $n > N$ implies $|\beta_n| < \frac{\epsilon}{\alpha}$, where we implicitly assume that $\alpha > 0$. For $n > N$, we have

$$
\left| \sum_{k=0}^{n} a_k \beta_{n-k} \right| \leq \sum_{k=0}^{n} | a_{n-k} \beta_k |
$$

$$
= \sum_{k=0}^{N} | a_{n-k} \beta_k | + \sum_{k=N+1}^{n} | a_{n-k} \beta_k |
$$

$$
\leq \beta \sum_{k=0}^{N} | a_{n-k} | + \frac{\epsilon}{\alpha} \sum_{k=N+1}^{n} | a_{n-k} |
$$

$$
\leq \beta \sum_{k=0}^{N} | a_{n-k} | + \epsilon.
$$

To complete the proof, notice that as $n \to \infty$, by the Cauchy Criterion,

$$
\beta \sum_{k=0}^{N} | a_{n-k} | \to 0.
$$

Since ϵ was arbitrary, we are done. $\square$

Discussion. In Exercise 21, The reader is asked to supply justifications for all the displayed equations in the above proof.

There are two important steps in this argument. The first is the manipulation of C_n into a form involving $A_n b$ and other terms. The second key step in this argument is noticing that by splitting the sum of the other terms into two parts as follows,

$$
\sum_{k=0}^{n} | a_{n-k} \beta_k | = \sum_{k=0}^{N} | a_{n-k} \beta_k | + \sum_{k=N+1}^{n} | a_{n-k} \beta_k |
$$

that both parts can then be made arbitrarily small. Another example that requires this type of argument is given in Exercise 22 (see also Exercise 1.2.14).

This theorem provides a means for ensuring that the product of two series converges, and further, that the product series has the right sum. As has already been noted, if neither series is absolutely convergent, then the product does not have to converge. This leaves a small, but important, question. Suppose that one starts with two convergent, but not absolutely convergent, series. Could we have a situation in which the Cauchy product converges to other than the required sum? The next theorem deals with this question. $\square$

Theorem 6.3.7 (Abel). *Suppose $\sum_{n=0}^{\infty} a_n$ converges to a, $\sum_{n=0}^{\infty} b_n$ converges to b, and the Cauchy product is convergent. Then the Cauchy product converges to ab.*

Proof. Let A_n, B_n, and C_n be the partial sums of the three series. Observe that

$$
\sum_{k=0}^{n} C_k = \sum_{k=0}^{n} A_k B_{n-k}.
$$

It follows that

$$\frac{1}{n} \sum_{k=0}^{n} C_k = \frac{1}{n} \sum_{k=0}^{n} A_k B_{n-k}.$$

Now $A_n \to a$ and $B_n \to b$, whence $\frac{1}{n} \sum_{k=0}^{n} C_k \to ab$ (Exercise 24). $\square$

There remains one additional question, namely, if two series are absolutely convergent, what about their product?

Theorem 6.3.8. *If two series are absolutely convergent, so is their Cauchy product, and the product of the sums is the sum of the products.*

Proof. Exercise 25. $\square$

Cauchy products are a special example of double series. Recall that a double sequence, $\{\sigma(n, m)\}$, is a function from $\mathbf{N} \times \mathbf{N}$ into $\mathbf{R}$ and that a double sequence has a limit, a exactly if for every positive ϵ there exists $N \in \mathbf{N}$ such that

$$|\sigma(n, m) - a| < \epsilon \text{ whenever } n, m > N.$$

(See Exercises 3.3.31–3.3.33.) We want to develop a notion of convergence for double series, $\sum_{n,m=1}^{\infty} \sigma(n, m)$. As with ordinary series, the vehicle is the sequence of partial sums.

Definition. Let $\{\sigma(n, m)\}$ be a double sequence and consider the formal infinite series $\sum_{n,m=1}^{\infty} \sigma(n, m)$. We define the sequence of partial sums of the double series by

$$s(n, m) = \sum_{k=1}^{n} \sum_{j=1}^{m} \sigma(k, j).$$

We will say that the double series **converges** to a sum, s, provided

$$\lim_{n,m \to \infty} s(n, m) = s.$$

If the limit of the sequence of partial sums does not exist, we say the series is **divergent**.

Discussion. This definition is completely consistent with previous definitions of convergence for series. The notions are based on the basic definition of double sequences and the convergence of same. If the reader has not completed Exercises 3.3.31–3.3.33 of Section 3.3, he would be well advised to do so before continuing with this material.

If we have two sequences $\{a_n\}$ and $\{b_m\}$ $n, m \geq 0$, we can define a double sequence (series) by setting $\sigma(n, m) = a_n b_m$. We can then ask about the convergence properties of the series, $\sum_{n,m=0}^{\infty} \sigma(n, m)$. Further, we can ask whether this double series bears any relation to the Cauchy product of the two series. $\square$

Theorem 6.3.9. *Let f be a one-to-one function from $\mathbf{N}$ onto $\mathbf{N} \times \mathbf{N}$ and let $\sum_{n,m=1}^{\infty} \sigma(n, m)$ be a double series. Then $\sum_{k=1}^{\infty} \sigma(f(k))$ converges absolutely exactly if $\sum_{n,m=1}^{\infty} \sigma(n, m)$ converges absolutely.*

Proof. Exercise 31. $\square$

Exercises

1. Verify directly that the alternating form of the harmonic series, $\sum \frac{(-1)^{n+1}}{n}$, fails the absolute convergence property, the rearrangement convergence property, and the subseries convergence property.

2. Let p_n and q_n be obtained from a_n as in the definition of positive and negative parts of a series. Prove all claims made about p_n and q_n in the discussion.

3. Complete the proof of Theorem 6.3.2.

4. Let $\sum b_n$ be a rearrangement of $\sum a_n$. Show the positive part of $\sum b_n$ is a rearrangement of the positive part of $\sum a_n$.

5. Give a direct proof of Theorem 6.3.3 that does not apply Theorem 6.3.2.

6. Let $\sum a_n$ be a convergent series, and let $\sum a_{f(n)}$ be a rearrangement of same. Show $\lim\limits_{n \to \infty} a_{f(n)} = 0$.

7. Complete the missing details in the proof of Theorem 6.3.4.

8. Rearrange the alternating form of the Harmonic series to converge to 0, to 1, to -1.

9. Show the expression for the general term of t_{3n} that is given in Example 1 is correct.

10. Show if the alternating form of the Harmonic series is rearranged so that each time p positive terms are followed by q negative terms, $p, q \in \mathbf{N}$, then the resultant series is convergent. Also find its sum.

11. Let $\{f_n\}$ be a sequence of functions satisfying:
    (a) $f_n : \mathbf{N} \to \mathbf{N}$ is one-to-one for all $n \in \mathbf{N}$; (b) $A_n \subseteq \operatorname{Rng} f_n$ for each n;
    (c) if $i \neq j$, then $A_i \cap A_j = \emptyset$; (d) $\bigcup_j A_j = \mathbf{N}$.

12. Let $\sum a_n$ be an absolutely convergent series, and set $b_n^k = a_{f_k(n)}$ for each n and k. Show

    (a) For each fixed k, $\sum_{n=1}^{\infty} b_n^k$ is an absolutely convergent subseries of $\sum a_n$.

    (b) If $s_k = \sum_{n=1}^{\infty} b_n^k$, then $\sum s_k$ is absolutely convergent and has the same sum as $\sum a_n$.

13. If $\sum a_n$ is absolutely convergent, which of the following are absolutely convergent?
    (a) $\sum \frac{a_n}{1+a_n}$; (b) $\sum \frac{a_n^2}{1+a_n^2}$; (c) $\sum \sqrt{a_n a_{n+1}}$; (d) $\sum \frac{|a_n a_{n+1}|}{|a_n|+|a_{n+1}|}$.

14. Let $\sum a_n$ be a absolutely convergent series and $\{b_n\}$ a bounded sequence. Is $\sum a_n b_n$ absolutely convergent?

15. Let a denote the sum of the alternating harmonic series. Rearrange the series by taking p positive terms followed by q negative terms and repeating indefinitely. Find an expression for the sum of the rearranged series.

16. Show it is possible to rearrange the series $\sum \frac{(-1)^{n+1}}{n}$ so that the resultant sequence of partial sums is bounded but does not converge.

17. Let $\sum a_n$ be a convergent series. Give a condition that will ensure that $\sum a_n^2$ will converge.

18. Suppose that $\sum a_n^2$ and $\sum b_n^2$ are both convergent. Is the same true for $\sum a_n b_n$?

19. We have shown that $\sum \frac{1}{n^2}$ is convergent. Let its sum be denoted by s. (In fact $s = \frac{\pi^2}{6}$.) Show $\sum_{n=0}^{\infty} \frac{1}{(2n+1)^2} = \frac{3}{4}s$. Find the sums of the following series:

(a) $\frac{1}{2^2} + \frac{1}{4^2} + \frac{1}{6^2} + \cdots$; (b) $1 + \frac{1}{5^2} + \frac{1}{7^2} + \frac{1}{11^2} + \frac{1}{13^2} + \cdots$;

(c) $1 - \frac{1}{2^2} - \frac{1}{4^2} + \frac{1}{5^2} + \frac{1}{7^2} - - + \cdots$.

20. Let $S = \sum_{j=1}^{\infty} (-1)^{j+1} a_j$, where $\{a_j\}$ is an eventually monotonically decreasing sequence of strictly positive terms, tending to 0. Define S^{pq} as the rearrangement (no signs changed) obtained from S by taking groups of p positive terms followed by a group of q negative terms. Prove that if f is a continuous and eventually monotone function such that $f(2n-1) = a_{2n-1}$ and $p \geq q > 0$, then

$$S^{pq} = S + \frac{1}{2} \lim_{n \to \infty} \int_{qn}^{pn} f$$

(with the understanding that both sides may be infinite or may diverge by oscillation). Compute S^{pq} for the following series S:

(a) $1 - \frac{1}{2} + \frac{1}{3} - + - \cdots$; (b) $1 - \frac{1}{2^k} + \frac{1}{3^k} - + \cdots$;

(c) $1 - \frac{1}{3} + \frac{1}{5} - + \cdots$; (d) $\sum_{n=1}^{\infty} \frac{(-1)^{n+1} \cos(\ln n) + 2}{n}$.

21. Prove the definition of c_m in Cauchy product has the property of being the coefficient on x^m in the product of two finite polynomials of degree n, where $n \geq m$.

22. Verify all displayed equations in the proof of Theorem 6.3.6.

23. Let $\{s_n\}$ be a sequence. Show if $s_n \to 0$, then $\frac{1}{n} \sum_{k=1}^{n} s_k$ converges to 0 as $n \to \infty$.

24. Verify the displayed equations in the proof of Theorem 6.3.7. (Hint: use Exercise 1.2.14 and Exercise 1.2.16.)

25. Let $\{a_n\}$ and $\{b_n\}$ be two sequences converging to a and b, respectively. Define $t_n = \frac{1}{n} \sum_{k=1}^{n} a_k b_{n+1-k}$. Show $\{t_n\}$ converges to ab.

26. Prove Theorem 6.3.8.

27. Consider $a_1 = 1, a_n = 2$ for $n \geq 2$ and $b_1 = 1, b_n = (-1)^{n+1}2$ for $n \geq 2$. Show the Cauchy product of $\sum a_n$ and $\sum b_n$ is convergent.

28. Let $\sum_{i=1}^{\infty} a_i$ be a series and $\{s_n\}$ its sequence of partial sums. Define σ_n by

$$\sigma_n = \frac{1}{n} \sum_{k=1}^{n} s_k.$$

Show if $\{s_n\}$ converges, then $\{\sigma_n\}$ converges to the same sum. Give a counterexample to show that the converse is not true.

29. Let s_n and σ_n be as in Exercise 27. Show if for every $M \in \mathbf{N}$, there exists $N \in \mathbf{N}$ such that $n > N$ implies $\sigma_n > M$, then s_n has the same property. Show the converse fails, namely, if s_n is unbounded, then σ_n may be bounded, or indeed, converge to 0.

30. For a series, $\sum_{n=1}^{\infty} a_n$, the sequence, $\{\sigma_n\}$ is called the **sequence of arithmetic means**. A series is called **Cesaro summable**, or **(C,1) summable**, exactly if $\{\sigma_n\}$ is convergent. Show for series with nonnegative terms, convergence and summability are identical concepts. Discuss the Cesaro summability of the following series:

(a) $1 - 1 + 1 - 1 + \cdots$;   (b) $1 - 1 + 0 + 1 - 1 + 0 + 1 + \cdots$;

(c) $1 + 0 - 1 + 1 + 0 - 1 + \cdots$;   (d) $\frac{1}{2} - 1 + \frac{1}{2} + \frac{1}{2} + \frac{1}{2} - 1 + \frac{1}{2} + \frac{1}{2} + \cdots$;

(e) $\frac{1}{2} - \frac{1}{2} + \frac{1}{2} - \frac{1}{2} + \cdots$;   (f) $1 - 1 + 0 + 0 + 1 - 1 + 0 + 0 + \cdots$;

(g) $\frac{1}{k} - \frac{1}{k} + \frac{1}{k} - \frac{1}{k} + \cdots$, $k \in \mathbf{N}$.

31. Establish the Cauchy-Schwartz and Minkowski inequalities for series. Let $\sum_{n=1}^{\infty} a_n^2$ and $\sum_{n=1}^{\infty} b_n^2$ converge. Show $\sum_{n=1}^{\infty} a_n b_n$ and $\sum_{n=1}^{\infty} (a_n + b_n)^2$ also converge and that

$$\left[\sum_{n=1}^{\infty} a_n b_n \right]^2 \leq \left[\sum_{n=1}^{\infty} a_n^2 \right] \cdot \left[\sum_{n=1}^{\infty} b_n^2 \right],$$

and

$$\left[\sum_{n=1}^{\infty} (a_n + b_n)^2 \right]^{\frac{1}{2}} \leq \left[\sum_{n=1}^{\infty} a_n^2 \right]^{\frac{1}{2}} + \left[\sum_{n=1}^{\infty} b_n^2 \right]^{\frac{1}{2}}.$$

32. Prove Theorem 6.3.9.

33. Show there is a one-to-one function f from $\mathbf{N} \cup \{0\}$ onto $(\mathbf{N} \cup \{0\}) \times (\mathbf{N} \cup \{0\})$ that will generate the Cauchy product of two series. What does this imply for absolutely convergent series?

34. Suppose that $\sum_{n,m=1}^{\infty} \sigma(n, m)$ is a double series. Consider the associated partial sums $A_m = \sum_{n=1}^{\infty} \sigma(n, m)$ and $B_n = \sum_{m=1}^{\infty} \sigma(n, m)$, as well as the iterated sums, $\sum_{m=1}^{\infty} A_m$ and $\sum_{n=1}^{\infty} B_n$. Under what conditions will these various series converge? Under what conditions will they all yield the same sums?

35. Prove that a double series of positive terms converges if and only its sequence of partial sums is bounded.

36. Show that an absolutely convergent double series converges. What can you say about the converse?

37. Let $\sum_{n,m} \sigma(n, m)$ be a double series. The series defined by A_m, and B_n of Exercise 33 are called the **column series** and **row series**, respectively. Similarly, a **diagonal series** is defined by $\sum_{k=1}^{\infty} C_k$, where $C_k = \sum_{j=1}^{k} \sigma(j, k - j)$. There are various other ways (infinitely many, in fact) of forming a single series from $\sum_{n,m} \sigma(n, m)$. Prove that if a double series of positive terms converges, then any single series formed out of it converges. What can you say about the converse?

38. Discuss the convergence of the following double series. In particular, find the sums of each row series, column series, and diagonal series. Also discuss the two iterated sums $\sum_{m=1}^{\infty} \sum_{n=1}^{\infty} \sigma(m, n)$, $\sum_{n=1}^{\infty} \sum_{m=1}^{\infty} \sigma(m, n)$, in each case.

(a) $\sigma(1, m) = \sigma(n, 1) = 1$, $\sigma(2, m + 1) = \sigma(n + 1, 2) = -1$, $\sigma(n, m) = 0$, otherwise;

(b)

$$1 + 2 + 4 + 8 + \cdots$$
$$-\frac{1}{2} - 1 - 2 - 4 - \cdots$$
$$-\frac{1}{4} - \frac{1}{2} - 1 - 2 - \cdots$$
$$-\frac{1}{8} - \frac{1}{4} - \frac{1}{2} - 1 - \cdots$$
$$\cdots ;$$

(c)

$$\sigma(m,n) = \begin{cases} 1, & m = n+1, \ n = 1, 2, \dots, \\ -1, & m = n-1, \ n = 1, 2, \dots, \\ 0, & 0, \ \text{otherwise;} \end{cases}$$

(d) $\sigma(n,n) = 2$, $\sigma(n, n+2) = -1$, $\sigma(m+2, m) = 1$, $\sigma(n,m) = 0$, otherwise;

(e) $\sigma(n,n) = 2$, $\sigma(n, n+1) = -1$, $\sigma(m+1, m) = 1$, $\sigma(n,m) = 0$, otherwise;

(f) (Cesaro's example)

$$\frac{1}{2} - \frac{1}{4} + \frac{1}{4} - \frac{1}{8} + \frac{1}{8} - \frac{1}{16} + \frac{1}{16} - \cdots$$

$$\frac{1}{2^2} - \frac{3}{4^2} + \frac{3}{4^2} - \frac{7}{8^2} + \frac{7}{8^2} - \frac{15}{16^2} + \frac{15}{16^2} - \cdots$$

$$\frac{1}{2^3} - \frac{3^2}{4^3} + \frac{3^2}{4^3} - \frac{7^2}{8^3} + \frac{7^2}{8^3} - \frac{15^2}{16^3} + \frac{15^2}{16^3} - \cdots$$

$$\frac{1}{2^4} - \frac{3^3}{4^4} + \frac{3^3}{4^4} - \frac{7^3}{8^4} + \frac{7^3}{8^4} - \frac{15^3}{16^4} + \frac{15^3}{16^4} - \cdots$$

$$\cdots$$

39. Show that any single series formed out of an absolutely convergent double series also converges absolutely. In particular, in an absolutely convergent double series, the sums by rows, by columns and by diagonal are all equal.

40. Give examples to illustrate that the conclusions of Exercise 38 need not be true, if the double series is conditionally convergent.

41. Show that the Cauchy product of two series can be obtained as the diagonal series of a suitable double series. Use this idea to give an alternate proof of Theorem 6.3.8.

6.4 Series with Arbitrary Terms

As pointed out in the last section, all the tests that we have developed so far require the series to be nonnegative. Actually, this can be weakened to a requirement that eventually the terms of the series be all of one sign, but this is not a significant addition to the power of the tool. The net result is that our tests may be used to decide whether an arbitrary series will converge absolutely, but none can detect conditional convergence. For this reason we would like to have some tests that can be applied to an arbitrary series.

One of the most easily dealt with forms of a series with both positive and negative terms is the one in which terms are alternately positive and negative. Such a series is called an **alternating series** for obvious reasons. Taking all the $a_n > 0$, an alternating series could be represented by $\sum_{n=1}^{\infty} (-1)^{n-1} a_n$. The following elegant test decides the convergence of such a series.

Theorem 6.4.1 (Leibnitz's test).   *If* $\{a_n\}$ *is a nonincreasing sequence of positive terms converging to* 0, *then the alternating series* $\sum_{n=1}^{\infty}(-1)^{n-1}a_n$ *is convergent.*

Proof. Consider the odd partial sums, namely, those ending in positive terms. Since

$$s_{2n+1} - s_{2n-1} = a_{2n+1} - a_{2n} \leq 0,$$

the odd partial sums form a monotonic decreasing sequence. The analogous computation for the even partial sums shows that $s_{2n+2} \geq s_{2n}$, whence the even partial sums form a monotonic increasing sequence. Now, $s_{2n+1} = s_{2n} + a_{2n+1}$, whence $s_{2n+1} \geq s_{2n} \geq a_2$. Consequently, the odd partial sums form a decreasing sequence bounded below, and so must converge, say, to s. Similarly, the even partial sums form an increasing sequence bounded above and so must converge, say to t. But

$$0 = \lim_{n\to\infty} a_{2n+1} = \lim_{n\to\infty} (s_{2n+1} - s_{2n}) = s - t,$$

whence $s = t$. $\square$

Discussion. This is a nice theorem because it is simple to apply and the proof is conceptually easy to understand. It does, however, require that the sequence $\{a_n\}$ is monotonically decreasing and that the associated series is alternating. Both are strong conditions, and so reduce the utility of the theorem. The condition that $\{a_n\}$ is monotone can be weakened to a requirement that the sequence be eventually monotonically decreasing to 0; however, this is not a significant weakening of the hypothesis. Finally, note that completeness in the form of Theorem 1.3.1 was vital for the proof. $\square$

Example 1. Discuss the convergence of the series $\sum \frac{(-1)^{n-1}}{n}$.

Solution. The sequence $\{\frac{1}{n}\}$ is monotonically decreasing to 0, whence we may apply Leibnitz's test to see that it converges. $\square$

Our next two tests are similar to Leibnitz's test in that they consider situations where the terms of a series are composed of the term by term product of corresponding terms of two sequences. To establish these results, we shall require the following useful identity:

Lemma (Abel).   *Let* $\{a_n\}$, $\{b_n\}$ *be two arbitrary sequences and let* $s_n = \sum_{i=1}^{n} a_i$. *Then*

$$\sum_{i=1}^{n} a_i b_i = s_n b_{n+1} - \sum_{i=1}^{n} s_i(b_{i+1} - b_i).$$

Discussion. The reader will be invited to check the above identity in Exercise 2. An immediate consequence of the identity is that the series $\sum a_i b_i$ converges exactly if both the sequence $\{s_n b_{n+1}\}$ and the infinite series $\sum s_n(b_{n+1} - b_n)$ converge. $\square$

Theorem 6.4.2 (Dirichlet's test).   *Let* $\sum a_n$ *be a series whose partial sums form a bounded sequence, and let* $\{b_n\}$ *be a sequence of decreasing terms converging to zero. Then* $\sum a_n b_n$ *is convergent.*

Proof. It is given that for some M, the partial sums s_n of $\sum a_n$ satisfy the inequality $|s_n| < M$ for all n. Thus, $\lim s_n b_{n+1} = 0$ (**WHY?**). Also,

$$|s_n(b_{n+1} - b_n)| \le M(b_n - b_{n+1}),$$

whence

$$\sum_{i=1}^{n} |s_i(b_{i+1} - b_i)| \le M \sum_{i=1}^{n} (b_i - b_{i+1}) = M(b_1 - b_{n+1}),$$

and consequently, $\sum a_n b_n$ converges (**WHY?**). $\square$

Theorem 6.4.3 (Abel's test). *If $\sum a_n$ is convergent and if $\{b_n\}$ is monotonic convergent sequence, then $\sum a_n b_n$ is convergent.*

Proof. The existence of $\lim s_n b_{n+1}$ is guaranteed by the convergence of $\sum a_n$ and of $\{b_n\}$. An argument similar to the Dirichlet's test completes the proof (Exercise 3). $\square$

Discussion. It is easy to see that Dirichlet's test is a generalization of Leibnitz's test, since we have only to note that the setting $a_n = (-1)^{n-1}$ will produce a series with a bounded sequence of partial sums.

Abel's test may be thought of in a somewhat different way. If we take the series $\sum a_n$ and multiply it term by term by a constant sequence $\{b_n\}$, then the resultant series $\sum a_n b_n$ will be convergent. We can ask: How can we generalize this result? There are several ways to go. We could require only that $\{b_n\}$ be bounded, or we could require only that $\sum b_n$ be convergent. It turns out that neither of these requirements is enough, and we will explore this further in the exercises. $\square$

Example 2. Determine the convergence properties of $\sum_{n=2}^{\infty} \frac{(n^3+1)^{1/3} - n}{\ln n}$.

Solution. For $n \ge 2$, the sequence $\{\frac{1}{\ln n}\}$ is monotone and converges to 0. Thus, if $\sum (n^3 + 1)^{1/3} - n$ is convergent, we can use Abel's test to conclude convergence for the original series. Using algebra we have

$$\sum \left[(n^3 + 1)^{1/3} - n \right] = \sum \frac{1}{(n^3 + 1)^{2/3} + (n^3 + 1)^{1/3}n + n^2}.$$

The series on the right-hand side can now be shown to be convergent by an appropriate application of the Comparison test. It is left to the reader to complete the details (Exercise 4). $\square$

Discussion. The basic technique employed is to rationalize the numerator. We cannot overly stress the importance of algebra as a tool for attacking these problems. It is used implicitly throughout all the manipulations. $\square$

Two of the very useful and frequently employed tests are given next. The first, called the Ratio test and due to D'Alembert, draws conclusions about the behavior of a series by studying the sequence of ratios of consecutive terms of the series. The second, the Root test, due to Cauchy, draws conclusions based on the behavior of the sequence formed by taking the nth root of the nth term of the given series.

Theorem 6.4.4 (Ratio test). *Given the series* $\sum a_n$, *let*

$$r = \underline{\lim} \left| \frac{a_{n+1}}{a_n} \right| \quad and \quad R = \overline{\lim} \left| \frac{a_{n+1}}{a_n} \right|.$$

Then

  (i) *the series* $\sum a_n$ *converges absolutely if* $R < 1$;

  (ii) *the series diverges if* $r > 1$;

  (iii) *the test gives no information if* $r \leq 1 \leq R$.

Proof. For (i) assume that $R < 1$, and choose x such that $R < x < 1$. By definition of R, there exists an N such that $|\frac{a_{n+1}}{a_n}| < x$ whenever $n \geq N$. It is immediate that $|a_{n+1}| < |a_n|x$, which together with the above inequality can be used to show that

$$\sum_{k=N}^{N+p} |a_k| \leq \sum_{k=1}^{p+1} |a_N||x^{k-1}|.$$

The sum on the right-hand side is geometric with ratio $x < 1$ and so is convergent as $p \to \infty$. Consequently, the sum on the left-hand side is also convergent, and so $\sum a_n$ converges as well.

For (ii), since $r > 1$, we have $|a_{n+1}| > |a_n|$ for all $n > N$. Evidently, this prevents the terms of the series from approaching 0, and so the series cannot converge.

For (iii), the series $\sum \frac{1}{n}$ is divergent, while the series $\sum \frac{1}{n^2}$ is convergent. Both satisfy $R = r = 1$ (Exercise 6). $\square$

Discussion. The initial thrust of the argument in this theorem asserts that there is an x and an $N \in \mathbf{N}$ such that $n \geq N$ implies

$$\left| \frac{a_{n+1}}{a_n} \right| < x < 1.$$

The truth of this inequality is derived from the basic properties of the limit inferior of a sequence. The reader should be able to supply all the reasons as to why this inequality is valid, or lacking this, review the basic properties of limit inferior (Exercise 5).

In the special case that the sequence $\{|\frac{a_{n+1}}{a_n}|\}$ converges, $R = r$, whence we obtain convergence for $R < 1$, and divergence for $R > 1$, and the test fails when $R = 1$.

This test answers the question: Does a given series eventually have its terms go to 0 with the rapidity of a convergent geometric series or, on the other hand, diverge with the rapidity of a divergent geometric series. If the answer is 'yes' in the first instance, we conclude convergence; if 'yes' in the second, we conclude divergence. If the limiting ratio is 1, we get no conclusion.

While the test applies to all series, if we conclude convergence with this test, the series will converge absolutely, as well. If the series diverges by this test, no amount of changing signs of individual terms will ever generate a convergent series.

Finally, this test has particular utility, since it is totally self-contained. By this we mean that to apply this test, one need only consider the properties of the terms of the series whose convergence properties are required. No additional series are required for comparisons. The Root test, which is given below, also has this feature. $\square$

Example 3. Discuss the convergence properties of the series $\sum_{n=0}^{\infty} \frac{x^n}{n!}$ for $x \in \mathbf{R}$.

Solution. Let x be fixed, and set $a_n = \frac{x^n}{n!}$. We apply the Ratio test as follows:

$$\left| \frac{a_{n+1}}{a_n} \right| = \left| \frac{\frac{x^{n+1}}{(n+1)!}}{\frac{x^n}{n!}} \right| = \left| \frac{x^{n+1}}{(n+1)!} \frac{n!}{x^n} \right| = \left| \frac{x}{n+1} \right|.$$

Since x is fixed, it is immediate that $\frac{x}{n+1}$ converges to 0 for all x, whence the series converges (absolutely) for all values of x. $\square$

Discussion. The computations are straightforward and illustrate the ease with which the Ratio test may be applied.

The series to which we have applied the test may be well known to the reader. It sums to e^x, where e is the base for the natural logarithms. This series is tremendously useful and shows up in almost every branch of mathematics. $\square$

Theorem 6.4.5 (Root test).   *Given the series* $\sum a_n$, *let*

$$R = \overline{\lim} |a_n|^{\frac{1}{n}}.$$

Then,

(i) $\sum a_n$ *converges absolutely if* $R < 1$;

(ii) $\sum a_n$ *diverges if* $R > 1$;

(iii) *no information is obtained if* $R = 1$.

Proof. For (i), if $R < 1$, choose x such that $R < x < 1$. The definition of R guarantees the existence of an N such that $|a_n|^{\frac{1}{n}} < x$ for $n > N$ so that $|a_n| < x^n$ after a stage. By the Comparison test, $\sum a_n$ converges (**WHY?**).

For (ii), notice that $a_n > 1$ infinitely often, since $R > 1$, so that $\lim a_n$ is different from zero. Hence, the series must diverge.

For (iii), the same examples that were used in the proof of the Ratio test will also do the job here (Exercise 6). $\square$

Discussion. As in the case of Ratio test, if the sequence $\{|a_n|^{\frac{1}{n}}\}$ converges to R, then we conclude that the series converges for $R < 1$, and diverges for $R > 1$, and the test fails for $R = 1$.

Since taking ratios and proceeding to limits is generally easier than a root extraction (of order n), the Ratio test has the advantage that it is simpler to operate. On the other hand, Root test is certainly more powerful than the Ratio test in the sense that whenever the Ratio test succeeds, Root test also succeeds and delivers the same conclusions. But there are instances where the Root test is decisive, while the Ratio test fails to give any conclusion whatsoever. The key to this remark is the simple inequality (see Exercise 3.4.22) if $a_n > 0$,

$$\underline{\lim} \frac{a_{n+1}}{a_n} \le \underline{\lim} \, a_n^{\frac{1}{n}} \le \overline{\lim} \, a_n^{\frac{1}{n}} \le \overline{\lim} \frac{a_{n+1}}{a_n}. \quad \Box$$

The next example illustrates a situation where the Ratio test fails but the Root test prevails.

Example 4. Under what conditions will the series defined by

$$1 + a + ab + a^2 b + a^2 b^2 + a^3 b^2 + a^3 b^3 + a^4 b^3 + \cdots$$

converge?

Solution. It is easily seen that the ratio of any two consecutive terms alternate between a and b so that the ratio test will fail if $a < 1 < b$. However, the root test when applied (counting the first term 1 as a_0), yields

$$\lim \left(a^n b^n \right)^{\frac{1}{2n}} = \lim \left(a^{n+1} b^n \right)^{\frac{1}{2n+1}} = \sqrt{ab},$$

which is decisive provided $ab \ne 1$. The reader can complete the solution for the case where $ab = 1$ (Exercise 7). $\Box$

The question of whether a given series is convergent or divergent is in general a difficult one. There is no single, universal test that will deal with all possible cases. The foremost method is to verify the Cauchy criteria, but we have discussed several other useful tests above, including the popular ones like the Ratio and the Root tests, as well as the elegant test for an alternating series. Most of these have been derived in some fashion from one of the forms of the Comparison test. We include here some more delicate tests that may be applied when all the earlier tests fail. In particular, some of these will be useful when the Ratio and Root tests fail, in other words, when the limit of $\frac{a_{n+1}}{a_n}$ and $a_n^{\frac{1}{n}}$ is equal to 1.

The following characterization of convergence, due to Kummer, yields a large number of powerful tests, including the Ratio and Root tests, as special cases.

Theorem 6.4.6 (Kummer's test).   *Let $\sum a_n$ be a series of positive terms, $\{D_n\}$ a sequence of positive terms, and set*

$$p_n = D_n \frac{a_n}{a_{n+1}} - D_{n+1}.$$

Then $\sum a_n$ is convergent if and only if for some sequence $\{D_n\}$, $\underline{\lim} \, p_n > 0$, and $\sum a_n$ diverges if and only if for some sequence $\{D_n\}$, $\sum \frac{1}{D_n}$ is divergent and $p_n \le 0$.

Proof. We first show the existence of the sequence $\{D_n\}$ with the given property implies convergence or divergence; the case where $\lim p_n > 0$ is treated first. Fix $x > 0$ such that there exists $N \in \mathbf{N}$ such that $n \ge N$ implies $p_n > x$, and fix N with the asserted property. For $n = N, N+1, N+2, \cdots$, we have the following string of inequalities:

$$
\begin{aligned}
D_N a_N - D_{N+1} a_{N+1} &> x a_{N+1} \\
D_{N+1} a_{N+1} - D_{N+2} a_{N+2} &> x a_{N+2} \\
\cdots \quad \cdots \quad \cdots \\
\cdots \quad \cdots \quad \cdots \\
\cdots \quad \cdots \quad \cdots \\
D_{N+p-1} a_{N+p-1} - D_{N+p} a_{N+p} &> x a_{N+p}.
\end{aligned}
$$

If we let $\{s_n\}$ be the sequence of partial sums of $\sum a_n$ and add the inequalities, we obtain

$$
x(s_{N+p} - s_N) < D_N a_N - D_{N+p} a_{N+p} \le D_N a_N,
$$

with the validity of the last inequality being due to the fact that we are dealing with sequences of positive terms. Mere manipulation of the last inequality yields

$$
s_{N+p} < s_N + \frac{1}{x} D_N a_N,
$$

whence $\{s_{N+p}\}$, $p \in \mathbf{N}$ is a bounded sequence. It now follows that $\sum a_n$ converges.

For divergence, suppose $\sum \frac{1}{D_n}$ diverges and

$$
p_n a_{n+1} = D_n a_n - D_{n+1} a_{n+1} \le 0.
$$

This inequality is equivalent to saying that $\{D_n a_n\}$ is monotone increasing and yields

$$
D_1 a_1 \frac{1}{D_n} \le a_n
$$

for all $n \in \mathbf{N}$. Since the series $\sum \frac{1}{D_n}$ is divergent and $D_1 a_1$ is a fixed, positive real number, the Comparison test establishes the divergence of $\sum a_n$.

Now we show that if $\sum a_n$ is either convergent or divergent, then a sequence $\{D_n\}$ with the requisite properties must exist. Suppose first $\sum a_n$ converges to a and set

$$
D_n = \frac{a - \sum_{i=1}^{n} a_i}{a_n}.
$$

Since $a_i > 0$ for all $i \in \mathbf{N}$, $D_n > 0$ and

$$
\begin{aligned}
p_n &= D_n \cdot \frac{a_n}{a_{n+1}} - D_{n+1} \\
&= \frac{a - \sum_{i=1}^{n} a_i}{a_{n+1}} - \frac{a - \sum_{i=1}^{n+1} a_i}{a_{n+1}} \\
&= \frac{a_{n+1}}{a_{n+1}} = 1 = \lim p_n > 0
\end{aligned}
$$

as required.

Suppose now $\sum a_n$ diverges and set

$$
D_n = \frac{\sum_{i=1}^{n} a_i}{a_n}.
$$

Since $a_i > 0$ for all $i \in \mathbf{N}$, $D_n > 0$ and

$$
\begin{aligned}
p_n &= D_n \cdot \frac{a_n}{a_{n+1}} - D_{n+1} = \frac{\sum_{i=1}^n a_i}{a_{n+1}} - \frac{\sum_{i=1}^{n+1} a_i}{a_{n+1}} \\
&= -\frac{a_{n+1}}{a_{n+1}} = -1 \le 0.
\end{aligned}
$$

To complete the proof we must show $\sum \frac{1}{D_n}$ diverges. Observe

$$
\begin{aligned}
\sum_{k=m+1}^{n} \frac{1}{D_k} &= \frac{a_{m+1}}{\sum_{k=1}^{m+1} a_k} + \frac{a_{m+2}}{\sum_{k=1}^{m+2} a_k} + \cdots + \frac{a_n}{\sum_{k=1}^{n} a_k} \\
&> \frac{\sum_{k=m+1}^{n} a_k}{\sum_{k=1}^{m} a_k + \sum_{k=m+1}^{n} a_k}.
\end{aligned}
$$

Since $\sum_{k=m+1}^{\infty} a_k$ is divergent, we may choose $n > m$ such that

$$
\frac{\sum_{k=m+1}^{n} a_k}{\sum_{k=1}^{m} a_k + \sum_{k=m+1}^{n} a_k} > \frac{1}{2},
$$

whence $\sum \frac{1}{D_n}$ must diverge. $\square$

Discussion. Kummer's test provides a characterization for convergence and/or divergence of series of positive terms; thus, previous comments to the contrary, Kummer's test can be used to decide the issue of convergence for all positive series. The difficulty is that Kummer's test requires the user to find the sequence $\{D_n\}$. Obviously, we would like to have a single sequence $\{D_n\}$ that would serve to test all series of positive terms. Such is not possible, as we ask the reader to show in Exercise 31. Because of this, Kummer's test is not a single test, rather it is an infinitude of tests, one for each sequence, $\{D_n\}$. Some of the useful choices for the sequence $\{D_n\}$ are $\{1\}$, $\{n\}$, $\{n \ln n\}$, $\{n \ln n \ln(\ln n)\}$, and so on. For series thought to be divergent, the proof also suggests a construction of $\{D_n\}$ that may be useful. While the $\{D_n\}$ constructed in the case of convergence must work for any convergent series, it requires knowledge of the sum to be applied in any particular case. Evidently, if the sum is known, one would not be trying to prove convergence, so this construction cannot be generally applied in practice.

Portions of the proof given follow Tong [1994], and a brief discussion of the history surrounding Kummer's test together with historical references are given there.[1]

If we set $D_n = 1$, it is easily seen that Kummer's test reduces to the Ratio test, while the choice $D_n = n$ gives rise to Raabe's test given below. $\square$

Corollary (Raabe's test).   *Let $\sum a_n$ be a positive series. Then*

(i) $\sum a_n$ *is convergent if* $\varliminf n(\frac{a_n}{a_{n+1}} - 1) > 1$;

(ii) $\sum a_n$ *is divergent if* $\varlimsup n(\frac{a_n}{a_{n+1}} - 1) < 1$.

---

[1] J. Tong, "Kummer's test Gives Characterizations for Convergence of All Positive Series," *American Mathematical Monthly*, May 1994, 450–452.

Proof. The proof of this result is left as Exercise 8. □

Example 5. Test for convergence the series whose nth term is given by

$$a_n = \left[\frac{1 \cdot 4 \cdot 7 \cdots (3n - 2)}{3 \cdot 6 \cdot 9 \cdots 3n} \right]^2.$$

Solution. We may apply Kummer's test directly by setting $D_n = n$, or we can apply Raabe's test. Using the latter, a simple computation shows that

$$n \left(\frac{a_n}{a_{n+1}} - 1 \right) = n \left(\left[\frac{3n + 3}{3n + 1} \right]^2 - 1 \right)$$

$$= \frac{12n^2 + 8n}{9n^2 + 6n + 1}.$$

Since the limit of this sequence exceeds 1, the series is convergent by Raabe's test. □

Discussion. In many instances, the nth term of a series involves powers, factorials or huge products as above. Such series are particularly suited to tests that involve the ratio of successive terms, since a good deal of cancellation can occur. One should always start with the Ratio test, since this is by far the simplest to apply and, moreover, to apply any of the more delicate tests, such as Kummer's or Raabe's test, the ratio $\frac{a_n}{a_{n+1}}$ must be calculated. In the instance above, it is easy to see that the ratio of successive terms of the series has limit 1, and so we are forced to use a test other than the Ratio test. As with the Ratio test, Raabe's test will not give any information if the generated sequence has limit 1. □

The last test presented again illustrates the application of Kummer's test to generate a test with a specific sequence $\{D_n\}$. This test is of even greater sensitivity than those already given. However, the list of such subtler tests that can be derived from Kummer's test is by no means exhaustive, and the reader will be asked to establish additional tests in the exercises.

Theorem 6.4.7 (Gauss's test). *Let $\sum a_n$ be a series of positive terms and suppose there is a bounded sequence $\{b_n\}$ and a constant k such that*

$$\frac{a_n}{a_{n+1}} = 1 + \frac{k}{n} + \frac{b_n}{n^2}.$$

Then $\sum a_n$ is convergent if $k > 1$ and divergent if $k \leq 1$.

Proof. We treat two cases, namely, $k \neq 1$ and $k = 1$. For the former, observe that we may rewrite

$$\frac{a_n}{a_{n+1}} = 1 + \frac{k}{n} + \frac{b_n}{n^2}$$

as

$$n \left(\frac{a_n}{a_{n+1}} - 1 \right) = k + \frac{b_n}{n}.$$

The sequence on the left is exactly the expression in Raabe's test. Moreover, since k is a constant and $\{b_n\}$ is bounded, this sequence (on the left) has a limit, namely k. It is immediate that if $k > 1$, then $\sum a_n$ is convergent, whereas if $k < 1$, the series must diverge.

For the case where $k = 1$, we apply Kummer's test with $D_n = n \ln n$. Now, $\sum \frac{1}{D_n}$ is divergent (**WHY**?); hence, we have only to show that

$$\lim \left(D_n \frac{a_n}{a_{n+1}} - D_{n+1} \right) < 0.$$

Using the hypothesis regarding the ratio $\frac{a_n}{a_{n+1}}$, we can rewrite the expression above to be

$$n \ln n \left(1 + \frac{1}{n} + \frac{b_n}{n^2} \right) - (n+1) \ln(n+1).$$

We leave it to the reader to show (Exercise 9) that the limit of this expression is in fact, -1, whence the conclusion follows by Kummer's test. $\square$

Discussion. Gauss's test is very useful in practice. Unlike the Ratio, Root, or Raabe's test, this test gives conclusions even when the limit $k = 1$. Gauss's test should be tried whenever the Ratio and Raabe's tests fail. This is particularly true when the expression $\frac{a_n}{a_{n+1}}$ is a ratio of polynomials, since the required expression in $\frac{1}{n}$ may be very easy to obtain. This is illustrated in our last example. $\square$

Example 6. Test the convergence of the series whose nth term is

$$a_n = \frac{1}{5^n n!} \cdot \frac{(1 \cdot 2)(6 \cdot 7)(11 \cdot 12) \cdots ((5n - 4) \cdot (5n - 3))}{3 \cdot 8 \cdots (5n - 2)}.$$

Solution. First, we observe both Ratio test and Raabe's test fail. We apply Gauss's test. Now

$$\frac{a_n}{a_{n+1}} = \frac{5(n+1)(5n+3)}{(5n+1)(5n+2)} = 1 + \frac{1}{n} + \frac{b_n}{n^2},$$

where

$$b_n = \frac{-(1 + \frac{1}{n})}{1 + \frac{15}{2n} + \frac{1}{n^2}}. \qquad (\textbf{HOW}?)$$

Also, $\lim b_n = -1$, the sequence $\{b_n\}$ is bounded. Since k, the coefficient of $\frac{1}{n}$ is 1, the series diverges. $\square$

Discussion. The reader may wonder how the expression for b_n on the second equation was generated. The answer is simple, long division, which once again reminds us about the power of simple algebraic techniques. $\square$

Exercises

1. Let $\sum (-1)^{n-1} a_n$ be a series that satisfies the hypotheses of Leibnitz's test. Show that $s_{2n-1} \geq s_j \geq s_{2n}$ for $j > 2n \geq 2$.

2. Let $\{a_n\}$, $\{b_n\}$ be two arbitrary sequences with $s_n = \sum_{i=1}^{n} a_i$. Show for all n,

$$\sum_{i=1}^{n} a_i b_i = s_n b_{n+1} - \sum_{i=1}^{n} s_i (b_{i+1} - b_i).$$

3. Complete the details in the proof of Abel's test.

4. Complete the details of Example 2.

5. Give a careful derivation of the inequality that forms the basis of the proof of case (i) of the Ratio test.

6. Check that the series $\sum \frac{1}{n}$ and $\sum \frac{1}{n^2}$ both have the property asserted in the proof of the Ratio and Root tests.

7. Complete the solution of Example 4.

8. Prove Raabe's test.

9. Complete the proof of Gauss's test.

10. Show that Gauss's test remains true when the the expression for $\frac{a_n}{a_{n+1}}$ is $1 + \frac{k}{n} + \frac{b_n}{n^{1+\lambda}}$, for some $\lambda > 0$.

11. Let $\sum a_n$ be a convergent series and $\{b_n\}$ be a sequence.

    (a) Show that if $\{b_n\}$ is bounded, then $\sum a_n b_n$ will not necessarily converge.

    (b) Show that if $\{b_n\}$ is Cauchy, then $\sum a_n b_n$ will not necessarily converge.

    (c) Is there any condition other than monotonicity, which when added to the Cauchy condition will guarantee convergence?

12. Let $\sum a_n$ be a convergent series and $\{b_n\}$ a monotonic bounded sequence. Show $\sum a_n b_n$ is convergent.

13. Verify the inequalities that lead to the conclusion that the Root test is a more sensitive test than the Ratio test (see discussion following the root test).

14. Show if $\sum a_n^2$ is convergent, then $\sum \frac{|a_n|}{n}$ converges.

15. Give a careful proof (independent of Kummer's test) of Raabe's test.

16. Verify the calculations in Example 3.

17. Verify the calculations in Example 4.

18. Test for convergence and absolute convergence the series whose nth term is given by

(a) $\frac{\ln n}{n}$;

(b) $\frac{\ln n}{n^2}$;

(c) $\ln \frac{n}{n+1}$;

(d) $\frac{(-1)^{n-1}}{n^{1+\frac{1}{n}}}$;

(e) $\frac{(-1)^n}{\ln(n+1)}$;

(f) $\frac{1}{\ln(n+1)^p}$, $p \geq 1$;

(g) $\frac{n^3 + 2^n}{2^n n^3}$;

(h) $\left(\frac{n}{1+n^2}\right)^n$;

(i) $\frac{\sqrt{n+1} - \sqrt{n}}{n^p}$, $p \geq 1$;

(j) $\frac{2^n + n^{1000}}{3^n + 1}$;

(k) $\frac{2 \cdot 4 \cdot 6 \cdots 2n}{1 \cdot 3 \cdot 5 \cdots (2n+1)}$;

(l) $\frac{n! e^n}{n^n}$;

(m) $(n^{\frac{1}{n}} - 1)^n$;

(n) $\frac{(-1)^n}{[\ln(n+1)]^{\frac{1}{n}}}$;

(o) $\frac{n^5 (n+1)^n}{3 n^n}$;

(p) $\frac{1 \cdot 3 \cdot 5 \cdots (2n-1)}{2 \cdot 4 \cdot 6 \cdots 2n} \frac{1}{n}$;

(q) $\left[\frac{1 \cdot 3 \cdot 5 \cdots (2n-1)}{2 \cdot 4 \cdot 6 \cdots 2n}\right]^p$, $p \geq 1$;

(r) $\frac{n!}{3 \cdot 5 \cdot 7 \cdots (2n+1)}$;

(s) $\frac{(n!)^2 100^n}{3 n^2}$;

(t) $(-1)^n \left[\frac{1 \cdot 3 \cdot 5 \cdots (2n-1)}{2 \cdot 4 \cdot 6 \cdots 2n}\right]^p$, $p \geq 1$;

(u) $\frac{e^n n!}{3 \cdot 5 \cdot 7 \cdots (2n-1)}$;

(v) $\frac{2^n + 1}{3^n + 2n}$;

(w) $\frac{1}{x^n - y^n}$ $x > y > 0$;

(x) $\frac{1}{a^n + 1}$, $a > -1$;

(y) $\frac{(-1)^{n-1} n^3}{2^n - 1}$;

(z) $\frac{\sin n}{n}$;

(a') $(-1)^n \frac{|\sin n|}{n}$;

(b') $\frac{|\sin n|}{n}$;

(c') $(-1)^n \frac{\sin(\frac{n\pi}{2})}{n}$;

(d') $\frac{(2n!)^3}{2^{6n} (n!)^6}$.

19. Test for convergence the series defined by

$$1 + 2r + r^2 + 2r^3 + r^4 + 2r^5 + r^6 + 2r^7 + \cdots,$$

where $r \in \mathbf{R}$ is arbitrary.

20. Discuss the convergence of the hypergeometric series given by

$$1 + \frac{ab}{cd} + \frac{a(a+1)b(b+1)}{c(c+1)d(d+1)} + \frac{a(a+1)(a+2)b(b+1)(b+2)}{c(c+1)(c+2)d(d+1)(d+2)} + \cdots,$$

where a, b, c, $d \in \mathbf{R}^+$.

21. Let $a, b, c > 0$ with $b < a$. Test for convergence:

$$\frac{a}{b} + \frac{a(a+c)}{b(b+c)} + \frac{a(a+c)(a+2c)}{b(b+c)(b+2c)} + \cdots.$$

22. Test for convergence:

(a) $a + 1 + a^3 + a^2 + a^5 + a^4 + a^7 + \cdots$, $a \in \mathbf{R}$;

(b) $1 - \frac{1}{3 \cdot 2^2} + \frac{1}{5 \cdot 3^2} - \frac{1}{7 \cdot 4^2} + \cdots$;

(c) $0 - \frac{1}{2} + \frac{1}{2^2} - \frac{1}{3} + \frac{2}{3^2} - \frac{1}{4} + \frac{3}{4^2} - \cdots$.

23. Let $\sum (-1)^{n-1} a_n$ satisfy the hypothesis of Leibnitz's test. If s_n denotes the nth partial sum of the series and s the sum, find a bound for $|s_n - s|$.

24. The series

$$1 - \frac{1}{2^2} + \frac{1}{3} - \frac{1}{4^2} + \frac{1}{5} - \frac{1}{6^2} + \cdots$$

is divergent. Does this contradict Leibnitz's test?

25. Consider the series $\sum \frac{(-1)^n}{n}$. For what values of n will s_n be accurate to six decimal places?

26. Let $\{a_n\}$ be a sequence of terms that alternate signs and such that $|a_n| \to 0$ monotonically. Let $\sum b_n$ be formed by taking p positive terms from a_n followed by q negative terms and continuing this pattern. Is $\sum b_n$ convergent?

27. If the alternating series $\sum (-1)^{n-1} a_n$ converges by applying Leibnitz's test, with sum A, show there exists n_0 such that for $n \geq n_0$,

(a) $|A - s_n| < |a_{n+1}|$; (b) $|A - s_n - \frac{1}{2} a_{n+1}| < \frac{1}{2} |a_{n+1}|$.

Use this to compute the sum of $\sum_{n=1}^{\infty} \frac{(-1)^{n+1}}{n^4}$ to three decimal places. Moreover, by considering the series

$$1 - \frac{1}{2} + \frac{1}{3^2} - \frac{1}{2^2} + \frac{1}{3^4} - \frac{1}{2^4} + \cdots$$

demonstrate that if the series converges, but Leibnitz's test does not work, then the above conclusions are not valid.

28. Prove **logarithmic test**: Let $\sum a_n$ be a series of nonnegative terms, and let

$$L_n = \frac{\ln \frac{1}{a_n}}{\ln n}.$$

The series converges if $\underline{\lim} L_n > 1$ and diverges if $L_n \leq 1$ for all n. Use this test to discuss the convergence or divergence of $\sum x^{\ln n}$ and $\sum x^{\ln \ln n}$.

29. Prove **second logarithmic test**: Let $\sum a_n$ be a series of nonnegative terms, and let
$$M_n = \frac{\ln \frac{1}{n a_n}}{\ln \ln n}.$$
The series converges if $\varliminf M_n > 1$ and diverges if $M_n \leq 1$ for all n. Use this test to discuss the convergence of $\sum \frac{1}{n^{1+\frac{1}{\ln n}}}$.

30. Prove **Bertrand's test**: If $a_n \geq 0$ and $B_n = (n-1)\ln n - \frac{a_{n+1}}{a_n} n \ln n$, then $\sum a_n$ converges if $\varliminf B_n > 1$ and diverges if $\varlimsup B_n < 1$.

31. Show that there is no universal sequence $\{D_n\}$ that can be used to test all series of positive terms by the conditions given in Kummer's test.

Chapter 7

Sequences of Functions

In the previous chapter, we studied the notion of the convergence of an infinite series whose terms are real numbers and discussed the concept of the 'sum' of an infinite series of constant terms. In this chapter, we shall see that we can also consider infinite sequences and series whose terms are not mere constants, but real-valued functions defined on certain subsets of $\mathbf{R}$. Thus, we shall study the behavior of a sequence $\{f_n\}$ or an infinite series $\sum f_n$, where the f_n's are functions defined on suitable subsets of real numbers. We shall discuss two types of convergence for sequences of functions, namely, 'pointwise convergence' and 'uniform convergence', respectively. Pointwise convergence is a natural extension of previous convergence concepts. But as we shall see, the pointwise limit function corresponding to a sequence of functions does not, in general, inherit the 'nice' properties of the individual functions comprising the sequence. We would like the limit function to share any nice properties common to the functions in the sequence, such as continuity, differentiability, and integrability. To achieve this end, pointwise convergence is not the answer, and we are led instead to the notion of uniform convergence of a sequence of functions. We shall study both these properties in detail, and see how uniform convergence of sequences of functions guarantees that all the nice properties possessed by the individual members are also shared by the limit function. In the end, we shall see that in the presence of uniform convergence, certain limiting operations can be interchanged freely without affecting any of the limit properties. The chapter concludes with some consequences of uniform convergence.

7.1 Pointwise Convergence of a Sequence of Functions

In Chapter 1, a thorough study of the convergence of a sequence $\{a_n\}$ of real numbers was presented. Such sequences, whose members are real numbers will be referred to as **sequences of constant terms** and it is essential that the reader be totally familiar with all the theory of such sequences prior to attempting this chapter.

Definition. A **sequence of functions defined on** A is a function f, having domain $\mathbf{N}$ and whose range is a subset of the collection of all functions having domain A and range a subset of $\mathbf{R}$. If f is a sequence of functions on A, we will denote $f(n)$ by f_n and denote the sequence itself by $\{f_n\}$.

Discussion. The definition of a sequence of functions as a function having domain $\mathbf{N}$ and range in the collection of functions from A into $\mathbf{R}$ is the formal concept. However, the focus of the concept is in the denotation of these sequences, $\{f_n\}$, which emphasizes the nature of the sequence as a collection of functions ordered by $\mathbf{N}$. Thus, each **term** of the sequence is a function having domain A and range a subset of $\mathbf{R}$.

As the reader may have guessed, our intention in discussing the convergence of such sequences is to build on previous concepts and knowledge. Thus, if we now fix x such that $x \in A$, this creates a sequence of constant terms, $\{f_n(x)\}$. Since $\{f_n(x)\}$ is a sequence of real numbers, we are in a position to apply all of our previously developed theory to the present task.

It should also be clear why we require the various functions, f_n, to have a common domain, A. Were this not the case, for a particular k, $x \in A$ might fail to be in the domain of f_k, whence some terms of the sequence, $\{f_n(x)\}$, would fail to exist. This would create needless difficulties, and for this reason, we have given a definition that avoids the problem. Thus, we stress that a sequence of functions defined on A refers to a collection of functions that all share the same domain, namely, A. □

Definition. The sequence $\{f_n\}$ of functions defined on $A \subseteq \mathbf{R}$ is said to **converge pointwise** on A, provided there exists a function f having domain A, and for each $x \in A$ and every $\epsilon > 0$, there exists an $N(x) \in \mathbf{N}$ such that

$$| f_n(x) - f(x) | < \epsilon$$

whenever $n > N(x)$. In case a function, f, exists with the required properties, we say f is the **pointwise limit** of the sequence $\{f_n\}$ of functions on the set A, and we write $f_n \to f$ (**pointwise**) on A.

Discussion. The set A is a common domain for all the functions f_n in the sequence and each point x in this common domain, A, manufactures a sequence, $\{f_n(x)\}$, of constant terms. If for each $x \in A$ this sequence is a convergent sequence of real numbers in the usual sense, then a unique limit exists, and it makes sense to call this limit $f(x)$. Indeed, the function, f, whose existence is asserted in the definition is completely defined by the equation

$$f(x) \equiv_{df} \lim_{n \to \infty} f_n(x).$$

This equation captures the essence of pointwise convergence, since it requires convergence of each of the individual sequences, $\{f_n(x)\}$, x fixed, and defines the limit function in terms of the pointwise limit at x. We use $N(x)$, rather than the simpler $N \in \mathbf{N}$, to emphasize that the value that will enable the definition to be satisfied depends on the point x. Since this value also depends on the given ϵ, we should have used $N(x, \epsilon)$. We refrained since all readers are well aware of the dependence of $N(x)$ on ϵ.

If the definition of pointwise convergence for a sequence of functions is compared with the definition of convergence for a sequence of constants, the similarities are striking. Specifically,

(i) Convergent sequences of constants require the existence of a real number, L; convergent sequences of functions require the existence of a function, f, on the common domain A;

(ii) Convergent sequences of constants require that to each positive ϵ there exits $N \in \mathbf{N}$; convergent sequences of functions require that to each positive ϵ and each $x \in A$ there exits $N(x) \in \mathbf{N}$;

(iii) Convergent sequences of constants require that

$$|a_n - L| < \epsilon \text{ whenever } n > N;$$

convergent sequences of functions require that

$$|f_n(x) - f(x)| < \epsilon \text{ whenever } n > N(x).$$

The first key difference relates to the fact that $N(x)$ depends on the value of x as well as on the value of ϵ. The second difference is that convergence of the sequence $\{f_n(x)\}$ must be validated for each x in A. This is why the latter type of convergence is said to be pointwise on A. $\square$

We give one essential theorem and then illustrate pointwise convergence concept by means of some concrete examples.

Theorem 7.1.1. *Let $\{f_n\}$ be a sequence of functions on A, which converges pointwise on A. Then the pointwise limit, f, of the sequence is unique.*

Proof. The proof of this theorem is implicit in the discussion above. We leave the generation of the formal details to the reader as Exercise 1. $\square$

Example 1. Discuss the pointwise convergence of the sequence of functions, $\{f_n\}$, where $f_n : (0, 1] \to \mathbf{R}$ is defined by

$$f_n(x) = \frac{1}{nx}, \ x \in (0, 1], \ n \in \mathbf{N}.$$

Solution. Fix $\epsilon > 0$ and $x_0 \in (0, 1]$. Now, $f_n(x_0) = \frac{1}{nx_0}$, whence choose $N(x_0)$ such that $N(x_0) > \frac{1}{\epsilon x_0}$. It follows that if $n > N(x_0)$, then

$$|f_n(x_0) - 0| = \left| \frac{1}{nx_0} - 0 \right| < \left| \frac{1}{N(x_0)x_0} \right| < \left| \frac{\epsilon x_0}{x_0} \right| = \epsilon.$$

Thus $\lim_{n \to \infty} f_n(x_0) = 0$ for each $x_0 \in (0, 1]$. We conclude that the sequence $\{f_n\}$ converges pointwise to the function, f, defined by $f(x) = 0$ for all $x \in (0, 1]$. $\square$

Discussion. The reader should study this argument from the point of view of examining how the difficulty of constructing an argument that will deal with all of the various sequences that can arise for the different values of $x \in (0, 1]$. The difficulty is eliminated by picking an arbitrary $x_0 \in (0, 1]$ and then proceeding with the argument while keeping x_0 fixed. It is critical that the reader understand this technique, since all arguments in this chapter will employ it in some way. It should not be totally unfamiliar to the reader, since similar arguments were used when dealing with uniform continuity and the reader may choose to review some of those arguments as an aid in understanding these.

The crucial fact used above is that no matter how close we take x_0 to 0, $\lim_{n \to \infty} \frac{1}{n x_0} = 0$. This is because x_0 is fixed, while n is increasing without bound. The existence of a unique limit for each x_0 permits the definition of f in terms of this limit. Thus, we set $f(x) = 0$ on $(0, 1]$ and write $f_n \to 0$ pointwise on $(0, 1]$. It is to be clearly understood that the 0 appearing on the right side of the arrow is not the number 0, but is to be interpreted as the function having domain $(0, 1]$ and satisfying $0(x) = 0$ for all $x \in (0, 1]$.

In Figure 7.1.1 we present the graphs of several of the f_n's. We have fixed x_0 and graphed the line $x = x_0$. Notice that the sequence $\{f_n(x_0)\}$ can be pictured as a sequence of real numbers lying on the line determined by $x = x_0$. The individual members of the sequence are the points of intersection between the graph of particular f_n's and the graph of $x = x_0$. In this example, these points on the line $x = x_0$ clearly converge to a point on the x-axis. The reader may find that looking at pointwise convergence from this geometric perspective increases his insight into the topic. $\square$

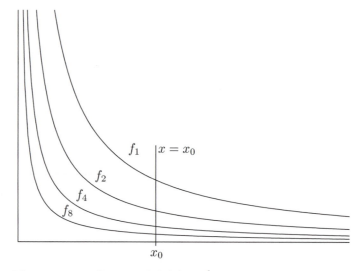

Figure 7.1.1 Graphs of $f_n(x) = \frac{1}{nx}$, for $n = 1$, 2, 4, and 8.

Example 2. Discuss the pointwise convergence of the sequence of functions, $\{f_n\}$, where $f_n : [0, 1] \to \mathbf{R}$ is defined by

$$f_n(x) = \frac{x}{n}, \ x \in [0, 1], \ n \in \mathbf{N}.$$

Solution. Fix $\epsilon > 0$ and $x_0 \in [0, 1]$. In this case $f_n(x_0) = \frac{x_0}{n}$, whence set

$N(x_0) = \frac{1}{\epsilon}$. It follows that if $n > N(x_0)$, then

$$|f_n(x_0) - 0| = \left| \frac{x_0}{n} - 0 \right| \le \left| \frac{x_0}{N(x_0)} \right| \le \left| \frac{1}{N(x_0)} \right| = \epsilon.$$

Once again $\lim_{n \to \infty} f_n(x_0) = 0$ for each $x_0 \in [0, 1]$. We conclude that the sequence $\{f_n\}$ converges pointwise to the function, f, defined by $f(x) = 0$ for all $x \in [0, 1]$. $\square$

Discussion. In this example, there is an apparent lack of dependence of $N(x)$ on the choice of x. It is important that the reader understand that this results from the fact that x occurs in the numerator of $f_n(x) = \frac{x}{n}$ and that $0 \le x \le 1$. For this reason, it is natural to make use of the inequality, $f_n(x) \le \frac{1}{n}$, which eliminates the dependence on x. However, for a particular x, say $x = \frac{1}{2}$, this results in a choice of $N(x)$ that is larger than necessary. Last, the reader should compare the computations in Example 2 with those in Example 1 to see if it is possible to eliminate the dependence of $N(x)$, in Example 1, on x. $\square$

Example 3. Discuss the pointwise convergence of the sequence, $\{f_n\}$, where $f_n : [0, 1] \to \mathbf{R}$ is defined by $f_n(x) = x^n$, $x \in [0, 1]$, $n \in \mathbf{N}$.

Solution. If $x = 1$, then for each n, we have $f_n(1) = 1^n = 1$, so the sequence, $\{f_n(1)\}$, has all its terms equal to 1, and hence the limit is also 1. On the other hand, if $0 \le x < 1$, we know that $f_n(x) = x^n$, which has limit 0. Hence, for each $x < 1$, the sequence $\{f_n(x)\}$ converges to 0. Thus, in either case, we have that $\{f_n(x)\}$ is a convergent sequence of real numbers. We therefore conclude that the sequence, $\{f_n\}$, converges pointwise to the function f defined on $[0, 1]$ by

$$f(x) = \begin{cases} 1, & \text{if } 0 \le x < 1 \\ 0, & \text{if } x = 1. \end{cases} \square$$

Discussion. The most striking feature of this example is that each of the functions, f_n, is continuous, indeed, uniformly continuous, the sequence of functions converges pointwise to a limit, and yet the limit function is not continuous! It is one of the simplest examples demonstrating that pointwise convergence is the wrong tool for ensuring that a nice property, shared by all members of a sequence of functions, should be shared by the limit function, as well. Figure 7.1.2 presents the geometry of this example and should permit the reader to see why this sequence results in a discontinuous limit.

The fact that the limit f is discontinuous is evident in the argument for the existence of the limit. Its presence is in the fact that two arguments must be presented. The first is for the case $x = 1$, and the second for the case $0 \le x < 1$. For these two cases, the limits of $\{x^n\}$ were, respectively, 1 and 0. Thus, the resulting pointwise limit function, f, takes the value 0 throughout the interval [0,1) while at $x = 1$, $f(1) = 1$.

The argument presented was not an $\epsilon - N(x)$ argument, but depended heavily on previously established facts, which are presumed to be well known. Nevertheless, it is essential that the reader be able to write out $\epsilon - N(x)$ arguments, and we ask the reader to generate such an argument for this example in Exercise 2. $\square$

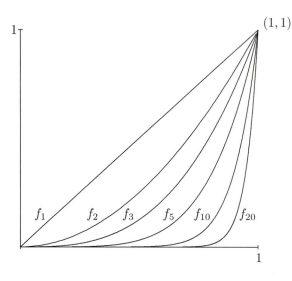

Figure 7.1.2 Graphs of $f_n(x) = x^n$, for $n = 1,\ 2,\ 3,\ 5,\ 10,\ 20$, in the interval $0 \le x \le 1$.

Example 4. Discuss the pointwise convergence of the sequence $\{f_n\}$, where f_n is defined on $\mathbf{R}$ by

$$f_n(x) = \begin{cases} 1, & \text{if } -n \le x \le n \\ 0, & \text{otherwise.} \end{cases}$$

Solution. Define f on $\mathbf{R}$ by $f(x) = 1$ for all $x \in \mathbf{R}$. For any fixed x_0 and positive ϵ, let $N = N(x_0)$ be chosen such that $N > |x_0|$. If $n > N$, we have

$$|f_n(x_0) - f(x_0)| = |f_N(x_0) - 1| = |1 - 1| = 0 < \epsilon,$$

whence $\{f_n(x_0)\}$ converges to $f(x_0) = 1$. Thus, $f_n \to f$ pointwise on $\mathbf{R}$. $\square$

Discussion. Note that f_N vanishes only outside $[-N, N]$ and takes the value 1 in that interval. Thus, the N manufactured in the above solution depends very much on the point x_0 in $\mathbf{R}$. What we observe here intuitively is that as n gets larger, the functions, f_n, assume the value 1 over larger intervals, and so ultimately the pointwise limit becomes the function that is identically 1 on all of $\mathbf{R}$.

Curiously enough, even though each f_n's was a discontinuous function on $\mathbf{R}$, the pointwise limit function is a continuous function (see Figure 7.1.3). $\square$

Example 5. Discuss the pointwise convergence of the sequence $f_n : [0, 1] \to \mathbf{R}$ be defined by

$$f_n(x) = \begin{cases} n, & \text{if } x \in (0, \frac{1}{n}) \\ 0, & \text{otherwise.} \end{cases}$$

Solution. Fix $x \in (0, 1]$. If $x = 0$, then $\{f_n(0)\}$ is a constant sequence consisting of 0's, and so converges. Thus, fix x such that $0 < x \le 1$. Then there exists $N(x)$ such that $\frac{1}{N(x)} < x$. It is immediate that

$$f_{N(x)}(x) = f_{N(x)+1}(x) = \cdots = f_{N(x)+k}(x) = \cdots = 0, \ (k \in \mathbf{N}),$$

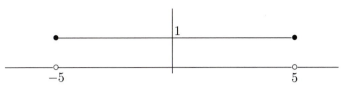

Figure 7.1.3 The function, f_5, from the sequence of functions in Example 4.

whence $\lim_{n \to \infty} f_n(x) = 0$. Thus, for each $x \in [0, 1]$, $\{f_n(x)\}$ converges; hence, $\{f_n\}$ converges pointwise to the function, f, which is identically 0 on $[0, 1]$. $\square$

Discussion. Again we stress that the $N(x)$ that is chosen depends on the particular point x in $(0, 1]$. As a result, $f_k(x) = 0$ only for those subscripts, k, satisfying $x > \frac{1}{k}$.

An important feature of this example becomes evident if we compute $\int_0^1 f_n$ for each n. As the reader can check, the value of any of these integrals is 1, whereas the integral of the pointwise limit function is $\int_0^1 f = \int_0^1 0 = 0$. Consequently, we have the unpleasant situation that even though all limits exist,

$$\lim_{n \to \infty} \left(\int_0^1 f_n \right) \neq \int_0^1 (\lim_{n \to \infty} f_n),$$

where the limit on the right side refers to the pointwise limit. This example employs the canonical convergent sequence of integrable functions, for which it is impossible to interchange the order in which the limits are computed. That is, the limit of the integrals, as a convergent sequence of real numbers, is not equal to the integral of the limit function, as a real number.

The reader is urged to draw the graphs of the various f_n's and the limit function f on the same set of axes so as to more completely understand this example. $\square$

Examples 3 and 5 illustrate major difficulties with the pointwise convergence concept. In the exercises, there are additional examples where the pointwise limit fails to have some property that is shared by all the functions in the sequence. For this reason, we are forced to seek a more powerful notion of convergence for sequences of functions, which will ensure that properties shared by all the members of a sequence will be shared by the limit function as well. This will be accomplished in Section 7.2.

We close this section with a discussion of the negation of pointwise convergence.

Negation of the Definition of Pointwise Convergence. Let $\{f_n\}$ be a sequence of functions defined on A. Then $\{f_n\}$ **fails to converge (pointwise)** on A provided for every function f having domain A, there exists $x \in A$ and $\epsilon > 0$ such that for all $N \in \mathbf{N}$ there exists $n > N$ such that

$$| f_n(x) - f(x) | \geq \epsilon.$$

Discussion. We want to extract the essence of this negation. First, consider that the negation requires that we examine every function f defined on A. Since if the limit of a given sequence exists, it is unique, a given convergent sequence of functions, $\{f_n\}$, can fail to converge to a given candidate, say, g, because g is not the unique limit function. Quite apparently, most candidate functions will not be the right limit function, and their failure will reflect this fact rather than the fact that the sequence doesn't converge. (To make this concrete, consider the sequence of Example 1 and the candidate limit function, g, on the interval $[0, 1]$ defined by $g(x) = 1$.) Thus, we want to make sure that the failure to converge is due to the properties of $\{f_n\}$ and not due to a property of the candidate limit function.

Recall that if the limit of the sequence of functions exists, then we have a surefire method for finding the correct candidate for the limit, namely, the limit function is given by

$$f(x) = \lim_{n \to \infty} f_n(x).$$

From this we see that the limit exists exactly if each of the sequences, $\{f_n(x)\}$, converges for each $x \in A$. Thus, the sequence, $\{f_n\}$, will fail to converge, exactly if we can find a particular $x \in A$ such that the sequence of real numbers, $\{f_n(x)\}$, fails to converge.

In Chapter 1, much effort was spent determining why a sequence of real numbers could fail to converge. Two reasons were found. First, the sequence was unbounded. Second, the sequence oscillated. Thus, a sequence of functions will fail to converge pointwise on A, exactly if there is a particular $x \in A$ such that the sequence, $\{f_n(x)\}$, is either unbounded or oscillates. This is exactly what the above negation expresses. $\square$

Example 6. Discuss the pointwise convergence properties of the sequence of functions defined on $[-1, 1]$ by

$$f_n(x) = \begin{cases} n, & \text{if } x \in (-\frac{1}{n}, \frac{1}{n}) \\ 0, & \text{otherwise.} \end{cases}$$

Solution. The reader can check, as shown in Example 5, that if $x \neq 0$, then $\lim_{n \to \infty} f_n(x) = 0$. However, $f_n(0) = n$, for every $n \in \mathbf{N}$, so that $\{f_n(0)\}$ is an unbounded sequence of real numbers, and $\{f_n\}$ is not convergent on $[-1, 1]$ pointwise. $\square$

Discussion. We have established lack of convergence by finding a divergent sequence of real numbers. We have not established the details as specified in the statement of Negation above. The reader will be asked to do this as Exercise 4.

It is suggested that the reader carefully compare Examples 5 and 6. They employ very similar sequences, with rather different results. $\square$

In previous chapters when a new notion of limiting process has been introduced, a number of standard questions have arisen. Basically, these questions related to how the limiting process behaved with respect to the standard arithmetic operations on the reals, or in those cases where the limiting process dealt with functions, the arithmetic of functions. Similar questions arise with respect to pointwise convergence of sequences of functions.

Consider then two sequences of functions, $\{f_n\}$ and $\{g_n\}$, that are defined on a common domain A. Clearly, we can create a new sequence of functions by employing one of the operations for combining functions, $\{f_n + g_n\}$, for example. Obviously, one would like to have some standard theorems that deal with these questions. The next theorem is one such. Because the proof is generally straightforward, it is left to the exercises. As well, in the exercises the reader will be asked to work out some of the theory.

Theorem 7.1.2. *Let $\{f_n\}$ and $\{g_n\}$ be two sequences of functions defined and pointwise convergent on a common domain A. Then $\{f_n \pm g_n\}$, $\{f_n \cdot g_n\}$, and $\{|f_n|\}$ are pointwise convergent on A.*

Proof. Exercise 8. $\square$

The reader should recall that bounded, monotone sequences have very nice properties, after all, we have invoked Theorem 1.3.1 with great frequency. As a result, the reader may guess that bounded, monotone sequences of functions are also well behaved.

Theorem 7.1.3. *Let $\{f_n\}$ be a sequence of functions on a common domain, A, and suppose that $f_n \leq f_{n+1}$ for all $n \in \mathbf{N}$. If, for each $x \in A$, there exists $K_x \in \mathbf{R}$ such that $|f_n(x)| < K_x$, then $\{f_n\}$ converges pointwise on A.*

Proof. Exercise 15. $\square$

Exercises

1. Write out the details of the proof of Theorem 7.1.1.

2. Give an $\epsilon - N(x)$ argument to establish the properties of the sequence of functions in Example 3.

3. Discuss the pointwise convergence of the following sequences $\{f_n\}$ of functions on the given domains[1] and illustrate graphically wherever possible:

   (a) $f_n(x) = 1 - \frac{1}{nx}$, $x \in (0, 1]$;　　(b) $f_n(x) = \frac{\sin nx}{\sqrt{n}}$, $x \in [0, 1]$;

   (c) $f_n(x) = (\sin x)^n$, $x \in [0, \pi]$;　　(d) $f_n(x) = \frac{x}{n}$, $x \in [0, \infty)$;

   (e) $f_n(x) = \frac{nx}{1+nx}$, $x \in \mathbf{R}$;　　(f) $f_n(x) = \frac{nx}{1+n^2x^2}$, $x \in \mathbf{R}$;

   (g) $f_n(x) = \frac{x^{2n}}{1+x^{2n}}$, $x \in \mathbf{R}$;　　(h) $f_n(x) = n^2x^n(1-x)$, $x \in [0, 1]$;

   (i) $f_n(x) = \frac{x^2+nx}{n}$, $x \in \mathbf{R}$;　　(j) $f_n(x) = (1 - |x|)^n$, $x \in (-1, 1)$;

   (k) $f_n(x) = \frac{xe^{-\frac{x}{n}}}{n}$, $x \in [0, \infty)$;　　(l) $f_n(x) = nxe^{-n^2x^2}$, $x \in \mathbf{R}$;

   (m) $f_n(x) = \begin{cases} 1, & \text{if } x = \frac{p}{q}, \ (p, q) = 1, \text{ with } p + q = n \\ 0, & \text{otherwise}; \end{cases}$　Dmn $f_n = [0, 1]$;

   (n) $\chi_n(x) = \begin{cases} 1, & \text{if } x \in [-n, n] \\ 0, & \text{otherwise}. \end{cases}$　Dmn $\chi_n = \mathbf{R}$;

   The function χ_n is called the **characteristic function** of the closed interval, $[-n, n]$.

---

[1] In all instances for the domain of f_n, take the largest subset of the indicated domain set.

4. Give a detailed $\epsilon - N(x)$ argument showing that the sequence in Example 6 fails to converge on $[-1, 1]$.

5. Let $\{r_n\}$ be a sequence that contains each rational number in $[0,1]$ exactly once and let $f_n : [0, 1] \to \mathbf{R}$ defined by

$$f_n(x) = \begin{cases} 0, & \text{if } x \text{ is irrational, or if } x = r_i, \, i > n \\ 1, & \text{if } x = r_i, \, i \leq n. \end{cases}$$

Discuss the pointwise convergence of the sequence $\{f_n\}$. For what values of x does the infinite series $\sum f_n(x)$ converge?

6. Give examples (different from the ones given in this section) in support of the following statements:

   (a) each f_n is integrable, but the pointwise limit is not integrable;

   (b) each f_n is bounded, but the pointwise limit is unbounded;

   (c) each f_n is continuous on $[a, b]$ and differentiable on (a, b), and the pointwise limit is continuous on $[a, b]$ but not differentiable on (a, b);

   (d) each f_n is uniformly continuous on A, but the pointwise limit is not uniformly continuous on A;

   (e) $\int_a^b (\lim_{n \to \infty} f_n) \neq \lim_{n \to \infty} \int_a^b f_n$;

   (f) $\int_a^b (\lim f_n') \neq (\lim f_n)'$.

7. For each x in the domain A of a sequence $\{f_n\}$ of functions, the sequence $\{f_n(x)\}$ has an upper and lower limit point in $\mathbf{R}^* = \mathbf{R} \cup \{\infty\}$. Define $\overline{\lim} f_n$ to be the function defined by $\overline{\lim} f_n(x)$ and a similar definition for lower limits. Prove

   (a) $\underline{\lim} f_n \leq \overline{\lim} f_n$;   (b) $-\underline{\lim} f_n = \overline{\lim} (-f_n)$;

   (c) $\underline{\lim} f_n + \underline{\lim} g_n \leq \underline{\lim} (f_n + g_n)$.

   What is the corresponding relation to that in (c) for upper limits? Prove it.

8. If $f_n \to f$ and $g_n \to g$ (pointwise on A), prove the following:
   (a) $f_n \pm g_n \to f \pm g$ (pointwise on A);   (b) $f_n \cdot g_n \to f \cdot g$ (pointwise on A);
   (c) $|f_n| \to |f|$ (pointwise on A).   (d) What can you say about $\frac{f_n}{g_n}$?

9. Give an example where $|f_n|$ converges pointwise, but f_n does not have a pointwise limit.

10. Obtain each of the following functions as a pointwise limit of a sequence of functions, each member of which is continuous and bounded (differentiable in the case of (c)):

   (a) $f(x) = x, \, x \in \mathbf{R}$;   (b) $f(x) = x^2, \, x \in \mathbf{R}$;   (c) $f(x) = |x|, \, x \in [-1, 1]$;

   (d) $f(x) = \begin{cases} 0, & \text{if } x \in [0, 1] \sim \mathbf{Q} \\ \frac{1}{q}, & \text{if } x = \frac{p}{q} \in [0, 1] \cap \mathbf{Q} \text{ in lowest terms}.; \end{cases}$

   (e) $f(x) = \begin{cases} x, & \text{if } x \in [0, 1] \cap \mathbf{Q} \\ 1 - x, & \text{if } x \in [0, 1] \sim \mathbf{Q}. \end{cases}$

11. A sequence $\{f_n\}$ of functions is said to be **uniformly bounded** on a domain $D \subseteq \mathbf{R}$, provided there exists a positive number K such that $|f_n(x)| \leq K$ for each $n \in \mathbf{N}$ and for each $x \in D$. Determine whether each of the sequences in Exercise 3 is uniformly bounded.

12. Suppose $\{f_n\}$ is a sequence functions that is uniformly bounded on D. What can be said about the limit function?

13. Give an example of a sequence of continuous functions that is *not* uniformly bounded, but converges to a bounded function.

14. Suppose $f_n \to f$ pointwise on A and $f_n, f : A \to B$. Further, suppose $g_n \to g$ pointwise on B. What can be said about the convergence of $g_n \circ f_n$?

15. Prove Theorem 7.1.3.

16. Let $\{f_n\}$ satisfy the hypothesis of Theorem 7.1.3. Must the pointwise limit, f, be bounded above? Must the pointwise limit be bounded below? Must it be universally bounded?

17. State a Cauchy Criterion for pointwise convergence of a sequence of functions. Prove that this criterion is equivalent to the original definition of pointwise convergence.

18. A sequence, $\{f_n\}$, defined on $[a, b]$ is said to converge **in the mean** to a function, f, provided

$$\lim_{n \to \infty} \left\{ \int_a^b [f_n(x) - f(x)]^2 \right\} = 0.$$

Which of the following functions converge in the mean?

(a) $f_n(x) = 1 - \frac{1}{nx}$, $x \in (0, 1]$;

(b) $f_n(x) = \frac{x^{2n}}{1+x^{2n}}$, $x \in [0, 10]$;

(c) $f_n(x) = \frac{x}{n}$, $x \in [0, 10]$;

(d) $f_n(x) = (\sin x)^n$, $x \in [0, \pi]$;

(e) $f_n(x) = \frac{nx}{1+nx}$, $x \in [0, 10]$;

(f) $f_n(x) = \frac{nx}{1+n^2x^2}$, $x \in [0, 10]$;

(g) $f_n(x) = \frac{\sin nx}{\sqrt{n}}$, $x \in [0, 1]$;

(h) $f_n(x) = n^2 x^n (1 - x)$, $x \in [0, 1]$;

(i) $f_n(x) = \frac{xe^{-\frac{x}{n}}}{n}$, $x \in [0, 8]$;

(j) $f_n(x) = (1 - |x|)^n$, $x \in (-1, 1)$;

(k) $f_n(x) = \frac{x^2 + nx}{n}$. $x \in \mathbf{R}$;

(l) $f_n(x) = nxe^{-n^2x^2}$, $x \in [0, 2]$.

19. Give an example of a sequence of functions on $[0, 1]$ that converges in the mean to a function f but converges pointwise to a different function, g.

20. Discuss and compare the pointwise and mean convergence of the following sequences of functions:

(a) $f_n(x) = \cos^n(x)$, $x \in [0, \pi]$;

(b) $f_n(x) = n^{3/2} x e^{-n^2 x^2}$, $x \in [-1, 1]$;

(c) $f_n(x) = \begin{cases} n, & \text{if } x \in (0, \frac{1}{n}) \\ 0, & \text{if } x \in [0, 1] \sim (0, \frac{1}{n}). \end{cases}$

7.2 Uniform Convergence of Sequences of Functions

In the previous section, we studied the convergence of a sequence of functions by determining convergence at each point of the common domain. Thus, pointwise convergence could aptly be termed as a 'local property'. We also saw by means of several examples that this sort of convergence may destroy many pleasant properties possessed by all of the individual functions of the sequence. This defect will now be remedied by considering a new type of convergence, defined 'globally' on the entire domain under consideration, and called uniform convergence over a set.

Recall the sequence $\{f_n\}$ of functions converges pointwise to the function f provided for each x in the domain A, we have the relation: $\lim_{n\to\infty} f_n(x) = f(x)$. In other words, the necessary and sufficient condition for this to hold is that at each point, x, the constant sequence, $\{f_n(x)\}$, converges to the real number, $f(x)$; namely, given $\epsilon > 0$, at each point x, we can come up with an integer $N(x)$ (more precisely, $N(x, \epsilon)$), just to emphasize the fact that $N(x)$ depends on the ϵ given in advance, as well as the point x in question such that for all $n > N(x)$, we have $|f_n(x) - f(x)| < \epsilon$. The important thing to note here, as was stressed in the last section, is that the integer $N(x)$ is a function of x as well as ϵ, and different points x in the domain may well yield different integers $N(x)$ satisfying this inequality.

It would be very nice, if instead of having to find an $N(x)$ for each x, one could find a single $N = N(\epsilon)$, depending only on ϵ, that served the purpose 'uniformly' for all x in the domain. This is the property that we will insist on in generating the notion of uniform convergence. These observations lead to the following important definition.

Definition. A sequence $\{f_n\}$ of functions defined on a subset $A \subseteq \mathbf{R}$ is said to be **uniformly convergent on the set** A provided there exists a function, f, having domain A such that given $\epsilon > 0$ we can find an integer $N > 0$ such that

$$|f_n(x) - f(x)| < \epsilon, \text{ whenever } n > N \text{ and for all } x \in A.$$

In the case where such a function exists, we will say that f is the **uniform limit of the sequence, $\{f_n\}$, on** A and write $f_n \to f$ **(uniformly)** on A.

Discussion. We want to explore the distinction between pointwise and uniform convergence very carefully:

(i) pointwise convergence of sequences of functions requires the existence of a function, f, defined on A; uniform convergence of a sequence of functions also requires the existence of a function, f, on the common domain A;

(ii) pointwise convergence of sequences of functions requires that to each positive ϵ, and each $x \in A$, there exits $N(x) \in \mathbf{N}$; uniform convergence of a sequence of functions requires that to each positive ϵ, there exists an $N \in \mathbf{N}$, depending only on ϵ;

(iii) pointwise convergence of sequences of functions requires require that

$$|f_n(x) - f(x)| < \epsilon, \text{ whenever } n > N(x);$$

uniform convergence of a sequence of functions requires

$$\text{for all } x \in A, \ |f_n(x) - f(x)| < \epsilon, \text{ whenever } n > N.$$

Examination of the three points of comparison reveals that the significant difference in the two concepts is completely related to the nature of $N(x)$ as opposed to N. In the former case, $N(x)$ is permitted to depend on x; one for each x, whence pointwise. But N, in the uniform case, must be independent of x and must depend only on ϵ.

Within the definitions proper, this amounts to interchanging the order of quantification. Specifically, the 'pointwise' definition reads

'for each $\epsilon > 0$ and for each x, there exists $N(x)$.'

The 'uniform' definition reads

'for each $\epsilon > 0$, there exists $N \in \mathbf{N}$ such that for all x.'

The effect of interchanging the order of quantification in this manner is to eliminate the dependence of N on x. Thus, in the uniform case, once an ϵ has been given, one must search for an N that will force the inequality

$$|f_n(x) - f(x)| < \epsilon$$

to hold whenever $n > N$ and no matter what value of $x \in A$ is considered. There can be no exceptions!

As described, uniform convergence occurs in a 'global' way over an entire domain. Pointwise convergence occurs in a 'local' way, point by point throughout the domain. As a result, for some values of x, the rate of convergence can be arbitrarily slow, whereas for uniform convergence, there must be a minimum rate of convergence that holds globally over the whole domain.

It should be evident from the discussion above that a uniformly convergent sequence of functions will converge pointwise, as well. This fact, which we prove below, permits us to observe that if a sequence of functions converges uniformly, the natural candidate for the 'uniform' limit is none other than the 'pointwise-limit' function. In other words, the uniform limit, if it exists, cannot be different from the pointwise limit. $\square$

Theorem 7.2.1. *Let $\{f_n\}$ be a sequence of functions that converges uniformly on A. Then $\{f_n\}$ converges pointwise on A, and the pointwise limit is the uniform limit.*

Proof. Since $\{f_n\}$ converges uniformly on A, it has a limit on A, which we call f. Now let $\epsilon > 0$ be given, and fix $x_0 \in A$. By uniform convergence, there exists $N \in \mathbf{N}$ such that for all $x \in A$, $|f_n(x) - f(x)| < \epsilon$ whenever $n > N$. This implies $|f_n(x_0) - f(x_0)| < \epsilon$ whenever $n > N$, whence $\{f_n(x_0)\}$ converges to $f(x_0)$. Since x_0 was arbitrary, $f_n \to f$ pointwise. Since the pointwise limit is unique, the uniform limit must also be unique, whence the uniform limit and the pointwise limit coincide. $\square$

Discussion. All the details in the argument have been presented. The essence of the argument is in noticing that if N works for all the x's, it must work for each particular x. $\square$

We illustrate the concept with a simple example.

Example 1. Show that the sequence of functions, $\{f_n\}$, where $f_n : [0, 1] \to \mathbf{R}$ is defined by

$$f_n(x) = \frac{x}{n}, \ x \in [0, 1], \ n \in \mathbf{N}$$

is uniformly convergent on $[0, 1]$.

Solution. Let f be the function that is identically 0 on $[0, 1]$. Fix $\epsilon > 0$ and set $N = \frac{1}{\epsilon}$. It follows that if $n > N$, then

$$|f_n(x) - 0| = \left| \frac{x}{n} - 0 \right| \leq \frac{1}{n} \leq \frac{1}{N} = \epsilon,$$

whence $f_n \to f$ uniformly on $[0, 1]$. $\square$

Discussion. The sequence of functions presented in this example was first discussed as Example 7.1.2. In the discussion, the essence of why the sequence is uniformly convergent was presented, and the reader who does not fully recall that discussion should review it now. $\square$

Let us now consider the problem of why a sequence of functions on A can fail to converge uniformly.

Negation of Uniform Convergence. The sequence $\{f_n\}$ is **not uniformly convergent** on the set A, provided for every function, f, having domain A, there exists an $\epsilon > 0$ such that given any integer $N > 0$, there exists $x \in A$ and an $n > N$ such that $|f_n(x) - f(x)| \geq \epsilon$.

Discussion. Our intent is to identify the reasons why a sequence of functions can fail to converge uniformly. The contrapositive of Theorem 7.2.1 asserts that if a function fails to converge pointwise on A, then it cannot converge uniformly on A. We have already identified why a sequence of functions can fail to converge pointwise on A, namely, there must exist an $x \in A$ such that the sequence of real numbers, $\{f_n(x)\}$, fails to converge. These ideas, together with their meaning in terms of the $\epsilon - N$ definition were thoroughly discussed at the end of the last section, and the reader should review that discussion prior to continuing further with this one.

Given the above, in the remainder of this discussion we assume that the sequence of functions converges pointwise on A. Thus, in denying that $\{f_n\}$ converges uniformly on A, we are asserting that f does not converge uniformly to its pointwise limit, f. To prove this assertion, we must, once and for all, declare a number, $\epsilon > 0$, which will universally witness the condition that whatever the choice of $N \in \mathbf{N}$ we will always be able to find a specific point x in the common domain A such that the distance between $f_n(x)$ and $f(x)$ is at least ϵ. Notice that since $\{f_n\}$ converges pointwise, for a given fixed x, there will always be an $N = N(x)$ that will guarantee that $|f_n(x) - f(x)| < \epsilon$. Thus, we see that each time we increase N, we will be required to find a new value of x to witness the failure of the uniform convergence inequality. In this sense, the choice of x depends on the choice of N, and no single x will do the job for all values of N.

Let us now examine the geometry of uniform convergence. Our aim will be to develop further intuition about these abstract definitions. It is evident from the discussions thus far, that the sequence, $\{f_n\}$, converges uniformly to f on A provided for $\epsilon > 0$, we can find $N \in \mathbf{N}$ such that

$$\sup_{x \in A} |f_n(x) - f(x)| < \epsilon, \text{ whenever } n > N.$$

Uniform convergence is a concept that can be clearly understood from a geometrical point of view. Given an arbitrary positive number, ϵ, consider two functions, $f + \epsilon$ and $f - \epsilon$ that are defined by

$$(f + \epsilon)(x) = f(x) + \epsilon \quad \text{and} \quad (f - \epsilon)(x) = f(x) - \epsilon.$$

The graphs of these functions run 'parallel' to the graph of the (pointwise-limit) function, f. This creates a 'tube-shaped' region of width 2ϵ centered around the graph of the function f (see Figure 7.2.1). The definition of uniform convergence demands that for this 2ϵ-tube, we must reach a certain stage (specified by a number $N \in \mathbf{N}$) such that the graphs of the functions, f_{N+1}, f_{N+2}, f_{N+3}, $\cdots$, must all be completely captured inside this tube. The significance of the number, N, is to tell us the maximum number of functions in the sequence that can (possibly) escape this tube.

Once the geometrical perspective on uniform convergence is grasped, then the negation of uniform convergence becomes very clear. Here, the value of ϵ is declared in advance around the pointwise-limit function, f, and this defines a tube of width 2ϵ that remains fixed for the remainder of the discussion. We must then convince ourselves that no matter how large we choose N, the graph of some f_n will escape the tube even though $n > N$! Again we stress that as N changes, the place in the domain, A, at which the graph of an f_n escapes the tube will change.

As a final note, one could view the definition and the negation of uniform convergence in terms of the two-person game discussed earlier in the context of limit of a sequence and limit of a function. The diligent reader may wish to do this as a further aid to understanding of the uniform convergence concept. $\square$

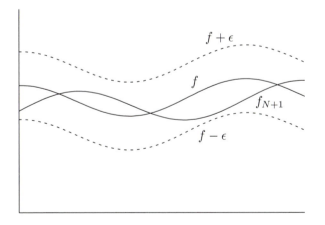

Figure 7.2.1 A graphical illustration of the uniform convergence concept. Note the dashed curves (defined by $f \pm \epsilon$) are centered on the graph of f, and the graph of f_{N+1} and all f_m with $m \geq N + 1$ are contained between the dashed curves.

We illustrate the concept of uniform convergence as well as its negation by means of further examples.

Example 2. Discuss the uniform convergence of the sequence of functions, $\{f_n\}$, where $f_n : (0, 1] \to \mathbf{R}$ is defined by

$$f_n(x) = \frac{1}{nx}, \ x \in (0, 1], \ n \in \mathbf{N}.$$

Solution. This sequence of functions was shown to be pointwise convergent to the limit function, f, which is identically 0, on $(0, 1]$ in Example 7.1.1. We show that this sequence is not uniformly convergent. Set $\epsilon = 1$, and fix N. Choose $x_0 \in (0, 1]$ such that $x_0 < \frac{1}{N+1}$; then

$$|f_{N+1}(x_0) - f(x_0)| = \left| \frac{1}{x_0(N+1)} - 0 \right| \geq 1,$$

whence f_n does not converge uniformly to f on $(0, 1]$. $\square$

Discussion. The basic reason why this sequence of functions cannot converge uniformly to f is that f is bounded on $(0, 1]$, whereas each f_n is unbounded on $(0, 1]$. Thus, no matter how large we had taken ϵ, we could always find a portion of the graph of f_n that escaped a 2ϵ-tube about f. The geometry of this situation was presented in Figure 7.1.1. $\square$

Example 3. Discuss the uniform convergence of the sequence, $\{f_n\}$, where $f_n : [0, 1] \to \mathbf{R}$ is defined by $f_n(x) = 1 - \frac{x}{n}, \ n \in \mathbf{N}.$

Solution. As in Example 7.1.2, it is easy to see that f_n converges pointwise to the function f, which is identically 1 on $[0, 1]$. We claim that this convergence is uniform as well. To this end, let $\epsilon > 0$ be given. We can find $N > 0$ satisfying $\frac{1}{N} < \epsilon$. Now, if $n > N$,

$$\begin{aligned}
|f_n(x) - f(x)| &= \left| \left(1 - \frac{x}{n} \right) - 1 \right| \\
&= \left| \frac{x}{n} \right| < \frac{1}{n}, \ \text{for all } x \in [0, 1] \\
&< \frac{1}{N} < \epsilon
\end{aligned}$$

by our choice of N. Thus, $\{f_n\}$ converges uniformly to 1 on $[0, 1]$. $\square$

Discussion. Note the manner in which we came up with the magic number, N. First, we compute $|f_n(x) - f(x)|$, which reduces to $|\frac{x}{n}|$, which, unfortunately, still involves x. The key to making this expression less than ϵ for all x lies in the fact that x is in $[0, 1]$ so that $x \leq 1$. Thus we simplify the problem to that of making $\frac{1}{n}$ smaller than ϵ, and we only have to invoke the Archimedean Property of the real number system to find the N that will accomplish this. Note that what is really behind this argument is the powerful tool of the Completeness Axiom for the real numbers disguised as the Archimedean Property!

The geometry of this example is presented in Figure 7.2.2. The pointwise limit is the graph of $y = 1$ from $x = 0$ to $x = 1$. Let us take a small strip around the line $y = 1$ (the limit function) of width ϵ on either side. The issue here is: Can we capture the graphs of all the f_n's after certain stage (say, $n > N$) in this strip of width 2ϵ? That this can be done at each point of $[0, 1]$ is the

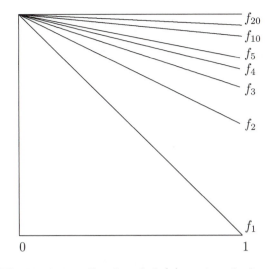

Figure 7.2.2 Graphs of $f_n(x) = 1 - \frac{x}{n}$, for $n = 1, 2, 3, 5, 10,$ and 20.

notion of pointwise convergence. But we want the entire graph to be in this strip irrespective of which point in the domain is considered. Clearly, it is possible to find an N that ensures that the f_n's with $n > N$ have their graphs enclosed in the strip of width 2ϵ around the limit function. Of course, the stage N depends on the size of ϵ.

The reader should see that the sort of graphical considerations presented above can serve as an intuitive aid in deciding whether the convergence of a sequence of functions is uniform. $\square$

Example 4. Discuss the uniform convergence of $\{f_n\}$, where $f_n : [0,1] \to [0,1]$ are defined by $f_n(x) = x^n$, $n = 1, 2, \ldots$.

Solution. The pointwise limit of this function was found in Example 7.1.3 and shown to be

$$f(x) = \begin{cases} 0, & \text{if } x < 1 \\ 1, & \text{if } x = 1. \end{cases}$$

Let us choose $\epsilon = \frac{1}{2}$ and let N be fixed. Consider the function f_{N+1}. The reader should have long since been aware of the fact that $\lim_{x \to 1^-} x^{N+1} = 1$. Hence, there exists $x_0 \in (0,1)$ such that $x_0^{N+1} \geq \frac{1}{2}$. It is immediate that

$$|f_{N+1}(x_0) - f(x_0)| = |x^{N+1} - 0| \geq \epsilon,$$

whence f_n does not converge uniformly on $[0,1]$. $\square$

Discussion. In Example 2, we remarked that any value of ϵ would have sufficed for the argument. In this example, not every positive number will do as a choice for ϵ. Indeed, the requirement on the choice of ϵ is that $\epsilon < 1$. Any larger choice will not permit the construction of the argument (see Exercise 1).

The intuition behind this example can be seen from Figure 7.1.2. Specifically, the graph of each f_n must eventually curve upward, away from the x-axis, so as to attain the value 1 when $x = 1$. The argument simply builds on the fact that to get to 1, the graph of f_n must go outside of a 2ϵ-tube around the x-axis, provided ϵ is sufficiently small. □

Example 5. Discuss the uniform convergence of the sequence $\{f_n\}$ of functions where $f_n : [0, \infty) \to \mathbf{R}$ is defined by

$$f_n(x) = \frac{x}{1 + nx}.$$

Solution. The first step in the solution is to identify the pointwise limit, if it exists. This limit is easily computed for each x, and is the function that is identically zero on $[0, \infty)$. This will be confirmed by the argument that the convergence is uniform. Given $\epsilon > 0$, pick N satisfying $\frac{1}{N} < \epsilon$. If $n > N$ and $x > 0$, we have

$$|f_n(x) - f(x)| = \left| \frac{x}{1 + nx} - 0 \right| = \frac{x}{1 + nx} < \frac{x}{nx} = \frac{1}{n} < \frac{1}{N} < \epsilon,$$

and for $x = 0$, the inequality is obvious. Thus, $f_n \to f$ uniformly on $[0, \infty)$. □

Discussion. The crucial steps in the above reasoning are that from the given expression for f_n, we identify the pointwise limit and then simplify the expression $|f_n(x) - f(x)|$. We must exploit our basic knowledge of inequalities to reduce this expression to one that is free of x, but depends on n, so that this quantity can be made small by choosing n large. The domain of the functions plays a very important role in this sort of calculation, which again illustrates the global nature of the concept of uniform convergence. □

Example 6.   Discuss the uniform convergence of the sequence $\{f_n\}$, where $f_n : [0, \infty) \to \mathbf{R}$ is defined by

$$f_n(x) = \frac{nx}{1 + n^2 x^2}.$$

Solution. For each fixed x, $f_n(x)$ clearly has the limit 0. Hence, for $x \geq 0$, $f_n \to f$ pointwise, where $f(x) = 0$ for all x. We claim that f_n cannot converge uniformly on $[0, \infty)$. To see this, note that $f_n(\frac{1}{n}) = \frac{1}{2}$ and take $\epsilon = \frac{1}{2}$. Then

$$\left| f_n \left(\frac{1}{n} \right) - 0 \right| = \frac{1}{2} \geq \epsilon, \text{ for all } n \in \mathbf{N}. \ \square$$

Discussion. How were the choice of ϵ and the point x, which were used to deny uniform convergence, obtained? The considerations are clear if we look at the graphs of members of the sequence of functions. Each f_n has a continuous graph commencing at origin and rising to a maximum height of $\frac{1}{2}$ at the point $x = \frac{1}{n}$ (use Theorem 4.3.1 to see this). The graph then falls rapidly and becomes asymptotic to the x-axis, that is, $\lim_{x \to \infty} f_n(x) = 0$ (see Figure 7.2.3). Thus, once the details of the graphs are available, the construction of the argument follows.

However, instead of $[0, \infty)$, if the common domain is taken to be $[a, \infty)$, for $a > 0$, the convergence becomes uniform. The reason is that the maximum height, $\frac{1}{2}$, is arrived at by each of the graphs f_n, at $\frac{1}{n}$, and as n gets larger, this phenomenon occurs closer and closer to origin. Thus, if n is sufficiently large, then $\frac{1}{n} < a$, whence the difficulty is eliminated. The reader should present an $\epsilon - N$ argument to show uniform convergence in $[a, \infty), a > 0$. $\square$

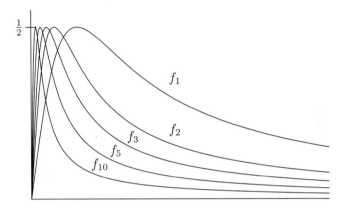

Figure 7.2.3 Graphs of $f_n(x) = \dfrac{nx}{1 + n^2 x^2}$, for $n = $ 1, 2, 3, 5, and 10.

The reader should by now be asking a number of questions regarding the notion of uniform convergence. Some of these questions are explored in the theorems below, and some in the exercises.

Theorem 7.2.2. *Let $\{f_n\}$ and $\{g_n\}$ be uniformly convergent on A. Then $\{f_n \pm g_n\}$ is uniformly convergent on A.*

Proof. Exercise 3. $\square$

Discussion. To construct this proof, the reader should examine the proof of the analogous result for sequences of real numbers. Then recognize that if an inequality holds for all $x \in A$, it must hold for each particular $x \in A$.

The shrewd reader will have noticed that we have avoided mentioning certain other well-known arithmetic operations. Is this oversight, or is there a reason? $\square$

Theorem 7.2.3. *Let $\{f_n\}$ be a monotone increasing sequence of functions on A, and suppose $\{f_n\}$ has a subsequence that is uniformly convergent on A. Then $\{f_n\}$ is uniformly convergent on A.*

Proof. Let f be the limit of the subsequence. It can be checked (see Exercise 5) that for each n, $f_n \le f$. It follows that $f_n \to f$ pointwise and uniformly. $\square$

The next theorem provides a useful test for uniform convergence of a sequence of functions.

Theorem 7.2.4. *Let $\{f_n\}$ be pointwise convergent on A to a function, f. Define*

$$M_n = \sup_{x \in A} |f_n(x) - f(x)|.$$

Then $f_n \to f$ uniformly on A if and only if $\lim_{n \to \infty} M_n = 0$.

Proof. Exercise 7. □

Discussion. This theorem can be quite useful. For example, in Example 6, one would merely have had to note that $\frac{1}{2} \leq M_n$ to complete the argument. The difficulty in applying this result is in identifying the required supremum. But since many sequences comprise differentiable functions, the task is not insurmountable, as the next example shows. □

Example 7. Discuss the pointwise and uniform convergence of the sequence $\{2nxe^{-nx^2}\}$ on $[a, \infty)$, where $a \geq 0$.

Solution. It can be checked (Exercise 8) that $f_n \to 0$ pointwise on $[0, \infty)$. It follows that $M_n = \sup_{x \in A} |f_n(x)|$. Further, each f_n is differentiable on $(0, \infty)$, and its derivative is given by

$$f_n'(x) = (2n - 4n^2x^2)e^{-nx^2}.$$

Theorem 4.3.1 establishes that the relative extrema of f_n on the interval $[0, \infty)$ occurs at $x = \dfrac{1}{\sqrt{2n}}$, whence $M_n = \sqrt{\dfrac{2n}{e}}$ provided that $a = 0$. Since M_n does not converge to 0, we have that f_n does not converge uniformly on $[0, \infty)$. It is left to the reader to show that the convergence is uniform on $[a, \infty)$ for $a > 0$. □

Discussion. The functions in the sequence are presented graphically in Figure 7.2.4. It is obvious from the graphs of the individual functions that the convergence could not possibly be uniform on $[0, \infty)$. However, since the 'hump' gets squeezed in toward zero as n increases, the graphs demonstrate why the sequence can converge uniformly on the smaller domain. □

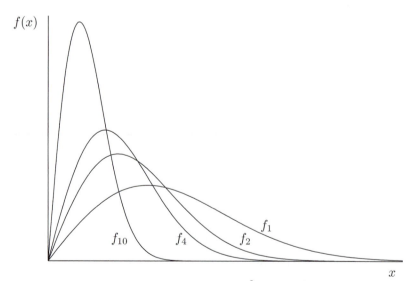

Figure 7.2.4 Graphs of $f_n(x) = 2nxe^{-nx^2}$ for $n = 1$, 2, 4, and 10 in the interval $0 \leq x \leq 2.5$. Notice how the absolute maximum increases and moves closer to the y-axis with increasing index.

The next theorem states the Cauchy Criterion for the uniform convergence of a sequence of functions.

Theorem 7.2.5 (Uniform Cauchy Criterion).   *Let $\{f_n\}$ be a sequence of functions defined on A. Then $\{f_n\}$ is uniformly convergent on A if and only if for positive ϵ there exists an $N \in \mathbf{N}$ such that for all $x \in A$,*

$$|f_n(x) - f_m(x)| < \epsilon, whenever\ n, m > N.$$

Proof. We prove uniform Cauchy implies uniform convergence leaving the rest to the reader as Exercise 14. Thus, let $\{f_n\}$ be uniformly Cauchy on A. Then, for each fixed $x \in A$, $\{f_n(x)\}$ is a Cauchy sequence and so has a limit that we call $f(x)$, thus defining the limit function f for each $x \in A$. Fix $\epsilon > 0$, and for each $x \in A$, choose $N_x \in \mathbf{N}$ such that $m > N_x$ implies $|f_m(x) - f(x)| < \frac{\epsilon}{2}$. By the uniform Cauchy Criterion, choose $N \in \mathbf{N}$ such that for all $x \in A$ $n, m > N$ imply $|f_n(x) - f_m(x)| < \frac{\epsilon}{2}$. Now let $n > N$. Then for any given $x \in A$, if $m > \max\{N, N_x\}$, we have

$$|f_n(x) - f(x)| \leq |f_n(x) - f_m(x)| + |f_m(x) - f(x)| < \frac{\epsilon}{2} + \frac{\epsilon}{2} = \epsilon,$$

whence the sequence converges uniformly on A. $\square$

Discussion. Cauchy criteria have been discussed for sequences in Section 3.3, for infinite series in Section 6.1. They are important tools for testing convergence, and if the reader does not feel totally able to construct the required proofs, he should review the mentioned sections. $\square$

Theorem 7.2.6.   *Let $\{f_n\}$ be a uniformly convergent sequence of functions defined on $[a, b]$. If each f_n is bounded on $[a, b]$, then the uniform limit, f is bounded on $[a, b]$.*

Proof. Exercise 20. $\square$

Discussion. This is one of the many useful results that illustrate that a property shared by all the members of a uniformly convergent sequence of functions is retained by the limit function. We shall see several other results of the same nature in the next section. $\square$

Exercises

1. Show for the sequence of Example 4, an argument can be constructed for any value of ϵ that is less than 1.

2. Discuss the uniform convergence of the following sequences of functions on the domains mentioned against each. Where possible, describe the limit function.

   (a) $f_n(x) = 1 - nx$, $x \in \mathbf{R}$;

   (b) $f_n(x) = 1 - \frac{x}{n}$, $x \in [0, \infty)$;

   (c) $f_n(x) = \frac{n^2 x}{1 + n^3 x^2}$, $x \in \mathbf{R}$;

   (d) $f_n(x) = \frac{e^{-nx}}{n}$, $x \geq 0$;

(e) $f_n(x) = nxe^{-nx}$, $x \in [0, 1]$;

(f) $f_n(x) = \frac{x^n}{1+x^n}$, $x \in [0, 1]$;

(g) $f_n = nx(1 - x^2)^n$, $x \in [0, 1]$;

(h) $f_n(x) = \frac{1}{n} \sin n^2 x$, $x \in \mathbf{R}$;

(i) $f_n(x) = \frac{nx}{1+nx}$, $x \geq 0$;

(j) $f_n(x) = \frac{\sin nx}{\sqrt{n}}$, $x \in [0, 2\pi]$;

(k) $f_n(x) = (\sin x)^n$, $0 \leq x \leq \pi$;

(l) $f_n(x) = n \ln(1 + \frac{x}{n})$, $x \geq 0$;

(m) $f_n(x) = n(x^{\frac{1}{n}} - 1)$, $x \geq a > 0$;

(n) $f_n(x) = x(1 + \frac{1}{n})$, $x \in \mathbf{R}$;

(o) $f_n(x) = g(x + \frac{1}{n})$, where g is uniformly continuous on $\mathbf{R}$;

(p) $f_n(x) = g(nx)$ where $g(x) = \begin{cases} x, & \text{if } x \in [0, \frac{1}{2}) \\ 1 - x, & \text{if } x \in [\frac{1}{2}, 1] \\ 0, & \text{if } x > 1; \end{cases}$

(q) $f_n(x) = \begin{cases} nx, & \text{if } x \in [0, \frac{1}{n}] \\ -nx + 2, & \text{if } x \in (\frac{1}{n}, \frac{2}{n}] \\ 0, & \text{if } x \in (\frac{2}{n}, 1]; \end{cases}$

(r) $f_n(x) = \begin{cases} n^2 x^2 - 2nx, & \text{if } x \in [0, \frac{2}{n}] \\ 0, & \text{if } x \in (\frac{2}{n}, 2]; \end{cases}$

(s) $f_n(x) \begin{cases} \frac{1}{n}, & \text{if } x = 0 \text{ or } x \in \mathbf{R} \sim \mathbf{Q} \\ q + \frac{1}{n}, & \text{if } x = \frac{p}{q} \in \mathbf{Q} \text{ is in lowest terms, and } q > 0. \end{cases}$

3. Prove if $\{f_n\}$ and $\{g_n\}$ are both uniformly convergent on a set A, so are $\{f_n \pm g_n\}$ (Theorem 7.2.2).

4. Suppose that in addition both sequences in Exercise 3 consist of functions that are bounded on A. Show $\{f_n g_n\}$ converges uniformly on A. What can be said if the boundedness condition fails?

5. Complete the proof of Theorem 7.2.3.

6. Let $\{f_n\}$ be a uniformly convergent sequence of functions on A. Show $\{f_n^+\}$ and $\{f_n^-\}$ are uniformly convergent sequences. What can be concluded about $\{|f_n|\}$?

7. Prove Theorem 7.2.4.

8. Show the sequence in Example 7 tends to 0 pointwise on $[0, \infty)$. Further show that the sequence converges uniformly on $[a, \infty)$ for $a > 0$.

9. Give an example of a uniformly convergent sequence $\{f_n\}$ such that the sequence of its derivatives fails to be uniformly convergent in the same domain.

10. Let $\{f_n\}$ be a sequence of continuous functions on $[0, 1]$ that converges uniformly. Show there exists $K > 0$ such that $|f_n(x)| < K$ for each $n \in \mathbf{N}$ and each $x \in [0, 1]$. Does the result still hold if we replace uniform convergence by pointwise convergence? What if we assume the limit function is bounded?

11. Prove that if $\{f_n\}$ is uniformly convergent to a bounded function, then there exists $K, N > 0$ such that $|f_n(x)| < K$ for all $n > N$ and all x in the domain. Give an example to show why we cannot replace $n > N$ by 'for all $n \in \mathbf{N}$'?

12. If $\{f_n\}$ converges uniformly on an open interval (a, b) and converges at the points a and b, must it converge uniformly on $[a, b]$?

13. Give an example where $f_n \to f$ uniformly, but f_n^2 does not converge uniformly to f^2.

14. Complete the proof of Theorem 7.2.5.

15. Let $\{f_n\}$ converge uniformly on A. Show that every subsequence of $\{f_n\}$ also converges uniformly on A. Is the converse true?

16. Let $\{f_n\}$ be uniformly convergent on A_i $(i = 1, 2, \ldots, k)$. Show that $\{f_n\}$ is uniformly convergent on $\bigcup_{i=1}^{k} A_i$. Show that this result cannot be extended to the union of a countable number of sets.

17. Let each f_n be continuous on a set A and B be a dense subset of A. If $\{f_n\}$ is uniformly convergent on A, is it true that $\{f_n\}$ is uniformly convergent on B? What can you say about the converse?

18. Let $\{f_n\}$ and f be defined on A. Then $\{f_n\}$ fails to converge uniformly to f on A if and only if there exists $\epsilon > 0$, a subsequence $\{f_{n_k}\}$, and a sequence $\{x_k\}$ of elements of A such that $|f_{n_k}(x_k) - f(x_k)| \geq \epsilon$ for all $k \in \mathbf{N}$.

19. Is it possible for a sequence of continuous functions to be uniformly convergent to a discontinuous limit function?

20. Prove Theorem 7.2.6. Is the result still valid if we have only pointwise convergence?

21. If $\{f_n\}$ and $\{g_n\}$ converge uniformly to f and g, respectively, in $[a, b]$, prove the following:

    (a) if each f_n is bounded, and f is bounded, then $\{f_n\}$ is uniformly bounded (Exercise 7.1.11);

    (b) if $\{f_n\}$ and $\{g_n\}$ are both uniformly bounded, then $\{f_n \cdot g_n\}$ is uniformly convergent to fg;

22. If $\{f_n\}$ is a sequence of continuous functions on a compact set A such that for each x in A, $\{f_n(x)\}$ is monotonically increasing, and if $\{f_n\}$ converges pointwise to a continuous function f, show that the convergence is uniform on A (Dini's Theorem). Will the result still hold if we relax the compactness of A? What if we drop the monotonicity condition?

23. For the functions in Exercise 7.1.10, is it possible to obtain a sequence $\{f_n\}$ of functions that converge uniformly to f in the indicated domains?

24. The sequence $\{f_n\}$ is defined for $x \geq 0$ as follows: set $f_1(x) = \sqrt{x}$, and for $n > 1$, $f_{n+1}(x) = \sqrt{x + f_n(x)}$. Show if $0 < a < b < \infty$, then $\{f_n\}$ converges uniformly on $[a, b]$. Is the convergence uniform on $[0, 1]$?

25. The sequence $\{f_n\}$ is defined by $f_0(x) = 1$, and for $n \geq 1$, $f_n(x) = \sqrt{x \cdot f_{n-1}(x)}$. Prove that $\{f_n\}$ converges uniformly in $[0, 1]$. What is the limit?

26. Suppose that $f_n \to f$ uniformly on A and that it is uniformly bounded (Exercise 7.1.11) by K. If g is continuous on $[-K, K]$, then $\{(g \circ f_n)\}$ converges uniformly on A. What is the limit function?

27. Recall that the sequence given by $f_n(x) = x^n$ was shown to converge pointwise, but not uniformly on $[0, 1]$. Let g be continuous on $[0, 1]$. Find a necessary and sufficient condition on g that will guarantee that the product $\{g \circ f_n\}$ converges uniformly.

28. Suppose $f_n \to f$ uniformly on A, that each f_n is continuous and that $\{x_n\}$ is a sequence of elements of A that tends to $x \in A$. Show $\{f_n(x_n)\}$ is convergent. What is its limit?

29. Let $\{f_n\}$ be a sequence of continuous functions that converges pointwise to a limit, f, on a compact set A. Show that the convergence is uniform exactly if

(a) f is continuous and

(b) for every $\epsilon > 0$ there exists $m \in \mathbf{N}$ and $\delta > 0$ such that $n > m$ and $|f_k(x) - f(x)| < \delta$ imply $|f_{k+n}(x) - f(x)| < \epsilon$ for all $x \in A$ and $k \in \mathbf{N}$.

30. Let $\{f_n\}$ and $\{g_n\}$ be the sequences of Exercises 2(r) and 2(s), respectively. Prove that both sequences converge uniformly on every bounded interval. Prove $\{h_n\}$, defined by $h_n = f_n g_n$, does not converge uniformly on any bounded interval.

31. A sequence of functions $\{f_n\}$ with a common domain D is **equicontinuous** on $A \subseteq D$ if and only if for each positive ϵ, there exists a $\delta = \delta(\epsilon) > 0$ such that $x, y \in A$, and $|x - y| < \delta$ implies that $|f_n(x) - f_n(y)| < \epsilon$ for each $n \in \mathbf{N}$. Let $\{f_n\}$ be a sequence of continuous functions on D which converges uniformly on a compact subset $A \subseteq D$. Prove $\{f_n\}$ is uniformly bounded and equicontinuous on A.

32. Which of the following sequences of functions are (i) uniformly bounded, (ii) equicontinuous on the specified domains:

(a) $f_n(x) = \frac{x^n}{n}$, $x \in [-1, 1]$;   (b) $f_n(x) = \sin nx$, $x \in [0, 1]$;

(c) $f_n(x) = \frac{\sin nx}{x}$, $x \in (0, 1)$.

33. Show that a subsequence of an equicontinuous sequence is equicontinuous.

7.3   Consequences of Uniform Convergence

We have seen from earlier examples that the pointwise limit function, f, of a sequence, $\{f_n\}$, of functions will not, in general, share all the nice properties of each of the f_n's. Among the sequences considered were examples where each f_n was continuous, (respectively bounded, integrable, differentiable), but the pointwise-limit function was not. As Theorem 7.2.6 shows, what was lacking in these examples was uniform convergence of the sequence. Because uniform convergence ensures the transfer of many properties from individual functions of a sequence to the limit function, we will conclude that the concept of uniform convergence is superior to that of pointwise convergence. Our first theorem treats continuity.

Theorem 7.3.1. *If each f_n is continuous at a point, $a \in A$, and the sequence $\{f_n\}$ is uniformly convergent on the domain A to the function, f, then f is continuous at the point a.*

Proof. Let $\epsilon > 0$ be given. By uniform convergence, there exists $N \in \mathbf{N}$ such that for all $x \in A$ and $n \geq N$, we have

$$|f_n(x) - f(x)| < \frac{\epsilon}{3}.$$

Since f_N is continuous at the point a, we can find a $\delta > 0$ such that for each $x \in A$,

$$|f_N(x) - f_N(a)| < \frac{\epsilon}{3}, \text{ whenever } |x - a| < \delta.$$

Finally, if $x \in A$ and $|x - a| < \delta$, we have

$$
\begin{aligned}
| f(x) - f(a) | &\leq | f(x) - f_N(x) | + | f_N(x) - f_N(a) | + | f_N(a) - f(a) | \\
&< \frac{\epsilon}{3} + \frac{\epsilon}{3} + \frac{\epsilon}{3} = \epsilon,
\end{aligned}
$$

proving that f is continuous at the point a. $\square$

Discussion. Note the technique adopted in the proof. Starting with the given ϵ we first arrive at the single N that satisfies the condition for uniform convergence for all x. We then use this N, consider f_N at a general point x close enough to the given point a, and employ the the triangle inequality by adding and subtracting the appropriate candidates. The reader should be well familiar with such techniques. This argument is the famous $\frac{\epsilon}{3}$ argument, which we have seen earlier.

Observe that the proof will break down if we had only pointwise convergence instead of uniform convergence. This is what happened in Example 7.1.3, and the reader would do well to attempt to construct the present argument for that sequence of functions to see exactly where and why the argument breaks down. Another noteworthy feature of this theorem is that it can profitably be used to negate uniform convergence of a sequence $\{f_n\}$ of continuous functions that has a discontinuous pointwise-limit function. This is of great practical value as the next example shows. However, the reader should be aware that the relationship is not one of equivalence. The fact that a sequence of continuous functions converges pointwise on A to a continuous limit does not ensure that the convergence is uniform. $\square$

Corollary. *If $\{f_n\}$ is a sequence of continuous functions defined on a set A that converges uniformly to the function f on A, then f is continuous on the set A.*

Example 1. Discuss the uniform convergence of the sequence of functions, $\{f_n\}$, where $f_n(x) = \arctan nx$, $(x \in \mathbf{R})$.

Solution. It is known that $\arctan 0 = 0$, that

$$
y > 0 \text{ implies } \lim_{y \to \infty} \arctan y = \frac{\pi}{2}
$$

and that

$$
y < 0 \text{ implies } \lim_{y \to -\infty} \arctan y = -\frac{\pi}{2}.
$$

Thus, for any fixed $x \in \mathbf{R}$, it follows that

$$
\lim_{n \to \infty} \arctan nx = \frac{\pi}{2} \operatorname{sgn} x,
$$

whence $\{\arctan nx\}$ converges pointwise to a discontinuous function. Since $\arctan nx$ is continuous, the convergence is not uniform. $\square$

The next theorem presents a somewhat more general version of the previous theorem.

Theorem 7.3.2. *Let $\{f_n\}$ be a sequence of continuous functions defined on a set A that converges uniformly to the function f on A. Further suppose that a is* *a limit point of A and that for each $n \in N$, $\lim_{x \to a} f_n(x) = a_n$, then*

$$\lim_{x \to a} f(x) = \lim_{n \to \infty} a_n,$$

or equivalently,

$$\lim_{x \to a} \lim_{n \to \infty} f_n(x) = \lim_{n \to \infty} \lim_{x \to a} f_n(x).$$

Proof. Exercise 1. □

Discussion. The proof of this form of the theorem is very similar to the proof of Theorem 7.3.1. The point of stating the theorem in this form is that it illustrates the fact that uniform convergence permits one to interchange the order in which limits are computed. This is particularly evident from

$$\lim_{x \to a} \lim_{n \to \infty} f_n(x) = \lim_{n \to \infty} \lim_{x \to a} f_n(x).$$

On the left-hand side, we first find the limit function, f, and then let $x \to a$ for the limit function. On the right-hand side, the individual limits, $\lim_{x \to a} f_n(x)$, are computed for each n, thereby forming the sequence $\{a_n\}$. Finally, the limit of the sequence $\{a_n\}$ is computed. Notice that in this computation, the requirement on a is that a is a limit point of A, not that $a \in A$. □

Next we apply the uniform convergence concept to Riemann integrability. It was already established in Example 7.1.5 that pointwise convergence was insufficient to preserve the value of the limit of the integrals as the integral of the limit. However, as the next example shows, the total situation is even worse.

Example 2. Let $\{r_n\}$ be an enumeration of the rationals in $[0, 1]$. Define $\{f_n\}$ on $[0, 1]$ by

$$f_n(x) = \begin{cases} 1, & \text{if } x = r_m \text{ and } m \leq n \\ 0, & \text{otherwise.} \end{cases}$$

What can be said about the integration properties of the sequence and its pointwise limit?

Solution. The reader can verify that the pointwise limit on $[0, 1]$ is given by

$$f(x) = \begin{cases} 1, & \text{if } x \in \mathbf{Q} \cap [0, 1] \\ 0, & \text{if } x \in [0, 1] \sim \mathbf{Q}. \end{cases}$$

Since each f_n has a discontinuity only at a finite number of points, it is integrable on $[0, 1]$ for each n. The pointwise limit function, f, is the canonical example of a function that is not Riemann integrable (Example 5.2.4). □

Discussion. The present example shows just how ugly the pointwise limit can be. It is nowhere continuous and fails to be integrable on any subinterval of its domain. Yet each of the functions f_n has only finitely many discontinuities, is integrable, and the integral of each f_n yields the same value, 1. The requirement of uniform convergence will turn this around. □

Theorem 7.3.3. *If each f_n, $n \in \mathbf{N}$, is Riemann integrable on $[a, b]$ and the sequence $\{f_n\}$ converges uniformly to f on $[a, b]$, then f is Riemann integrable on $[a, b]$; furthermore,*

$$\lim_{n \to \infty} \int_a^b f_n = \int_a^b \lim_{n \to \infty} f_n.$$

Proof. Let $\epsilon > 0$ be given. Since each f_n is integrable in $[a, b]$, there exists a partition, P_n, of $[a, b]$ such that

$$|\overline{S}(f_n, P_n) - \underline{S}(f_n, P_n)| < \frac{\epsilon}{3}.$$

Also by uniform convergence of the given sequence of functions, there exists $N \in \mathbf{N}$ such that for all $x \in [a, b]$ and for $n \geq N$, we have

$$|f_n(x) - f(x)| < \frac{\epsilon}{3(b-a)}.$$

Now if $m_{n,r}$, and m_r denote the infima of f_n and f, respectively, in the rth subinterval of the partition, P_n, and d_r, the length of the rth subinterval, then in particular, we have

$$|m_{n,r} - m_r| < \frac{\epsilon}{3(b-a)} \quad \text{for} \quad n \geq N,$$

because failure of this inequality leads to an immediate contradiction. Therefore,

$$
\begin{aligned}
|\underline{S}(f_N, P_N) - \underline{S}(f, P_N)| &= \left| \sum m_{N,r} d_r - \sum m_r d_r \right| \\
&= \left| \sum (m_{N,r} - m_r) d_r \right| < \frac{\epsilon}{3(b-a)} \sum d_r = \frac{\epsilon}{3}
\end{aligned}
$$

so that

$$\underline{S}(f_N, P_N) \leq \underline{S}(f, P_N) + \frac{\epsilon}{3}.$$

Similarly, considering the upper sums, we have the inequality

$$\overline{S}(f, P_N) \leq \overline{S}(f_N, P_N) + \frac{\epsilon}{3}.$$

Thus, we obtain

$$\overline{S}(f, P_N) - \underline{S}(f, P_N) < \overline{S}(f_N, P_N) - \underline{S}(f_N, P_N) + 2\frac{\epsilon}{3} \leq \epsilon.$$

Now, by Theorem 7.2.6, f is bounded, whence we apply Theorem 5.2.4, and the function f is integrable on $[a, b]$. Finally, note that

$$\left| \int_a^b f_n - \int_a^b f \right| \leq \int_a^b |f_n - f| < \frac{\epsilon}{b-a}(b-a) = \epsilon \qquad \textbf{(WHY?)},$$

proving that the limit of the integral is the integral of the uniform limit. $\square$

Discussion. The proof of this theorem is founded on the application of Theorem 5.2.4. The hypothesis of Theorem 5.2.4 requires that the function under discussion be bounded. Recall that for a function to be Riemann integrable, it must be bounded. Thus, each f_n is bounded. This permits the application of Theorem 7.2.6, to obtain that f is bounded, whence Theorem 5.2.4 applies.

To establish integrability of f, we have to produce a partition, P, of $[a, b]$, which satisfies the condition $\overline{S}(f, P) - \underline{S}(f, P) < \epsilon$ for any preassigned ϵ. The way we obtain this partition is by using the uniform convergence of the sequence to find a value of N that will guarantee that all f_n's having subscript exceeding N are inside a $\frac{2\epsilon}{3(b-a)}$-tube around f. This tube is then used in an extremely powerful way, namely, to conclude that

$$m_{n,r} - m_r < \frac{\epsilon}{3(b-a)} \qquad \text{for} \quad n \geq N,$$

where $m_{n,r}$, and m_r are infima over various subintervals of arbitrary partitions for f_n and f, respectively. Notice that these infima do not have to occur at the same value of x for f_n and f; thus, it is not at all obvious why this inequality should hold. It is essential that the reader work this out (Exercise 2) so as to completely understand it. Once this inequality has been obtained, it becomes possible to replace statements about lower sums relative to f, by lower sums relative to f_N, leading to

$$\underline{S}(f_N, P_N) \leq \underline{S}(f, P_N) + \frac{\epsilon}{3}.$$

Obtaining the analogous inequalities for upper sums is straightforward, and the proof is essentially complete.

The theorem also asserts that in the presence of uniform convergence, we have the pleasant feature that we can interchange the integral and the limit, namely, the limit of the integrals is equal to the integral of the (uniform) limit. Again, this theorem is useful for negating uniform convergence of a sequence of functions each of which is Riemann integrable, and for which the pointwise limit function is either not Riemann integrable, or takes a different value from the limit of the integrals.

Several other types of integrals exist, two of which have been developed elsewhere in this text. The obvious question arises as to whether it is possible to prove analogous results for Stieltjes integrals and improper Riemann Integration. These ideas will be explored further in the exercises. $\square$

Example 3. Is the sequence of functions $\{2(n+1)x(1-x^2)^n\}$ uniformly convergent to its pointwise limit on $[0,1]$?

Solution. It is left to the reader to verify that the pointwise limit of the f_n's is the zero function (Exercise 3). Further, the value of the integral of each f_n is quickly seen to be 1 by the Fundamental Theorem. Thus,

$$\int_0^1 f = 0 \neq \lim_{n \to \infty} \int_0^1 f_n = 1.$$

Thus, the convergence to the pointwise limit on $[0,1]$ cannot be uniform. $\square$

Next, we consider the effect of uniform convergence on differentiation applied to each member of the sequence. Here, again, the issue is obtaining the conditions under which the process of differentiation can be interchanged with the process of computing the limit of the sequence of functions. In thinking about this problem, it would seem that the conditions are likely to be restrictive since we are dealing with two sequences of functions, and two limit functions, that is, $f_n \to f$ and $f'_n \to f'$.

Theorem 7.3.4. *Let $\{f_n\}$ be a sequence of functions defined on $[a, b]$. If each $\{f_n\}$ has a continuous derivative on $[a, b]$ and the sequence of derivatives, $\{f_n'\}$, converges uniformly on $[a, b]$, and there exists $x_0 \in [a, b]$ such that $\{f_n(x_0)\}$ converges, then $f_n \to f$ uniformly on $[a, b]$ and*

$$\lim_{n \to \infty} f_n' = f'.$$

Proof. We first show that $\{f_n\}$ satisfies the Cauchy Criterion for uniform convergence. Thus, fix $\epsilon > 0$, use the fact that f_n' converges uniformly to choose N such that $n, m > N$ imply

$$|f_n'(t) - f_m'(t)| < \frac{\epsilon}{2(b - a)}, \quad \text{for all} \quad t \in [a, b].$$

Again, since $f_n(x_0)$ converges, we can choose N large enough, so that for $m, n > N$, we have

$$|f_n(x_0) - f_m(x_0)| < \frac{\epsilon}{2}.$$

Now we apply the Fundamental Theorem of Calculus in the form of Theorem 5.5.2 to observe that for each $x \in [a, b]$,

$$f_n(x) = \int_{x_0}^{x} f_n'(t)\, dt + f_n(x_0).$$

It follows that

$$f_n(x) - f_m(x) = \int_{x_0}^{x} [f_n'(t) - f_m'(t)]\, dt + f_n(x_0) - f_m(x_0),$$

whence we obtain

$$
\begin{aligned}
|f_n(x) - f_m(x)| \ &\leq\ \left| \int_{x_0}^{x} f_n'(t) - f_m'(t)\, dt \right| + |f_m(x_0) - f_n(x_0)|, \\
&<\ \frac{\epsilon(b - a)}{2(b - a)} + \frac{\epsilon}{2} = \epsilon.
\end{aligned}
$$

Thus, the sequence, $\{f_n\}$, converges uniformly on $[a, b]$, and we denote its uniform limit by f. To complete the proof, let $f_n' \to g$ on $[a, b]$. By the corollary to Theorem 5.5.2, we have

$$f_n(x) - f_n(a) = \int_{a}^{x} f_n'(t)\, dt.$$

Since the convergence of $\{f_n'\}$ is uniform, Theorem 7.3.3 applies and letting n tend to infinity, we obtain

$$f(x) - f(a) = \int_{a}^{x} g(t)\, dt.$$

But g is continuous, whence for each $x \in (a, b)$ we have, by Theorem 5.5.3,

$$f'(x) = g(x). \qquad \square$$

Discussion. The key to understanding the proof of this theorem begins with an understanding of how the Fundamental Theorem of Calculus is applied in this situation.

The first application is in the form of Theorem 5.5.2 to f_n', which is continuous on $[a, b]$, hence integrable there, and for which f_n is a primitive. The distinctive feature of the application is that in the present situation, we observe

$$f_n(x) = \int_{x_0}^{x} f_n'(t)\, dt + f_n(x_0),$$

where $x, x_0 \in [a, b]$. This application requires the introduction of a minus sign if $x < x_0$, but other than that, it is straightforward. While this theorem could be applied using any member of $[a, b]$ as the lower limit of integration, we must use x_0 as the lower limit since subtraction of f_m from f_n leaves the quantity, $f_m(x_0) - f_n(x_0)$ as a residual, and the only way to eliminate it is by virtue of convergence of the sequence $\{f_n(x_0)\}$, whence it becomes arbitrarily small. The final outcome of this application of the Fundamental Theorem is the fact that $f_n \to f$ uniformly on $[a, b]$.

The second application of the Fundamental Theorem is in the form of Theorem 5.5.3, which states that for a continuous function, h, the function, H on $[a, b]$ defined by

$$H(x) = \int_{a}^{x} h(t)\, dt$$

satisfies $H' = h$ on $[a, b]$. In this case, the theorem can be applied with the lower limit on the integral being a, since it is known in advance that $f_n(a)$ converges to $f(a)$. Of course, to complete this section of the proof, Theorem 7.3.3 must be applied, which is why the sequence of derivatives must be uniformly convergent. The use of Theorem 5.5.3 makes clear why each of the f_n''s must be continuous.

It is worth pointing out that the hypothesis requires convergence of f_n only at one point. One can construct examples to show that if any of the conditions in the hypothesis is dropped, the conclusion of the theorem may fail. These ideas are explored further in the exercises. $\square$

Exercises

1. Prove Theorem 7.3.2.

2. Write out a complete argument establishing the truth if the inequality

$$m_{n,r} - m_r < \frac{\epsilon}{3(b-a)} \text{ for } n \geq N$$

    of Theorem 7.3.3.

3. Verify the pointwise limit in Example 3 is the zero function.

4. Consider the sequence of functions, f_n, given by $f_n = \frac{\sin nx}{n}$ in $[0, 1]$. Show $\{f_n\}$ converges uniformly to 0, while f_n' does not converge even pointwise in $[0, 1]$. What do you conclude from this?

5. Show $f_n = nx(1-x)^n$ is integrable in $[0,1]$ for each n and $\lim\limits_{n\to\infty} \int_0^1 f_n = \int_0^1 \lim\limits_{n\to\infty} f_n$, yet the sequence is not uniformly convergent.

6. Compare $\lim\limits_{n\to\infty} \int_0^1 f_n$ and $\int_0^1 \lim\limits_{n\to\infty} f_n$, where $f_n = \frac{nx}{1+nx}$ on $[0,1]$.

7. Give an example of a nonuniformly convergent sequence $\{f_n\}$ in $[0,1]$ with pointwise limit f, for which $\lim\limits_{n\to\infty} \int_0^1 f_n$ and $\int_0^1 \lim\limits_{n\to\infty} f_n$ are equal.

8. Show although $f_n(x) = xe^{-nx^2}$ is uniformly convergent in $[-1,1]$ to a differentiable function, the limit and differentiation process cannot be interchanged.

9. Let g be continuous on $[a,b]$, and let $\{f_n\}$ be a sequence of continuous functions converging uniformly to f on $[a,b]$. Prove $\lim\limits_{n\to\infty} \int_a^b f_n g = \int_a^b fg$.

10. Let f_n be such that f_n' is defined and continuous on $[a,b]$, $\{f_n'\}$ converges uniformly on $[a,b]$, and $\{f_n\}$ converges pointwise to f. Show

$$f' = \lim\limits_{n\to\infty} f_n'.$$

11. Consider the sequence defined by $f_n(x) = \frac{2x}{1+n^2x^2}$ for $x \geq 0$ and its limit f. Check whether the equation $\lim f_n'(x) = f'(x)$ is valid.

12. Let $f_n : [0,1] \to \mathbf{R}$ be defined by $f_n = n$ if $x = \frac{1}{n}$ and 0 otherwise. Compare $\int_0^1 f_n$ with $\int_0^1 f$.

13. Discuss the pointwise and uniform convergence of $\{f_n\}$ where

$$f_n(x) = \begin{cases} -1, & \text{if } x < -\frac{\pi}{n} \\ \sin\frac{nx}{2}, & \text{if } -\frac{\pi}{n} \leq x \leq \frac{\pi}{n} \\ 1, & \text{if } x > \frac{\pi}{n}. \end{cases}$$

14. Give an example showing $\{f_n\}$ may converge to f uniformly on $[a,b]$, f_n' may exist and be continuous on $[a,b]$, but $f'(x_0)$ fails to exist at some point $x_0 \in [a,b]$.

15. If $\{f_n\}$ is a sequence of functions that converge uniformly to a continuous function f on $\mathbf{R}$, show

$$\lim\limits_{n\to\infty} f_n\left(x + \frac{1}{n}\right) = f(x), \quad x \in \mathbf{R}.$$

16. Let $\{f_n\}$ be a uniformly bounded sequence of continuous functions on $[a,b]$ and suppose f to be a function on $[a,b]$ to which $\{f_n\}$ converges uniformly on $[a,c]$ for each $c \in (a,b)$. Prove $\lim\limits_{n\to\infty} \int_a^b f_n = \int_a^b f$.

17. Prove **Arzela's Theorem**: if $\{f_n\}$ is a uniformly bounded sequence of Riemann integrable functions converging to a Riemann integrable function on $[a,b]$, then

$$\lim\limits_{n\to\infty} \int_a^b f_n = \int_a^b f.$$

18. Let $\{f_n\}$ be a sequence of functions that converges uniformly on $[a,b]$. If each f_n is improperly Riemann integrable on $[a,b]$, what can be said about f?

19. Repeat Exercise 18, except that the interval is $(0, \infty)$.

20. State and prove a suitable theorem about Stieltjes integration and uniformly convergent sequences of functions.

21. Let g and f_0 be continuous on $[a, b]$. Define the sequence, $\{f_n\}$, on $[a, b]$ by

$$f_n(x) = \int_a^x g(t) f_{n-1}(t) dt, n \geq 1.$$

Show $\{f_n\}$ converges uniformly to f, and f satisfies $f'(x) = g(x) f(x)$. What can be said about the form of f?

22. Consider the sequence of functions defined by $f_n(x) = \frac{1}{n} e^{-n^2 x^2}$ defined on $[-1, 1]$. Give a complete discussion of the convergence properties of this sequence and its associated sequence of derivatives.

23. Let f be defined on $[0, 1]$ and define $S(f)$ by

$$S(f) = \lim_{n \to \infty} \frac{1}{n} \sum_{k=1}^n f\left(\frac{k}{n}\right) \quad \text{whenever this limit exists.}$$

Show if $f_n \to f$ uniformly on $[0, 1]$ and $S(f)$ and $S(f_n)$ exist, then the two limiting process may be exchanged. Can this hypothesis be weakened?

24. Suppose f and g are continuous on $[a, b]$ and $f \geq 0$ and $g > 0$. Let M denote the maximum value of f on the interval and define M_n by

$$M_n = \int_a^b g(x)[f(x)]^n.$$

Show $(M_n)^{1/n} \to M$. Further show if $M > 0$, then $\frac{M_{n+1}}{M_n} \to M$.

25. Let g be a function of bounded variation in $[a, b]$. Let $\{f_n\}$ be a sequence of R-S integrable functions with respect to g on $[a, b]$, and let $\{f_n\}$ converge uniformly to f. Prove f is R-S integrable with respect to g on $[a, b]$, and $\{\int_a^x f_n dg\}$, converges uniformly on $[a, b]$ to $\int_a^x f dg$.

26. Let $\sigma(n, m)$ be a double sequence. For each $n \in \mathbf{N}$, define $\{f_n\} : \mathbf{N} \to \mathbf{R}$ by $f_n(m) = \sigma(n, m)$. If f_n converges uniformly f on $\mathbf{N}$, and if the iterated limit $\lim_{m \to \infty} \lim_{n \to \infty} \sigma(n, m)$ exists, then the double sequence converges, and all three values coincide.

Chapter 8

Infinite Series of Functions

In the last chapter, the concept of a sequence of functions was developed and various notions of convergence studied. In this chapter, the notion of an infinite series of functions is defined and analogous notions of convergence are developed. In thinking about this chapter, the reader should recall the previously established relationship between sequences and series of constants. Specifically, the basic theory of infinite series was developed by employing the sequence of partial sums as the foundation concept. This approach, which proved so successful in the earlier case, has obvious merit and should likely bear fruit in the present case. This then is the approach we will take, and with this in mind, the ambitious reader may want to try to develop some of the main ideas himself.

8.1 Convergence of Series of Functions

Recall the manner in which we constructed an infinite series $\sum a_n$ from a given sequence $\{a_n\}$ of real numbers and discussed the convergence of such a series by using the auxiliary sequence, $\{s_n\}$, of its partial sums. We can imitate the same technique to consider infinite series whose members are functions defined on a subset A of $\mathbf{R}$.

Definition. Let $\{f_n\}$ be a sequence of functions defined on a set $A \subseteq \mathbf{R}$. By the nth **partial sum** of this sequence, we mean the function, s_n, having domain A and defined by

$$s_n(x) = \sum_{i=1}^{n} f_i(x), \ x \in A.$$

The sequence of functions, $\{s_n\}$, is called the **associated sequence of partial sums**. By an **infinite series of functions** , we mean the pair $(\{f_n\},\{s_n\})$, where $\{f_n\}$ is a sequence of functions on A and $\{s_n\}$ is its associated sequence of partial sums.

Notation. We abbreviate the infinite series by the symbol $\sum_{n=1}^{\infty} f_n$. This will generally be further abbreviated to $\sum f_n$.

Discussion. Infinite series of functions will always be specified in terms of the formal collection of symbols, $\sum f_n$. To be able to work with this concept, it must be replaced by previously defined terms that capture the intent of the underlying notion. Thus, the moment an infinite series $\sum f_n$ of functions is given, we immediately replace it with the sequence of partial sums, $\{s_n\}$. This sequence consists of functions that are defined on the same subet A as the f_n's, and it is this sequence that will be the focus of our interest.

   Once the series has been replaced by a sequence of functions, natural questions arise as to whether the sequence, $\{s_n\}$, of functions converges pointwise or uniformly on the subset A. These questions will occupy the remainder of this section. $\square$

Definition. The series $\sum f_n$ of functions defined on a set A is said to **converge pointwise** on A provided there exists a function, f, having domain, A, and the sequence, $\{s_n\}$, of associated partial sums converges pointwise to f on A. In this case, we write

$$\sum f_n = f \text{ pointwise on } A$$

and refer to f as the pointwise limit of $\sum f_n$ on A.

Discussion. The statement $\sum f_n = f$ pointwise means that for each x in the common domain, A, we have the relationship

$$\lim_{n \to \infty} s_n(x) = f(x),$$

which in turn is equivalent to the statement that

$$\sum_{i=1}^{\infty} f_i(x) = f(x)$$

for each $x \in A$. This last assertion is a statement about an infinite series of constants. Thus, pointwise convergence of an infinite series of functions reduces to a statement about an infinite series of constants, in the same way that pointwise convergence for a sequence of functions reduced to a statement about convergence a sequence of constants. For this reason, we are able to bring to bear on questions of convergence for infinite series of functions, not only the previously developed theory for sequences of functions, but also the previously developed theory related to infinite series of constants.

   We have not stated the definition of pointwise convergence for series as an $\epsilon - N$ definition, and this is left to the reader as Exercise 1. $\square$

   The first concern with any limiting process is the uniqueness of the limit.

Theorem 8.1.1.   *Let $\sum f_n$ converge pointwise on A.   Then the limit, f, is unique.*

Proof. Exercise 2. $\square$

Example 1. Discuss the pointwise convergence of the geometric series $\sum_{n=1}^{\infty} x^{n-1}$.

Solution. Here, $f_n(x) = x^{n-1}$ for each $x \in \mathbf{R}$, and so

$$
\begin{aligned}
s_n(x) &= 1 + x + x^2 + \cdots + x^{n-1} \\
&= \begin{cases} \frac{1-x^n}{1-x}, & \text{if } x \neq 1 \\ n, & \text{if } x = 1. \end{cases}
\end{aligned}
$$

If $|x| < 1$, we know that $x^n \to 0$, and so $\{s_n(x)\}$ converges pointwise to the function, $f(x) = \frac{1}{1-x}$. For $|x| > 1$, it is clear that $\{x^n\}$ is divergent. Consequently, $\{s_n\}$ also diverges. For $x = 1$, $s_n(1) = n$, and hence the series diverges. Finally, if $x = -1$, we have the alternating series, $1 - 1 + 1 - 1 + \cdots$, which is also divergent. Thus, the above series converges pointwise to the function $\frac{1}{1-x}$ if and only if $|x| < 1$. $\square$

Discussion. Alternatively, the solution could proceed by fixing x and applying the results of Example 6.1.3 for the infinite series of constants, $\sum_{n=1}^{\infty} x^{n-1}$.

A noteworthy feature of this example is that each of the functions, f_n, is defined on the totality of $\mathbf{R}$. However, as is apparent from the solution, the series converges only on a portion of $\mathbf{R}$, namely, $(-1, 1)$. This is a very important feature of series of functions. Often the sequences of functions comprising the series will be defined on all of $\mathbf{R}$, but the series will converge only on an subset, or an interval, of $\mathbf{R}$. Indeed, it will often be the case that identifying the interval of convergence for a particular series is the crux of the problem. $\square$

Example 2. Discuss the pointwise convergence of the series $\sum \frac{x^n}{n!}$, $x \in \mathbf{R}$.

Solution. Fix $x \in \mathbf{R}$. Then $\sum \frac{x^n}{n!}$ is a series of constants, and we can apply the Ratio test (Theorem 6.4.4) with $a_n = \frac{x^n}{n!}$ to obtain

$$
\lim \left| \frac{\frac{x^{n+1}}{(n+1)!}}{\frac{x^n}{n!}} \right| = \lim \left| \frac{x^{n+1}}{(n+1)!} \times \frac{n!}{x^n} \right| = \lim \left| \frac{x}{n+1} \right| = 0.
$$

By Theorem 6.4.4 the series converges at x, and since x was arbitrary, the series converges pointwise on all of $\mathbf{R}$. $\square$

Discussion. The argument in this example illustrates how all of the previous theory developed for series of constants can be brought to bear on the issue of pointwise convergence of a series. All of the sophisticated tests for convergence are based on the behavior of the terms, and the rate at which the terms get small. As such, they can be applied to series of functions simply by holding x fixed.

The suggestion that the theory from series of constants has application to series of functions suggests a number of questions. Among them are what analogies exist between the theory of series of constants and series of functions, and which definitions from the theory of series of constants have natural extensions to the theory of series of functions? $\square$

Example 3. Discuss the pointwise convergence of the series, $\sum f_n$, of functions defined on $\mathbf{R}$ by $f_1(x) = \frac{1}{x^2+1}$ and for $n > 1$,

$$
f_n(x) = \left(\frac{x^{2n-2}}{x^{2n}+1} - \frac{x^{2n-4}}{x^{2n-2}+1} \right).
$$

Solution. From the expression for f_n as a difference of two terms, we notice that this is an example of a telescoping series, whence, as the reader can easily verify (Exercise 3),

$$s_n(x) = \frac{x^{2n-2}}{x^{2n} + 1}.$$

The reader can check (Exercise 3) that if x satisfies the inequality $|x| < 1$, then $\{s_n(x)\}$ converges to 0, whereas for $|x| > 1$, $\{s_n(x)\}$ has the limit $\frac{1}{x^2}$. But for $x = \pm 1$, $\lim_{n \to \infty} s_n(x) = \frac{1}{2}$. Hence, the pointwise limit function is given by

$$f(x) = \begin{cases} 0, & \text{if } |x| < 1 \\ 1/2, & \text{if } x = \pm 1 \\ 1/x^2, & \text{if } |x| > 1. \end{cases}$$

Thus, $\sum f_n = f$ (pointwise), for all $x \in \mathbf{R}$. $\square$

Discussion. The manner in which we obtained the expression for $s_n(x)$, by virtue of the fact that the series telescopes, was first discussed in Example 6.1.5. Since the resultant expression for $s_n(x)$ involves powers of x, we have to treat various cases and employ the facts about the geometric series.

As with series of constants, this example and Example 1 both demonstrate the utility of obtaining an expression for the nth partial sum. Using the expression for the partial sum, we are able to obtain a functional specification for the limit function, f.

Unlike Example 1, the present series converges for all $x \in \mathbf{R}$. However, the limit function is not continuous, even though the individual f_n's are continuous, as are the s_n's. While this result should not be surprising, given the theory for sequences of functions, it does suggest that the stronger form of convergence is required. $\square$

Definition. Let $\{f_n\}$ is a sequence of functions defined on a subset $A \subseteq \mathbf{R}$. We say that the infinite series of functions, $\sum f_n$, **converges uniformly** on the set A provided there exists a function, f, defined on A such that for every $\epsilon > 0$, there exists an $N \in \mathbf{N}$ such that for all $x \in A$,

$$\left| \sum_{i=1}^{n} f_i(x) - f(x) \right| = |s_n(x) - f(x)| < \epsilon \text{ whenever } n > N.$$

In the event that such an f exists, we say that $\sum f_n$ **converges to f uniformly on A**, and write $\sum f_n = f$ **uniformly on A**, or $\sum f_n \to f$ uniformly on A and refer to f as the **uniform limit** of $\sum f_n$ on A.

Discussion. We have given the $\epsilon - N$ definition of uniform convergence. We could as easily have given a definition of convergence by appealing to the concept $s_n \to f$ uniformly on A. Such a definition would have been completely equivalent to that given here (Exercise 4).

The key concept underlying uniform convergence for series is the same as that for sequences. Namely, the idea is create a tube of width 2ϵ, and require all partial sums having subscripts exceeding N to lie inside this tube. Notice that unlike the case for sequences, where the f_n's themselves had to lie inside the tube, for series it is the partial sums that must be contained inside the tube.

Indeed, in almost all cases, the functions, f_n, will not lie inside the 2ϵ-tube for any value of n. $\square$

Theorem 8.1.2. *Let $\sum f_n$ be a series of functions defined on A. If $\sum f_n$ converges uniformly to f on A, then f is the pointwise limit on A and, as such, is unique.*

Proof. Exercise 4. $\square$

From this definition it is easy to frame the familiar Cauchy Criterion for uniform convergence of an infinite series $\sum f_n$ as follows:

Theorem 8.1.3 (Cauchy Criterion). *A necessary and sufficient condition for the series $\sum f_n$ to converge uniformly on the set A is that given $\epsilon > 0$, we can find an $n \in \mathbf{N}$ such that for all $m > n > N$ and all $x \in A$, we have*

$$|s_m(x) - s_n(x)| < \epsilon,$$

or equivalently,

$$|f_{n+1}(x) + f_{n+2}(x) + \cdots + f_m(x)| < \epsilon.$$

Proof. Exercise 6. $\square$

Discussion. The reader is already familiar with Cauchy Criterion in several contexts. We again stress that the utility of this condition is that we do not have to know in advance that the limit function, f, in order to test for uniform convergence. $\square$

Theorem 8.1.4. *Let $\sum_{n=1}^{\infty} f_n$ and $\sum_{n=1}^{\infty} g_n$ be two infinite series that are uniformly convergent on A. If f and g are the respective limits, and a, b are two real constants, then*

$$\sum_{n=1}^{\infty} (af_n \pm bg_n) = af \pm bg \quad \text{uniformly on } A.$$

Proof. Exercise 8. $\square$

Next, we analyze the impact of uniform convergence on continuity.

Theorem 8.1.5. *If each f_n is continuous on A, and the series $\sum f_n$ converges uniformly to the function f on A, then f is continuous on A. Further, suppose that a is a limit point of A, and that for each $n \in N$, $\lim_{x \to a} f_n(x) = a_n$. Then*

$$\lim_{x \to a} \sum_{n=1}^{\infty} f_n(x) = \sum_{n=1}^{\infty} a_n.$$

Proof. If each member, f_n, of the series, $\sum f_n$, is continuous, certainly s_n has to be continuous, being a finite sum of continuous functions. Now, if $\sum f_n$ converges uniformly to f on A, then $\{s_n\}$ converges uniformly to f on A. Thus, by Theorem 7.3.1, we know that f is necessarily continuous. The remainder of the theorem follows from Theorem 7.3.2. $\square$

Discussion. The proof given above depends completely on the fact that $\sum f_n \to f$ uniformly on A exactly if the sequence of partial sums s_n tends uniformly to f on A. While this is apparently true, still, it must be verified (Exercise 4).

We have avoided giving an $\epsilon - N$ proof of this theorem, not because we dislike such proofs, but rather because we wanted to preserve that pleasure for the reader as Exercise 9. Generating such a proof is a useful exercise, since it will force the reader to come to grips with many of the inequalities employed in these types of arguments. $\square$

Example 4. Discuss the uniform convergence of the series, $\sum_{n=0}^{\infty} x(1 - x)^n$, on $[0, 1]$.

Solution. A simple computation (Exercise 11) shows that

$$s_n(x) = \sum_{k=0}^{n} x(1-x)^k = x\left[\frac{1 - (1 - x)^{n+1}}{1 - (1 - x)}\right].$$

Now, if $x \in (0, 1)$, then

$$\lim_{n \to \infty} s_n(x) = x\left(\frac{1}{1 - 1 + x}\right) = 1.$$

For $x = 0, 1$, clearly $f_n(x) = 0$, whence $s_n(x) = 0$ for all $n \in \mathbf{N}$. Thus the pointwise-limit function f is defined by

$$f(x) = \begin{cases} 1, & \text{if } x \in (0, 1) \\ 0, & \text{if } x = 0, \ 1. \end{cases}$$

Consequently, we write $\sum x(1 - x)^n = f(x)$. Since the sum function f is discontinuous, while each component is continuous, by Theorem 8.1.5 the convergence cannot be uniform. $\square$

Discussion. Note the ease with which we exploited Theorem 8.1.5 to negate uniform convergence, once we knew that the sum was a discontinuous function. This method should be tried before trying the definition using the $\epsilon - N$ technique. Because of its utility in this respect, Theorem 8.1.5 can be viewed as a test for the failure of uniform convergence of a series of functions. $\square$

There is a very useful positive test for uniform convergence of a series of functions that arises by relating the terms of the series to those of a series with constant terms. This is known as Weierstrass's M-test and is given in the next theorem.

Theorem 8.1.6 (Weierstrass's M-test). *If $\sum f_n$ is a series of functions defined on A and $\{M_n\}$ is a sequence of real numbers such that $\sum M_n$ is convergent, and the relation $|f_n(x)| \le M_n$ holds for all $x \in A$, and for each $n \in \mathbf{N}$, then $\sum f_n$ converges uniformly and absolutely on the set A.*

Proof. Let $\epsilon > 0$ be given. Since $\sum M_n$ converges we can find an $N \in \mathbf{N}$ such that for $m, n > N$, we have $\sum_{i=n+1}^{m} M_i < \epsilon$. Now for each $x \in A$,

$$|s_m(x) - s_n(x)| = \left| \sum_{i=n+1}^{m} f_i(x) \right|$$

$$\leq \sum_{i=n+1}^{m} |f_i(x)| \leq \sum_{i=n+1}^{m} M_i < \epsilon,$$

whenever $m > n > N$. Hence, by Cauchy Criterion, we obtain uniform convergence on A. Absolute convergence is implicit in the inequalities generated. □

Discussion. The reader will have noticed that we have adopted a very relaxed style in the presentation of proofs. In the present argument, we have not mentioned any of the basic elementary facts, such as the triangle inequality, which are used in the proof. It is assumed that the reader is so familiar with these facts that they are by now second nature, and that the reader will invoke them as required. This assumption may be incorrect. For this reason, the reader is invited to test its truth in Exercise 12.

As well, we did not define the notion of absolute convergence for series of functions. The original definition of absolute convergence is in Section 6.3. If one thinks about how to generalize this definition, the natural generalization would appear to be that $\sum f_n$ converges **absolutely** on A, provided $\sum |f_n|$ converges pointwise on A. This definition raises obvious questions. For example, if $\sum f_n$ converges absolutely and uniformly on A, will the absolute series converge uniformly? If a series converges absolutely, will the convergence be uniform? And so forth. Some of these questions are answered in the remaining examples of this section, others by examples in the exercises.

Weierstrass's M-test is one of the most practical methods for establishing uniform convergence in cases for which one does not have a specification for the pointwise limit. The only tool needed is a proper choice of a convergent series, $\sum M_n$, which dominates the terms f_n over the whole of A. Note also that it suffices that the condition $|f_n(x)| \leq M_n$ hold after a certain stage, and not necessarily for all n without exception. □

Example 5. Discuss the uniform convergence of $\sum \frac{\sin nx}{n^2}$, $x \in \mathbf{R}$.

Solution. Here, $f_n(x) = \frac{\sin nx}{n^2}$. Since the maximum value of $\sin nx$ is 1, $\sum \frac{1}{n^2}$ converges, and $|\frac{\sin nx}{n^2}| \leq \frac{1}{n^2}$, by Weierstrass's M-test, we conclude that $\sum f_n$ converges uniformly and absolutely. □

Discussion. Weierstrass's test for uniform convergence automatically guarantees that the series is also absolutely convergent. Thus, if a series is uniformly but not absolutely convergent, then its uniform convergence cannot be derived by the application of the M-test. While this might suggest that an example of a series that converged uniformly, but not absolutely, was very ill-behaved, such is not the case. If the reader cannot create such an example on his own, an example can be found among the series in Exercise 13.

Note the manner in which we hit upon the candidate for the M_n series, namely, $\frac{1}{n^2}$, which converges. The success here is due to the fact that we can bound the sine function by a maximum value, 1, thus eliminating the variable x from the expression and obtaining a constant series. Finally, from the absolute convergence of $\sum f_n$, we conclude that $\sum f_n(x)$ converges for each x. $\square$

Example 6. Does the series $\sum_{n=0}^{\infty} \frac{x^n}{n!}$ converge uniformly and/or absolutely on **R**?

Solution. If Theorem 6.4.4 is applied to the computations in Example 6.4.3, it is seen that the series $\sum \frac{x^n}{n!}$ converges absolutely on **R**. We claim that the convergence is not uniform. To see this, observe that $\sum_{n=0}^{\infty} \frac{x^n}{n!}$ converges pointwise on **R** to the limit function e^x, as we will show in Chapter 9. Given this is the case, the reader can show that for any fixed n, we can find x such that $|e^x - s_n| \geq 1$ (Exercise 14). Thus, the convergence is not uniform on **R**. $\square$

Discussion. This example shows that absolute convergence on **R** does not imply uniform convergence. Further examples can be constructed to show that absolute convergence on $[a, b]$ does not imply uniform convergence there (Exercise 15). It may be more surprising that uniform convergence does not imply absolute convergence. The reader can construct an example or find an example in the exercises. $\square$

There are several other delicate tests, more powerful than the M-test, that are useful to test the uniform convergence of a conditionally convergent series of functions. Two such tests are appended below. We start with a definition.

Definition. The infinite series, $\sum_{n=1}^{\infty} f_n$, is said to be **uniformly bounded** on a set A provided there exists a constant $K > 0$ such that $|\sum_{k=1}^{n} f_k(x)| < K$ for all $n \in \mathbf{N}$ and for all $x \in A$.

Discussion. The uniformity of the boundedness concept is spelled out by the requirement that we need to produce a single constant K for which

$$\left| \sum_{k=1}^{n} f_k(x) \right| < K \text{ for all } x \in A \text{ and for all } n \in \mathbf{N}.$$

The reader should note the difference between this concept and a concept of pointwise boundedness.

The definition of a uniformly bounded sequence of functions has previously appeared in Exercise 7.1.11. In essence, the present definition asserts that the sequence of partial sums, $\{s_n\}$, is uniformly bounded on A. Both concepts will be required in the next several theorems. $\square$

The following simple lemma is needed for the next two tests.

Lemma. *Let $\sum_{n=1}^{\infty} a_n$ converge to a, $s_n = \sum_{i=1}^{n} a_i$, and let $t_n = \sum_{i=n+1}^{\infty} a_i$. If $\{b_n\}$ is an arbitrary sequence, then*

(i) $a_n b_n = s_n[b_n - b_{n+1}] - s_{n-1}b_n + s_n b_{n+1};$

(ii) $a_n b_n = -t_n[b_n - b_{n+1}] + t_{n-1}b_n - t_n b_{n+1}.$

Proof. A straightforward computation, using $s_n = a_1 + a_2 + \ldots + a_n$, yields (i), and (ii) follows easily from (i) by substituting $t_n = a - s_n$. $\square$

Discussion. These identities are extremely useful, in the discussion of the convergence of a series of the form $\sum f_n g_n$, as the next two tests clearly illustrate. Convergence of the series $\sum a_n$ was not required in the proof of (i). $\square$

Theorem 8.1.7 (Abel's test). *If $\sum f_n$ is uniformly convergent on a set A, $\{g_n\}$ is a uniformly bounded sequence on A such that for each $x \in A$, $\{g_n(x)\}$ is a monotonic (increasing or decreasing) sequence, then the series $\sum f_n g_n$ is uniformly convergent on A.*

Proof. If $s_n(x)$ denotes the partial sum function of $\sum f_n$, we can define $t_n = \sum_{i=n+1}^{\infty} f_i(x)$, so that $\sum f_n = s_n + t_n$ (**WHY?**). Using the identity (ii) of the above lemma and summing $f_i g_i$ from $i = n+1$ to $n+k$, we obtain

$$\sum_{i=n+1}^{n+k} f_i g_i = \sum_{i=n+1}^{n+k} [-t_i(g_i - g_{i+1})] + t_n g_{n+1} - t_{n+k} g_{n+k+1},$$

so

$$\left| \sum_{i=n+1}^{n+k} f_i g_i \right| \leq \sum_{i=n+1}^{n+k} [|t_i| |g_i - g_{i+1}|] + |t_n| |g_{n+1}| + |t_{n+k}| |g_{n+k+1}|.$$

Let $\epsilon > 0$ be given. By uniform boundedness of $\{g_n\}$ there is $K > 0$ such that $|g_n(x)| < K$ for all x in A and for all n. Also, there exists $N \in \mathbf{N}$ such that for all $x \in A$ and all $n > N$, $|t_n(x)| < \frac{\epsilon}{4K}$ (**WHY?**). Again, by the monotonicity of the sequence $\{g_n\}$, we have

$$\sum_{i=n+1}^{n+k} |g_i(x) - g_{i+1}(x)| = |g_{n+1}(x) - g_{n+k+1}(x)| < 2K.$$

Putting all these facts together, for $k \geq 0$,

$$\left| \sum_{i=n+1}^{n+k} f_i(x) g_i(x) \right| < 2K \cdot \frac{\epsilon}{4K} + 2K \cdot \frac{\epsilon}{4K} = \epsilon.$$

The desired conclusion follows, once we invoke the Cauchy Criterion. $\square$

Discussion. The proof of this theorem is quite complicated. One would like to construct a simple proof, that would be founded on the following string of inequalities:

$$\left| \sum_{k=n}^{m} f_k g_k \right| \leq \sum_{k=n}^{m} |f_k g_k| = \sum_{k=n}^{m} |f_k| |g_k| \leq \sum_{k=n}^{m} |f_k| K.$$

One would then apply the Cauchy Criterion to the last series and be done. The difficulty with this argument is that the last series is not, in general, convergent. The canonical example of a resultant series, that is, at the far right, which would fail to converge is the series, $\sum \frac{x}{n}$ defined on $[0, 1]$. The reader may think that this situation is avoided because its occurrence requires the introduction of absolute values. Such is not the case, for $\{f_n\}$ and $\{g_n\}$ may be chosen so as to guarantee $f_n g_n \geq 0$.

The question therefore arises: What in the hypothesis prevents such a situation from arising? The answer is monotonicity of $\{g_n\}$. If this assumption is dropped, then the divergent series $\sum \frac{x}{n}$ may result as the product, $f_n g_n$. $\square$

Theorem 8.1.8 (Dirichlet's test).   *If $\{f_n\}$ and $\{g_n\}$ are such that $\sum f_n$ is uniformly bounded on A, the sequence $\{g_n\}$ is uniformly convergent on A, and for each $x \in A$, the sequence $\{g_n(x)\}$ is monotonic (increasing or decreasing) and tends to 0, then $\sum f_n g_n$ is uniformly convergent on A.*

Proof. We use the identity (i) in the lemma (where s_n has the usual meaning) and sum $f_i g_i$ from $i = n + 1$ to $n + k$ to obtain

$$\left| \sum_{i=n+1}^{n+k} f_i g_i \right| \leq \sum_{i=n+1}^{n+k} [|s_i||g_i - g_{i+1}|] + |s_{n+k}||g_{n+k+1}| + |s_n||g_{n+1}|.$$

Let $\epsilon > 0$ be given and find $K > 0$ satisfying $|s_n(x)| < K$ for all $x \in A$, and all $n > N$. By uniform convergence of $\{g_n\}$, there exists $N > 0$ such that for all $x \in A$, and $n > N$, we have $|g_n(x)| < \frac{\epsilon}{4K}$. Hence, if $k > 0$, $n > N$, and $x \in A$, we have

$$\left| \sum_{i=n+1}^{n+k} f_i(x) g_i(x) \right| \leq K \sum_{i=n+1}^{n+k} |g_i(x) - g_{i+1}(x)| + 2K \frac{\epsilon}{4K}$$

$$< K|g_{n+1}(x) - g_{n+k+1}(x)| + \frac{\epsilon}{2} \text{ by monotonicity of } g_n$$

$$\leq \epsilon,$$

proving the uniform convergence of $\sum f_n g_n$. $\square$

Discussion. Similar comments apply to the role of monotonicity in the hypothesis of Dirichlet's test. $\square$

Example 7. Discuss the uniform convergence of $\sum_{n=1}^{\infty} \frac{\sin nx}{n^p}$, $0 < p \in \mathbf{R}$.

Solution. We can take $f_n(x) = \sin nx$, and by elementary trigonometry, it can be shown (Exercise 19, also see Exercise 5.3.21) if $x \neq \pm 2k\pi$, $k = 1, 2, \ldots$, then

$$s_n(x) = \sum_{k=1}^{n} f_k(x) = \frac{\sin \frac{nx}{2} \sin \frac{(n+1)x}{2}}{\sin \frac{x}{2}}.$$

If $x \in [d, 2\pi - d]$, then $|s_n(x)| < \frac{1}{\sin \frac{d}{2}}$, so that $\{s_n(x)\}$ is uniformly bounded in $[d, 2\pi - d]$. Letting $g_n(x) = \frac{1}{n^p}$, the sequence $\{g_n(x)\}$ is free of the variable x, and so is trivially uniformly convergent to 0 in the same interval, if $0 < k \leq 1$. So the conditions for Dirichlet's test are fulfilled, and by applying the test, we conclude that the given series is uniformly convergent in $[d, 2\pi - d]$. $\square$

Discussion. It is worth comparing the solution for this example to that for Example 5. The solution for Example 5 employed Weierstrass's M-test. Why not do the same in this case? The difficulty with the present example is the failure of $\sum \frac{1}{n^p}$ to converge for $p \leq 1$. Thus, more subtle methods are required. $\square$

Next, we provide a short discussion of the term-by-term integration and term-by-term differentiation of an infinite series of functions. The considerations are very similar to those of sequences of functions discussed earlier. The following theorem justifies term-by-term integration in the presence of uniform convergence.

Theorem 8.1.9. *If each f_n is Riemann integrable in $[a, b]$, and the series $\sum f_n$ converges uniformly in $[a, b]$ to f, then f is also Riemann integrable in $[a, b]$ and further $\sum \int_a^b f_n = \int_a^b f$. (In other words, the integral of the sum is the sum of the integrals.)*

Proof. Let $\{s_n\}$ denote the sequence of partial sums of the series. Then, the sequence, s_n, converges uniformly in $[a, b]$ to f. Clearly, s_n is integrable, being a finite sum of integrable functions. Hence by Theorem 7.3.3, f is Riemann integrable in $[a, b]$ and

$$\int_a^b f = \lim \int_a^b s_n.$$

Using the fact that

$$\int_a^b s_n = \sum_{i=1}^n \int_a^b f_i,$$

and taking limit as n approaches infinity, we obtain

$$\int_a^b f = \sum_{n=1}^\infty \int_a^b f_n. \quad \square$$

The final theorem of this section gives conditions under which term-by-term differentiation of a series of function is valid.

Theorem 8.1.10. *If each f_n possesses a derivative throughout $[a, b]$, each f_n' is continuous on $[a, b]$ for each n, $\sum f_n$ converges to f on $[a, b]$, and $\sum f_n'$ converges uniformly on $[a, b]$, then*

$$f'(x) = \sum f_n'(x)$$

for $x \in [a, b]$.

Proof. The partial sum s_n has a derivative throughout $[a, b]$, $\{s_n\}$ converges to f on $[a, b]$, and since $s_n' = f_1' + f_2' + \cdots + f_n'$, it is clear that $\{s_n'\}$ converges uniformly to a function g on $[a, b]$. So Theorem 7.3.4 applies to guarantee that $f' = g$ throughout $[a, b]$, whence $f'(x) = \sum f_n'$ for each $x \in [a, b]$. $\square$

Discussion. The proof of both of these theorems depend for their truth on finite additivity. That is, a finite sum of integrable (differentiable), functions is integrable (differentiable) and the sum of the integrals (derivative) is the integral (derivative) of the sum. These theorems delineate the circumstances in which this property, finite additivity, can be extended to infinite sums. $\square$

We conclude this section with an example of an everywhere continuous, nowhere differentiable function, which we promised in Chapter 4. The first example was constructed by Weierstrass in 1872, but it is believed that Bolzano knew of such a construction as early as 1830. The version we present here is essentially that of Van der Waerden [1930].

Example 8. There exists a function $f : \mathbf{R} \to \mathbf{R}$, which is continuous on $\mathbf{R}$, but does not possess a derivative at any point of $\mathbf{R}$.

Solution. We start with the function $g(x) = |x|, -2 \le x \le 2$, and extend it to the entire real line, periodically, by setting $g(x + 4p) = g(x)$, $x \in \mathbf{R}$, $p \in \mathbf{Z}$. If $S = \{4m : m \in \mathbf{Z}\}$, it is then clear that

$$g(x) = \inf\{|x - s| : s \in S\}. \tag{1}$$

Also, note that if there is no even integer in the open interval (a, b), then $|g(a) - g(b)| = |a - b|$. For each n, set $f_n(x) = \frac{g(4^n x)}{4^n}$, $x \in \mathbf{R}$. Let $f(x) = \sum_{n=1}^{\infty} f_n(x)$. We claim that the function f has the desired properties.

Clearly, each f_n is continuous on $\mathbf{R}$ (**WHY?**). Also, $0 \le f_n(x) \le \frac{2}{4^n}$. Since $\sum \frac{2}{4^n}$ is a convergent series of constants, by Weierstrass's M-test, it follows that the series $\sum f_n$ is uniformly convergent to f on $\mathbf{R}$. Also, since each f_n is continuous on $\mathbf{R}$, we invoke Theorem 8.1.5 to conclude that f is continuous on $\mathbf{R}$.

We now show that f is not differentiable at any point on $\mathbf{R}$. Let $a \in \mathbf{R}$ be arbitrary, but fixed. We demonstrate that $f'(a)$ does not exist. To this end, for each $k \in \mathbf{N}$, define $\delta_k = +1$ or -1, depending on k, such that there is no even integer in the open interval $(4^k a, 4^k a + \delta_k)$. In the ambiguous case where $4^k a$ happens to be an integer, either choice will do. For $1 \le n \le k$, we observe that there is no even integer between $4^n a$ and $4^n a + 4^{n-k} \delta_k$. For, if an even integer $2p$ were between them, then $4^{k-n} 2p$ would be between $4^k a$ and $4^k a + \delta_k$, which is a contradiction. Hence, by (1) we have

$$\left| f_n \left(a + \frac{1}{4^k} \delta_k \right) - f_n(a) \right| = \frac{1}{4^n} \left| g(4^n a + 4^{n-k} \delta_k) - g(4^n a) \right|$$

$$= \frac{1}{4^k} \ (1 \le n \le k) \tag{2}$$

On the other hand, if $n > k$, it is clear from (1) and the first equality in (2) that

$$f_n \left(a + \frac{1}{4^k} \delta_k \right) = f_n(a). \tag{3}$$

Call $h_k = \frac{\delta_k}{4^k}$, and use (2) and (3) to obtain

$$\frac{f(a + h_k) - f(a)}{h_k} = \sum_{n=1}^{k} \frac{f_n(a + h_k) - f_n(a)}{h_k} = \sum_{n=1}^{k} (\pm 1).$$

The last expression is an even integer if k is even and an odd integer if k is odd. When $k \to \infty$, $h_k \to 0$, and so

$$\lim_{h_k \to 0} \frac{f(a + h_k) - f(a)}{h_k}$$

does not exist. This proves that $f'(a)$ does not exist. $\square$

Discussion. This basic fact employed in this ingenious construction is that the absolute value function is not differentiable at its corner point, $x = 0$, even though it is continuous. The function, g, described above is composed of several of these graphs arrayed together on the real line, so as to form an infinite number of 'roofs', a roof constructed on each interval $(4m, 4m + 4)$, with height 2 at $4m + 2$, where $m \in \mathbf{Z}$. Such a graph is called a **saw-tooth function**. The function, f_1, has the same shape as that of g, except that each roof in the function g gets replaced by 4 smaller roofs, with $\frac{1}{4}$ times the width and height of the roofs in g. Thus, the graph of f_n consists of 4^n roofs in any interval of the form $(4m, 4m + 4), m \in \mathbf{Z}$, and the height of each roof is $\frac{1}{4^n}$. Note that each roof contributes three points of nondifferentiability, one at its maximum and two points at either ends, common to the two adjacent roofs. So, as n increases, we have more and more points of nondifferentiability, since there are more and more sharp corners. Specifically, if $f_n(x) = 0$ for some x, it follows that $f_m(x) = 0$ for all $m > n$; also the 'peak point' of f_n reduces to a 'zero point' for f_{n+1}. Thus, it is conceivable that when we take the sum of all these functions, and then take limit as n tends to infinity, we must have destroyed the differentiability at a huge collection of point. The above proof formalizes this intuition. The graphs of g and f_1 are drawn in Figure 8.1.1.

The continuity part of the proof is straightforward and employs only the fact that uniform convergence preserves continuity. The differentiability part needs a delicate argument, which is not hard. Rather than proving $\lim_{h \to 0} \frac{f(a+h)-f(a)}{h}$ does not exist, we choose a sequence $\{h_k\}$ approaching 0, so that the corresponding difference quotient does not approach a limit. $\square$

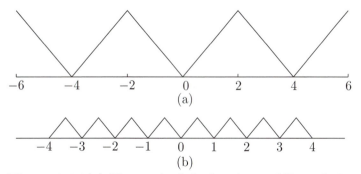

Figure 8.1.1(a) The graph of the function g of Example 8.
(b) The graph of the function f_1 defined by $f_1(x) = \frac{1}{4}g(4x)$.

Exercises

1. Give an $\epsilon - N$-type definition for the pointwise convergence of an infinite series of functions.

2. Prove Theorem 8.1.1.

3. Verify the formula $s_n(x)$ in Example 3. Verify the statements about the limit of the sequence.

4. Give a definition for the uniform convergence of a series of functions in terms of uniform convergence of the associated sequence of partial sums. Show that this definition is completely equivalent to the definition given. Conclude that the uniform limit must coincide with the pointwise limit, and hence must be unique.

5. Write a discussion of the geometric interpretation of uniform convergence for an infinite series of functions.

6. Prove Theorem 8.1.3.

7. Make a list of the various forms of Cauchy criteria that have been developed in this text. Write a discussion that illustrates their essential similarities.

8. Prove Theorem 8.1.4. State and prove an similar theorem related to pointwise convergence.

9. Give an $\epsilon - N$ proof of Theorem 8.1.5.

10. Write a discussion of the proof of Theorem 8.1.5. What is the role of the triangle inequality in the proof? What is the geometric intuition underlying the proof?

11. Verify the formula for s_n in Example 4.

12. What is the role of the Triangle inequality in Theorem 8.1.6? Why does the proof imply that the series converges absolutely?

13. Discuss the uniform and absolute convergence of the following series of functions on the indicated domains:

(a) $\sum_{n=1}^{\infty} n^2 x^n$, $x \in [\frac{-1}{2}, \frac{1}{2}]$;

(b) $\sum_{n=1}^{\infty} \frac{x^n}{n^2(1+x^n)}$, $x \in [0, 1]$;

(c) $\sum_{n=1}^{\infty} \frac{\cos nx}{5^n}$, $x \in \mathbf{R}$;

(d) $\sum_{n=1}^{\infty} \frac{(-1)^n}{n^x}$, $x \geq 1$;

(e) $\sum_{n=1}^{\infty} ne^{-nx}$, $x \geq 2$;

(f) $\sum_{n=1}^{\infty} \frac{x}{n^\alpha(1+nx^2)}$, $\alpha > \frac{1}{2}$, $x \in \mathbf{R}$;

(g) $\sum_{n=1}^{\infty} \frac{(-1)^n x^{2n}}{n^p(1+x^{2n})}$, $x \in \mathbf{R}$, $p \in \mathbf{R}$;

(h) $\sum_{n=1}^{\infty} \frac{x^2}{(1+x^2)^n}$, $x \in \mathbf{R}$;

(i) $\sum_{n=1}^{\infty} e^{-nx} x^n$, $x \in [0, 100]$;

(j) $\sum_{n=1}^{\infty} \frac{(-1)^{n+1} x^2}{n}$, $x \in [0, 5]$;

(k) $\sum_{n=1}^{\infty} \frac{1}{n^2+x^2}$, $x \in \mathbf{R}^+$.

(l) $\sum_{n=1}^{\infty} \frac{nx^2}{n^3+x^3}$, $x \in [0, a]$, $a > 0$;

(m) $\sum_{n=1}^{\infty} \frac{(-1)^n}{x+n}$, $x \in \mathbf{R}^+$;

(n) $\sum_{n=1}^{\infty} \frac{1}{x^n+1}$, $x \in \mathbf{R}^+$;

(o) $\sum_{n=1}^{\infty} \frac{x^{2n}}{(x+n)^2}$, $x \in [0, 1]$;

(p) $\sum_{n=1}^{\infty} \frac{x^n}{n^2}$, $x \in [5, \infty)$;

(q) $\sum_{n=1}^{\infty} \frac{1}{1+n^2x^2}$, $x \in [1, \infty)$;

(r) $\sum_{n=1}^{\infty} n^{-x}$, $x \in [\sqrt{3}, \infty)$;

(s) $\sum_{n=1}^{\infty} \frac{1}{(nx)^2}$, $x \neq 0$;

(t) $\sum_{n=1}^{\infty} (\frac{x^{2n+1}}{2n+1} - \frac{x^{n+1}}{2n+2})$, $x \in [0, 1]$;

(u) $\sum_{n=1}^{\infty} n^{-x}$, $[1 + \delta, \infty)$, $\delta > 0$;

(v) $\sum_{n=1}^{\infty} \frac{\sin nx}{n2^n}$, $x \in \mathbf{R}$;

(w) $\sum_{n=1}^{\infty} \sin^n x$, $x \in [0, \frac{\pi}{2} - \delta]$, $\delta > 0$;

(x) $\sum_{n=1}^{\infty} \frac{1}{n} \sqrt{\frac{|\sin nx|}{n}}$, $x \in \mathbf{R}$;

(y) $\sum_{n=1}^{\infty} \frac{x^n}{n} e^{nx}$, $x \in [0, 12]$;

(z) $\sum_{n=1}^{\infty} (1 - x^2) x^n$, $x \in [0, 1]$;

(a') $\sum_{n=1}^{\infty} \frac{x}{1+nx^p}$, $x \in \mathbf{R}$ $p \in \mathbf{R}$;

(b') $\sum_{n=1}^{\infty} \frac{1}{n^p+|x|^p}$, $x \in \mathbf{R}$ $p \in \mathbf{R}$;

(c') $\sum_{n=1}^{\infty} \frac{x}{n^p+x^2n^q}$, $x \in \mathbf{R}$ $p, q \in \mathbf{R}$;

(d') $\sum_{n=1}^{\infty} \frac{x^n}{1+x^{n+1}}$, $x \in [0, a]$, $0 < a < 1$ and $x \in [0, 1)$;

(e') $\sum_{n=1}^{\infty} \frac{(-1)^n \sin(1+\frac{x}{n})}{\sqrt{n}}$, any compact subset of $\mathbf{R}$;

(f') $\sum_{n=1}^{\infty} \frac{x}{(1+(n-1)x)(1+nx)}$, $x \in [a, b]$, $[0, b)$, $b > a > 0$;

(g') $\sum_{n=1}^{\infty} f_n(x)$, where for $x \in \mathbf{R}$,

$$f_n(x) = \begin{cases} 1, & \text{if } x = \frac{m}{n}, \ m, n \in \mathbf{Z}, \ (n, m) = 1 \\ 0, & \text{otherwise.} \end{cases}$$

14. Under the assumption that $\sum_{n=0}^{\infty} \frac{x^n}{n!}$ converges pointwise on $\mathbf{R}$ to $f(x) = e^x$, show the convergence cannot be uniform.

15. Construct an example of a series that is absolutely convergent on $[0, 1]$, but that fails to be uniformly convergent there.

16. Prove the identities (a) and (b) of the lemma that precedes Abel's test (Theorem 8.1.7).

17. If $\{f_n\}$, $\{g_n\}$ be such that $|f_n| \le |g_n|$ on A, and if $\sum g_n$ is uniformly convergent on A, prove the same applies to $\sum f_n$.

18. If $\sum f_n$ converges uniformly on A, prove that the sequence of functions $\{f_n\}$ converges uniformly to the zero function. Is the converse true?

19. If $\{f_n\}$ is a decreasing sequence of nonnegative functions that converge uniformly to 0 on A, prove that the alternating series $\sum (-1)^n f_n$ converges uniformly on A.

20. Consider $\sum f_n$, where $f_{2n}(x) = -f_{2n+1}(x) = (x^n - x^{n+1})$, $(x \in \mathbf{R})$. Show that $\sum f_n$ is uniformly and absolutely convergent in $\mathbf{R}$, but $\sum |f_n|$ is not uniformly convergent on $\mathbf{R}$.

21. Consider the trigonometric identity required for the solution of Example 7. Using any of the basic trigonometric identities that can be found in an elementary calculus book, prove this identity for $n = 2$. Now prove it for arbitrary values of n.

22. If $\sum a_n$ converges absolutely, prove that $\sum_{n=0}^{\infty} a_n \sin nx$ and $\sum_{n=0}^{\infty} a_n \cos nx$ are both uniformly convergent.

23. Let $\{a_n\}$ be a decreasing sequence of positive terms. Prove the series $\sum a_n \sin nx$ converges uniformly on $\mathbf{R}$ if and only if $na_n \to 0$ as $n \to \infty$.

24. If $\{a_n\}$ is monotonic sequence converging to 0, discuss the uniform convergence of $\sum a_n \sin nx$ and $\sum a_n \cos nx$ in $[d, 2\pi - d]$.

25. Show $\sum \frac{\sin nx}{n^2}$ converges uniformly on $\mathbf{R}$ but the differentiated series $\sum \frac{\cos nx}{n}$ diverges for all x.

26. If $f(x) = \sum \frac{x^n}{n}$ for $x \in [0, 1)$, show $\int_0^1 f = \sum_{n=1}^{\infty} \frac{1}{n(n+1)}$.

27. Give examples in support of the following:

(a) a series may converge uniformly, but not absolutely;

(b) a series may converge to an integrable limit without being uniformly convergent;

(c) a series of continuous functions with a continuous limit function need not be uniformly convergent;

(d) $\sum f_n$ converges pointwise to f, term-by-term differentiation is valid, but the convergence is not uniform;

(e) $\sum f_n$ converges pointwise to f, term-by-term integration is valid, but the convergence is not uniform;

(f) the converse of Weierstrass's M-test is not true, that is, $\sum f_n$ converges uniformly and absolutely on A, but $\sum M_n$ diverges, where

$$M_n = \sup\{f_n(x) : x \in A\}.$$

28. Comment on the effect of the following operations on a uniformly convergent series $\sum f_n$ on a domain A:

(a) adding a finite number of terms;

(b) deleting a finite number of terms;

(c) adding an infinite number of terms;

(d) deleting an infinite number of terms;

(e) adding an infinite number of terms, each of which is the zero function.

29. Prove **Dini's theorem**: If each f_n is nonnegative and continuous on $[a, b]$ and $\sum f_n$ converges to a continuous function on $[a, b]$, then the convergence is uniform in $[a, b]$.

30. Prove the series $\zeta(x) = \sum_{n=1}^{\infty} \frac{1}{n^x}$ (known as **Riemann's zeta function**) converges uniformly on every interval of the form $1 + h \leq x < \infty$, where $h > 0$. Justify the validity of the equation

$$\zeta'(x) = -\sum_{n=1}^{\infty} \frac{\ln n}{n^s}$$

for $x > 1$. Obtain a similar series for the kth derivative $\zeta^{(k)}(x)$.

31. Given a convergent series $\sum a_n$, prove the series, $\sum_{n=1}^{\infty} \frac{a_n}{n^x}$, converges uniformly on the interval $0 \leq x < \infty$. Use this to show

$$\lim_{x \to 0+} \sum_{n=1}^{\infty} \frac{a_n}{n^x} = \sum_{n=1}^{\infty} a_n.$$

32. If $\sum |a_n|$ converges, prove

$$\int_0^1 \left(\sum_{n=1}^{\infty} a_n x^n \right) = \sum_{n=1}^{\infty} \frac{a_n}{n+1}.$$

33. Given $f(x) = \sum_{n=1}^{\infty} \frac{1}{n^3 + n^4 x^2}$, justify the equation

$$f'(x) = -2x \sum_{n=1}^{\infty} \frac{1}{n^2(1+nx^2)^2}.$$

34. Justify the equation

$$\lim_{x \to 1} \left[\sum_{n=1}^{\infty} \frac{nx^2}{n^3 + x^3} \right] = \sum_{n=1}^{\infty} \frac{n}{n^3 + 1}.$$

35. Find the derivative f' for the following functions:

(a) $f(x) = \sum_{n=-1}^{\infty} \frac{\sin nx}{e^{nx}}$, $x > 0$;   (b) $f(x) = \sum_{n=1}^{\infty} \frac{\sin nx}{n^2}$, $x \in [0, \pi]$;

(c) $f(x) = \sum_{n=1}^{\infty} \frac{n}{x^n}$, $|x| > 1$.

36. Compute the following:

(a) $\int_0^{\pi} \sum_{n=1}^{\infty} \frac{\sin nx}{n^2}$;   (b) $\int_1^2 \sum_{n=1}^{\infty} \frac{n}{e^{nx}}$;

(c) $\int_a^b \sum_{n=1}^{\infty} \frac{\sin nx}{e^{nx}}$, where $0 < a < b$.

37. If $\sum |a_n|$ converges, prove that $\sum a_n f_n$ also converges uniformly, where f_n is any one of the following:

(a) $\cos nx$;   (b) $\sin nx$;   (c) $\cos a_n x$;

(d) $\sin a_n x$;   (e) $\frac{|x|^n}{1+x^{2n}}$;   (f) $\frac{x^{2n}}{1+x^{2n}}$.

38. If $\{f_n\}$ is bounded on A, and $\sum a_n$ is absolutely convergent, is it true that $\sum a_n f_n$ is uniformly convergent on A?

39. Let $\sum f_n \to f$ uniformly on A, a be a limit point of the common domain, and for each $n \in \mathbf{N}$, $\lim_{x \to a} f_n = a_n$. Prove that $\sum a_n$ converges to $\lim_{x \to a} f$.

40. Prove **du Bois-Reymond test**: If $\sum f_n$ and $\sum g_n$ are such that $\sum f_n$ and $\sum |g_n - g_{n+1}|$ are uniformly convergent on A, and $\{g_n\}$ is uniformly bounded on A, then $\sum f_n g_n$ is uniformly convergent on A.

41. Prove **Dedekind's test**: If the sequence $\{s_n\}$ of partial sums of $\sum f_n$ is uniformly bounded, the sequence $\{g_n\}$ converges uniformly to 0, and if $\sum |g_n - g_{n+1}|$ is uniformly convergent, then $\sum f_n g_n$ converges.

42. If either

(a) $\{a_n\}$ is monotonic and decreases to 0 or

(b) $\{a_n\} \to 0$, and $\sum |a_n - a_{n+1}| < \infty$,

prove $\sum a_n \sin nx$ and $\sum a_n \cos nx$, both converge uniformly in $[\delta, 2\pi - \delta]$, $\delta > 0$.

43. Let $h : [0, 2] \to \mathbf{R}$ be defined by

$$h(t) = \begin{cases} 0, & \text{if } t \in [0, \frac{1}{3}] \cup [\frac{5}{3}, 2] \\[2mm] 1, & \text{if } t \in [\frac{2}{3}, \frac{4}{3}] \\[2mm] 3t - 1, & \text{if } t \in (\frac{1}{3}, \frac{2}{3}) \\[2mm] 5 - 3t, & \text{if } t \in (\frac{4}{3}, \frac{5}{3}) \end{cases}$$

and extend the definition of h to $\mathbf{R}$ by periodicity by setting $h(t + 2) = h(t)$. Also set

$$f(t) = \sum_{n=1}^{\infty} \frac{h(3^{2n-2}t)}{2^n}, \quad g(t) = \sum_{n=1}^{\infty} \frac{h(3^{2n-1}t)}{2^n}.$$

Prove the following:

(a) f and g are continuous on $\mathbf{R}$;

(b) if $\Delta = \{(f(t), g(t) : t \in [0, 1]\}$, then $\Delta \subseteq [0, 1] \times [0, 1]$;

(c) if $(a, b) \in [0, 1] \times [0, 1]$, let

$$a = \sum_{n=1}^{\infty} \frac{a_n}{2^n}, \quad b = \sum_{n=1}^{\infty} \frac{b_n}{2^n}, \quad c = 2 \sum_{n=1}^{\infty} \frac{c_n}{2^n},$$

where each a_n and each b_n is either 0 or 1 and $c_{2n-1} = a_n$, $c_2 n = b_n$, $n \in \mathbf{N}$, then $f(c) = a$, $g(c) = b$.

This is the classical example of a **space-filling curve** (that is, Δ 'fills' the unit square).

8.2   Power Series

In the first section of this chapter, we considered the basic properties of series of functions. Specifically, we considered many questions, for instance, when such a series could be integrated term by term, and so forth. The reason underlying our interest in these questions is that in many instances, a given function can be represented by a series, and under suitable conditions, that is, requirements on uniform convergence, the series may be easy to deal with, whereas the original function is not. If one thinks about the type of functions that would make up a series representing a given unpleasant function, it is evident that if the series consisted of polynomials, then it would be easy to deal with in the extreme. This is the notion that motivates our discussion of power series.

Definition. Let $\{a_n\}$ be a sequence of real numbers for $n = 0, 1, \ldots$. The infinite series of functions

$$\sum_{n=0}^{\infty} a_n x^n$$

is called a **power series** in x, centered at the origin. More generally, one can consider power series of the form

$$\sum_{n=0}^{\infty} a_n (x - a)^n,$$

and this will be known as a power series in x **centered at the point** a.

Discussion. We shall be concerned mainly with power series centered at origin, since the methodology for dealing with power series centered at a results from a minimal modification of the methodology for dealing with series centered at the origin. The questions that we are primarily interested in are when does such a series converge, and moreover, when is the convergence uniform? Also, can we perform term by term operations such as differentiation and integration on such a series? The reason why we are interested in these questions is that many functions can be represented by power series, and for the representation to be useful, we must have a complete understanding of its properties.

We want to adopt a convention for indices of power series. Namely, for power series, the notation, $\sum a_n x_n$, means $\sum_{n=0}^{\infty} a_n x^n$. The reason for this convention is that a power series includes the constant term $a_0 = a_0 x^0$.

Trivially, every power series $\sum a_n x^n$ converges for $x = 0$. It may happen that this is the only point at which it converges, as it happens in the case of $\sum (n!) x^n$. On the other extreme, it is possible that the power series converges for every value of x, as in the example of $\sum \frac{x^n}{n!}$. The third possibility is that for certain values of x, we obtain convergence, and for certain others, we have divergence. This is illustrated by the geometric series $\sum x^n$. Our aim is to precisely nail down a stage that separates the values of x for which the series converges and those for which the series diverges. We approach this problem with a series of results, culminating in the precise formulation of what is known as the 'interval of convergence' of a power series. $\square$

Theorem 8.2.1. *If the power series $\sum a_n x^n$ converges for a value $x = x_0$, then it converges absolutely for all x satisfying the inequality $|x| < |x_0|$.*

Proof. Since $\sum a_n x_0^n$ converges as a series of constant terms, the sequence $\{a_n x_0^n\}$ converges to 0 and hence is bounded, say, by the real number M. Thus, for any x, we have

$$|a_n x^n| = |a_n x_0^n| \left| \frac{x}{x_0} \right|^n \leq M r^n,$$

where $r = |\frac{x}{x_0}| < 1$. Since the geometric series $\sum M r^n$ is convergent, the result follows by the Comparison test. $\square$

Discussion. The power of the theorem rests in the fact that by the mere knowledge that the series converges at a single point, x_0, we are able to conclude absolute convergence in an entire open interval $(-|x_0|, |x_0|)$. We stress that the theorem does not even require the convergence at x_0 be absolute. The theorem says nothing at all about the behavior of the power series outside this interval (including the two endpoints). Thus, the series may or may not converge outside the open interval. $\square$

For the divergence case, we have the following theorem:

Theorem 8.2.2. *If $\sum a_n x^n$ diverges for a value $x = x_0$, then it diverges for all x satisfying $|x| > |x_0|$.*

Proof. If for some x satisfying $|x| > |x_0|$, the series $\sum a_n x^n$ converges, then by the previous theorem, $\sum a_n x_0^n$ must converge, which is a contradiction. $\square$

Discussion. As remarked in the previous discussion, divergence at a single point enables us to conclude divergence outside an entire interval $[-|x_0|, |x_0|]$, but the theorem gives no information whatsoever as to what happens inside the interval. The series may converge or diverge. We cannot obtain any conclusion about the behavior at the two endpoints. $\square$

As stated, a prime matter of concern for power series is delineating the points of convergence. Let us define A by

$$A = \{x : x \in \mathbf{R} \text{ and } \sum |a_n x^n| < \infty\}.$$

We have already noted that $0 \in A$ for every power series. Theorem 8.2.1 requires that A be connected and exhibit a degree of symmetry about 0; that is, $a \in A$ implies $(-|a|, |a|) \subseteq A$. If A is not bounded, then A must be all of $\mathbf{R}$. Otherwise,

A is bounded, whence it has a supremum. These considerations are summarized in the following possibilities about the power series $\sum a_n x^n$:

(i) it converges for all $x \in \mathbf{R}$;

(ii) it converges for no value of x other than $x = 0$;

(iii) there is a certain point, say, $x = R$, so that for all x satisfying $|x| < R$, the series converges absolutely and diverges for those x with $|x| > R$.

It is our aim to spell out precisely how to arrive at this magic number R. This bring us to our next theorem.

Theorem 8.2.3. *Let $\sum_{n=0}^{\infty} a_n x^n$ be a power series. If*

$$A = \{x : x \in \mathbf{R} \quad and \quad \sum |a_n x^n| < \infty\}$$

is bounded and $R = \sup A$, then $\sum a_n x^n$ converges absolutely for $|x| < R$ and diverges for $|x| > R$. The series may converge or diverge for $x \pm R$.

Proof. If $|x| < R$, we can choose x_0 such that $|x| < |x_0| < R$ **(WHY?)**. Thus $x_0 \in A$, and so by Theorem 8.2.1, the series converges (absolutely). If $|x| > R$, then $x \notin A$ and so $\sum a_n x^n$ diverges. $\square$

Discussion. The above theorem separates the real line into two parts, one an open interval, where we are guaranteed to have absolute convergence, and the other consisting of two half-rays where there is divergence. Both these regions have the points $x = \pm R$ as boundaries. The theorem draws no conclusion about the nature of the series at these endpoints. In fact all possibilities can occur at $\pm R$ (see Examples 1, 2, 3 below). $\square$

Definition. The number R, defined in Theorem 8.2.3 above, is called the **radius of convergence** of the power series $\sum a_n x^n$. The **interval of convergence** of the power series is the set of all values for which the power series converges.

Discussion. If R is defined as above, and $x \in (-R, R)$, then $|x| < R$, and hence by Theorem 8.2.1, we have (absolute) convergence. On the other hand, if $x \notin [-R, R]$, then $|x| > R$, so we have divergence. Thus, the interval of convergence must include the open interval $(-R, R)$. Because the results we have had so far do not tell us anything about the behavior of the series at $x = \pm R$, these two points may or may not belong to the interval of convergence.

If $R = 0$, then the set $A = \{0\}$, and so the radius of convergence is 0, and the interval of convergence is the singleton $\{0\}$. On the other hand, if A is unbounded, R becomes infinite, so the interval of convergence is the entire real line $\mathbf{R}$. $\square$

Example 1. Find the interval of convergence of $\sum x^n$.

Solution. We know that the geometric series converges for $|x| < 1$ diverges for $|x| > 1$, whence we see that $R = 1$. At $x = \pm 1$, by inspection, we have divergence, so we conclude that the interval of convergence is $(-1, 1)$. $\square$

Example 2. Find the interval of convergence of $\sum \frac{x^{n+1}}{n+1}$.

Solution. Since for $x = 1$, $\sum \frac{1}{n+1}$ diverges, and for $x = -1$, $\sum \frac{(-1)^n}{n+1}$ converges, we conclude from the previous theorems that $R = 1$. It is immediate that the interval of convergence is $[-1, 1)$. $\square$

Discussion. The solution was immediate, since substitution of ± 1 for x in the power series yielded the Harmonic series, or its alternating form. We have used the properties of this series many times, and they should by now be completely familiar. In any case, the fact that the Harmonic series diverges forces $R \leq 1$, while the fact that its alternating form converges forces $R \geq 1$. $\square$

Example 3. Find the interval of convergence of $\sum \frac{x^n}{(n+1)^2}$.

Solution. Since $\sum \frac{1}{(n+1)^2}$ converges, $R \geq 1$. On the other hand, if we choose a value of x greater than 1, say, $1 + \delta$, $(\delta > 0)$, we can use Ratio or Root test to conclude that the series diverges for this value of x. So R must be less than $1 + \delta$ for every $\delta > 0$. Thus, $R = 1$. Also, since $\sum \frac{1}{(n+1)^2}$ is absolutely convergent, we have convergence at $\pm R$, so the interval of convergence is $[-1, 1]$. $\square$

Discussion. Once again, the strategy here is to look for some familiar points of convergence or divergence. In this example, both the points 1 and -1 are points of convergence. Luckily, for $x > 1$, we obtain divergence by using standard tests such as the Ratio or Root test. $\square$

The next important question is: How can the interval of convergence be found? Equivalently, what is R? To obtain R as the supremum of a set of real numbers is, in general, not easy. Ideally, we require a formula that will enable us to write down the value of R directly from the coefficients, a_n, of the power series by simple computations. That such formulae might exist is plausible, since the distinguishing feature between two power series is the coefficients, a_n. The next two theorems provide methods for obtaining the radius of convergence.

Theorem 8.2.4. *If $\sum a_n x^n$ is a power series with $a_n \neq 0$ for all n, and if*

$$\rho = \lim \left| \frac{a_{n+1}}{a_n} \right|$$

(finite or infinite), then the radius of convergence is given by $R = \frac{1}{\rho}$ (with the convention, $R = 0$ if $\rho = \infty$ and $R = \infty$ if $\rho = 0$).

Proof. Using the ratio test, we have

$$\lim \left| \frac{a_{n+1} x^{n+1}}{a_n x^n} \right| = |x| \lim \left| \frac{a_{n+1}}{a_n} \right| = |x| \rho = \frac{|x|}{R},$$

so that if $\left| \frac{x}{R} \right| < 1$, we have (absolute) convergence, and for $\left| \frac{x}{R} \right| > 1$, we have divergence. In other words, R has the properties listed in Theorem 8.2.3, and since such a number is unique (**WHY?**), we get the desired expression for R. $\square$

The formula for R given above was obtained by using the ratio test. We can generate a second formula for the calculation of R, by using Cauchy's Root test.

Theorem 8.2.5. *If $\sum a_n x^n$ is a power series and $\alpha = \overline{\lim} |a_n|^{1/n}$, then the radius of convergence is given by $R = \frac{1}{\alpha}$ (where $R = 0$ if $\alpha = \infty$ and $R = \infty$ if $\alpha = 0$).*

Proof. Since

$$\overline{\lim} \, |a_n x^n|^{\frac{1}{n}} = |x| \cdot \overline{\lim} \, |a_n|^{\frac{1}{n}} = |x|\alpha,$$

by the Root test, $\sum |a_n x^n|$ converges for $|x|\alpha < 1$ and diverges for $|x|\alpha > 1$. If $|x|\alpha > 1$, then $|a_n x^n| > 1$ for infinitely many values of n, then, $a_n x^n$ cannot tend to 0, so the series, $\sum a_n x^n$ cannot converge. The remainder of the proof amounts to checking details for the various cases (Exercise 1). □

Discussion. These two theorems enable us to compute the radius of convergence directly in terms of the coefficients a_n. All we have to do is to compute the limit of either the ratio $|\frac{a_{n+1}}{a_n}|$ or the root $|a_n|^{1/n}$, whichever is easier to obtain, and the reciprocal of the number .15 yields the radius of convergence. To obtain the interval of convergence, we have to check separately the convergence or divergence at the two endpoints of the interval in question. □

Example 4. Find the interval of convergence of $\sum \frac{x^n}{n!}$.

Solution. We have

$$\frac{a_{n+1}}{a_n} = \frac{n!}{(n+1)!} = \frac{1}{n+1} \to 0$$

showing $\rho = 0$; hence, $R = \infty$ and the interval of convergence is **R**. □

Example 5. Find the exact interval of convergence of $\sum a_n x^n$ where $a_{2n-1} = \frac{1}{2^n}$ and $a_{2n} = \frac{1}{4^n}$.

Solution. The reader should check that the ratio approach is undesirable in this case. Applying the Root test yields

$$\overline{\lim} \, |a_n|^{\frac{1}{n}} = \frac{1}{\sqrt{2}} = \alpha = \frac{1}{R}.$$

Further, the reader can check (Exercise 2) that the series converges at both endpoints so the interval of convergence is $[-\sqrt{2}, \sqrt{2}]$. □

Determination of the radius of convergence provides an open interval on which the power series is guaranteed to converge absolutely. Unfortunately, as we have seen from examples in the previous section, there is little relation between concepts of absolute and uniform convergence. Moreover, we know that the power that preserves important properties is uniform convergence, not absolute convergence. For this reason, the next task is to consider the uniform convergence of the power series, $\sum a_n x^n$. We have the following result:

Theorem 8.2.6. *If R is the radius of convergence of the power series, $\sum a_n x^n$, then for every $\delta > 0$, the series converges uniformly in the interval $[-R+\delta, R-\delta]$.*

Proof. If $x \in [-R+\delta, R-\delta]$, then $|a_n x^n| \le a_n(R-\delta)^n$, and the series $\sum a_n(R-\delta)^n$ is convergent. By Weierstrass's M-test, we obtain uniform convergence in the interval $[-R+\delta, R-\delta]$. □

Discussion. This theorem establishes the utility of power series as a means of representing functions. Specifically, if the power series converges to f on an interval, we can almost be sure that the convergence will be uniform. The reason for this, which is employed in the proof, is the monotonicity of the various terms, $|a_n x^n|$.

The above theorem asserts that the power series is uniformly convergent in any closed interval that is contained in the open interval $(-R, R)$. One might wonder whether this is enough to guarantee uniform convergence in the entire closed interval $[-R, R]$, or even on the open interval $(-R, R)$. Uniform convergence on either of the intervals does not follow from our results, and is not true in general.

Uniform convergence in $[-R+\delta, R-\delta]$ implies the existence of an integer, N, independent of x in that interval and with its usual properties. But the choice of N may very well depend on δ, whence if δ assumes a sequence of values that decrease to 0, the sequence of corresponding values of N need not be bounded. In such a case no value of N will exist that is independent of x in the entire open interval $(-R, R)$. The next example illustrates this point. □

Example 6. Discuss the uniform convergence properties of the power series, $\sum x^n$.

Solution. Clearly, $R = 1$, and therefore the power series converges absolutely in $(-1, 1)$. By the above theorem, the convergence is uniform in $[-1 + \delta, 1 - \delta]$. We shall see that the convergence is not uniform in $(-1, 1)$. For, if it were, then the series would be uniformly Cauchy, so that given $\epsilon > 0$, there is N such that

$$|x^{n+1} + x^{n+2} + \cdots + x^{n+p}| < \epsilon$$

for $n > N$ and $p = 1, 2, \ldots$. Choosing $p = 1$, we get $|x^{n+1}| < \epsilon$ for $n > N$ and $x \in (-1, 1)$. But this is impossible since whatever be n, the function x^{n+1} on $(-1, 1)$ has the limit 1 as $x \to 1$, whence it follows that for some values of x, x^{n+1} must be closer to 1. Thus, the power series converges uniformly in all the intervals $[-1 + \delta, 1 - \delta]$, but not in $(-1, 1)$. □

From the example given above, the question naturally arises as to when we can extend the interval of uniform convergence right up to either of the endpoints. A possible suggestion would be to insist that the series converge at each point (which was not the case in the example cited above). That this is sufficient is the content of the next theorem due to Abel.

Theorem 8.2.7 (Abel). *If $\sum a_n x^n$ has radius of convergence R, and if the series converges at $x = R$, then it converges uniformly in $[-R + \delta, R]$, for any positive δ; also if it converges at $x = -R$, then the convergence is uniform in $[-R, R - \delta]$.*

Proof. Without loss of generality, we can assume that $R = 1$. To see this, observe that $\sum a_n x^n$ has radius of convergence R if and only if $\sum b_n x^n$ has radius of convergence 1, where $b_n = \frac{a_n}{R^n}$. We consider two cases.

Case 1. Assume $\sum a_n x^n$ converges at $R = 1$. For $x \in [0, 1]$, let $f_n(x) = a_n$ and $g_n(x) = x^n$. Note that $\sum f_n$ converges uniformly on $[0, 1]$. Also, $\{g_n\}$ is uniformly bounded on $[0, 1]$ and for each $x \in [0, 1]$, $\{g_n(x)\}$ is monotone decreasing. This verifies the hypothesis of Abel's test, whence the series converges uniformly on $[0, 1]$, as required. It is immediate that this implies uniform convergence on $[-1 + \delta, 1]$.

Case 2. Assume $\sum a_n x^n$ converges at $R = -1$. For $x \in [-1, 0]$, set $f(x) = (-1)^n a_n$ and $g_n(x) = |x^n|$. Again apply Abel's test to obtain uniform convergence on $[-1, 0]$ and extend to $[-1, 1 - \delta)$ to complete the proof. $\square$

Discussion. This theorem is very useful. If R is the radius of convergence of a known power series and if $\sum a_n R^n$ converges, then the series is uniformly convergent in $[-R + \delta, R]$ for any positive δ. Similarly, if $\sum (-1)^n a_n R^n$ is convergent, then the uniformity of convergence extends right up to $x = -R$. $\square$

Corollary. *Let $\sum_{n=0}^{\infty} a_n$ be any convergent series. Then the power series $\sum a_n x^n$ converges uniformly on $(-1, 1]$ to a limit function, f, and*

$$\lim_{x \to 1^-} f(x) = \sum a_n.$$

Proof. Exercise 14. $\square$

As one might expect, the uniform convergence properties of power series mean that the limit function will share the nice features of the polynomials making up the power series. The next several theorems explore the effects of uniform convergence.

Theorem 8.2.8. *Every power series represents a continuous function on its interval of convergence.*

Proof. Let $\sum a_n x^n$ converge to f on A, the interval of convergence of the power series. Fix $x \in A$. If x is an interior point of A, then there is a positive δ such that $\sum a_n x^n$ converges uniformly on $[x - \delta, x + \delta]$. Since each $f_n(x) = a_n x^n$ is continuous and the series converges uniformly, f is continuous on $[x - \delta, x + \delta]$. On the other hand, if x is an endpoint of the interval, there are two possibilities, namely, $x = 0$ or $x \neq 0$. If the former, then $f = \{(0, a_0)\}$ and continuity follows trivially. If the latter, then we have uniform convergence on an interval of the form, $[x, x + \delta]$ if x is the left-hand endpoint or $[x - \delta, x]$ if x is the right-hand endpoint. In either case, the continuity of f at x follows. $\square$

Discussion. The argument presented consists of assertions without reasons. Each of the assertions amounts to the application of a theorem. The reader should supply the required reasons (Exercise 4). $\square$

Theorem 8.2.9. *Let $\sum a_n x^n$ have A for its interval of convergence and suppose $[a, b] \subseteq A$. The $\sum a_n x^n$ is term-by-term Riemann integrable on $[a, b]$.*

Proof. Exercise 5. $\square$

With respect to term-by-term differentiation, we have the following pleasant result:

Theorem 8.2.10. *The power series $\sum a_n x^n$ and the series $\sum n a_n x^{n-1}$ obtained by term-by-term differentiation have the same radius of convergence.*

Proof. If R denotes the radius of convergence of $\sum a_n x^n$, then the radius of convergence of $\sum n a_n x^{n-1}$ is computed using 8.2.5 by

$$\alpha_1 = \overline{\lim} |n a_n|^{\frac{1}{n}} = \lim n^{\frac{1}{n}} \cdot \overline{\lim} |a_n|^{\frac{1}{n}} = \alpha,$$

whence $\frac{1}{\alpha_1} = \frac{1}{\alpha} = R$. In Exercise 6 the reader is asked to find examples to show that the differentiated series may or may not converge at the endpoints of the interval of convergence. $\square$

Corollary. *The power series* $\sum a_n x^n$ *and the term-by-term integrated series,* $\sum \frac{a_n x^{n+1}}{n+1}$, *have the same radius of convergence.*

Proof. Exercise 7. □

The next result, which justifies term-by-term differentiation of a power series, is a consequence of Theorem 8.2.6.

Theorem 8.2.11. *If* $\sum a_n x^n$ *has a nonzero radius of convergence, it can be differentiated term-by-term on the interior of its interval of convergence.*

Proof. Let x belong to the interior of the interval of convergence. Then there is a closed interval, $[a, b]$, such that $x \in [a, b]$ and $\sum a_n x^n$ converges uniformly on $[a, b]$ (Exercise 8). The reader may check that the hypothesis of Theorem 8.1.10 apply (Exercise 9), and the result follows. □

Discussion. Theorem 8.2.10 asserts that if

$$a_0 + a_1 x + a_2 x^2 + \cdots = f(x),$$

then, after term-by-term differentiation, we have

$$a_1 + 2a_2 x + \cdots = f'(x)$$

for each x in the interior of the interval of convergence. This theorem certainly hints at the fact that functions that can be represented by power series must have nice differentiation properties. □

An important consequence of this result, in conjunction with Theorems 8.2.8 and 8.2.9 is the following theorem, which concerns with the repeated term-by-term differentiation of a given power series any number of times.

Theorem 8.2.12. *A power series,* $\sum a_n x^n = f(x)$, *possesses derivatives of all orders on the interior of its interval of convergence; the power series for each derivative may be obtained by term-by-term differentiation of the power series for f, and all of them converge and are continuous on the common interior of the interval of convergence for f.*

Proof. Exercise 10. □

Corollary. *If the power series,* $\sum a_n x^n$, *has a nonzero radius of convergence, and if $f(x)$ denotes the sum function in the interval of convergence, then for each n,*

$$a_n = \frac{f^{(n)}(0)}{n!}.$$

Proof. If

$$f(x) = a_0 + a_1 x + a_2 x^2 + \cdots,$$

then, by repeated application of Theorem 8.2.12, we have

$$f^{(n)}(x) = n(n-1)(n-2) \cdots 3 \cdot 2 \cdot 1 (a_n + a_{n+1} x + a_{n+2} x^2 \cdots).$$

Since $x = 0$ belongs to the interval of convergence, substituting $x = 0$, we obtain

$$\frac{f^n(0)}{n!} = a_n$$

as required. $\square$

The next result shows that the limit function $f(x)$ is uniquely determines the coefficients of the power series $\sum a_n x^n$.

Theorem 8.2.13. *If $\sum a_n x^n$ and $\sum b_n x^n$ both converge to the same function $f(x)$ in some neighborhood of 0, then for all n, $a_n = b_n$; in other words, the two series are identical.*

Proof. If in some open interval containing 0

$$f(x) = \sum a_n x^n = \sum b_n x^n,$$

then by corollary to Theorem 8.2.12,

$$\frac{f^n(0)}{n!} = a_n = b_n$$

for all n. $\square$

Discussion. Theorem 8.2.13 is remarkable. To see why, consider an arbitrary function, f, that is defined on $[-1, 1]$. It seems plausible that there would be many distinct series of functions that converge to f. If one adds the requirement that the given series must consist of continuous functions and that the convergence much be uniform, then these requirements will have two effects. First, the two requirements limit the choice of f to continuous functions on $[-1, 1]$. Second, there will be fewer series that converge to f under these more stringent conditions. Nevertheless, there will still be an infinite number of such series.

Now consider the case of power series, and of a function, f, that is the sum of that power series, $\sum a_n x^n$. Suppose, for the sake of argument, that we changed the value of a_0 to be b_0. The new power series $b_0 + \sum_{n=1}^{\infty} a_n x^n$ will no longer converge to f, although it will converge. While this is not surprising, what is surprising is the fact that there is no way to adjust the values of the coefficients, a_n, $n = 1, 2, \ldots$, so as to obtain a new series, $\sum b_n x^n$, which does converge to f. Thus, for a power series to converge to f, one must get all the coefficients right, so to speak. In summary, then, given that f has a power series representation, f uniquely determines that representation.

The results contained in the last several theorems also impose rather strong conditions on f. Specifically, f must have derivatives of all orders on its interval of convergence. Thus, the class of functions that have power series representations must be restricted to very nice functions indeed.

This leaves us with several burning questions. What functions have power series representations? Given that a function has a power series representation, how can it be found? These questions will be considered in the next section. $\square$

We conclude with some results for combining power series.

Theorem 8.2.14. *Let $\sum a_n x^n$ and $\sum b_n x^n$ be two power series having intervals of convergence, A_1 and A_2, respectively. Then, the formula*

$$\sum a_n x^n + \sum b_n x^n = \sum (a_n + b_n) x^n$$

is valid on $A_1 \cap A_2$.

Proof. Exercise 11. □

In Section 6.3 we defined the Cauchy product of two series, $\sum_{n=0}^{\infty} a_n$ and $\sum_{n=0}^{\infty} b_n$, by setting $c_n = \sum_{k=0}^{n} a_k b_{n-k}$ and taking the product to be

$$\left(\sum_{n=0}^{\infty} a_n\right) \cdot \left(\sum_{n=0}^{\infty} b_n\right) \equiv_{df} \sum_{n=0}^{\infty} c_n,$$

where on the left we have the formal product of the two infinite series. For two power series, $\sum a_n x^n$ and $\sum b_n x^n$, we define the **Cauchy product** by extension as

$$\left(\sum a_n x^n\right) \cdot \left(\sum b_n x^n\right) \equiv_{df} \sum c_n x^n,$$

where on the left we have the formal product of the two infinite series. The reader can check that this definition is the correct extension, based on the discussion following the definition of the Cauchy product in Section 6.3.

Theorem 8.2.15. *Let $\sum a_n x^n$ and $\sum b_n x^n$ be two power series with a common radius of convergence, R, and limit functions, f and g, respectively. Then*

$$\sum c_n x^n = f(x) \cdot g(x)$$

for each $x \in (-R, R)$.

Proof. Without loss of generality, we may assume that the radius of convergence is nonzero. Fix $x \in (-R, R)$. Then $\sum a_n x^n$ converges absolutely to $f(x)$. Similarly, $\sum b_n x^n$ converges absolutely to $g(x)$. Now (Exercise 12), $\sum c_n x^n$ is the Cauchy product of the two power series evaluated at x and considered as series of real constants. By Theorem 6.3.8,

$$\sum c_n x^n = f(x) \cdot g(x)$$

as desired. □

Discussion. It would be really nice if the open interval, $(-R, R)$, could be replaced by the interval of convergence. Such a result seems very plausible given Theorems 8.2.6 and 8.2.7. The reader is invited to explore this suggestion in Exercise 13. □

It is also possible to treat division of two power series. This is explored in Exercise 33.

Exercises

1. Complete the proof of Theorem 8.2.5.

2. Complete the details of the solution to Example 5.

3. Complete the details of Theorem 8.2.7.

4. Supply all reasons to justify the assertions making up the proof of Theorem 8.2.8.

5. Prove Theorem 8.2.9.

6. Give examples to illustrate that the differentiated series obtained from a given power series $\sum a_n x^n$ may or may not converge at the endpoints of its interval of convergence.

7. Prove the corollary to Theorem 8.2.10.

8. Suppose that $\sum a_n x^n$ has a nonzero radius of convergence. Show if x belongs to the interval of convergence, then there is a closed interval, $[a, b]$, such that $x \in [a, b]$ and the series converges uniformly on $[a, b]$.

9. Complete the details of Theorem 8.2.11.

10. Prove Theorem 8.2.12.

11. Prove Theorem 8.2.14.

12. Let $\sum a_n x^n$ and $\sum b_n x^n$ be two power series. Show that for any fixed x, $\sum c_n x^n$ is the Cauchy product of the two series considered as series of real constants.

13. Can the interval $(-R, R)$ mentioned in Theorem 8.2.15 be replaced by the interval of convergence?

14. Prove the corollary to Theorem 8.2.7. Use this corollary to give a simple proof of Abel's test (Theorem 6.3.7).

15. Determine the radius of convergence and the exact interval of convergence of the following power series:

   (a) $\sum \frac{(-1)^n (nx)^n}{n!}$;

   (b) $\sum (-1)^n \frac{(x-1)^n}{n!}$;

   (c) $\sum \frac{(\ln n)(x-5)^n}{\sqrt{n}}$;

   (d) $\sum (\sin n) x^n$;

   (e) $\sum a^{n^2} x^n$, $a < 1$;

   (f) $\sum (1 - (-2)^n) x^n$;

   (g) $\sum \frac{(x+3)^n}{(n+2)2^n}$;

   (h) $\sum \frac{(-1)^n 2^{2n} x^{2n}}{2n}$;

   (i) $\sum \frac{n(x-2)^n}{3^n (n+1)}$;

   (j) $\sum \frac{(n!)^2 x^{n+1}}{(2n)!}$;

   (k) $\sum \frac{(-1)^n (n+1)^n x^n}{n^2 + 1}$;

   (l) $\sum (-1)^n \frac{x^{2n+1}}{(2n+1)!}$;

   (m) $\sum \frac{(n!)(x+2)^n}{n^n}$;

   (n) $\sum \frac{3^{\sqrt{n}} x^n}{\sqrt{n^2 + 1}}$;

   (o) $\sum \frac{3^{\sqrt{n}} x^{2n+1}}{\sqrt{n}}$;

   (p) $\sum [\frac{4 + (-1)^n}{5}]^n x^n$;

   (q) $\sum \frac{n^\alpha x^n}{n!}$;

   (r) $\sum \frac{n^n x^n}{n!}$;

   (s) $\sum \frac{x^n}{n \ln n}$;

   (t) $\sum \frac{x^n}{\ln n}$;

   (u) $\sum \binom{k}{n} x^n$;

   (v) $\sum \frac{3^n x^{2n+1}}{\sqrt{x}}$;

   (w) $\sum \frac{x^{2n+1}}{4^n}$;

   (x) $\sum \frac{(-1)^n 4^n x^{2n}}{(3n-4)}$;

   (y) $\sum \frac{(n!)^2 (2n+2)! x^n}{(2n!)[(n+1)!]^2}$;

   (z) $\sum \frac{(-1)^n x^{4n+1}}{(4n+1)!(2n)}$;

   (a') $\sum \frac{(-1)^n x^{2n}}{(n!)^2 4^n}$;

   (b') $\sum a_n x^n$, where $a_n = \begin{cases} 1 & \text{if } n = m^2, \, m \in \mathbf{N} \\ 0 & \text{otherwise}; \end{cases}$

   (c') $\sum a_n x^n$, where $a_n = \begin{cases} 1 & \text{if } n = m!, \, m \in \mathbf{N} \\ 0 & \text{otherwise}. \end{cases}$

16. Prove if a_n is an integer for each n, and if infinitely many of them are 0, then the radius of convergence, R, of the power series $\sum a_n x^n$ can never exceed 1. Can the same conclusion be obtained is $\overline{\lim} |a_n| > 0$?

17. If R and S are the radii of convergence of $\sum a_n x^n$ and $\sum b_n x^n$, respectively, what can you say about the radius of convergence of the following power series?

   (a) $\sum a_n x^{kn}$;

   (b) $\sum (a_n \pm b_n) x^n$;

   (c) $\sum (a_n b_n) x^n$;

   (d) $\sum a_n^k x^n$;

   (e) $\sum a_n^k x^{2n}$;

   (f) $\sum a_n x^{n^2}$;

   (g) the Cauchy product series $\sum c_n x^n$, where $c_n = \sum_{j=1}^{n} a_n b_{n-j}$.

18. Does there exist a power series $\sum a_n x^n$ that represents $|x|$ for all $x \in \mathbf{R}$?

19. Show the series $\sum (n+1)! x^n$ and $\sum n! x^n$ have the same interval of convergence and define the same function on the interval of convergence. Is there any contradiction to Theorem 8.2.13?

20. Construct a power series whose exact interval of convergence is $[-3, 5)$.

21. Given $a_0 = 1$, $a_n = 2$ for $n \geq 1$, $b_n = \frac{(-1)^n 8 - \frac{3}{4^n}}{5}$,

    (a) show $\sum a_n$, $\sum b_n$ both diverge, but the Cauchy product $\sum c_n$ is absolutely convergent;

    (b) find the radii of convergence of $\sum a_n x^n$, $\sum b_n x^n$, and $\sum c_n x^n$.

22. Find the Cauchy product of the power series $3 + \sum_{n=1}^{\infty} 3^n x^n$ and $-2 + \sum_{n=1}^{\infty} 2^n x^n$.

23. Prove, using the Cauchy product, that

$$\sum_{n=1}^{\infty} n x^{n-1} = \frac{1}{(1-x)^2}.$$

24. If $\sum a_n$ converges, but not absolutely, show that the radius of convergence of $\sum a_n x^n$ is 1.

25. We say $\sum a_n$ is **Abel summable** to L, if $\lim_{x \to 1-} f(x) = L$, where $f(x) = \sum a_n x^n$. In this case, we write $\sum a_n = L(A)$. Show the following series are Abel summable, and find their Abel sums.

    (a) $1 - 1 + 1 - 1 + \cdots$;   (b) $1 - 2 + 3 - 4 + \cdots$;
    (c) $1 - 3 + 6 - 10 + \cdots$;   (d) $1 - \frac{1}{3} + \frac{1}{5} - \frac{1}{7} + \cdots$.

26. If $\sum_{n=0}^{\infty} a_n = L(A)$ and $\sum_{n=0}^{\infty} b_n = M(A)$, show $\sum_{n=0}^{\infty} (a_n + b_n) = L + M(A)$. What can you say about the Abel sum of the following series?

$$0 + a_0 + 0 + a_1 + 0 + a_2 + \cdots + 0 + a_n + 0 + \cdots.$$

27. If $\sum a_n = L$, prove that $\sum a_n = L(A)$, but not conversely.

28. Justify the equation

$$\frac{\pi}{4} = 1 - \frac{1}{3} + \frac{1}{5} - \cdots.$$

[Hint: Consider $\int_0^x \frac{1}{1+t^2} dt$ in $|t| < 1$, and use Abel's Theorem.]

29. By integrating the power series for $\frac{1}{1+x+x^2}$, derive the identity

$$\frac{\pi}{3^{\frac{3}{2}}} = 1 - \frac{1}{2} + \frac{1}{4} - \frac{1}{5} + \cdots.$$

30. Let $\sum a_n x^n$ have radius of convergence $R > 0$ and $\sum \frac{a_n R^{n+1}}{n+1}$ be convergent. If $\sum a_n x^n = f(x), |x| < R$ show that f is integrable in $[0, R]$, and

$$\int_0^R f = \sum \frac{a_n R^{n+1}}{n+1},$$

irrespective of the way f is defined at R.

31. Starting from the geometric series $\sum x^{n-1}, -1 < x < 1$, derive the following:

    (a) $\sum_{n=0}^{\infty}(n+1)x^n = \frac{1}{(1-x)^2}$;
        (b) $\sum_{n=0}^{\infty}(n+1)x^{2n} = \frac{1}{(1-x^2)^2}$;

    (c) $\sum_{n=0}^{\infty}(-1)^n(n+1)x^n = \frac{1}{(1-x)^2}$;
        (d) $\sum_{n=0}^{\infty}(n+1)x^{n+2} = \frac{x^2}{(1-x)^2}$;

    (e) $\sum_{n=0}^{\infty}\frac{n+1}{n+3}x^{n+3} = \int_0^x \frac{t^2}{1-t^2}\,dt$.

32. If $a_n + a_{n-1} + a_{n-2} = 0$, $n = 2, 3, \ldots$, show that for each x for which the power series $\sum a_n x^n$ converges, the sum is $\dfrac{a_0 + (a_1 + a_0)x}{1 + x + x^2}$.

33. Let $\sum b_n x^n$ and $\sum c_n x^n$ be two power series that converge on $(-R, R)$ to f and g, respectively. Further suppose that $b_0 \neq 0$. Show there is an interval, $(-r, r) \subseteq (-R, R)$ such that $\frac{g}{f}$ is defined and bounded. Further show $\frac{g}{f}$ has a power series representation on this interval, and give an explicit method for finding the coefficients of this power series.

34. Suppose $f(x) = \sum_{n=0}^{\infty} a_n x^n$ on $(-R, R)$, $R > 0$ and $g(x) = \sum_{n=1}^{\infty} b_n x^n$ on $(-S, S)$ for $S > 0$. Show the composite function, $h(x) = f(g(x))$, is represented by a power series having a positive radius of convergence, and this power series can be obtained by substituting the series for g into the series for f and rearranging terms.

8.3   Taylor Series

In Section 8.2, we saw how a power series gives rise to a continuous function in its interval of convergence. In this section, the converse problem is investigated, namely, given an arbitrary function, $f(x)$, find a power series whose sum is precisely the given function.

    The idea that a given function might be represented as a power series was introduced in Section 4.4, where the formula

$$f(a+h) \approx f(a) + f'(a)h + \frac{1}{2}f''(a)h^2$$

was developed as part of Example 4.4.2. If one sets $a = 0$ and replaces h by x, the formula becomes

$$f(x) \approx f(0) + f'(0)x + \frac{1}{2}f''(0)x^2,$$

which looks suspiciously like the first three terms of the power series for f, should such exist. This approach to approximating a function with a polynomial was known, in a very primitive form, to Newton, around 1676, and was developed further by Taylor and Maclaurin in the 1700s. As such, this approach to the representation of functions predates any notion of power series, uniform convergence, and so forth.

Functions, which can be represented as power series over an interval, are very special. For example, to have a representation as a power series requires that the function have lots of derivatives, at least at $x = 0$. This suggests that as a class, this collection of function is worthy of study and leads to the following definition.

Definition. Let f be a function defined in a neighborhood of 0. We say that f is **analytic at** 0, provided there is a power series, $\sum a_n x^n$, that converges to f on an interval $(-R, R)$.

Discussion. We stress that for f to be analytic at 0, it must be representable by a power series over an open interval that includes 0.

Having defined this class of functions, one would like to have a nice test for membership in the class. The elements of such a test must relate to requirements on f to have a power series. If f has such a power series, then by Theorem 8.2.13, the series must be unique and its coefficients, a_n, calculated by the formula

$$a_n = \frac{f^{(n)}(0)}{n!}.$$

Thus, one immediate test is whether f has derivatives of all orders at the origin. Unfortunately, as we shall see, this is not sufficient. $\square$

Given that f is a function that has derivatives of all orders at the origin, or indeed, at a, it makes sense to give the following definition:

Definition. If $f(x)$ possesses derivatives of all orders in some neighborhood of origin, then the (formal) power series,

$$\sum_{n=0}^{\infty} \frac{f^{(n)}(0)}{n!} x^n$$

is called the **Taylor series expansion** of f about the point 0. If $f(x)$ possesses derivatives of all orders in some neighborhood of a, then the (formal) power series,

$$\sum_{n=0}^{\infty} \frac{f^{(n)}(a)}{n!} (x - a)^n$$

is called the **Taylor series expansion** of f about the point a.

Discussion. The definition requires that we start with a function $f(x)$ that can be differentiated infinitely often at a fixed point, which for purposes of discussion we take to be 0. We then compute the value of the nth derivative $f^{(n)}(x)$ at the point $x = 0$ and write down the formal infinite series $\sum_{n=0}^{\infty} \frac{f^{(n)}(0)}{n!} x^n$ (with the constant term equal to $f(0)$) without worrying about the convergence properties of the series.

The moment one writes down such an infinite series, several questions naturally arise about its convergence properties. First, there is no guarantee that the series converges for any value of x other than 0, or $x = a$ in the general case. Second, even if it converges for some x, there is still no guarantee that the sum of the series is exactly the function value, $f(x)$, for the function that generated the series. Thus, at this stage, the Taylor series expansion of the infinitely differentiable function, f, is just a formal infinite series.

Given our previous definition, if the Taylor expansion for f at 0 actually converges to f on some open interval, then f is analytic at 0. More generally, if the Taylor expansion of f at a converges to f on an open interval that includes a, then f is said to be **analytic at** a. Thus, all the questions posed above relate to whether f is analytic.

As a matter of historical interest, the Taylor expansion for a function about 0 is referred to as the **Maclaurin series**, or **Maclaurin expansion.** One would think that this is due to the fact that Maclaurin's work preceded Taylor's. But this is not the case. Taylor's more general work preceded Maclaurin's by almost 50 years, and in addition, it appears the Maclaurin was aware of Taylor's work. Nevertheless, Maclaurin's name stuck to the special case of Taylor's series! But as we have hinted, the basic theory of approximating a function by a polynomial having coefficients determined by the derivative of the functions was worked out in the late 1600s by Newton and others. $\square$

Example 1. Obtain the Taylor series generated by the function given below (usually referred to as **Cauchy's function**):

$$f(x) = \begin{cases} e^{-1/x^2}, & \text{if } x \neq 0 \\ 0, & \text{if } x = 0. \end{cases}$$

Solution. To write down the Taylor infinite series, one has to compute all the derivatives at origin. It is left as Exercise 1 to show that $f^{(n)}(0) = 0$ for each n. (The computations are not easy.) Thus, the corresponding Taylor series is given by

$$\sum \frac{0}{n!} x^n = 0 + 0 \cdot x + 0 \cdot x^2 + \cdots,$$

and the series (trivially) converges to the function, g, which is identically 0 for all $x \in \mathbf{R}$. $\square$

Discussion. In this example, the function, f, admits a nice Taylor series expansion, which certainly converges for each x, but unfortunately the sum function obtained is not the function, f, which we started with, except at the single point $x = 0$. The point of this example is that it shows that even though a function possesses all the derivatives one could want, it is not enough to guarantee that the function is analytic on even a minuscule interval.

The graph of this function is drawn in Figure 8.3.1, Observe that the graph has an 'infinite-order' contact with x-axis at origin. $\square$

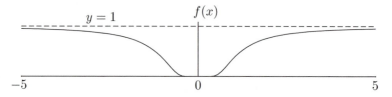

Figure 8.3.1 Graph of the Cauchy's function (Example 8.3.1). Note that the graph has an 'infinite-order' contact with the x-axis at 0.

Next, we consider an alternate approach. Rather than writing an infinite power series corresponding to a given function, let us begin by posing the question:

> Given a function f, can we find a sequence of polynomials that approximates the function in some region, that is, near some point, a?

This is the historical approach that began with Newton, Taylor, and others. The polynomials employed were developed by extending the formula

$$f(a + h) \approx f(a) + f'(a)h + \frac{1}{2}f''(a)h^2$$

to include higher powers of h. Thus, a sequence of polynomials, $\{P_n\}$, is generated each of which is hoped to be a better approximation to f.

Definition. Let f be infinitely differentiable in an open interval about a. Then, for each n, the polynomial

$$P_n(x) = \sum_{i=0}^{n} \frac{f^{(i)}(a)}{i!}(x - a)^i$$

is called the **Taylor polynomial of degree** n at a corresponding to the function, f. For $a = 0$, the Taylor polynomial of degree n at origin is given by

$$P_n(x) = \sum_{i=0}^{n} \frac{f^{(i)}(0)}{i!}x^i.$$

Discussion. For each n, $P_n(x)$ is just a polynomial of degree n consisting of $n+1$ terms, with the function value at a as the constant term. Indeed, this polynomial is just the nth partial sum of the Taylor series. However, Taylor polynomials do not arise from power series considerations. Rather, as shown in Section 4.4, they arise from attempts to approximate differentiable functions near points for which functional values are known. Even so, the process obviously generates a sequence of functions that are intended to approximate a given function, so the natural questions related to convergence immediately appear.

In summary, then, we have a sequence of polynomials, $\{P_n(x)\}$, and we want to know whether this sequence converges (uniformly) to some function, and if so, is the limit function the parent function, f? In Example 1, all the Taylor polynomials of all degrees turned out to be the zero polynomial, and hence, the sequence, though convergent at each point, was not convergent to the parent function, except at origin. So the answer to these questions is in general 'no'. But if 'no' were the complete picture, the Taylor series would not appear as a topic in every calculus course. $\square$

Definition. If $P_n(x)$ denotes the nth Taylor polynomial corresponding to $f(x)$ about a, then the function, R_{n+1}, defined by the difference

$$R_{n+1}(x) = f(x) - P_n(x)$$

is called the $(n + 1)$st **Taylor remainder** at a.

Discussion. The $(n+1)$st Taylor remainder at a, when evaluated at x, is precisely the value to be added to the value of the nth Taylor polynomial at x so as to equal the function value, $f(x)$. Thinking about Taylor polynomials in this way, via remainders, implies that the purpose of the nth Taylor polynomial is to approximate the function, f, on some interval. Thus, the remainder is, in effect, an error term. We stress that as an error term, its value depends on x. As a general rule, one would expect that the farther a particular x is from a, the point of expansion, the larger will be the magnitude of $R_{n+1}(x)$.

As a matter of notation, the reader may wonder why the subscript on the remainder is $n+1$, instead of n. If one writes the equation,

$$f(x) = P_n(x) + R_{n+1}(x),$$

then R_{n+1} is in fact the $(n+1)$st term in the expression on the right-hand side (if we do not count the constant term), since $P_n(x)$ is a polynomial of degree n, whose last term is $\frac{f^{(n)}(a)}{n!}(x-a)^n$. For this reason, it is common practice to assign it a subscript of $n+1$ instead of n. So it is obvious that when we refer to $R_n(x)$, we mean the Taylor remainder that corresponds to the Taylor polynomial $P_{n-1}(x)$ of degree $(n-1)$. In what follows, we shall be frequently using $R_n(x)$, as a matter of convenience.

Evidently, the sequence, $\{P_n\}$, converges (uniformly) to f on A, if and only if the Taylor remainder sequence of functions, $\{R_{n+1}\}$, converges (uniformly) to 0 (Exercise 2). Thus, adequate information about the nature of functions possessing a Taylor series (that converges precisely to the function at all points) can be obtained from the study of the sequence, $\{R_{n+1}\}$, of its Taylor remainders. For this reason, different authors have studied different forms of the Taylor remainder function. The object of the study is not that different remainders are possible, rather, it is that one would like to bound the function, R_{n+1}, over an interval. In the next few results, we will illustrate some of the forms of the Taylor remainders and their uses in determining the nature of the Taylor series corresponding to a given function. However, for clarity, we begin with an example. □

Example 2. Use graphical techniques to illustrate the nature of the convergence problem for the Taylor expansion of $f(x) = e^x$ at the origin.

Solution. The fifth-degree Taylor polynomial for e^x is given by

$$P_5(x) = 1 + x + \frac{x^2}{2!} + \frac{x^3}{3!} + \frac{x^4}{4!} + \frac{x^5}{5!}.$$

The formula for $R_6(x)$ is given by

$$R_6(x) = e^x - \left(1 + x + \frac{x^2}{2!} + \frac{x^3}{3!} + \frac{x^4}{4!} + \frac{x^5}{5!}\right).$$

In Figure 8.3.1 we present a graph of e^x, P_1, P_3, P_5, and R_6 on $[-4, 4]$ to illustrate these concepts. □

Discussion. In terms of the approximation of a function by a polynomial, the fifth Taylor polynomial does a remarkable job of approximating e^x. Indeed, in the interval, $[-2, 2]$, the magnitude of the remainder, $|R_6|$, is less than 0.1, and it is only 1.7 at $x = 3$.

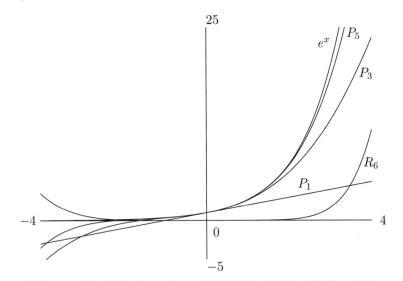

Figure 8.3.2 Taylor polynomials $P_1(x), P_3(x), P_5(x)$, and the remainder, $R_6(x)$, for the function $f(x) = e^x$.

In general terms, the remainder grows as x moves away from 0, the point about which the expansion is taken. If one is to understand this growth, it would be useful to have a formula for $R_{n+1}(x)$ that does not involve the original function, as does the defining equation for R_{n+1}. □

Theorem 8.3.1 (Schlomlich-Roche's form of the Taylor remainder). *Let f be defined and possess derivatives up to order n in some neighborhood U of 0. Then there exists a number θ such that $0 < \theta < 1$, and if $k < n$,*

$$R_n(x) = \frac{f^{(n)}(\theta x)(1 - \theta)^k}{(n - k)(n - 1)!} x^n.$$

Proof. Let $x_0 \neq 0$ be a member of U, and define the auxiliary function g by setting

$$g(t) = f(x_0) - f(t) - (x_0 - t)f'(t) - \cdots - \frac{(x_0 - t)^{n-1}}{(n - 1)!} f^{(n-1)}(t) - A(x_0 - t)^{n-k}$$

where the constant A is determined by the condition that $g(0) = 0$. Now the function g defined above is continuous and differentiable in $[0, x_0]$. Further, $g(0) = 0$ by the definition of A, and $g(x_0) = 0$ as the reader can check. Hence the Mean Value Theorem on differential calculus (Theorem 4.3.2) applies and there exists $\xi \in (0, x_0)$ satisfying $g'(\xi) = 0$. In fact, $\xi = \theta x_0$ for some θ satisfying $0 < \theta < 1$ (Exercise 3). A direct computation shows that

$$g'(t) = -\frac{(x_0 - t)^{n-1}}{(n - 1)!} f^{(n)}(t) + A(n - k)(x_0 - t)^{n-k-1}$$

so that the requirement that $g'(\xi) = 0$ together with the fact that $\xi = \theta x_0$ yields

$$A = \frac{(x_0 - \theta x_0)^k f^{(n)}(\theta x_0)}{(n - 1)!(n - k)} = \frac{x_0^k (1 - \theta)^k f^{(n)}(\theta x_0)}{(n - 1)!(n - k)}.$$

Substituting this value in the definition of A, we finally obtain

$$f(x_0) = f(0) + x_0 f'(0) + \frac{x_0^2 f^{(2)}(0)}{2} + \cdots + \frac{x_0^{n-1} f^{(n-1)}(0)}{(n-1)!} + \frac{x_0^n (1-\theta)^k f^{(n)}(\theta x_0)}{(n-1)!(n-k)}.$$

Since x_0 is an arbitrary point in the neighborhood, U of 0, replacing it by a general point x, we get the desired expression for $R_n(x)$. $\square$

Discussion. The key to this theorem is the Mean Value Theorem. First, it is another in the long string of consequences of the Mean Value Theorem (see Exercises in Section 4.3). More important, the proof of the Mean Value Theorem supplies the critical ideas required for generating the proof of this theorem. If the reader does not see the reasons that support this last statement, he should review the proof of the Mean Value Theorem and the rest of the material presented in Section 4.3. In Exercise 4 the reader will be asked to establish the general form of the Schlomlich-Roche remainder.

As we have stated, one purpose in obtaining expressions for the Taylor remainder is to obtain bounds on the error in $P_n(x)$. Ideally, one wants an expression that does not involve the original function, f, in any way. This is too much to hope for, and inspection reveals that f has been replaced by $f^{(n)}$ in

$$R_n(x) = \frac{f^{(n)}(\theta x)(1-\theta)^k}{(n-k)(n-1)!} x^n.$$

While this situation is less than the ideal, in many instances $f^{(n)}$ can be replaced by a maximum value, or by dint of cleverness, something even better, and a formula not involving f in any form results. $\square$

The Schlomlich-Roche form of the Taylor remainder is one of the most general forms. More useful and simpler forms of the remainder can be obtained from this form as special cases. For example, setting $k = 0$ we have the following corollary:

Corollary (Lagrange's form of the Taylor remainder). *If the function f possesses derivatives up to order n in some neighborhood of 0, and if $x \neq 0$ is an arbitrary point in that neighborhood, there exists a number $\theta \in (0,1)$ such that*

$$R_n(x) = \frac{f^{(n)}(\theta x)}{n!} x^n.$$

Discussion. This is the form of the Taylor remainder that is most common and widely used. In essence, it states that a function f that is differentiable n times in some neighborhood of origin can be expanded as a polynomial of degree n together with an error term, namely, the Taylor remainder after n terms as follows:

$$f(x) = f(0) + \frac{f'(0)x}{1!} + \frac{f''(0)x^2}{2!} + \cdots + \frac{f^{(n-1)}(0)x^{n-1}}{(n-1)!} + \frac{f^{(n)}(\theta x)x^n}{n!}$$

for some number $\theta \in (0,1)$. We stress that the number θ depends on the point x, as well as the stage, n, at which the expansion is terminated. Thus, for each x and each n, we have a number $\theta = \theta(x, n)$. $\square$

By setting $k = n - 1$ in Theorem 8.3.1 we obtain

Corollary (Cauchy's form of the Taylor remainder). *Under the conditions of Theorem 8.3.1, there exists* $\theta \in (0, 1)$ *satisfying*

$$R_n(x) = \frac{f^{(n)}(\theta x)(1 - \theta)^{n-1} x^n}{(n - 1)!}.$$

Example 3. Estimate the error involved in using P_5 to approximate e.

Solution. The sixth Taylor remainder for e^x evaluated at 1 is given by

$$R_6(1) = \frac{e^\theta 1^6}{6!},$$

where $\theta \in (0, 1)$. Since $e \leq 3$, it is immediate that the error is less than $\frac{1}{240}$. $\square$

Theorem 8.3.2. *If the function, f, admits derivatives of all orders in $(-b, b)$ and if there exists $K > 0$ such that $\frac{f^{(n)}(x)}{n!} < K^n$ for all n and all $x \in (-b, b)$, then the Taylor series, $\sum \frac{f^{(n)}(0)x^n}{n!}$, converges to the function, f, throughout $(-a, a)$ where $a = \min\{b, \frac{1}{K}\}$.*

Proof. Using Lagrange's form of the Taylor remainder, we have $R_n(x) = \frac{f^{(n)}(\xi)x^n}{n!}$ for some $\xi \in (-b, b)$. Consequently, by hypothesis,

$$|R_n(x)| < K^n |x|^n \text{ for all } x \in (-b, b).$$

If a is chosen as required in the statement of the theorem and $x \in (-a, a)$, then $|R_n(x)| < r^n$, where $r = K|x| < 1$. Since r^n converges to 0, we have $R_n(x) \to 0$, so the Taylor series converges to $f(x)$ in $(-a, a)$. $\square$

Discussion. In the light of the above theorem, let us further analyze Example 1. We have seen that all the derivatives vanish so that the Taylor remainder after n terms is still $f(x)$. Thus, for any nonzero x, $R_n(x) = e^{-1/x^2}$, which never approaches zero. Again, if $b > 0$, let $x_n = n^{-\frac{1}{2}}$. Then after a stage, say, $n > N$, $x_n \in (-b, b)$. Using Lagrange's form of remainder, now corresponding to each x_n, we have

$$f^{(n)}(x_n) = e^{-1/x_n^2} \cdot \frac{n!}{x_n^n} = \left(\frac{n}{e^2}\right)^{\frac{n}{2}} n!.$$

Since for sufficiently large n, we have $\frac{n}{e^2}$ arbitrarily large, Theorem 8.3.2. cannot apply for any value of b and K. $\square$

Example 4. Obtain a Taylor expansion of the function $(1 + x)^m$ in a neighborhood of 0, where m is any real number. Deduce the **binomial series**

$$(1 + x)^m = \sum_{n=0}^{\infty} \binom{m}{n} x^n.$$

[Here, for an arbitrary real number m, the expression $\binom{m}{n}$ is the nth **binomial coefficient**, which is the fraction $\frac{m(m-1)(m-2)\cdots(m-n+1)}{n!}$.]

Solution. The function $f(x) = (1 + x)^m$ is clearly defined in a neighborhood of 0, and it is readily seen that

$$f^{(n)}(x) = m(m - 1) \cdots (m - n + 1)(1 + x)^{m-n}.$$

So the Taylor infinite series corresponding to the function, f, about origin is the formal series

$$\sum_{n=0}^{\infty} \frac{f^{(n)}(0)x^n}{n!} = \sum_{n=0}^{\infty} \binom{m}{n} x^n.$$

We first study the convergence of this series. If m is a nonnegative integer, then for $n > m$, the coefficients, $\binom{m}{n} = 0$; hence, the series converges trivially for all real x. If $m \notin \mathbf{N}$, then the above is an infinite power series, whose radius of convergence is easily seen to be 1, so the infinite Taylor series converges for $|x| < 1$. However, we still do not know whether the sum is actually $(1 + x)^m$.

For $x \in (0, 1)$, we use Lagrange's form of the Taylor remainder and obtain the remainder

$$R_n(x) = \binom{m}{n}(1 + \theta x)^{m-n} x^n,$$

where $0 < \theta < 1$. If we choose $n > m$, then $(1 + \theta x)^{m-n} < 1$, so

$$|R_n(x)| \le \left| \binom{m}{n} x^n \right|.$$

Now, $\sum_{n=0}^{\infty} \binom{m}{n} x^n$ converges (Exercise 6), so $\lim_{n\to\infty} \binom{m}{n} x^n = 0$. Consequently, $R_n(x) \to 0$, proving that in $(0, 1)$, the infinite Taylor series actually converges to the sum $(1 + x)^m$.

For x in the interval $(-1, 0)$, we choose the Cauchy's form of the Taylor remainder and obtain

$$
\begin{aligned}
R_n(x) &= \frac{(1 - \theta)^{n-1}}{(n - 1)!} x^n m(m - 1)(m - 2) \cdots (m - n + 1)(1 + \theta x)^{m-n} \\
&= n \binom{m}{n} x^n \left(\frac{1 - \theta}{1 + \theta x} \right)^{n-1} (1 + \theta x)^{m-1},
\end{aligned}
$$

where $0 < \theta < 1$. Since $-1 < x < 0$, we have $0 < 1 - \theta < 1 + \theta x$, so $\frac{1-\theta}{1+\theta x} < 1$. Thus, $\left(\frac{1-\theta}{1+\theta x} \right)^{n-1}$ converges to 0. Next, for $|x| < 1$, by Ratio test, the series $\sum n \binom{m}{n} x^n$ is convergent, so $n \binom{m}{n} x^n$ tends to 0. Finally, $0 < (1 + \theta x)^{m-1} \le (1 + |x|)^{m-1}$ if $m > 1$, and if $m < 1$, then $0 < (1 + \theta x)^{m-1} = \frac{1}{(1+\theta x)^{1-m}} \le \frac{1}{(1-|x|)^{1-m}} = (1 - |x|)^{m-1}$. So, for all $x \in (-1, 0)$, $(1 + \theta x)^{m-1} \le (1 \pm |x|)^{m-1}$ hence is bounded. It is now immediate that $R_n(x) \to 0$ for $x \in (-1, 0)$.

Thus, for all x satisfying $|x| < 1$, the infinite series expansion

$$(1 + x)^m = \sum_{n=0}^{\infty} \binom{m}{n} x^n$$

is valid. $\square$

Discussion. There are several important features to be noted in the above example. Since $(1+x)^m$ can be differentiated any number of times in a neighborhood of 0, we could immediately write down the Taylor infinite series associated with the function about origin in an elegant form. We also show very easily, using either the Ratio or the Root test that the series converges for all x in $(-1, 1)$. But the real problem here is to show that the infinite Taylor series converges *precisely* to the function from which it was generated. This is where the role of the Taylor remainder is important. In $(0, 1)$, we chose the Lagrange's form, which was quite easy to compute, and we could easily show that the remainder converges to 0, by looking at each component in the expression. But, in the interval $(-1, 0)$, we preferred to use the Cauchy's form, because the Lagrange's form will only provide a very poor estimate for $R_n(x)$ in $(-1, 0)$. In the Cauchy form expression for $R_n(x)$, the real difficulty was to deal with the term $(1 + \theta x)^{m-1}$ (since θ depends on n). This was overcome, by considering the two cases $m > 1, m < 1$ separately, and generating the bounds, which are independent of n. □

Example 5. Show that if $f^{(n+1)}$ exists and is continuous and if $f^{(n+1)}(0) \neq 0$, then $\lim_{x \to 0} \theta_n(x) = \frac{1}{n+1}$, where θ_n represents the θ occurring in the nth Taylor remainder.

Solution. Since f satisfies the hypothesis of the corollary to Theorem 8.3.1 (Lagrange's form), we can write the Taylor expansion about origin corresponding to $f(x)$ stopping at the nth, as well as the $(n+1)$st stages, respectively, as follows, resulting in the Taylor remainders at these two stages:

$$f(x) = f(0) + \frac{f'(0)x}{1!} + \frac{f''(0)x^2}{2!} + \cdots + \frac{f^{(n)}(\theta_n x)x^n}{n!}$$

and

$$f(x) = f(0) + \frac{f'(0)x}{1!} + \frac{f''(0)x^2}{2!} + \cdots + \frac{f^{(n)}(0)x^n}{n!} + \frac{f^{(n+1)}(\theta_{n+1}x)x^{n+1}}{n!},$$

where θ_n, θ_{n+1}, respectively, denote the θ occurring in the statement of the corollary at the nth and the $(n+1)$st stages. Equating the above two expressions, we obtain

$$\frac{x^n}{n!}[f^{(n)}(\theta_n x) - f^{(n)}(0)] = \frac{x^{n+1}}{(n+1)!}f^{(n+1)}(\theta_{n+1}x),$$

whence

$$f^{(n)}(\theta_n x) - f^{(n)}(0) = \frac{x}{n+1}f^{(n+1)}(\theta_{n+1}x).$$

Now, $f^{(n)}$ satisfies the conditions for the Mean Value Theorem in differential calculus (Theorem 4.3.2), so there exists a ξ between 0 and $\theta_n x$ satisfying

$$\theta_n x f^{(n+1)}(\xi) = \frac{x}{n+1}f^{(n+1)}(\theta_{n+1}x).$$

Hence,

$$\theta_n = \frac{1}{n+1}\frac{f^{(n+1)}(\theta_{n+1}x)}{f^{(n+1)}(\xi)}.$$

The continuity of $f^{(n+1)}$ guarantees that as x approaches 0, since ξ also approaches 0, the term $\frac{f^{(n+1)}(\theta_{n+1}x)}{f^{(n+1)}(\xi)}$ approaches 1, and we obtain

$$\lim_{x \to 0} \theta_n = \frac{1}{n+1}.$$

□

Exercises

1. Using the well-known formula for the derivative of e^x, show $f^{(n)}(0) = 0$ for all $n \in \mathbf{N}$, where f is as in Example 1.

2. Let f be infinitely differentiable on $(a - R, a + R)$. Show the sequence of Taylor polynomials for f at a converges uniformly on $(a-R, a+R)$ exactly if the sequence of remainders converges uniformly to zero on the same interval.

3. Establish the claim about ξ in the proof of Theorem 8.3.1, namely, that $\xi = \theta x_0$ for some $\theta \in (0, 1)$.

4. State and prove a general form of the Schlomlich-Roche theorem to cover the case of a Taylor expansion about a.

5. Supply an independent proof of the validity of Lagrange's and Cauchy's form of the Taylor remainder, without using Theorem 8.3.1.

6. Show that $\sum_{n=0}^{\infty} \binom{m}{n} x^n$ is convergent for $x \in (0, 1)$ and $m \in \mathbf{R}$.

7. Obtain the Taylor remainder (all the three forms) after three terms of the following functions in $[0, x]$. Use whatever formulae are required to find derivatives.

    (a) $x^3 + 4x^2 - 3x + 5$;

    (b) $\sqrt{1 + x}$, $x > 0$;

    (c) $\sqrt{1 - x^2}$, $-1 < x < 1$;

    (d) $(1 + x)^{\frac{1}{3}}$, $-1 < x$;

    (e) $f(x) = \ln \cos x$;

    (f) $f(x) = e^{\cos x}$;

    (g) $f(x) = \frac{x}{e^x - 1}$;

    (h) $f(x) = e^{\arcsin x}$;

    (i) $f(x) = \dfrac{1}{\sqrt{1 - x^2}}$.

8. Obtain a Taylor expansion of the following functions about origin. In each case, obtain the interval of convergence of the infinite Taylor series associated with the function. Prove (where possible) that the series converges to the function in that interval:

    (a) $f(x) = \sin x + \cos x$;

    (b) $f(x) = \cosh x = \frac{e^x + e^{-x}}{2}$;

    (c) $f(x) = \tanh x = \frac{e^x - e^{-x}}{e^x + e^{-x}}$;

    (d) $f(x) = \ln \frac{1+x}{1-x}$;

    (e) $f(x) = \arcsin x$;

    (f) $f(x) = \int_0^x e^{-t^2}\, dt$;

    (g) $f(x) = \ln(1 + x)$ [Hint: Use Cauchy's form of remainder in $(-1, 0)$];

    (h) $f(x) = \arctan x$ [Hint: Consider the expansion of the derivative].

9. Obtain a Taylor expansion of the following functions about the point $a \in \mathbf{R}$:

    (a) e^{x+1};

    (b) $\sin x$;

    (c) $\cos x$;

    (d) $\ln(1 + x)$ $a > -1$.

10. Prove any power series is the Taylor series of its sum.

11. Let P be a polynomial on $[a, b]$. Show this polynomial is its own Taylor series.

12. Obtain a Taylor series expansion of $f(x) = x^6 + 14x^5 - 14x^3 + 21x^2 + x - 6$ about the point $x = 2$. Hence, obtain $f(2.01)$ correct to four decimal places.

13. Expand $f(x) = \frac{1}{x}$ in a Taylor series about $a = 1$. What can be said about the convergence properties of this series?

14. If $a, h > 0$, $n \in \mathbf{N}$, prove the existence of θ, $0 < \theta < 1$ satisfying
$$\frac{1}{a+h} = \frac{1}{a} - \frac{h}{a^2} + \frac{h^2}{a^3} - \cdots + \frac{(-1)^{n-1}h^{n-1}}{a^n} + \frac{(-1)^n h^n}{(a+\theta h)^{n+1}}.$$

15. Show the power series, $\sum_{n=0}^{\infty} \frac{x^n}{n^n}$, is convergent to a function f for all $x \in \mathbf{R}$. What is the Taylor series for f expanded about 0?

16. Suppose there is known to be a function that is continuous and differentiable on $\mathbf{R}$ and satisfies $f'(x) = f(x)$ for all $x \in \mathbf{R}$. Given this function satisfies $f(0) = 1$, find the Taylor series expansion for f at 0.

17. Starting with the identity $\int_a^x f'(t)dt = f(x) - f(a)$, obtain the **integral form of the Taylor remainder**: If f has derivatives through order n in some neighborhood of x_0 and $f^{(n)}$ is Riemann integrable on any interval in that neighborhood, then for all x in that neighborhood, the Taylor remainder after n terms is given by the integral
$$R_n(x) = \frac{1}{(n-1)!} \int_{x_0}^x (x-t)^{n-1} f^{(n)}(t)\,dt.$$
Obtain also the Lagrange's and the Schlomlich-Roche's forms of remainder in integral form.

18. If $f(x)$ possesses continuous derivatives up to order $(n+2)$, $f^{(n+1)}(0) \neq 0$, and θ_n is the term occurring in the nth Taylor remainder, prove
$$\theta_n(x) = \frac{1}{n+1} + \frac{n}{2(n+1)^2(n+2)} \left[\frac{f^{(n+2)}(0)}{f^{(n+1)}(0)} + \epsilon_x \right] x,$$
where $\epsilon_x \to 0$ as $x \to 0$.

19. Let f be n times differentiable at c, and let $P_n(x)$ be the nth Taylor polynomial at $x = c$. Show
$$\lim_{x \to c} \frac{f(x) - P_n(x)}{(x-c)^n} = 0.$$

20. Let $0 < a < 1$. Show the equation $\sin x = ax$ has a root near π. Further, show $\pi(1-a)$ and $\pi(1-a+a^2)$ are successively better approximations to this root. What about $\pi(1 - a + a^2 - a^3)$?

21. Expand $(1-x)^{n+\frac{1}{2}}$ using Lagrange's form of remainder at the nth stage, and show θ converges to $\frac{4n+1}{(2n+1)^2}$ as x approaches 1. More generally, if $f(x) = (1-x)^m$, where $m > 0$, then θ approaches $1 - \left(\frac{n}{m}\right)^{\frac{1}{m-n}}$ as x approaches 1.

22. Let f be an odd function, that is, $f(x) = f(-x)$, for all x, and suppose f can be expanded in an infinite Taylor series at 0. Show the terms of this series all have odd degree. Prove an analogous result for even functions. Why are these functions called odd and even?

23. Prove the following form of Taylor's Theorem (**Young' form**): If $f(x)$ and its successive $n-1$ derivatives are continuous at $x = a$ and $f^{(n)}(a)$ exists, then
$$f(a+h) = f(a) + \frac{hf'(a)}{1!} + \cdots + \frac{h^{n-1}f^{(n-1)}(a)}{(n-1)!} + \frac{h^n}{n!}M,$$
where $M \to f^{(n)}(a)$ as $h \to 0$.

24. The numbers B_n in the expansion $\frac{x}{e^x-1} = \sum_{n=0}^{\infty} \frac{B_n x^n}{n!}$ are called **Bernoulli numbers**.

    (a) Prove $B_{2n+1} = 0$ for all $n \in \mathbf{N}$.

    (b) Obtain the first six Bernoulli's Numbers;

    (c) Show $\sum_{k=0}^{n-1} \binom{m}{n} B_k = 0$, and use this identity to compute the first six even Bernoulli's Numbers.

25. If $\frac{xe^{xt}}{e^x-1}$ is expanded in a power series about 0 in the form $\sum_{n=0}^{\infty} P_n(t)\frac{x^n}{n!}$, show that

$$P_n(t) = \sum_{k=0}^{n} \binom{n}{k} P_k(0) t^{n-k}.$$

The $P_n(t)$'s are called **Bernoulli's Polynomials**, and $P_n(0)$'s can be identified with the Bernoulli's Numbers. Prove

    (a) $P_n(t+1) - P_n(t) = nt^{n-1}, n \in \mathbf{N}$;

    (b) $\frac{P_{n+1}(k+1)-P_{n+1}(0)}{n+1} = \sum_{i=1}^{k-1} i^n$, for $n = 2,\ 3,\ \dots$.

26. If f is a polynomial of degree n, show

$$\sum_{k=1}^{n} (-1)^{k-1} \binom{n}{k} f(x+kh) = \left(\sum_{k=1}^{n} \frac{1}{k} \right) f(x) + hf'(x).$$

27. Prove **Cauchy's Generalized Mean Value Theorem**. If f and g together with their successive $n-1$ derivatives are continuous in $[a,b]$ and $f^{(n)}, g^{(n)}$ both exist in (a,b), then there exists a number $\xi \in (a,b)$ satisfying

$$\frac{f(b) - f(a) - \frac{(b-a)}{1!}f'(a) - \cdots - \frac{(b-a)^{n-1}}{(n-1)!}f^{(n-1)}(a)}{g(b) - g(a) - \frac{(b-a)}{1!}g'(a) - \cdots - \frac{(b-a)^{p-1}}{(p-1)!}g^{(p-1)}(a)} = \frac{(p-1)!}{(n-1)!}(b-\xi)^{n-p}\frac{f^{(n)}(\xi)}{g^{(p)}F'(\xi)}.$$

28. If $f(x+h) = f(x) + hf'(x) + \cdots + \frac{h^n}{n!}f^{(n)}(x+\theta h)$, where $0 < \theta < 1$, prove the following:

    (a) $\theta \to \frac{1}{n+1}$ as $h \to 0$;

    (b) θ is independent of x and h provided $f(x)$ is of the form $A + Bx + \cdots + Kx^{n+1}$.

8.4   Weierstrass's Approximation Theorem

In the last section, we defined an analytic function as being a function, f, which could be represented by a power series, specifically, a Taylor series, on an open interval. We saw that to even have a Taylor series, in a formal sense, the function, f, would have to be very well behaved on that interval, in the sense that it would have to possess an infinite number of derivatives, each of which was continuous on the interval. We followed this definition by presenting a example of a function that dashed our hopes that any function that admitted a Taylor series expansion would have to be represented by that Taylor series on at least some small open interval about the point of expansion.

The negative result that mere possession of derivatives in an interval is not enough to guarantee that a function is analytic is certainly disheartening and must have come as a rude shock to those studying the problem of representing functions by power series. In fact, Cauchy's function has so many nice properties that it hardly seems plausible that an arbitrary continuous function might be representable by a sequence or series of polynomials. Nevertheless, mathematicians did not give up the cause and continued to pursue the goal of being able to represent an arbitrary continuous function by polynomials.

In this section we present the realization of the goal in the form of Weierstrass's Approximation Theorem, which dates back to 1885. This remarkable theorem asserts that every continuous function defined on a closed interval is not only representable as a sequence of polynomials, indeed, it is uniformly representable as such a sequence! There are several constructive proofs of this statement, due to various mathematicians, Bernstein, Laguerre, Tchebychev, Lebesgue, Fourier, to mention a few. These proofs provide an explicit construction that will witness the truth of the theorem by spelling out the sequence of polynomials. Our proof will adopt this approach using the Bernstein Polynomials. We will define these polynomials and study their basic properties before proving the main theorem.

Definition. Let f be a real-valued function defined on $[0, 1]$. The nth **Bernstein Polynomial associated with** f is defined by

$$B_{f,n}(x) = \sum_{k=0}^{n} \binom{n}{k} x^k (1-x)^{n-k} f\left(\frac{k}{n}\right), n \in \mathbf{N}, x \in [0, 1].$$

Discussion. Recall from Chapter 0 that the numbers $\binom{n}{k}$ are the familiar binomial coefficients, namely,

$$\binom{n}{k} = \frac{n!}{k!(n-k)!}, \ n \in \mathbf{N}, \ k = 0, \ 1, \ 2, \ \dots.$$

There are several facts to be observed from this definition.

First, each $B_{f,n}$ is a polynomial of degree n in the variable x. For example, $B_{f,3}$ is

$$B_{f,3}(x) = (1-x)^3 f(0) + 3x(1-x)^2 f\left(\frac{1}{3}\right) + 3x^2(1-x)f\left(\frac{2}{3}\right) + x^3 f(1).$$

As such, $B_{f,3}(0) = f(0)$ and $B_{f,3}(1) = f(1)$.

Second, each Bernstein Polynomial depends on f. Indeed, the nth polynomial uses $n+1$ functional values. Moreover, each functional value that is used is used infinitely often in the sequence of polynomials, $\{B_{f,n}\}$. Thus, the polynomials depend ever more strongly on the given function, f.

As another example, consider f defined by $f(x) = 1$ for all x. Then the Bernstein Polynomials reduce to the constant polynomial, 1, for each n, since from the Binomial Theorem, we obtain

$$1 = (x + 1 - x)^n = \sum_{k=0}^{n} \binom{n}{k} x^k (1 - x)^{1-k} = B_{1,n}(x).$$

This representation of the constant polynomial, 1, will prove useful in the proof of Theorem 8.4.1. $\square$

Before beginning the proof of the Weierstrass Approximation Theorem, we present a further identity in the form of an example that will be useful in the proof of the theorem.

Example 1. Show that $x = \sum_{k=0}^{n} \frac{k}{n} \binom{n}{k} x^k (1 - x)^{n-k}$.

Solution. We have

$$1 = B_{1,n-1} = \sum_{k=0}^{n-1} \binom{n-1}{k} x^k (1 - x)^{n-1-k}.$$

Thus, if we simply multiply by x, we obtain

$$x = x B_{1,n-1} = \sum_{k=0}^{n-1} \binom{n-1}{k} x^{k+1} (1 - x)^{n-1-k}.$$

Now, $\binom{n-1}{k} = \frac{k+1}{n} \times \binom{n}{k+1}$ (Exercise 1), whence

$$
\begin{aligned}
x &= \sum_{k=0}^{n-1} \frac{k+1}{n} \binom{n}{k+1} x^{k+1} (1 - x)^{n-(k+1)} \\
&= \sum_{j=1}^{n} \frac{j}{n} \binom{n}{j} x^j (1 - x)^{n-j} \\
&= \sum_{j=0}^{n} \frac{j}{n} \binom{n}{j} x^j (1 - x)^{n-j},
\end{aligned}
$$

where we may include the term with $j = 0$ in the last sum, since it is 0. Replacing j by k completes the proof. $\square$

Discussion. The calculations presented, while not particularly difficult, are nevertheless not obvious, nor easily motivated. Yet it is calculations like these which are the foundation of the proof of Weierstrass's Theorem by means of Bernstein Polynomials. In Exercises 1–3, there are further identities of a like nature that one should complete prior to reading the proof Theorem 8.4.1. $\square$

Theorem 8.4.1 (Weierstrass). *If f is continuous on $[0, 1]$, then the sequence Bernstein Polynomials $\{B_{f,n}\}$ associated with f converge uniformly to f in $[0, 1]$.*

Proof. Since f is continuous on $[0, 1]$, there exists a constant, $K > 0$, such that $|f(x)| < K$ for all $x \in [0, 1]$. Since $[0, 1]$ is compact, the function, f, is uniformly continuous on $[0, 1]$. Hence, given $\epsilon > 0$, there exists $\delta > 0$ such that for all $x, y \in [0, 1]$, whenever $|x - y| < \delta$, we have

$$|f(x) - f(y)| < \frac{\epsilon}{2}.$$

Now, fix an arbitrary $x \in [0, 1]$. For this fixed x we have

$$
\begin{aligned}
|f(x) - B_{f,n}(x)| &= |f(x) \cdot 1 - B_{f,n}(x)| \\
&= \left| \sum_{k=0}^{n} \binom{n}{k} x^k (1-x)^{n-k} \left(f(x) - f\left(\frac{k}{n}\right) \right) \right| \\
&\leq \sum_{k=0}^{n} \binom{n}{k} x^k (1-x)^{n-k} \left| f(x) - f\left(\frac{k}{n}\right) \right|. \quad (1)
\end{aligned}
$$

Recall that x is fixed. Thus, for each value of $k = 0, 1, \dots, n$, the inequality, $|x - \frac{k}{n}| < \delta$, is either satisfied or fails. Thus, we decompose the sum (1) into two parts, $\sum_1$ and $\sum_2$, where the first sum includes all terms from (1) for which the inequality, $|x - \frac{k}{n}| < \delta$, is satisfied and the second include all terms from (8.4.1) for which the inequality fails. For $\sum_1$, we have,

$$
\begin{aligned}
\sum_1 &= \sum_{k:|x-\frac{k}{n}|<\delta} \binom{n}{k} x^k (1-x)^{n-k} \left| f(x) - f\left(\frac{k}{n}\right) \right| \\
&< \sum_{k:|x-\frac{k}{n}|<\delta} \binom{n}{k} x^k (1-x)^{n-k} \frac{\epsilon}{2} \\
&= \left[\sum_{k:|x-\frac{k}{n}|<\delta} \binom{n}{k} x^k (1-x)^{n-k} \right] \cdot \frac{\epsilon}{2} \\
&\leq \left[\sum_{k=0}^{n} \binom{n}{k} x^k (1-x)^{n-k} \right] \cdot \frac{\epsilon}{2} \\
&= \frac{\epsilon}{2}, \quad (2)
\end{aligned}
$$

where (2) is by virtue of the identity for $B_{1,n}$. The second identity in Exercise 3 asserts that i

$$
\sum_{k=0}^{n} \binom{n}{k} x^k (1-x)^{n-k} \left(x - \frac{k}{n} \right)^2 = \frac{x(1-x)}{n}
$$

$$
\leq \frac{1}{4n},
$$

since $x(1-x) \leq \frac{1}{4}$ for all $x \in [0, 1]$ (Exercise 4). Now take $N \geq \frac{K}{\epsilon\delta^2}$, then

$$
\begin{aligned}
\sum_2 &\leq \sum_{k=0}^{n} \left(\frac{(x-\frac{k}{n})^2}{(x-\frac{k}{n})^2} \right) \binom{n}{k} x^k (1-x)^{n-k} \left(|f(x)| + |f\left(\frac{k}{n}\right)| \right) \\
&= \sum_{k=0}^{n} \left(x - \frac{k}{n} \right)^2 \binom{n}{k} x^k (1-x)^{n-k} \left(\frac{|f(x)| + |f\left(\frac{k}{n}\right)|}{(x-\frac{k}{n})^2} \right)
\end{aligned}
$$

$$\leq \sum_{k=0}^{n} \left(x - \frac{k}{n}\right)^2 \binom{n}{k} x^k (1-x)^{n-k} \left(\frac{2K}{\delta^2}\right)$$

$$< \frac{1}{4n} \cdot \frac{2K}{\delta^2}$$

$$< \frac{\epsilon}{2}, \text{ whenever } n > N. \tag{3}$$

Thus, for the given ϵ, we have chosen $N \in \mathbf{N}$, such that if $n \geq N$, then

$$|f(x) - B_{f,n}| \leq \sum\nolimits_1 + \sum\nolimits_2 < \tfrac{\epsilon}{2} + \tfrac{\epsilon}{2} = \epsilon,$$

proving that the sequence $\{B_{f,n}\}$ converges uniformly to f on $[0,1]$. $\square$

Discussion. This proof is difficult, and we review it step by step. To establish uniform convergence of the sequence, $\{B_{f,n}\}$, we have to show that for every positive ϵ there exists an $N \in \mathbf{N}$ such that $|f(x) - B_{f,n}(x)| < \epsilon$ for all $x \in [0,1]$ whenever $n > N$. Let us see that the proof actually accomplishes this requirement.

Consider $|f(x) - B_{f,n}(x)|$. This quantity is bounded by the sum, (1). The computations leading to this bound are based on a fixed x, but in no way depend on that x. Thus, the result captured in (1) is valid for all $x \in [0,1]$.

The next important step in the proof is the decomposition of the sum in (1) into $\sum_1$ and $\sum_2$. It is important to realize that the particular terms from the sum in (1) that end up as part of $\sum_1$ will depend on the value of x, the value of δ, and the value of n. Similarly, which particular terms end up as part of $\sum_2$ also depends on the value of x, the value of δ, and the value of n. For this reason, it is essential to make sure that the arguments that make up the remainder of the proof are not affected by changing that particular terms of the sum in (8.4.1) end up in $\sum_1$, and that end up in $\sum_2$.

Consider then the computations with $\sum_1$ leading to (2). The first step in this string depends on the fact that

$$\left|x - \frac{k}{n}\right| < \delta \text{ implies } \left|f(x) - f\left(\frac{k}{n}\right)\right| < \frac{\epsilon}{2}.$$

The remainder of these computations are algebraic in nature and do not depend on the value of x or the value of n or which terms are included in $\sum_1$. Thus, these computations will remain valid, provided we do not change ϵ or δ. Indeed, the inequality captured in (2) holds for all the Bernstein Polynomials, $B_{f,n}$, because in essence, it is not a statement about Bernstein Polynomials at all; rather, it is a statement that depends for its truth on the continuity of f. As a result of the computations leading to (2), we obtain an ϵ bound on $\sum_1$.

The next step in the proof amounts to generating an ϵ bound on $\sum_2$. This bound is generated from the identity

$$\sum_{k=0}^{n} \binom{n}{k} x^k (1-x)^{n-k} \left(x - \frac{k}{n}\right)^2 = \frac{1}{n} x(1-x), \tag{*}$$

which is the second identity requested in Exercise 3. Arriving at such an identity would result only after considerable experimentation with Bernstein Polynomials.

In any case, once one has this identity, one can notice that by introducing a fraction equivalent to 1 into $\sum_2$ and an application of the Triangle inequality, it becomes possible to replace $\sum_2$ by the quantity on the left-hand side of (*) multiplied by the constant, $\frac{2K}{\delta^2}$. Since the quantity in (*) becomes arbitrarily small as n becomes large, we are done as long as the previous computations do not depend on the choice of n. Since they do not, we have the required bound on $\sum_2$.

Further insight into this proof can be gained from looking again at (1), which in summary form asserts that

$$|f(x) - B_{f,n}(x)| \leq \sum_{k=0}^{n} \binom{n}{k} x^k (1-x)^{n-k} \left| f(x) - f\left(\frac{k}{n}\right) \right|.$$

If one poses the question as to why the sum on the right might be small, one concludes that the individual terms can be small for one of two reasons. First, $|f(x) - f(\frac{k}{n})|$ can be small, whence its product with another term that is bounded would also be small. Second, $\binom{n}{k} x^k (1-x)^{n-k}$ could be small, whence its product with a term that is bounded would also be small. To obtain the first, we apply continuity. To obtain the second, we use the fact that f is bounded on $[0,1]$ and the identity in (*).

It is at this point that one sees how the function, f, is employed in the proof. Specifically, f determines the dependence of δ on ϵ and the value of K, the universal bound. The number, $\frac{K}{\epsilon\delta^2}$, then determines the value of N, or how many terms of the sequence must be taken to obtain the required approximation.

Finally, the crucial fact on which the proof rests is that a continuous function on a closed interval is uniformly continuous. It is uniform continuity which permits us to obtain a single δ that does not depend on x and a single bound over the entire interval. Without these very powerful theorems, there would be no hope of a result like this. $\square$

Corollary. *Let f be a continuous function on $[a,b]$. Then there is a sequence of functions, $\{P_n\}$, such that P_n is a polynomial of degree, n, that converges uniformly to f on $[a,b]$.*

Proof. Let g be defined on $[0,1]$ by $g(x) = a + (b-a)x$. Then $f \circ g$ is a continuous function with domain $[0,1]$. As such there is a sequence of Bernstein Polynomials, $\{B_{f\circ g,n}\}$, that converges uniformly to $f \circ g$ on $[0,1]$. We claim, Exercise 5, that $\{B_{f\circ g,n}(a + (b-a)x)\}$ is a sequence of polynomials which converges uniformly to f on $[a,b]$. $\square$

Discussion. Theorem 8.4.1 and its corollary assert that every continuous function is almost a polynomial. With this in mind, the reader must surely be wondering about how this result and Cauchy's function could both appear in the same book, much less be products of the same axiom scheme.

Cauchy's function reveals that not every function can be obtained as the sum of its power series. Weierstrass's Approximation Theorem says that even Cauchy's function can be uniformly obtained as a sequence of polynomials, and in consequence, as a series, each member of which is a polynomial. The point is that the series of polynomials that approximates the Cauchy function is not a power series. This seems a simple enough answer, but the situation is still a bit more complex.

Let us consider the Cauchy function, f, on the interval from $[-1, 1]$. For any fixed $\epsilon > 0$, there is a tube about f of width, ϵ, such that there is a polynomial, P, that lies completely inside this ϵ-tube. By Exercise 8.3.11, this polynomial is a power series! Thus, there is a power series that approximates the Cauchy function arbitrarily closely. Why is this not a contradiction? The answer is that as ϵ is made smaller, the power series changes. In particular, each smaller ϵ will, in general, generate a new values for the constant term, the coefficient on x, and the other terms in the power series approximation to f. (If Bernstein Polynomials are used, the constant term will be fixed.) This is completely different from the Taylor series approach, where each addition to the approximation has no effect on the coefficients of terms of lower degree. Indeed, this fact about the Taylor series coefficients is at the heart of the properties of Taylor series. □

We leave the reader with a final example.

Example 2. Find the first five Bernstein Polynomials for $\cos 6x$. Illustrate how these polynomials approximate $\cos 6x$.

Solution. The polynomials in order are

$$
\begin{aligned}
B_{f,1} &= (1-x)\cos 0 + x\cos 6, \\
B_{f,2} &= (1-x)^2 \cos 0 + 2x(1-x)\cos 3 + x^2 \cos 6, \\
B_{f,3} &= (1-x)^3 \cos 0 + 3x(1-x)^2 \cos 2 + 3x^2(1-x)\cos 4 + x^3 \cos 6, \\
B_{f,4} &= (1-x)^4 \cos 0 + 4x(1-x)^3 \cos\frac{6}{4} + 6x^2(1-x)^2 \cos 3 \\
&\quad + 4x^3(1-x)\cos\frac{18}{4} + x^4 \cos 6, \\
B_{f,5} &= (1-x)^5 \cos 0 + 5x(1-x)^4 \cos\frac{6}{5} + 10x^2(1-x)^3 \cos\frac{12}{5} \\
&\quad + 10x^3(1-x)^2 \cos\frac{18}{5} + 5x^4(1-x)\cos\frac{24}{5} + x^5 \cos 6,
\end{aligned}
$$

where $f(x) = \cos 6x$. Figure 8.4.1 presents graphs of the Bernstein Polynomials together with that of $\cos 6x$. As can be seen from inspection, the convergence in not nearly so rapid as for the power series for e^x. But $\cos 6x$ is a more complicated function. □

Discussion. In relation to the previous discussion the reader can check the following. First, the constant term of each of the Bernstein Polynomials for $\cos 6x$ will be $\cos 0 = 1$. Second, the coefficients on all positive powers of x change as n changes. For example, the first coefficient on x is $\cos 6 - \cos 0$, while the second coefficient on x is $2(\cos 3 - \cos 0)$, and so forth. Thus, it is easy to see that the sequence of polynomials cannot possibly be part of a single power series. □

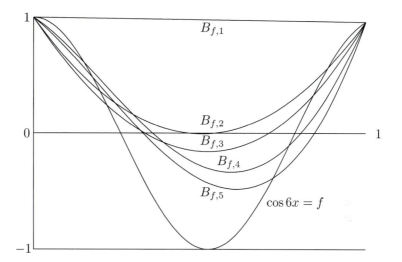

Figure 8.4.1 The Bernstein Polynomials $B_{f,1}(x)$, $B_{f,2}(x)$, $B_{f,3}(x)$, $B_{f,4}(x)$, and $B_{f,5}(x)$ for the function $f(x) = \cos(6x)$.

Exercises

1. Show $\binom{n-1}{k-1} = \frac{k}{n} \times \binom{n}{k}$ and $\binom{n-2}{k-2} = \frac{k(k-1)}{n(n-1)} \times \binom{n}{k}$.

2. Show $(n^2 - n)x^2 = \sum_{k=0}^{n} (k^2 - k)\binom{n}{k}x^k(1-x)^{n-k}$.

3. Show $(1 - \frac{1}{n})x^2 + \frac{1}{n}x = \sum_{k=0}^{n}(\frac{k}{n})^2\binom{n}{k}x^k(1-x)^{n-k}$. Use this to show

$$\frac{1}{n}x(1-x) = \sum_{k=0}^{n}(x - \frac{k}{n})^2 \binom{n}{k}x^k(1-x)^{n-k}.$$

4. Show $x(1-x) \le \frac{1}{4}$ for all $x \in [0, 1]$.

5. Complete the proof of the corollary to Theorem 8.4.1.

6. Obtain the first six Bernstein polynomials for the following functions in the indicated interval:
    (a) $f(x) = x$, $[0, 1]$;   (b) $f(x) = x^2$, $[0, 1]$;   (c) $f(x) = x^3 + 2x^2$, $[0, 1]$;
    (d) $f(x) = |x|$, $[-1, 1]$;   (e) $f(x) = \sin x$, $[0, \pi]$;   (f) $f(x) = e^x$, $[-1, 1]$.

7. In view of Exercise 6 and the results in Sections 8.3 and 8.4, discuss the relative merits of Taylor and Bernstein polynomials as methods for approximating functions.

8. Let $\epsilon = 0.2$ and $f(x) = \cos 6x$. Find a Bernstein polynomial that approximates f to within ϵ.

9. If f is continuous on $[a, b]$ such that $\int_a^b x^n f(x)dx = 0$ for all nonnegative integers n, prove f is the zero function on $[a, b]$.

10. How large must n be, so that the nth Bernstein Polynomial, B_n, associated with x^2 satisfies $|f(x) - B_n(x)| \le 0.001$ for all $x \in [0, 1]$?

11. Let f be a bounded function on $[0, 1]$, say, $|f(x)| \le K$ for all $x \in [0, 1]$. Show all the Bernstein Polynomials, $B_n(f)$, are bounded by K.

12. Let f be continuous on $[a, b]$. Show there exists a sequence, $\{p_n\}$, of polynomials converging to f uniformly on $[a, b]$, such that $p_n(a) = f(a)$ for all n.

13. Show there does not exist a sequence of polynomials converging uniformly on $\mathbf{R}$ to f, where

  (a) $f(x) = e^x$;   (b) $f(x) = \sin x$.

14. Show if f is continuous on $\mathbf{R}$, then there exists a sequence $\{p_n\}$ of polynomials converging uniformly to f on each bounded subset of $\mathbf{R}$.

15. If $f : [a, b] \to \mathbf{R}$ is continuous, show that f can be uniformly approximated on $[a, b]$ by a sequence $\{s_n\}$ of step functions.

16. A function $h : [a, b] \to \mathbf{R}$ is **piecewise linear** on $[a, b]$ if there exist points $\{t_i\}$ and real numbers $\{a_i\}$, $\{b_i\}$, $i = 0,\ 1,\ \ldots,\ n$, with $t_0 = a$, $t_n = b$ and such that

$$h(x) = a_k x + b_k,\ x \in [t_{k-1}, t_k],\ 0 \le k \le n - 1.$$

Prove that every continuous function on $[a, b]$ can be uniformly approximated by a sequence of piecewise linear functions.

17. Give another proof of Weierstrass's Approximation Theorem using the sequence of **Legendre's Polynomials**, $\{P_n\}$, where $P_0(t) = 0$ and

$$P_{n+1}(t) = P_n(t) + \frac{1}{2}\left[t - (P_n(t))^2\right],\ (n \ge 0).$$

[Hint: Show by induction

$$0 \le \sqrt{t} - P_n(t) \le \frac{2\sqrt{t}}{2 + n\sqrt{t}},$$

whence $\{P_n\}$ approaches $\sqrt{t}$ uniformly on $[0, 1]$.]

18. Let $D_n(x) = \alpha_n(1 - x^2)^n, n \in \mathbf{N}$, where α_n is defined by $\int_{-1}^{1} D_n = 1$ for each $n \in \mathbf{N}$. Let $f : [0, 1] \to \mathbf{R}$ be continuous. Prove the following:

  (a) $\alpha_n < \sqrt{n}$;

  (b) $D_n(x) \le \sqrt{n}(1 - \delta^2)^n$ for $\delta \le |x| \le 1$;

  (c) $D_n(x)$ converges to 0 uniformly for $\delta \le |x| \le 1$;

  (d) The polynomials $P_n(x)$ defined by

$$P_n(x) = \int_{-1}^{1} f(x + t)D_n(t)\,dt = \int_{0}^{1} f(t)D_n(t - x)\,dt$$

  converge uniformly to f in $[0, 1]$.

(This is Landau's proof of Theorem 8.4.1, and D_n is called **Landau's kernel**.)

19. A **Dirac sequence** is a sequence, $\{K_n\}$, of functions on $\mathbf{R}$ satisfying:
  (a) $K_n(x) \ge 0$ for all n and all x;   (b) each K_n is continuous;   (c) $\int_{-\infty}^{\infty} K_n(t)\,dt = 1$;
  (d) given ϵ and δ, there exists n such that if $n \ge N$,

$$\int_{-\infty}^{-\delta} K_n + \int_{\delta}^{\infty} K_n < \epsilon.$$

20. If f is a piecewise linear and bounded function, we define the **convolution** to be

$$f_n(x) = K_n * f(x) = \int_{-\infty}^{\infty} f(t) K_n(x - t)\, dt.$$

Prove the sequence, $\{f_n\}$, converges to f uniformly on each compact subset, $S \subseteq \mathbf{R}$.

21. If f is continuous on $[0, \pi]$, prove that given $\epsilon > 0$, there exists an even trigonometric polynomial T such that

$$|f(x) - T(x)| < \epsilon \text{ for all } x \in [0, \pi].$$

[Hint: Set $g(y) = f(\arccos y)$, $y \in [-1, 1]$. Approximate $g(y)$ as a polynomial in y, and substitute back $y = \cos x$.]

Chapter 9

Transcendental Functions

Even though we have used trigonometric, logarithmic, and exponential functions to furnish appropriate examples and exercises, we have, at this point, only defined powers of the function $f(x) = x$ in a completely formal way. Recall that for other than integral powers we made essential use of powerful facts from the theory. For example, the supremum principle was used to guarantee the existence of nth roots of real numbers, and continuity was used to obtain values of x^p for p irrational. So the existence of even these simple functions is not a trivial matter.

As well, the natural arithmetic of the real numbers was used to obtain more complicated functions, the simplest examples of which are the polynomials:

$$f(x) = a_0 + a_1 x + \cdots + a_n x^n = \sum_{k=0}^{n} a_k x^k,$$

where the a_i's are real numbers and $n \in \mathbf{N}$. In this way, starting with $f(x) = x$ and applying standard algebraic manipulations such as addition, subtraction, multiplication, division, root-extraction, and so forth, we are lead to a huge collection of functions, called **algebraic functions**. But there are many functions that cannot be manufactured by this procedure and many of these are in common use in mathematics and science, for example, the beta and gamma functions obtained by using the integral in Section 5.6.

In this chapter, we will formally introduce some of the the so-called **transcendental functions**, by which we mean those functions that cannot be obtained by standard algebraic procedures (addition, subtraction, multiplication, division, root-extraction, etc.) applied to polynomials. Explicitly, we shall define the exponential and logarithmic functions, the trigonometric functions, and the hyperbolic functions. There are several possible methods for introducing these new functions. We could use the concept of uniform convergence and define new functions by means of power series or as uniform limits of certain sequences of functions. Alternatively, we could introduce the new functions as indefinite integrals of some known functions. Or we could seek a function satisfying certain functional equations, such as $f(xy) = f(x) + f(y)$ or $f(x + y) = f(x)f(y)$. Still another possibility is to obtain the new functions as solutions to some differential equations, and this is our method of choice.

9.1   Exponential Functions

In this section, we establish the existence of a unique function that answers the demand that it be its own derivative. Then we will establish the basic properties of such a function and, finally, recognize this new function as the familiar exponential function.

Definition. The sequence $\{E_n\}$ of functions, $E_n : \mathbf{R} \to \mathbf{R}$, is defined inductively by

$$
\begin{aligned}
E_0(x) &= 1, \\
E_{n+1}(x) &= 1 + \int_0^x E_n(t)\, dt, \quad (n \in \mathbf{N},\ x \in \mathbf{R}).
\end{aligned}
$$

Discussion. The above definition is inductive. It is easily seen that $E_1(x) = 1 + x$, and more generally, using induction, one can prove (Exercise 1) that

$$
E_n(x) = 1 + \frac{x}{1!} + \frac{x^2}{2!} + \cdots + \frac{x^n}{n!}, \quad x \in \mathbf{R}.
$$

Thus, for each n, $E_n(x)$ is a polynomial of degree n in the variable x. Evidently, E_n is continuous for each n and satisfies $E_n(0) = 1$. Moreover, for a fixed $x > 0$, the sequence of real numbers, $\{E_n(x)\}$, is monotonically increasing and the polynomial E_n is strictly increasing on $[0, \infty)$. As well, these polynomials enjoy the property that the derivative of E_{n+1} is E_n, for all $n \in \mathbf{N}$, as can be checked readily. Thus, this sequence consists of functions that are the natural candidates from which to create the ultimate function with the property that it is its own derivative. $\square$

Our aim now is to use this sequence to obtain a unique function, $E(x)$, that has the property that $E'(x) = E(x)$ for all $x \in \mathbf{R}$. This is the purpose of our first theorem in this section.

Theorem 9.1.1. *There exists a function $E : \mathbf{R} \to \mathbf{R}$ satisfying the following properties:*

(i) $E(0) = 1$;

(ii) $E'(x) = E(x)$ *for all $x \in \mathbf{R}$.*

Proof. First, we show that the sequence, $\{E_n\}$, defined above, converges uniformly on each interval, $[-K, K]$, $K > 0$. For $m \geq n > 2K$ and $|x| \leq K$, we have

$$
\begin{aligned}
|E_m(x) - E_n(x)| &= \left| \frac{x^{n+1}}{(n+1)!} + \cdots + \frac{x^m}{m!} \right| \\
&\leq \frac{K^{n+1}}{(n+1)!} + \cdots + \frac{K^m}{m!} \\
&\leq \frac{K^{n+1}}{(n+1)!} \left[1 + \frac{K}{n} + \cdots + \left(\frac{K}{n} \right)^{m-n} \right] \\
&< 2\frac{K^{n+1}}{(n+1)!}.
\end{aligned}
$$

Since the last expression converges to 0, we can apply Theorem 7.2.5 to conclude that the sequence, $\{E_n\}$, converges uniformly on each interval, $[-K, K]$, $K > 0$. In particular, it follows that for each $x \in \mathbf{R}$, the sequence, $\{E_n(x)\}$, converges. We may therefore define the function $E : \mathbf{R} \to \mathbf{R}$ by

$$E(x) = \lim_{n \to \infty} E_n(x), x \in \mathbf{R}.$$

The fact that E is continuous follows from the corollary to Theorem 7.3.1, since for each x, there is a $K > 0$ satisfying $x \in [-K, K]$. Moreover, $E_n(0) = 1$ for all n (Exercise 1), and hence $E(0) = 1$. Again applying Theorem 7.3.4, we see that E is differentiable, and for all $x \in [-a, a]$,

$$E'(x) = \lim_{n \to \infty} E_n'(x).$$

Since $K > 0$ is arbitrary, this equality is satisfied for all $x \in \mathbf{R}$. Last, since $E_{n+1}'(x) = E_n(x)$, it follows that

$$E'(x) = \lim_{n \to \infty} E_n'(x) = \lim_{n-1 \to \infty} E_{n-1}(x) = E(x). \;\; \square$$

Discussion. The polynomial functions, E_n, serve as natural approximations to the desired function. They have the key property that $E_n' = E_{n-1}$. This fact, together with the differentiability of each of the polynomials in the sequence and the powerful tool of uniform convergence, enables us to invoke Theorem 7.3.4 and arrive at the conclusion that

$$\lim_{n \to \infty} E_n'(x) = \lim_{n-1 \to \infty} E_{n-1}(x).$$

Even though the sequence on the left side of the equality has one term less than the one on the right, this distinction disappears when taking the limits. This idea is the heart and soul of the proof.

It is almost an anticlimax that one finally obtains: The function, $E(x)$, is defined for all $x \in \mathbf{R}$, is continuous and differentiable and satisfies $E' = E$.

The definition of the sequence, $\{E_n\}$, makes it almost trivial to establish additional facts, for example, for $x > 0$, $E(x) > E_n(x)$ for all $n \in \mathbf{N}$. $\square$

Corollary 1. *For all $x \in \mathbf{R}$,*

$$E(x) = 1 + \frac{x}{1!} + \frac{x^2}{2!} + \cdots + \frac{x^n}{n!} + \cdots = \sum_{n=1}^{\infty} \frac{x^n}{n!}.$$

Proof. Exercise 2. $\square$

Corollary 2. *The function $E(x)$ possesses derivatives of all orders, and $E^{(n)}(x) = E(x)$ for all $n \in \mathbf{N}$, $x \in \mathbf{R}$.*

Proof. Exercise 3. $\square$

Next, we address the question: Are there functions other than $E(x)$, which possess properties (i) and (ii) of Theorem 9.1.1? In other words, we are interested in determining whether these two properties characterize a function. The next theorem provides the answer.

Theorem 9.1.2. *If $F : \mathbf{R} \to \mathbf{R}$ is a function with the properties*

(i) $F(0) = 1$ *and*

(ii) $F'(x) = F(x)$ *for all $x \in \mathbf{R}$,*

then $F(x) = E(x)$ for all $x \in \mathbf{R}$.

Proof. Set $G = E - F$. Then $G' = E' - F' = E - F = G$, so $G'(x) = G(x)$ for all $x \in \mathbf{R}$. Further,

$$G(0) = E(0) - F(0) = 1 - 1 = 0.$$

Thus, G possesses derivatives of all orders and $G^{(n)}(x) = G(x)$ for all $x \in \mathbf{R}$. Hence we can expand G in a Taylor series about the point 0 with a Lagrange remainder, and for any fixed $x \in \mathbf{R}$ we obtain

$$G(x) = G(0) + \frac{G'(0)}{1!}x + \cdots + \frac{G^{(n-1)}(0)}{(n-1)!}x^{n-1} + \frac{G^{(n)}(\xi_n)}{n!}x^n,$$

where ξ_n lies between 0 and x. Since $G^{(k)}(0) = 0$ for all k, it follows that $G(x) = \frac{G(\xi_n)x^n}{n!}$. Since x is fixed, there exists a constant K such that

$$|G^{(n)}(x)| = |G(x)| < K.$$

Successively integrating both sides of $|G^{(n)}(x)| \leq K$ n times yields $|G(x)| \leq K\frac{|x|^n}{n!}$. But this inequality is valid for all $n \in \mathbf{N}$. Since $\lim_{n\to\infty} \frac{|x|^n}{n!} = 0$, it follows that $G(x) = 0$ or, equivalently, that $E(x) - F(x) = 0$. Since x was arbitrary, we have $E(x) = F(x)$ for all $x \in \mathbf{R}$. $\square$

Discussion. Theorems of this nature are characterization theorems. In essence they establish the remarkable fact that there is exactly one function, f, with the two properties:

(i) $f' = f,$ and   (ii) $f(0) = 1.$

This function is none other than the exponential function, although we have yet to establish that the function, E, has the required properties to identify it as the exponential function.

It is really quite surprising that the two conditions are enough to totally determine a function, although this result should not surprise readers familiar with differential equations who will recognize Theorems 9.1.1 and 9.1.2 as existence and uniqueness theorems. In any case, if either of conditions (i) or (ii) were omitted, then we may obtain many functions possessing the the remaining properties, a fact pursued in Exercise 4. $\square$

Definition. The real number e defined by $e = E(1)$ is called the **exponential constant**.

Discussion. From Corollary 1 to Theorem 9.1.1, we immediately see that

$$e = \lim_{n \to \infty} \left(1 + \frac{1}{1!} + \frac{1}{2!} + \cdots + \frac{1}{n!} \right).$$

As we have long since known, by virtue of various convergence tests, the infinite series $\sum_{n=1}^{\infty} \frac{1}{n!}$ is convergent, and this definition gives the sum a name, e. Since $E(x) > 1 + x$ for $x > 0$, we see that $e > 2$. On the other hand,

$$\sum_{n=0}^{k} \frac{1}{n!} < 1 + \sum_{n=0}^{k-1} \frac{1}{2^n}.$$

By the definition, and careful analysis, we see

$$e = \sum_{n=0}^{\infty} \frac{1}{n!} < 1 + \sum_{n=0}^{\infty} \frac{1}{2^n} = 1 + 2 = 3.$$

Thus, the real number, e, lies between 2 and 3 and can be computed to any degree of accuracy by using the partial sums of the infinite series for e. A reasonable approximation for e is 2.71828182846.

The series for e converges very rapidly as the reader can check by reviewing Example 8.3.2. Computationally, we can estimate the rate of convergence as follows. Let s_n denote the nth partial sum of the series; then

$$
\begin{aligned}
e - s_n &= e - \sum_{k=0}^{n} \frac{1}{k!} = \sum_{k=n+1}^{\infty} \frac{1}{k!} \\
&= \frac{1}{(n+1)!} + \frac{1}{(n+2)!} + \frac{1}{(n+3)!} + \cdots \\
&= \frac{1}{(n+1)!} \left(1 + \frac{1}{n+2} + \frac{1}{(n+2)(n+3)} + \cdots \right) \\
&< \frac{1}{(n+1)!} \left(1 + \frac{1}{n+1} + \frac{1}{(n+1)^2} + \cdots \right) \\
&= \frac{1}{n!n},
\end{aligned}
$$

where the last inequality is obtained by once again applying the formula for the sum of a geometric series.

The last inequality provides a simple means for showing e is irrational. Suppose for the sake of argument that e were rational and $e = \frac{p}{q}$ for $p, q \in \mathbf{N}$. Then, by assumption and monotonicity for the series,

$$0 < q!(e - s_q) = q! \left(\frac{p}{q} - s_q \right) < \frac{1}{q} < 1.$$

Notice that $q!(e - s_q) = q! \left(\frac{p}{q} - 1 - \frac{1}{1!} - \frac{1}{2!} - \cdots - \frac{1}{q!} \right)$ is a positive integer that, by virtue of the above, must be between 0 and 1! It follows that e is irrational.

This argument, while it establishes that e is irrational, does not exclude the possibility that e is algebraic, that is, the solution to a polynomial equation where the polynomial has finite degree and the coefficients are integers. Numbers that are not algebraic are called transcendental, a term coined by Euler in the early 1700s.[1] Hermite established the transcendence of e in 1873, but this fact is more difficult to prove and is beyond the scope of this book. $\square$

Our next theorem summarizes the properties of the function, $E(x)$, and in consequence identifies it as the exponential function.

Theorem 9.1.3. *The function, E, defined on $\mathbf{R}$, has the following properties:*

(i) $E(x) > 0$ *for all* $x \in \mathbf{R}$;

(ii) $E(x)$ *is strictly increasing on* $\mathbf{R}$;

(iii) *the range of $E(x)$ is the set* $\mathbf{R}^+$;

(iv) $\lim\limits_{x \to \infty} E(x) = \infty$; $\lim\limits_{x \to -\infty} E(x) = 0$;

(v) $E(x+y) = E(x)E(y)$ *for all* $x, y \in \mathbf{R}$;

(vi) $E(\frac{m}{n}) = (e^m)^{\frac{1}{n}} = e^{m/n}$, *for all* $n, m \in \mathbf{Z}$, $n \neq 0$.

Proof. We shall prove (v) and (vi) leaving the rest as exercises.

For (v), fix $y \in \mathbf{R}$ and note $E(y) > 0$, by (i). Define a function $F : \mathbf{R} \to \mathbf{R}$ by setting $F(x) = \frac{E(x+y)}{E(y)}$, $x \in \mathbf{R}$. Now, $F(0) = 1$ and $F'(x) = F(x)$. Thus, by the uniqueness property of $E(x)$, we conclude that $F(x)$ must be $E(x)$ for all x. Thus, $E(x+y) = E(x)E(y)$ as desired.

To prove (vi), first note $E(nx) = (E(x))^n$, for all $n \in \mathbf{N}$ (Exercise 10) from which it follows that

$$e = E(1) = E\left(n \cdot \frac{1}{n}\right) = E\left(\frac{1}{n}\right)^n.$$

Now $E(\frac{1}{n}) > 0$. So by definition of nth root, we see $E(\frac{1}{n}) = e^{1/n}$. Next we observe,

$$E(n)E(-n) = E(n + (-n)) = E(0) = 1,$$

which implies

$$E(-n) = \frac{1}{E(n)} = \frac{1}{e^n} = e^{-n}.$$

In summary, these equations imply the following: if $m, n \in \mathbf{N}$, then

$$E\left(\frac{m}{n}\right) = E\left(\frac{1}{n}\right)^m = (e^{\frac{1}{n}})^m = e^{m/n},$$

establishing (vi). $\square$

Discussion. This theorem summarizes the familiar and essential properties of the function, $E(x)$, and relates them to the number e. In particular, the

---

[1]According to Kline (1972), p. 593, Euler asserted that numbers such as e 'transcended the power of algebra'. A proof the e is transcendental and not algebraic can be found in *Transcendental Number Theory* by A. Baker (1974).

properties asserted in (v) and (vi) almost justify our referring to $E(x)$ as the **exponential** function.

The reason underlying our reluctance to make the claim that E is, in fact, the exponential function is that it is only for rational numbers, t, that we know the equation, $E(t) = e^t$, is valid. Specifically, as noted in Section 2.5, we have not even assigned a meaning to the expression, e^x where x is not rational. The next definition deals with this problem. $\square$

Definition. If x is any real number, the xth **power of** e is defined by the equation

$$e^x = E(x).$$

Discussion. If x is a rational number of the form $\frac{m}{n}$, for some $m, n \in \mathbf{N}$, this definition agrees with the definition of $e^{\frac{m}{n}}$ given in Theorem 9.1.3 (vi). To see this, let us review the steps in the argument. First, it was shown that for every $n \in \mathbf{N}$, $E\left(\frac{1}{n}\right)$, was an nth root of e. Since $e^{\frac{m}{n}} = \left(e^{\frac{1}{n}}\right)^m$ is a definition (Exercise 0.5.14) to complete the argument, one has only to apply Theorem 9.1.3 (v) to $E(\frac{m}{n})$. Thus, all positive rational powers of the number e are defined. It can easily be deduced that $E(\frac{p}{q}) = e^{\frac{p}{q}}$ holds for negative rationals $\frac{p}{q}$ as well.

How should we go about defining an irrational power of e? For example, what is the meaning of $e^{\sqrt{2}}$? The definition supplies an answer, namely, $E(\sqrt{2})$. But is it the right answer?

Let us recall how powers were obtained. First, for any real number, x, and any positive integer n, x^n was defined (Section 0.4). Next, the definition was extended to other integers by setting $x^{-n} = \frac{1}{x^n}$, for $x \neq 0$ and $x^0 = 1$, also for $x \neq 0$. As well, it was established that this extended definition satisfied the basic law of exponents expressed in Theorem 9.1.3 (v) (see Exercise 0.4.24).

Once integral powers were defined, the next step was defining the nth root of a nonnegative real number. This step was taken in Section 0.5 by obtaining the required root from

$$x^{\frac{1}{n}} = \sup\{y : y^n \leq x \text{ and } y \in \mathbf{R}\}.$$

Rational powers for all positive real numbers were then obtained via Exercise 0.5.14.

This suggest two tests for the definition of e^x. First, for all $x, y \in \mathbf{R}$, we should have $e^x e^y = e^{x+y}$, which the reader should note is implicit in what has already been established. Second, we should have that

$$e^x = \sup\{e^{\frac{m}{n}} : m, n \in \mathbf{Z} \text{ and } n \neq 0 \text{ and } \frac{m}{n} \leq x\}. \tag{1}$$

The verification of these two tests for the extended definition of e^x is left to Exercise 11.

A completely different approach to the development of the function, e^x, would have been to begin with the function e^q, for $q \in \mathbf{Q}$. One could then extend the definition via (1) above. While this would give rise to the same function, one would be left with having to establish the basic properties listed in Theorem 9.1.3, plus the fact that the derivative of e^x is e^x. To illustrate the difficulties of this approach, we explore it in the exercises.

Having managed to obtain e^x for all $x \in \mathbf{R}$, one now wonders about a^x where $a \in \mathbf{R}^+$. It would appear plausible that a^x could be defined in terms of e^x. This is indeed the case and will be explored in the next section. $\square$

Exercises

1. Verify all claims made in the discussion following the definition of the sequence of functions, $\{E_n\}$, and Theorem 9.1.1. In particular, show that $E'_{n+1} = E_n$.

2. Prove Corollary 1 to Theorem 9.1.1.

3. Prove Corollary 2 to Theorem 9.1.1.

4. Let $f : \mathbf{R} \to \mathbf{R}$ be such that $f'(x) = f(x)$ for all $x \in \mathbf{R}$. Show $f(x) = \lambda E(x)$ for some constant, $\lambda \in \mathbf{R}$. Does this violate the uniqueness result in Theorem 9.1.2?

5. Show if $x \in [0, a]$ and $n \in \mathbf{N}$, then

$$\sum_{k=1}^{n} \frac{x^k}{k!} \leq e^x \leq \left[\sum_{k=1}^{n-1} \frac{x^k}{k!}\right] + \frac{x^n e^a}{n!}.$$

6. Prove if $n \geq 2$,

$$0 < en! - \left[\sum_{k=0}^{n} \frac{1}{k!}\right] n! < \frac{e}{n+1} < 1.$$

7. Calculate e accurate to seven decimal places and estimate the error involved.

8. If $x > 0$ and $x < \frac{n}{2}$, show

$$\left|\left(\sum_{k=0}^{n} \frac{x^n}{n!}\right) - e^x\right| < \frac{2}{(n+1)!} x^{n+1}.$$

Hence deduce $\frac{8}{3} < e < \frac{11}{4}$.

9. Prove parts (i)—(iv) of Theorem 9.1.3.

10. Show for all $n \in \mathbf{N}$, $E(nx) = (E(x))^n$. Extend this result to all $n \in \mathbf{Z}$.

11. Verify that the extended definition of e^x satisfies the two claims made for it in the discussion following the definition.

12. Let us suppose the function e^q is defined for all $q \in \mathbf{Q}$. Extend this definition to all reals by using the supremum. Now prove a suitable version of Theorem 9.1.3.

13. Show $\lim_{n \to \infty} x^n e^{-x} = 0$ for all $n \in \mathbf{N}$.

14. Supply detailed reasons for all steps in the proof of Theorem 9.1.2.

15. Define the function, $\exp : \mathbf{R} \to \mathbf{R}$, directly by

$$\exp(x) = \sum_{n=0}^{\infty} \frac{x^n}{n!}.$$

Prove $\exp(x)$ is well defined and has all the properties claimed for $E(x)$. In short, develop the exponential function directly from the power series definition. Also, show $\exp(x + y) = \exp(x)\exp(y)$ for $x, y \in \mathbf{R}$.

16. Show
$$e^x = \lim_{n \to \infty} \left(1 + \frac{x}{n}\right)^n, \quad x \in \mathbf{R}.$$

17. Consider the compound interest problem. Namely, let an amount of money, P, be invested for 1 year at an interest rate r. At simple interest, one is entitled to collect $P(1 + r)$ at the end of the year. If the interest is compounded twice, at the end of $\frac{1}{2}$ year, an amount of interest equal to $\frac{Pr}{2}$ is added to the account. At the end of the year, a further amount of interest is paid, $\frac{r}{2}\left(P + \frac{Pr}{2}\right)$. Similarly, interest could be compounded thrice, four times or n times. What rate of interest will result as $n \to \infty$?

18. If $f : \mathbf{R} \to \mathbf{R}$ is continuous and satisfies the identity $f(x + y) = f(x)f(y)$, $x, y \in \mathbf{R}$, prove $f(x) = e^{kx}$, for some constant, k.

19. What can you say about f in Exercise 4 if it satisfies the condition $f'(x) = kf(x)$, where k is a constant?

20. If $f'(x) \geq xf(x)$, $x \in \mathbf{R}$, show there exists a constant k such that $f(x) \geq ke^x$.

21. Prove the following inequalities:

    (a) $\left(1 + \frac{1}{n-1}\right)^{n-1} < e < \left(1 + \frac{1}{n-1}\right)^n$;

    (b) $\frac{n^n}{n!} < e^{n-1} < \frac{n^{n+1}}{n!}$;

    (c) $e\left(\frac{n^n}{e^n}\right) < n! < en\left(\frac{n^n}{e^n}\right)$;

    (d) $0 < e - \left(1 + \frac{1}{1!} + \frac{1}{2!} + \cdots + \frac{1}{n!}\right) < \frac{3}{(n+1)!}$.

22. Let $n \geq 1$ and P_n be polynomials such that
$$P_n(x)e^{nx} + P_{n-1}(x)e^{(n-1)x} + \cdots + P_1(x)e^x + P_0(x) = 0$$
for arbitrarily large x. Show $P_n = 0$ for all n.

23. Let F be differentiable and such that $F'(x) = -2xF(x)$ for all x. Show $F(x) = \lambda e^{-x^2}$ for some $\lambda \in \mathbf{R}$.

24. Show that e^q is irrational for all $q \in \mathbf{Q}$, $q \neq 0$.

25. Evaluate $\lim_{x \to 2} \frac{1}{x - 2} \int_2^x e^{\sqrt{1+t^2}} \, dt$.

26. Find a Taylor expansion of $\int_0^x e^{-t^2} \, dt$ in a neighborhood of 0.

27. Find the first four nonvanishing terms in the Taylor expansion for e^{e^x}.

28. Express the following functions $f(x)$ as a power series in x.

    (a) $f(x) = \int_0^x e^{-t^2}$; (b) $f(x) = e^{-x^2}$;

    (c) $f(x) = \frac{1}{1+e^x}$; (d) $f(x) = \frac{e^{-x}}{1+x^2}$.

29. Expand the following functions in a Taylor series about the point a:

    (a) e^{2x}; (b) $e^{x/3}$; (c) e^{x^2}.

30. If $a > 1$, prove that the sequence defined by
$$s_1 = a, s_{n+1} = a^{s_n}, \quad n \geq 1,$$
converges, provided $a \leq e^{\frac{1}{e}}$.

9.2   Logarithmic Function and Power Function

In the last section, we introduced the exponential function and studied its various remarkable properties; for example, it is strictly increasing, differentiable, and has the set $\{x : x \in \mathbf{R} \text{ and } x > 0\}$ for its range. These facts guarantee the exponential function is one-to-one and hence possesses a continuous, inverse function that will be called the **logarithmic function**. We shall study the basic properties of this function and use it to define arbitrary powers of any real number. Note that we have already defined arbitrary powers of the unique real number e.

Definition. The **natural logarithmic function**, $L : (0, \infty) \to \mathbf{R}$, is the inverse of the exponential function, $E(x)$.

Discussion. That $L = E^{-1}$ is a function, as opposed to a relation, follows from E being one-to-one. The domain, range, and continuity of L follow from Theorem 2.6.4. Also, since L and E are inverses of each other, we have

$$(L \circ E)(x) = x \text{ for all } x \in \text{Dmn } E$$

and

$$(E \circ L)(x) = x \text{ for all } x \in \text{Dmn } L.$$

Calling $E(x)$ by its standard name, e^x, and $L(x)$ by its standard name, **ln x**, these equations take the form

$$\ln(e^x) = x \quad \text{and} \quad e^{\ln y} = y.$$

This equation captures the fundamental relationship between the logarithmic and exponential functions. Although the initial development of logarithms by Napier in 1594 was not as exponents of a fixed positive number base, this approach followed shortly thereafter by Briggs in 1615, and it is the equation $e^{\ln y} = y$ that supplies the fundamental motivation for the logarithmic function. The utility of logarithms as a tool that aids the conduct of numerical computations arises from the fact that the use of logarithms permits the substitution of addition for the operation of multiplication. While this use has all but disappeared with the introduction of calculators, the use of logarithmic transforms for data remains as important as ever. $\square$

The following theorem summarizes the basic properties of the logarithmic function:

Theorem 9.2.1. *The function, L, has the following properties:*

(i) *$L(x)$ is strictly increasing, has domain $\{x : x \in \mathbf{R} \text{ and } x > 0\}$ and range* $\mathbf{R}$*;*

(ii) *L is differentiable and $L'(x) = \frac{1}{x}$, $x \in \mathbf{R}^+$;*

(iii) *For $x, y > 0, L(xy) = L(x) + L(y)$;*

(iv) *$L(1) = 0$; $L(e) = 1$;*

(v) $\lim\limits_{x\to 0^+} L(x) = -\infty$ *and* $\lim\limits_{x\to\infty} L(x) = \infty$.

Proof. To prove (ii), note that the fact that L is differentiable follows from Theorem 4.2.3. Since $E(L(x)) = x$ for all $x \in \mathbf{R}^+$, it follows from differentiation using the chain rule that $E'(L(x))L'(x) = 1$, whence

$$L'(x) = \frac{1}{E'(L(x))} = \frac{1}{E(L(x))} = \frac{1}{x}.$$

To prove (iii), let $a = L(x)$ and $b = L(y)$, so that $E(a) = x$ and $E(b) = y$. Now $xy = E(a)E(b) = E(a+b)$, so that

$$L(xy) = a + b = L(x) + L(y).$$

The other parts are left as Exercise 1. $\square$

Discussion. Compare this theorem with Theorem 9.1.3. These are simply the properties reflected from those of the exponential function by virtue of the fact that L is the inverse of E. The reader is advised to sketch the graphs of $E(x)$ and $L(x)$ on the same set of axes and compare them.

Part (iii) contains an extremely useful property of logarithms. Here the computation of L of a product $x \cdot y$ is rendered easy, since it is just the addition of two numbers $L(x)$ and $L(y)$. This feature was exploited in computational problems to multiply two numbers. For example, common logarithm and antilogarithm tables were routinely found as appendices to most university-level introductory physics and chemistry texts published prior to 1970. This use has disappeared with the advent of hand-held calculators.

Property (ii) is equally remarkable. The simplest of all polynomials are the constant 1 and the function $f(x) = x$. Their ratio, $\frac{1}{x}$, is a rational function. As the reader will recall, while the function, $\frac{1}{x}$, $x > 0$ is continuous and bounded on any interval, $[a, b]$, where $a > 0$, and so is clearly integrable on $[a, b]$, it was impossible to evaluate this integral, other than numerically, for lack of an antiderivative. This situation is now remedied, and antiderivatives have been supplied for all functions of the form x^q, $q \in \mathbf{Q}$. Property (ii) also provides an alternate method of defining the function, $L(x)$, in terms of the definite integral $\int_1^x \frac{1}{t}\,dt$, a matter that is pursued in Exercise 13 (also see Exercise 5.5.19). $\square$

As discussed in the last section, for $x > 0$, a meaning has been assigned to x^r for all rational numbers, r, or for all real numbers, r, subject to the condition, $x = e$. We want now to complete the definition of x^r so that a unique real number is specified for all $r \in \mathbf{R}$ and all $x \in \mathbf{R}^+$. To accomplish this goal, we exploit the logarithmic function.

Definition. If $\xi \in \mathbf{R}$ and $x > 0$, we define x^ξ by

$$x^\xi = E(\xi L(x)) = e^{\xi L(x)}.$$

The function $P_\xi : (0, \infty) \to \mathbf{R}$ defined by $P_\xi(x) = x^\xi$ is called the **ξth power function**.

Discussion. Since $x > 0$, this definition makes sense. However, it requires further validation, as discussed following the definition of e^x. This validation will be provided in the next several theorems, the proofs of which are left to the reader.

As initial validation, the reader can show by induction for each $n \in \mathbf{N}$, $n \cdot L(x) = L(x^n)$, so by the inverse nature of the function, E, $x^n = E(nL(x))$. Thus, the new definition extends the definition for powers that are natural numbers. This type of argument can be further extended to show for all $q \in \mathbf{Q}$, $x^q = e^{qL(x)}$ (Exercise 2), whence the new definition extends the earlier ones for rational powers also.

The question that remains is: Are we justified in using the definition for irrational powers? As in the case of e^ξ, what we want to verify is that

$$x^\xi = \sup\{x^q : q \in \mathbf{Q} \text{ and } q < \xi\}.$$

This problem will be explored below. $\square$

We now state some standard properties of the power function. Most of the facts are immediate from the well-known properties of the exponential and logarithmic functions and, hence, are left as exercises.

Theorem 9.2.2. *If $x, y > 0$ and $\xi \in \mathbf{R}$, then,*

(i) $1^\xi = 1$;

(ii) $x^\xi > 0$;

(iii) $(xy)^\xi = x^\xi y^\xi$;

(iv) $\left(\dfrac{x}{y}\right)^\xi = \dfrac{x^\xi}{y^\xi}$.

Proof. Exercise 3. $\square$

Theorem 9.2.3. *If $\xi, \eta \in \mathbf{R}$ and $x > 0$,*

(i) $x^{\xi+\eta} = x^\xi x^\eta$;

(ii) $x^{-\xi} = \dfrac{1}{x^\xi}$;

(iii) $x^{\xi\eta} = x^{\eta\xi}$;

(iv) $\xi < \eta$ *implies* $x^\xi < x^\eta$.

Proof. Exercise 4. $\square$

Theorem 9.2.4. *The function, $P_\xi : (0, \infty) \to \mathbf{R}$, defined by $P_\xi(x) = x^\xi$ is differentiable, and $P_\xi{}'(x) = \xi x^{\xi-1}$ for $x > 0$.*

Proof. $P_\xi{}'(x)$ is the derivative of the function specified by $P_\xi(x) = e^{\xi L(x)}$. An application of the Chain Rule yields

$$P_\xi'(x) = e^{\xi L(x)} \cdot [\xi\, L(x)]' = x^\xi \left(\frac{\xi}{x}\right) = \xi x^{\xi-1}. \quad \square$$

Discussion. These theorems summarize the properties of all functions of the form $f(x) = x^r$, where $r \in \mathbf{R}$ is fixed and $x \in \mathbf{R}^+$. While the properties presented go a long way toward justifying the definition of x^ξ, they do not quite show the equation

$$x^\xi = \sup\{x^q : q \in \mathbf{Q} \text{ and } q < \xi\}$$

is satisfied. To address this question, we need to look at functions of the form $f(x) = a^x$, where $a > 0$ is fixed. $\square$

Theorem 9.2.5. *Let $a \in \mathbf{R}^+$ be fixed and consider the function, $f : \mathbf{R} \to \mathbf{R}^+$, defined by $f(x) = a^x$. Then f has the following properties:*

(i) *f is continuous and differentiable on $\mathbf{R}$;*

(ii) *if $a > 1$, f is strictly increasing, if $a < 1$, f is strictly decreasing, if $a = 1$, $f = 1$;*

(iii) *if $a > 0$ and h is defined on $\mathbf{Q}$ by $h(q) = a^q$, then f is the continuous extension of h to all of $\mathbf{R}$.*

Proof. Exercise 5. $\square$

Discussion. Property (iii) has the effect of verifying the assertion that

$$x^\xi = \sup\{x^q : q \in \mathbf{Q} \text{ and } q < \xi\}. \quad \square$$

Exercises

1. Complete the proof of Theorem 9.2.1.

2. Show for all $q \in \mathbf{Q}$, $x^q = e^{qL(x)}$.

3. Prove Theorem 9.2.2.

4. Prove Theorem 9.2.3.

5. Prove Theorem 9.2.5.

6. Evaluate $\displaystyle\lim_{x \to \infty} \frac{(\ln x)^n}{x^\alpha}$, where α and n are positive real numbers.

7. If $a > 1$ and $a \neq 1$, define a function $\log_a : (0, \infty) \to \mathbf{R}$ by $\log_a(x) = \frac{L(x)}{L(a)}$, $x \in (0, \infty)$. This function is called the **logarithm of x to the base a**. Prove the following:

   (a) $\log_a$ is differentiable and $(\log_a)'(x) = \frac{1}{xL(a)}$;

   (b) $\log_a(xy) = \log_a(x) + \log_a(y)$;

   (c) $\log_a x = \frac{L(b)}{L(a)} \log_b x$, $x \in (0, \infty)$;

   (d) if $f(y) = a^y$ and $g(x) = \log_a x$, then f and g are inverses of each other;

   (e) $\log_{10} x = (\log_{10} e)L(x)$.

8. Show $\displaystyle\lim_{n \to \infty} (1 + \frac{1}{n})^n = e$. [Hint: consider $L'(1)$.]

9. Show

$$L(x+1) = x - \frac{x^2}{2} + \frac{x^3}{3} - \cdots + - \cdots + (-1)^{n-1}\frac{x^n}{n} + \int_0^x \frac{(-t)^n}{1+t}\, dt.$$

[Hint: expand $\frac{1}{1+x}$.] Hence evaluate $L(1.1)$ accurate to four decimal places.

10. Obtain a power series that represents the function $f(x) = \frac{1}{x+1}$ for $|x| < 1$ and use it to derive the fact that

$$\ln(1 + x) = \sum_{n=0}^{\infty} \frac{(-1)^n x^{n-1}}{n}.$$

Hence, justify the equation

$$\ln 2 = 1 - \frac{1}{2} + \frac{1}{3} - \cdots .$$

11. Compute the limits, $\lim_{x \to 0} x^{\alpha}$, $\alpha > 0$, and $\lim_{x \to \infty} x^{\alpha}$, for $\alpha < 0$.

12. Obtain a Taylor expansion of $(1 + x)^{\alpha}$ for $\alpha \in \mathbf{R}$.

13. Define $l(x) = \int_1^x \frac{1}{t} dt$, $x > 0$. Prove all the statements in Theorem 9.2.2 with $L(x)$ replaced by $l(x)$ and also show $l(x) = L(x)$ (c.f., Exercise 5.5.19).

14. Evaluate the following limits:

(a) $\lim_{x \to 0^+} x^x$;   (b) $\lim_{x \to \infty} x^{\frac{1}{x}}$;   (c) $\lim_{x \to 0^+} x^{\ln x}$;

(d) $\lim_{x \to 0^+} \frac{10^x - 7^x}{x^2}$;   (e) $\lim_{x \to 0^+} (\ln x)^x$;   (f) $\lim_{x \to 0^+} (\ln x)^{\ln x}$;

(g) $\lim_{x \to 0^+} \frac{\ln x}{x^n}$;   (h) $\lim_{x \to \infty} \frac{\ln x}{x^n}$;   (i) $\lim_{x \to \infty} \frac{\ln x}{x^{0.00001}}$;

(j) $\lim_{x \to 0^+} x \ln x$;   (k) $\lim_{x \to \infty} x \left[\left(1 + \frac{1}{x}\right)^x - e \right]$;

(l) $\lim_{x \to \infty} x \left[\left(1 + \frac{1}{x}\right)^x - e \ln \left(1 + \frac{1}{x}\right)^x \right]$.

15. Derive the following inequalities:

(a) $1 - x < -\ln x < \frac{1}{x} - 1$, $0 < x < 1$;

(b) $\frac{x^2}{2} < x - \ln(1 + x) < \frac{x^2}{2(1+x)}$, $-1 < x < 0$;

(c) $x - \frac{x^2}{2} + \frac{x^3}{3(1+x)} < \ln(1 + x) < x - \frac{x^2}{2} + \frac{x^3}{3}$, $x > 0$;

(d) $x < -\ln(1 - x) < \frac{x}{1-x}$, $0 < x < 1$;

(e) $\frac{1}{x} < \frac{1}{\ln(1+x)} < 1 + \frac{1}{x}$, $x \in \mathbf{R}^+$;

(f) $\frac{\ln x}{x} \leq \frac{1}{\sqrt{x}} \int_1^x \frac{1}{t^{3/2}} dt$;

(g) $\frac{\ln x}{x} < \frac{2}{\sqrt{x}}$, $x > 1$.

16. Prove the sequence $\{a_n\}$ defined by $a_n = n(a^{\frac{1}{n}} - 1)$, $a > 0$ is monotonic and converges to $\ln a$. Obtain standard properties of $\ln x$ from this definition.

17. Let $g : \mathbf{R} \to \mathbf{R}$ be continuous, nonidentically zero, and satisfy the identity $g(x + y) = g(x)g(y)$ for $x, y \in \mathbf{R}$. Show $g(x) = a^x$, where $g(1) = a$.

18. Let $h : (0, \infty) \to \mathbf{R}$ be continuous and satisfy $h(xy) = h(x) + h(y)$ for $x, y \in \mathbf{R}$. What can you say about the function $h(x)$? What happens if, instead, h satisfies the identity $h(xy) = h(x)h(y)$?

19. Let $n \geq 1$, and f_n be polynomials such that

$$f_n(x)(\ln x)^n + f_{n-1}(x)(\ln x)^{n-1} + \cdots + f_0(x) = 0$$

for all $x > 0$. Show all the f_n's are identically zero (see Exercise 9.1.22).

20. Derive the expansion

$$\ln 2 = 2 \sum_{k=1}^{\infty} \frac{1}{3^{2k-1}(2k-1)}.$$

Also estimate the error in taking the first six terms of the expansion for $\ln 2$.

21. In $-1 \le x < 1$, prove

$$-\ln(1-x) = x + \frac{x^2}{2} + \frac{x^3}{3} + \cdots .$$

Hence or otherwise, obtain the sum of the series $\sum_{n=0}^{\infty} \frac{1}{(n+1)(n+2)}$.

22. Obtain a power series expansion for $\ln \frac{1+x}{1-x}$, stating the range of validity of the expansion.

23. Derive the following expansions in $-1 < x \le 1$:

(a) $\frac{\ln(1+x)}{1+x} = x - x^2 \left(1 + \frac{1}{2}\right) + x^3 \left(1 + \frac{1}{2} + \frac{1}{3}\right) - \cdots$;

(b) $[\ln(1+x)]^2 = 2 \left[\frac{x^2}{2} - \frac{x^3}{3}\left(1 + \frac{1}{2}\right) + \frac{x^4}{4}\left(1 + \frac{1}{2} + \frac{1}{3}\right) - \cdots \right]$.

24. Prove

$$\int_0^1 \frac{\ln(1+t)}{t} = \frac{\pi^2}{12}.$$

25. Express $\int_0^x \frac{\ln t}{1-t} \, dt$ as a power series in $x - 1$, and find its radius of convergence.

9.3 Trigonometric Functions

In this section, we shall introduce the familiar sine and cosine functions, and from our definition, we shall derive all the well-known properties of these functions. Our approach will be similar to that for the exponential function. During this process, we shall identify the unique real number, **pi**, and study its role in relation to the trigonometric functions.

Definition. The pair $\{C_n\}$ and $\{S_n\}$ of sequences of functions mapping $\mathbf{R}$ into $\mathbf{R}$ are defined by

$$C_1(x) = 1, \quad S_1(x) = x,$$

and for $n > 1$,

$$C_{n+1}(x) = 1 - \int_0^x S_n(t) \, dt; \quad S_n(x) = \int_0^x C_n(t) \, dt.$$

Discussion. As in the case of exponential functions, these definitions are simple and easy to understand, although the properties of the functions that result from these definitions are by no means obvious. A straightforward induction (Exercise 1) shows, for $n \geq 1$,

$$C_{n+1}(x) = 1 - \frac{x^2}{2!} + \frac{x^4}{4!} - + - \cdots + \frac{(-1)^n x^{2n}}{(2n)!} \qquad (1)$$

and

$$S_{n+1}(x) = x - \frac{x^3}{3!} + \frac{x^5}{5!} - + - \cdots + \frac{(-1)^n x^{2n+1}}{(2n+1)!}. \qquad (2)$$

For each n, these expressions are polynomials. So they are differentiable (hence continuous) and integrable. As the reader can easily check by induction,

$$S'_{n+1}(x) = C_n(x), \text{ and } C'_{n+1}(x) = -S_n(x), \text{ for all } x \in \mathbf{R}, \ n \in \mathbf{N}. \qquad (3)$$

The reader can easily see the parallelism in the approach to the definition these functions and the exponential function, $E_n(x)$, of Section 9.1. In the present case, our aim is to create two new functions, S and C, with the properties $S(0) = 0$, $C(0) = 1$, $S' = C$, and $C' = -S$, which are the familiar properties of the sine and cosine functions. To obtain these functions, we will use the same type of approach and arguments as were used in the exponential case. With this in mind, the industrious reader may wish to proceed on his own at this point to see how much of the theory he can develop. $\square$

Definition. For $x \in \mathbf{R}$, we define the **trigonometric functions**, S and C, mapping $\mathbf{R}$ into $\mathbf{R}$ as follows:

$$S(x) = \lim_{n \to \infty} S_n(x);$$

$$C(x) = \lim_{n \to \infty} C_n(x).$$

Discussion. We must first make sure that these two limits exist at all points x and, hence, that S and C are well defined. The key to this is the concept of uniform convergence.

In Figure 9.3.1 graphs of the functions C, C_2, C_4, and C_6 on the interval $[-2\pi, 2\pi]$ are presented. The function C is identified by its standard name, **cos**. The number, π, will be formally identified in Theorem 9.3.5. As can be seen, the approximation of C_6 to cos is very good on the interval from $-\pi$ to π. $\square$

Theorem 9.3.1. *For each $x \in \mathbf{R}$ the sequences $\{S_n(x)\}$ and $\{C_n(x)\}$ converge. Further, for each $K > 0$, the sequences $\{S_n\}$ and $\{C_n\}$ converge uniformly on $[-K, K]$. Also, the limit functions, S and C, are continuous on $[-K, K]$.*

Proof. By arguments presented in Section 9.1, the truth of the second assertion will imply the first. Thus, fix $K > 0$. If $|x| \leq K$ and $m > n > 2K$, then,

$$|C_m(x) - C_n(x)| = \left| \sum_{k=n}^{m-1} (-1)^k \frac{x^{2k}}{(2k)!} \right|$$

$$\leq \frac{K^{2n}}{(2n)!} \left[\sum_{k=0}^{m-n-1} \left(\frac{K}{2n} \right)^{2k} \right]$$

$$< \frac{16 K^{2n}}{15(2n)!}. \qquad \textbf{(WHY?)}$$

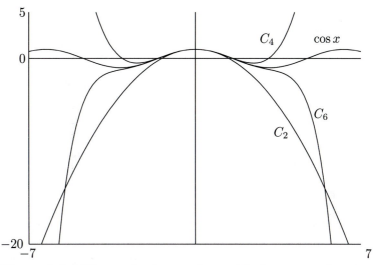

Figure 9.3.1 The graph of $\cos x$ along with three approximations $C_2(x)$, $C_4(x)$, $C_6(x)$ in $[-2\pi, 2\pi]$.

Since the last entry has limit 0 as n approaches infinity, we conclude that the sequence $\{C_n\}$ converges uniformly on each interval, $[-K, K]$. It follows immediately from the corollary to Theorem 7.3.1 that C is continuous on $[-K, K]$, for every $K > 0$.

The proof of the theorem for S is left to the reader as Exercise 2. □

Discussion. The techniques employed are completely analogous to those used in the development of the exponential function.

A further useful fact that follows directly from the definition is

$$C(0) = \lim_{n \to \infty} C_n(0) = 1.$$ □

Theorem 9.3.2. *The functions C and S possess derivatives of all orders; in particular, for all $x \in \mathbf{R}$,*

$$C'(x) = -S(x) \quad and \quad S'(x) = C(x).$$

Proof. The uniform convergence of the sequence $\{C_n\}$ coupled with Theorem 7.3.4 enable us to deduce that the limit function C is differentiable on $[-K, K]$. Further, if $x \in [-K, K]$,

$$C'(x) = \lim_{n \to \infty} C_n'(x) = \lim_{n \to \infty} (-S_{n-1}(x)) = -S(x).$$

Since $K > 0$ was arbitrary, it follows $C'(x) = -S(x)$, for all $x \in \mathbf{R}$. The remainder of the proof is left as Exercise 3. □.

Discussion. Evidently there are useful formulae that can be obtained for integrating C and S. These are left to the reader as Exercise 4.

Next, we show the functions C and S satisfy the so-called **Pythagorean Identity**.

Corollary. *The functions C and S satisfy the identity*

$$C^2(x) + S^2(x) = 1 \text{ for all } x \in \mathbf{R}.$$

Proof. Consider $F = C^2 + S^2$. Then, F is differentiable, and, further, for any $x \in \mathbf{R}$, we have

$$F'(x) = 2C(x)C'(x) + 2S(x)S'(x) = -2S(x)C(x) + 2S(x)C(x) = 0.$$

So it follows F must be a constant. But, then, $F(0) = 1 + 0 = 1$, and we obtain the desired identity. $\square$

Discussion. Instead of directly computing the value of expression, $C^2(x) + S^2(x)$, and checking it simplifies to 1 for every choice of x, we have used an indirect approach that avoids the direct computations and replaces them with the simple, but clever, technique of verifying the derivative is zero. Once this is known, it is trivial to find the constant value that F must be equal to by evaluating $F(0)$. $\square$

Our next aim is to establish the uniqueness of these functions. As in the case of exponential functions, the properties that characterize these functions are the form of the derivatives and the value of these functions at 0.

Theorem 9.3.3. *If A and B are two functions from $\mathbf{R} \to \mathbf{R}$ such that $A' = -B$, $B' = A$, $A(0) = 1$, and $B(0) = 0$, then $A = C$ and $B = S$.*

Proof. Call $A - C = F$ and $B - S = G$. Then F and G are continuous, possess derivatives of all orders, $F(0) = F'(0) = 0$, and $F''(0) = -F(0) = 0$ for all $n \in \mathbf{N}$. If $x \in \mathbf{R}$ is arbitrary, we can expand F in a Taylor series about origin as follows:

$$F(x) = F(0) + \frac{F'(0)x}{1!} + \cdots + \frac{F^{(n-1)}(0)x^{n-1}}{(n-1)!} + \frac{F^{(n)}(\xi_n)x^n}{n!},$$

where $\xi_n \in (0, x)$ or $(x, 0)$, depending on the sign of x. By virtue of the fact $A(0) = C(0)$ and $B(0) = S(0)$, it can be shown $F(x) = \frac{F^{(n)}(\xi_n)}{n!}x^n$. Observe $F^{(n)}(\xi_n)$ is either $\pm F(\xi_n)$ or $\pm G(\xi_n)$, whence in either case, by continuity there exists a $K > 0$ such that $|F^n(x)| \le K|x|^n$, and we immediately conclude that $F(x) = 0$, that is, $A = C$.

A similar argument proves that $B = S$ (Exercise 5). $\square$

Discussion. The proof is similar to the one used for the uniqueness of the exponential function (Theorem 9.1.2), and one may seek there for missing details, for example, the reason behind the last step.

Note that the above theorem shows if F is any function mapping $\mathbf{R} \to \mathbf{R}$ with the three properties

(i) $F'' = -F$,

(ii) $F(0) = 1$, and

(iii) $F'(0) = 0$,

then F is identical with the function, C. A similar set of properties can be found to characterize S (Exercise 6). $\square$

Notation. Having established the uniqueness of these functions, we now call $C(x)$ the **cosine** of x. The name of the function associated with C will be **cos**. Similarly, we call $S(x)$ the **sine** of x, and call S the **sin**.

Theorem 9.3.4. *The functions C and S satisfy the following equations:*

(i) $C(-x) = C(x)$ *and* $S(-x) = -S(x)$ *for all* $x \in \mathbf{R}$;

(ii) $C(x + y) = C(x)C(y) - S(x)S(y)$; $S(x + y) = S(x)C(y) + C(x)S(y)$, *for all* $x, y \in \mathbf{R}$.

Proof. Set $f(x) = C(-x)$. We see $f'' = -f$, $f(0) = 1$ and $f'(0) = 0$. Hence, by Theorem 9.3.3, $f = C$. A similar argument proves that $S(-x) = -S(x)$ (Exercise 7).

For (ii), we first claim if $f : \mathbf{R} \to \mathbf{R}$ satisfies the condition $f''(x) = -f(x)$, for $x \in \mathbf{R}$, then there exist constants a, b such that $f = aC + bS$. To see this, consider $g(x) = f(0)C(x) + f'(0)S(x)$. Then, it is readily seen that $g'' = -g$, $g(0) = f(0)$ and $g'(0) = f'(0)$ so that $f - g$ reduces to the zero function (Exercise 8). Thus, $a = f(0)$ and $b = f'(0)$. To complete the argument, fix $y \in \mathbf{R}$ and set $f(x) = C(x + y)$. One readily verifies $f'' = -f$, so there exist $a, b \in \mathbf{R}$ such that

$$f(x) = C(x + y) = aC(x) + bS(x)$$

and in consequence

$$f'(x) = -S(x + y) = -aS(x) + bC(x),$$

$x \in \mathbf{R}$. Letting $x = 0$, we obtain $a = C(y)$ and $b = -S(y)$, whence the first identity holds. The corresponding identity for $S(x + y)$ is proved similarly (Exercise 7). $\square$

Discussion. There are several features of the second part of the proof that are worthy of review. First, we employ the fact that if a function, f, satisfies the differential equation, $f'' = -f$, then f must be a scalar combination of the sine and cosine functions. It is not obvious this should be true. But if one considers the content of Theorems 9.3.1–9.3.3, the result becomes more plausible, particularly when one thinks of similar results such as the one for the exponential function.

Once the consequence of $f'' = -f$ is known, we have only to apply it in a clever way. Considering $C(x + y)$ as a function of x alone, that is, treating y as fixed for a moment, we verify that $C(x+y)$ has the property $C''(x+y) = -C(x+y)$. Thus, it is a scalar combination of $S(x)$ and $C(x)$. These scalars cannot involve x, which is a variable, but can be expressed in terms of y, which is fixed. By setting $x = 0$, these constants are seen to be $C(y)$ and $S(y)$, respectively.

This type of proof is an indirect proof in the sense that we did not directly evaluate the combination $S(x)C(y)+C(x)S(y)$, nor did we try to expand $S(x+y)$.

The differentiability properties of the functions C and S were the successful tools in this proof. As such, this argument and the other similar arguments in this chapter — the argument for part (i) of this theorem is another example — illustrate one of the truly beautiful features of mathematics, namely, that ideas from completely different areas, for example, differential equations, trigonometry, and various other areas of mathematics, can so fruitfully intermesh. □

Next, we are going to identify a real number that is of fundamental importance to the theory of trigonometry. Once we have delineated its properties, we shall recognize it as the familiar number π. The next theorem defines $\frac{\pi}{2}$ in a formal way.

Theorem 9.3.5.   *There exists a real number, $p \in [0,2]$, such that $C(p) = 0$, $C(x) > 0$ for $x \in [0,p]$, and p is the smallest number with these properties.*

Proof. We first claim $C(2) < 0$. To see this, first note $\frac{2^{2n}}{(2n)!} > \frac{2^{2n+2}}{(2n+2)!}$ for all $n > 1$. From equation (1) at the beginning of this section, we have

$$
\begin{aligned}
C(2) &= 1 - \frac{2^2}{2!} + \frac{2^4}{4!} - \frac{2^6}{6!} + \cdots \\
&= 1 - 2 + \frac{2}{3} - \cdots \\
&< -1 + \frac{2}{3} < 0. \qquad \textbf{(WHY?)}
\end{aligned}
$$

Since $C(0) = 1$ and C is continuous, there exists $y \in [0,2]$ such that $C(y) = 0$. Let $p = \inf\{y : y \in [0,2]$ and $C(y) = 0\}$. The reader can check (Exercise 9) the requirements for the existence of p are met and that $C(p) = 0$. □

Corollary.   *The real number, $2p$, is the least positive real number such that $S(2p) = 0$.*

Proof. From Theorem 9.3.4, $S(2x) = 2S(x)C(x)$, whence $S(2p) = 0$. From the same equation, $S(2y) = 0$ implies $C(y) = 0$. Thus, $2p$ is the least positive real such that $S(2p) = 0$. □

Discussion. The proof of this theorem invokes a very powerful theorem. As well the location of p, somewhere in $[0,2]$, is not particularly precise. An alternate proof, starting with the Pythagorean Identity leads to $\sqrt{2} < p < \sqrt{3}$. The reader is asked to develop such a proof in Exercise 10. □

Definition. The real number $2p$ of the above theorem, which is the smallest positive zero for $\sin x$, is called **pi** and is denoted by π.

Discussion. Our approach to the definition of π has been quite different than the historical approach. The number π was first identified geometrically as the ratio between the diameter and circumference of a circle, and values for π based on this approach are known from the Egyptian and Greek mathematics of 3000 years ago. The difficulty with the approach we have taken is the direct relationship of the trigonometric functions S and C with circles has to some degree been lost. Establishing this connection will be the reader's task in Exercises 11 and 12.

We shall show π is irrational, although at this stage, we will not be able to decide whether it is transcendental. In fact, it is transcendental as was shown by Lindemann in 1882. The method used by Lindemann is conceptually simple. He showed that if the p_k's and x_k's were all algebraic and the p_k's were not all zero, then

$$\sum_{k=1}^{n} p_k e^{x_k} \neq 0.$$

Since $e^{\pi i} + 1 = 0$, where $i^2 = -1$, π cannot be algebraic.

There are several methods to compute π to any number of decimal places, not all of which are equally effective. For example, we have already noted that $\frac{\pi^2}{6} = \sum_{n=1}^{\infty} \frac{1}{n^2}$. In 1674 Leibnitz obtained (see Exercise 26)

$$\frac{\pi}{4} = 1 - \frac{1}{3} + \frac{1}{5} - \frac{1}{7} + \frac{1}{9} - + \cdots$$

which after only 100,000 terms would yield an approximation of π as good as that known to Archimedes. In a paper in 1740, Euler developed several series approximations to powers of π, among them

$$\sum_{n=1}^{\infty} \frac{1}{(2n-1)^2} = \frac{\pi^2}{8}, \quad \sum_{n=1}^{\infty} \frac{(-1)^{n-1}}{(2n-1)^3} = \frac{\pi^3}{32}, \quad \text{and} \quad \sum_{n=1}^{\infty} \frac{1}{(2n-1)^4} = \frac{\pi^4}{96}.$$

While these facts may not be of tremendous ultimate significance, it is surprising that this number, π, which has clear geometric importance, should also be the result of such diverse calculations. Is it any wonder then, that mathematicians have viewed it with continued interest and fascination? $\square$

Theorem 9.3.6. *The functions, C and S, satisfy the following identities:*

(i) $C(x + 2\pi) = C(x); S(x + 2\pi) = S(x);$

(ii) $C(x) = S(\frac{\pi}{2} - x) = S(x + \frac{\pi}{2}); \quad S(x) = C(\frac{\pi}{2} - x) = -C(\frac{\pi}{2} + x).$

Proof. Observe that $S(\frac{\pi}{2}) = 1$. The rest follows by using the appropriate expansion formulas for $C(x + 2\pi)$ (Exercise 13). $\square$

Discussion. Theorem 9.3.6 (i) asserts that S and C are **periodic functions** with **period**, 2π. This fact will prove useful in Exercise 11. $\square$

Theorem 9.3.7. π *is irrational.*

Proof. We argue by contradiction. Thus, let $\pi = \frac{p}{q} \in \mathbf{Q}$. Next, for an arbitrary positive integer, n, set

$$f(x) = \frac{x^n (p - qx)^n}{n!}.$$

The reader can verify that for all $x \in \mathbf{R}$, $f(\frac{p}{q} - x) = f(x)$ (Exercise 14). Further, we note that $n! f(x)$ is a polynomial with integer coefficients and each individual term of this polynomial has degree at least n. In consequence, $f^{(k)}(0) \in \mathbf{Z}$ for $k = 0, 1, 2, \ldots$; in other words, if f or one of its derivatives is evaluated at $x = 0$, the result is an integer. Using the property, $f(\frac{p}{q} - x) = f(x)$, it can be shown that if f or one of its derivatives is evaluated at $x = \frac{p}{q}$, the result is also an integer. Finally, note that on the interval $[0, 2\pi]$, f attains an absolute maximum at $x = \frac{p}{2q}$, whence it follows that if $0 < x < \pi = \frac{p}{q}$ and $n > 1$, then

$$0 < f(x) \sin x \le f\left(\frac{p}{2q}\right) = \frac{(\pi p)^n}{2^{2n} n!} < \frac{1}{4}$$

provided n is sufficiently large.

Now we apply the properties of f in the following way. Define F by

$$F(x) = f(x) - f''(x) + f^{(4)}(x) - \cdots + (-1)^n f^{(2n)}(x).$$

Observe that $F(0)$ and $F(\pi)$ must be integers. A direct computation establishes

$$[F'(x) \sin x - F(x) \cos x]' = f(x) \sin x.$$

Or, equivalently, $F'(x) \sin x - F(x) \cos x$ is an antiderivative for $f(x) \sin x$. It follows that since $f(x) \sin x$ is positive on $(0, \pi)$, then

$$0 < \int_0^\pi f(x) \sin x \, dx = F(\pi) - F(0) \in \mathbf{Z},$$

for any choice of n. But we can choose n such that $f(x) \sin x < \frac{1}{4}$ on $(0, \pi)$, whence

$$0 < \int_0^\pi f(x) \sin x \, dx = F(\pi) - F(0) \le \frac{\pi}{4} < 1,$$

an evident contradiction. $\square$

Besides the functions discussed so far, there are many other transcendental functions that appear in many applications of mathematics. Some of them are the remaining four trigonometric functions ($\tan x, \cot x, \sec x, \csc x$), **inverse trigonometric functions** ($\arcsin x, \arccos x$, etc.), the **hyperbolic functions** ($\sinh x, \cosh x$, etc.), and the **inverse hyperbolic functions**. Many of these are explored in the exercises.

Exercises

1. Establish formulae (1), (2) and (3) for the functions C_{n+1} and S_{n+1}.

2. Prove the remaining assertions in Theorem 9.3.1. As well, show $S(0) = 0$.

3. Complete the proof of Theorem 9.3.2.

4. Prove a theorem about integrating C and S.

5. Complete the proof of Theorem 9.3.3 by showing in complete detail that $B = S$.

6. Find three properties that characterize the function S but that avoid any mention of C. Prove your answer.

7. Give detailed proofs of the identities in Theorem 9.3.4 for which no arguments have been given.

8. Give a detailed argument as to why the function $f - g$ of Theorem 9.3.4 must be the zero function.

9. Verify the p of Theorem 9.3.5 actually exists and has the required properties.

10. Use the Pythagorean Identity to show $\sqrt{2} < \frac{\pi}{2} < \sqrt{3}$.

11. Consider the collection of points in the plane, A, defined by

$$A = \{(x, y) : x = r \cdot C(t), y = r \cdot S(t), r \in \mathbf{R}^+, r \text{ fixed, and } 0 \le t \le 2\pi\}.$$

Show A is a circle of radius, r. Use this fact to show the functions C and S are the usual cosine and sine functions with their usual geometric interpretation relative to the unit circle.

12. Use the formula for arc length to show that the number π has its usual geometric significance relating the diameter of a circle to its circumference.

13. Supply detailed arguments to complete the proof of Theorem 9.3.6.

14. Verify all the facts mentioned about the function, f, in Theorem 9.3.7.

15. Verify all the facts mentioned about the function, F, in Theorem 9.3.7.

16. Use the fact that $\cos(x + y) = \cos x \cos y - \sin x \sin y$, to derive the formulae

$$\sin x = \sqrt{\frac{1 - \cos 2x}{2}} \quad \text{and} \quad \cos x = \sqrt{\frac{1 + \cos 2x}{2}}.$$

17. Show $\sin(\frac{\pi}{2}) = 1$. Obtain values for $\sin(\frac{\pi}{4})$ and $\cos(\frac{\pi}{4})$.

18. Define $\tan x = \frac{\sin x}{\cos x}$, $\cos x \ne 0$. Obtain the following results:

   (a) $\tan(\pi + x) = \tan x$;

   (b) $(\tan x)' = \frac{1}{\cos^2 x}$, $x \ne \frac{(2n+1)\pi}{2}$;

   (c) $\lim\limits_{x \to \frac{\pi}{2}^-} \tan x = \infty$, and $\lim\limits_{x \to \frac{\pi}{2}^+} \tan x = -\infty$.

19. Define $\sec x = \frac{1}{\cos x}$, $\cos x \ne 0$, $\csc x = \frac{1}{\sin x}$, $\sin x \ne 0$, and $\cot x = \frac{\cos x}{\sin x}$, $\sin x \ne 0$.

   (a) Prove $1 + \tan^2 x = \sec^2 x$.

   (b) Prove $1 + \cot^2 x = \csc^2 x$.

   (c) Obtain results similar to Theorems 9.3.4 and 9.3.6 for $\tan x$, $\cot x$, $\sec x$, and $\csc x$.

20. Show $\lim\limits_{x \to \frac{\pi}{2}} \sin x = 1$ and $\lim\limits_{x \to \frac{\pi}{2}} \cos x = 0$.

21. Obtain values of all the six trigonometric functions at the points 0, $\frac{\pi}{9}$, $\frac{\pi}{6}$, $\frac{\pi}{5}$, $\frac{\pi}{4}$, $\frac{\pi}{3}$, $\frac{\pi}{2}$, and π.

22. Prove the following inequalities:

   (a) $-1 \le \sin x \le 1$, $-1 \le \cos x \le 1$;

   (b) $-x \le \sin x \le x$, $x \ge 0$;

   (c) $1 - \frac{x^2}{2} \le \cos x \le 1$;

   (d) $x - \frac{x^3}{6} \le \sin x \le x$;

   (e) $1 - \frac{x^2}{2} \le \cos x \le 1 - \frac{x^2}{2} + \frac{x^4}{24}$;

   (f) $1 - \frac{x^2}{2} + \frac{x^4}{24} - \frac{x^6}{720} \le \cos x \le 1 - \frac{x^2}{2} + \frac{x^4}{24}$;

   (g) $\frac{2x}{\pi} \le \sin x$ for $x \in [0, \frac{\pi}{2}]$;

   (h) $\frac{2x}{\pi} < \sin x < \tan x$, $x \in (0, \frac{\pi}{2})$.

23. Evaluate the following limits:

$$\text{(a) } \lim_{x \to 0} \frac{\sin x}{x}; \quad \text{(b) } \lim_{x \to 0} \frac{\tan x}{x}; \quad \text{(c) } \lim_{x \to 0} \left(\frac{\sin x}{x}\right)^{\frac{1}{x^2}}; \quad \text{(d) } \lim_{x \to \infty} \left(\frac{\sin x}{x}\right)^{\frac{1}{x^2}}.$$

24. Let $S^{-1} : [-1, 1] \to \left[-\frac{\pi}{2}, \frac{\pi}{2}\right]$ be the inverse of the function $S(x)$. Show that S^{-1} is monotonic and differentiable in $(-1, 1)$ and

$$[S^{-1}(y)]' = \frac{1}{\sqrt{1 - y^2}}.$$

Deduce that

$$S^{-1}(y) = \int_0^y \frac{1}{\sqrt{1 - t^2}} \, dt,$$

and hence conclude that

$$\pi = 2 \int_0^1 \frac{1}{\sqrt{1 - t^2}} \, dt,$$

and obtain an infinite series expansion of π. [Note: S^{-1} is called the arcsin function.]

25. As in Exercise 24, define the arccos and arctan functions, and obtain similar results for their representation.

26. Derive the expansion

$$\arctan x = \sum_{k=0}^\infty (-1)^k \frac{x^{2k+1}}{2k + 1}, \qquad |x| < 1,$$

and deduce the following infinite series expansion of π:

$$\pi = 4 \sum_{k=0}^\infty \frac{(-1)^k}{2k + 1}.$$

27. Derive the equation

$$\arcsin x = \sum_{n=0}^\infty \frac{1 \cdot 3 \cdot 5 \, \cdots \, (2n - 1)}{2 \cdot 4 \, \cdots \, 2n} \cdot \frac{x^{2n+1}}{2n + 1}.$$

State the range of validity of the above equality.

28. Starting with the trigonometrical identity

$$\sin x = n \sin \frac{x}{n} \prod_{k=1}^{\frac{n-1}{2}} \left(1 - \frac{\sin^2 \frac{x}{n}}{\sin^2 \frac{k\pi}{n}}\right) \qquad n \text{ odd}, \ x \in \mathbf{R},$$

derive the following infinite product expansion of $\sin x$

$$\sin x = x \prod_{k=1}^\infty \left(1 - \frac{x^2}{k^2 \pi^2}\right).$$

Also, obtain the infinite product

$$\cos x = \prod_{k=1}^\infty \left(1 - \frac{4x^2}{(2k - 1)^2 \pi^2}\right).$$

29. Define two sequences of functions from $\mathbf{R}$ into $\mathbf{R}$ as follows. Set $\mathrm{ch}_1(x) = 1$, $\mathrm{sh}_1(x) = x$, and for $n > 1$, set

$$\mathrm{sh}_n(x) = \int_0^x \mathrm{ch}_n(t)\, dt \quad \text{and} \quad \mathrm{ch}_{n+1}(x) = 1 + \int_0^x \mathrm{sh}_n(t)\, dt.$$

Show there exist functions ch and sh from $\mathbf{R} \to \mathbf{R}$ satisfying $\mathrm{ch}(0) = \mathrm{sh}'(0) = 1$, $\mathrm{sh}(0) = \mathrm{ch}'(0) = 0$, $\mathrm{ch}'' = \mathrm{ch}$, and $\mathrm{sh}'' = \mathrm{sh}$. Moreover, $\mathrm{ch}' = \mathrm{sh}$ and $\mathrm{sh}' = \mathrm{ch}$. Show also that these functions are unique. [The functions sh and ch are respectively, called **hyperbolic sine** and **hyperbolic cosine** functions, and are abbreviated as $\sinh x$, $\cosh x$.]

30. Show
$$\sinh x = \frac{e^x - e^{-x}}{2}, \quad \cosh x = \frac{e^x + e^{-x}}{2} \quad (x \in \mathbf{R}).$$

31. Define $\tanh x$, $\coth x$, $\operatorname{sech} x$, $\operatorname{cosech} x$ in an obvious manner. Obtain the derivatives of all six hyperbolic functions.

32. The inverse of $\sinh x$, $\cosh x$, and so on are denoted by $\sinh^{-1} x$, $\cosh^{-1} x$, and so on. Derive the following formulae

    (a) $\sinh^{-1} x = \ln(x + \sqrt{x^2 + 1})$;

    (b) $\cosh^{-1} x = \ln(x + \sqrt{(x^2 - 1)})$, $(x \geq 0)$;

    (c) $\tanh^{-1} x = \frac{1}{2} \ln \frac{1+x}{1-x}$, $x \in (-1, 1)$.

33. Find the first four nonzero terms in the Taylor expansion of the following functions near the origin

$$\sqrt{\cos x}, \quad e^{\sin x}, \quad \sec x, \quad x \cot x, \quad (\arcsin x)^2, \quad \sinh^{-1} x, \quad \int_0^x \frac{\sin t}{t}\, dt.$$

34. The **gudermannian** is the function defined by $gd(x) = \arctan \sinh(x)$. Show the derivative of $\operatorname{sech} x$ is $gd(x)$.

35. If a function $f : \mathbf{R} \to \mathbf{R}$ satisfies the conditions $f(0) = f(\pi) = 0$ and $f''(x) = -cf(x)$ for some $c > 0$, show that $c = \frac{1}{n^2}$ and $f(x) = k \sin nx$ for some $k \in \mathbf{R}$ and $n \in \mathbf{N}$.

36. (**Wallis's product**) Let $a_n = \int_0^{\frac{\pi}{2}} \sin^n x\, dx$, $n = 1, 2, \dots$. Prove the following

    (a) the sequence $\{a_n\}$ is decreasing;

    (b) the products below define the terms of the sequence

$$a_{2n} = \left[\frac{1 \cdot 3 \cdot 5 \, \cdots \, (2n-1)}{2 \cdot 4 \cdot 6 \, \cdots \, (2n)} \right] \cdot \frac{\pi}{2}$$

and

$$a_{2n-1} = \left[\frac{2 \cdot 4 \cdot 6 \, \cdots \, (2n-2)}{3 \cdot 5 \cdot 7 \, \cdots \, (2n-1)} \right];$$

    (c) for all n,

$$\frac{2n}{2n+1} \cdot \frac{\pi}{2} < [2na_{2n-1}]^2 \cdot \frac{1}{2n+1} < \frac{\pi}{2};$$

    (d) the Wallis's product for π is

$$\frac{\pi}{2} = \lim_{n \to \infty} \left[\frac{2 \cdot 4 \, \cdots \, (2n)}{1 \cdot 3 \cdot 5 \, \cdots \, (2n-1)} \right]^2 \cdot \frac{1}{2n+1}.$$

37. Use Wallis's product to obtain **Stirling's formula**:

$$\sqrt{\pi} = \lim_{n \to \infty} \left(\frac{2^{2n}(n!)^2}{(2n)!\sqrt{n}} \right).$$

38. Prove the following special case of Stirling's approximation for $n!$:

$$n! \cong \sqrt{2\pi} e^{-n} n^{\frac{n+1}{2}}.$$

39. Show $\pi^e < e^\pi$.

40. Show

$$\pi = 16 \arctan \frac{1}{5} - 4 \arctan \frac{1}{239},$$

$$= 16 \sum_{n-0}^{\infty} \frac{(-1)^n}{(2n+1)5^{2n+1}} - 4 \sum_{n-0}^{\infty} \frac{(-1)^n}{(2n+1)(239)^{2n+1}}.$$

41. Let $a_1 = \frac{1}{\sqrt{2}}$ and $a_{n+1} = \sqrt{\frac{1+a_n}{2}}$ $(n > 1)$. Show that the infinite product $\prod_{n=1}^{\infty} a_n$ converges to $\frac{2}{\pi}$.

42. Show $\Gamma(\frac{1}{2}) = \sqrt{\pi}$.

43. By using standard methods, one can obtain the expansion of $\frac{x}{2}\frac{e^x+1}{e^x-1}$ as a power series. Performing this computation leads to:

$$\frac{x}{2}\frac{e^x+1}{e^x-1} = \frac{x}{2} \coth \frac{x}{2} = 1 + \sum_{n=2}^{\infty} \frac{B_n x^n}{n!}.$$

The nth **Bernoulli number**, B_n is defined by this equation, as shown. Calculate the first six Bernoulli numbers.

Obtain the following expansions (stating the range of validity):

(a) $x \coth x = \sum_{k=0}^{\infty} \frac{2^{2k} B_{2k} x^{2k}}{(2k)!}$;

(b) $\tanh x = \sum_{k=1}^{\infty} \frac{2^{2k}(2^{2k}-1)B_{2k} x^{2k-1}}{(2k)!}$;

(c) $\frac{x}{\sinh x} = -\sum_{k=0}^{\infty} \frac{(2^{2k}-2)B_{2k} x^{2k}}{(2k)!}$;

(d) $x \cot x = \sum_{k=0}^{\infty}(-1)^k \frac{2^{2k} B_{2k} x^{2k}}{(2k)!}$;

(e) $\tan x = \sum_{k=1}^{\infty}(-1)^{k+1} \frac{2^{2k}(2^{2k}-1)B_{2k} x^{2k-1}}{(2k)!}$.

44. The **Euler Numbers** E_n are defined by

$$\frac{1}{\cosh x} = \sum_{n=0}^{\infty} \frac{E_n x^n}{n!}.$$

Prove the following:

(a) $E_0 = 1, E_{2n-1} = 0, n \in \mathbf{N}$;

(b) $\sum_{k=0}^{n} \binom{2n}{2k} E_{2n-2k} = 0, n \geq 1$;

(c) $\sec x = \sum_{n=0}^{\infty}(-1)^n \frac{E_{2n} x^{2n}}{(2n)!}$.

Calculate the first 10 Euler Numbers.

45. Given $f(x) = \sin(p\arcsin x)$, prove

    (a) $(1 - x^2)f''(x) - xf'(x) + p^2 f(x) = 0$;

    (b) $(1 - x^2)f^{(n+2)}(x) - xf^{(n+1)}(x) + (p^2 - n^2)f^{(n)}(x) = 0$;

    (c) use (b) to obtain the first four terms in the Taylor expansion of $\sin(p\arcsin x)$ near 0.

46. Use the method suggested in Exercise 45 to obtain the Taylor expansion of the following functions near origin:

    (a) $\dfrac{\arcsin x}{\sqrt{1 - x^2}}$;   (b) $e^{\arctan x}$;   (c) $\arctan x$;   (d) $\sin\ln x$;   (e) $(\arcsin x)^2$.

Appendix A

Appendix on Set Theory

A.1 Introduction

This appendix contains only the briefest introduction to set theory. Any reader desiring a more thorough treatment should consult one of the texts mentioned in the references, for example, Monk (1969).

The approach presented is axiomatic, and a complete set of axioms on which to develop set theory is presented.[1] However, much of the theory is suppressed, and mainly those items of interest and importance to the development of analysis are discussed.

A.2 Axioms for Set Theory

The primitive notions of set theory are those of **class** and **membership**. Intuitively, classes are 'collections of objects'. Collecting objects into groups gives rise to the notion of 'being a member' of a group. Hence the two primitive notions whose behavior is captured and defined by the axioms are class and membership.

Notationally, capital letters $A, B, C, \ldots X, Y, Z$ will stand for classes, while the membership relation is denoted by $\in$. The formula $A \in B$ is read as 'the class A is a member of the class B'. The negation $A \notin B$ is read as 'the class A is not a member of the class B'.

Axiom 1 (Extensionality).

$$\forall A \forall B [\forall C (C \in A \iff C \in B) \implies A = B].$$

Remark. This axiom defines the conditions under which two classes are equal to one another, namely, when they have exactly the same members. $\square$

Definition. A class X is a **set** if there is a class B such that $X \in B$. A class that is not a set is called a **proper class**.

Remark. Lowercase letters are used to denote sets. Intuitively, sets are 'well-behaved classes'. They are well behaved because they are small enough to be

---

[1]Our axioms follow those of Monk (1969).

found as members of other classes. Note that for a class to be a set requires a witness; that is, to demonstrate A is a set, we must find a class B (the witness) such that $A \in B$. □

Formally, we want to discuss classes, sets, and the relations between them in the same way that we have discussed numbers, functions, and so forth. To accomplish this, we specify certain expressions as being of particular importance. Expressions of the form $A = B$, $A = C$, etc., are **set-theoretic formulae**. Expressions of the form $A \in A$, $X \in Y$, etc. are also set-theoretic formulas. Formulas generated from the above set-theoretic formulas by use of logical connectives, for example, $\vee$ (or), $\wedge$ (and), etc., or the universal ($\forall$) and existential ($\exists$) quantifiers are also set-theoretic[2]. In addition, in subjects such as analysis, one permits the symbols of analysis to appear, as shown below.

Axiom 2 (Class Building Axioms). *Let $\phi(X)$ be a set-theoretic formula not involving the letter A, then the following is an axiom:*

$$\exists A \forall X [X \in A \Longleftrightarrow X \text{ is a set } \wedge \phi(X)].$$

Remark. This axiom permits the generation of classes of objects having specific properties. Such constructions occur continually in mathematics, for example, in forming the interval, $[0, 1]$.

The notation $\{X : \phi(X)\}$ is commonly used to stand for the class A obtained from $\phi(X)$ by Axiom 2. Thus,

$$[0, 1] = \{x : 0 \le x \wedge x \le 1\}.$$

It is implicit that within our set theory we can find sets that will play the role of 0, 1 and all the other 'real numbers'. This is the case; however, a full discussion of modeling the real numbers within set theory is beyond the scope of this appendix.

If $\phi(X)$ is a set-theoretic formula not involving either the letter A or the letter B and if

$$\forall X [X \in A \Longleftrightarrow X \text{ is a set } \wedge \phi(X)]$$

and

$$\forall X [X \in B \Longleftrightarrow X \text{ is a set } \wedge \phi(X)],$$

then $A = B$. In short, this means that the classes constructed using Axiom 2 are unique. □

Definition. $A \subseteq B \Longleftrightarrow \forall C (C \in A \Longrightarrow C \in B)$.

Remark. The formula $A \subseteq B$ is read 'A is **subclass** of B' (or 'A is **contained in** B'). If a and b are sets, then we say 'a is a **subset of** b'.

The relationship of 'being contained in' is the other important relation that can hold between classes. It is essential for the reader to differentiate between the membership relation, $\in$, and the relation, $\subseteq$. These two relationships are fundamentally different, and they must not be confused. □

The first axiom defines equality, the second permits us to construct classes. However, we do not yet have any sets. Most of the rest of the axioms assert that certain constructions are guaranteed to yield sets, as opposed to classes.

---

[2] For a thorough discussion of the language of set theory and its relationship to logic, we again refer the reader Monk (1969).

Axiom 3 (Power Set). $\forall a \exists b \forall C (C \subseteq a \implies C \in b)$.

Remark. This axiom asserts that if one starts with a set, a, then another set, b, is guaranteed to exist that has the property that every subclass of a is a member of b. Among the obvious, but important, consequences of this axiom are that the collection of all subclasses of a set is, itself, a set and a subclass of a set is a set. $\square$

Axiom 4 (Pairing). $\forall a \forall b \exists c (a \in c \wedge b \in c)$.

Remark. This axiom guarantees that given two sets, a and b, a third set, c, can be found which has both these sets as members. Obviously, this principal extends to any finite collection. $\square$

Axiom 5 (Union). $\forall a \exists b \forall C (C \in a \implies C \subseteq b)$.

Remark. This axiom will guarantee the existence, as a set, of arbitrary unions of families of sets, so long as the family, itself, is a set. To see this, think of the set a as being an index set. What is being indexed is the members of a, and each member of a acts as its own index. Then the axiom guarantees the existence of a set b that collects together the members of all of the sets that are indexed, that is, all members of the sets that are members of a. $\square$

Definition. $\emptyset = \{x : x \neq x\}$.

Remark. $\emptyset$ is called the **empty class**. We cannot yet refer to it as the empty set, because, at this stage in the development, there is no way to prove $\emptyset$ is a set. That will follow from Axiom 7.

   The empty class is particularly useful for demonstrating the difference between the membership relation, $\in$, and the subclass relation, $\subseteq$. Observe, $\emptyset$ is not a member of every class, because we have $\forall x (x \notin \emptyset)$, whence $\emptyset \notin \emptyset$. However, $\emptyset$ is a subclass of every class, and in particular, $\emptyset \subseteq \emptyset$. Thus, the two relations must be different. $\square$

Definition. $A \cap B = \{x : x \in A \wedge x \in B\}$. $A \cap B$ is called the **intersection** of A and B.

Remark. Since $A \cap B$ is a subclass of A and B, no axiom will be needed to guarantee that $a \cap b$ will be a set. $\square$

Axiom 6 (Regularity). $\forall A [A \neq \emptyset \implies \exists X (X \in A \wedge X \cap A = \emptyset)]$.

Remark. This axiom, unlike the others, is aimed at ensuring that the relation $\in$ has no undesirable properties. One of the main properties being avoided is finite cyclical strings of classes of the form

$$A = B_0 \in B_1 \in \cdots \in B_{n-1} \in B_n = A.$$

The reader can show, for example, $\forall A (A \notin A)$, and so forth. $\square$

Definition. $\mathcal{S}A = \{x : x \in A \vee x = A\}$. $\mathcal{S}A$ is called the **successor** of A.

Remark. One should think of 'successor' as a unary operation. Given a set a, $\mathcal{S}a$ will be a new set that is distinct from a. $\square$

Axiom 7 (Infinity). $\exists a [\emptyset \in a \wedge \forall X (X \in a \implies \mathcal{S}X \in a)]$.

Remark. The axiom of infinity guarantees the existence of loads of sets. For example, $\emptyset$ is a set, $S\emptyset$ is a set, $SS\emptyset$ is a set, and so forth. Notice that any set with the property that it is nonempty and closed under successor must be infinite.

   The reader should think about this axiom and the successor operation in the context of the positive integers and the Peano Axioms. The 'smallest' set that has $\emptyset$ as a member and is closed under successor should look very much like the positive integers or, if one identifies 0 with $\emptyset$, the nonnegative integers. Indeed, this is how one can begin the process of modeling real analysis within set theory. That, however, is beyond the scope of this book. We will assume that all numbers can be represented as sets and that there is a set that represents **R**. $\square$

Definition. $\{A, B\} = \{x : x = A \vee x = B\}$. $\{A, B\}$ is called the **unordered pair** A, B.

Definition. $\{A\} = \{A, A\}$. $\{A\}$ is called **singleton** A.

Remark. Of course if a and b are sets, then $\{a, b\}$ is a set. It follows that all finite collections of sets will also be sets. But the critical fact about unordered pairs is contained in the next theorem. $\square$

Theorem 1. *If* $\{a, b\} = \{c, d\}$, *then either* $a = c$ *and* $b = d$, *or* $a = d$ *and* $b = c$.

Remark. The proof is a straightforward case analysis, but it is instructive. $\square$

Definition. $(A, B) = \{\{A\}, \{A, B\}\}$. (A, B) is called the **ordered pair** with **first coordinate** A and **second coordinate** B.

Remark. If one thinks about ordered pairs, one sees that two things are critical: it must be possible to distinguish coordinates, and ordered pairs must have the equality property stated in the next theorem. The first is not possible without the second, so it is really the second that one must have. $\square$

Theorem 2. *If* $(a, b) = (c, d)$, *then* $a = c$ *and* $b = d$.

Remark. This is the property of ordered pairs that is used repetitively throughout mathematics. We could not have functions or relations without it.

   The proof is case analysis and is instructive. $\square$

Definition. Let R be a class. Then

   (i) R is a **relation** if and only if $\forall A(A \in R \implies \exists c \exists d[A = (c, d)])$.

   (ii) Dmn $R = \{x : \exists y[(x, y) \in R]\}$. Dmn R is called the **domain** of R.

   (iii) Rng $R = \{y : \exists x[(x, y) \in R]\}$. Rng R is called the **range** of R.

   (iv) F is a **function** if and only if F is a relation and

$$\forall x \forall y \forall z[(x, y) \in F \wedge (x, z) \in F. \implies .y = z].$$

Axiom 8 (Substitution). *If* F *is a function and* Dmn F *is a set, then* Rng F *is a set.*

Remark. This axiom again specifies that certain objects are sets. In this case the image of a set under a function must again be a set. $\square$

Axiom 9 (Relational Axiom of Choice). *If R is a relation, then there is a function F such that F ⊆ R and* Dmn *F* = Dmn *R*.

Remark. This is a strong form of the Axiom of Choice, which is discussed in Section A.10. It is independent from the previous axioms and has many remarkable consequences. In analysis, one of these is the existence of a nonmeasurable set. □

The axioms presented are generally sufficient to do most of mathematics. Exploring their deeper consequences is far beyond the scope of this book, and again we recommend Monk (1969) for a concise and readable treatment.

Before continuing, we point out how the axioms permit the escape from Russell's Paradox, which results from considering the class

$$B = \{X : X \notin X\}.$$

The paradox results from asking whether $B \in B$. Ordinarily, one obtains

$$B \in B \quad \text{if and only if} \quad B \notin B.$$

The paradox is avoided, since for $B \in B$, it must be the case that B is a set. One concludes that B is not a set, but a proper class.

A.3   Boolean Algebra of Classes

Of particular interest and use to working mathematicians are the Boolean operations and the associated relation of containment, or subset. In this section, we discuss the basic theory associated with the containment relation, together with theorems related to finite unions and intersections.

Theorem 3. *Let A, B, and C be any classes. Then*

(i) $\emptyset \subseteq A$;

(ii) *if* $A \subseteq \emptyset$, *then* $A = \emptyset$;

(iii) $A \subseteq A$;

(iv) *if* $A \subseteq B$ *and* $B \subseteq A$, *then* $A = B$;

(v) *if* $A \subseteq B$ *and* $B \subseteq C$, *then* $A \subseteq C$.

Remark. The general method for showing equality of sets is specified in the extensionality axiom. The method is now refined to showing that containment goes in both directions, (iv). This, in fact, is a specification of an algorithm for showing equality between two sets, A and B. You must first show $A \subseteq B$ and then show $B \subseteq A$. *No method that does not establish these two facts is acceptable for establishing equality between sets!* □

Theorem 4. *Let A, B, C, and X be classes. Then*

(i) $A \cap B = B \cap A$;

(ii) $A \cap (B \cap C) = (A \cap B) \cap C;$

(iii) *if* $X \subseteq A$ *and* $X \subseteq B$, *then* $X \subseteq A \cap B$.

Definition. Classes A and B are called **disjoint** if $A \cap B = \emptyset$. A class A is called a **family of pairwise disjoint sets** if every two distinct members of A are disjoint, that is,

$$\forall x \forall y [x \in A \land y \in A \land x \neq y \implies x \cap y = \emptyset].$$

Theorem 5. *Let* A *be a class and* a *and* b *be any sets. Then*

(i) $\emptyset$ *and* A *are disjoint for any class* A;

(ii) *if* $a \notin A$, *then* A *and* $\{a\}$ *are disjoint;*

(iii) $\emptyset$ *and* $\{a\}$ *are families of pairwise disjoint sets;*

(iv) *for* $a \neq b$, $\{a, b\}$ *is a family of pairwise disjoint sets if and only if* $a \cap b = \emptyset$.

Definition. $A \cup B = \{x : x \in A \lor x \in B\}$. $A \cup B$ is called the **union** of A and B.

Theorem 6. *Let* A, B, *and* C *be classes and,* a, b, *and* x *be sets. Then*

(i) $A \cup (B \cup C) = (A \cup B) \cup C;$

(ii) $A \cup B = B \cup A;$

(iii) $A \cup (B \cap C) = (A \cup B) \cap (A \cup C);$

(iv) $a \cup b$ *is a set;*

(v) $\mathcal{S}x = x \cup \{x\}$ *and* $\mathcal{S}x$ *is a set.*

Remark. The union operation provides a simple means for defining finite sets of any size and for defining n-tuples for any positive integer n. $\square$

Definition. $\{A, B, C\} = \{A\} \cup \{B, C\}$; $(A, B, C) = ((A, B), C)$.

Definition. $\mathcal{V}^* = \{x : x = x\}$. $\mathcal{V}^*$ is called the **universe**.

Theorem 7. *Let* x *be any set and* A *be any class. Then*

(i) $\forall x (x \in \mathcal{V}^*);$

(ii) $\forall A (A \subseteq \mathcal{V}^*);$

(iii) $\forall A (A \cap \mathcal{V}^* = A);$

(iv) $\forall A (A \cup \mathcal{V}^* = \mathcal{V}^*);$

(v) $\mathcal{V}^*$ *is a proper class.*

Definition. (i) $A' = \{x : x \notin A\}$. A' is the **complement of** A.

(ii) $A \sim B = \{x : x \in A \land x \notin B\}$. $A \sim B$ is called the **complement of** B **relative to** A.

Theorem 8. *Let A and B be any classes and a any set. Then*

(i) $a \sim B$ *is a set;*

(ii) $A' = \mathcal{V}^* \sim A$;

(iii) $A \sim B = A \cap B'$;

(iv) $A'' = A$;

(v) $A \subseteq B$ *if and only if* $B' \subseteq A'$;

(vi) $(A \cap B)' = A' \cup B'$ *and* $(A \cup B)' = A' \cap B'$.

Remark. The facts contained in (vi) are known as **DeMorgan's Laws** for sets. $\square$

A.4 Infinite Boolean Operations

Arbitrary unions and intersections of families of sets occur throughout mathematics. This section contains the essential facts related to these more general operations.

Definition. $\bigcup A = \{x : \exists y[y \in A \land x \in y]\}$. We call $\bigcup A$ the **union** of the family A.

Remark. We often think of A as a function with domain I, called the index set. In such an instance, we set $A(i) = A_i$ and write

$$\bigcup \mathrm{Rng}\, A = \bigcup \{A_i : i \in I\} = \bigcup_{i \in I} A_i.$$

For this case we say that A is an **indexed family of sets** with domain I. $\square$

Theorem 9. *Let A and B be any classes and a and b be sets. Then*

(i) $\bigcup \emptyset = \emptyset$;

(ii) $\bigcup \{a\} = a$;

(iii) $\bigcup \{a, b\} = a \cup b$;

(iv) *if* $A \subseteq B$, *then* $\bigcup A \subseteq \bigcup B$;

(v) $\bigcup A \cup \bigcup B = \bigcup (A \cup B)$;

(vi) $\bigcup (A \cap B) \subseteq \bigcup A \cap \bigcup B$;

(vii) $\bigcup a$ *is a set.*

Definition. $\bigcap A = \{x : \forall y(y \in A \implies x \in y)\}$. $\bigcap$ is called the **intersection** of the family of sets, A.

Remark. $\bigcap A$ should be thought of as the intersection of all the sets that are members of the family A. $\square$

Theorem 10. *Let A and B be any classes and a and b be sets. Then*

(i) $\bigcap \emptyset = \mathcal{V}^*$;

(ii) $\bigcap\{a\} = a$;

(iii) $\bigcap\{a, b\} = a \cap b$;

(iv) $A \subseteq B$ *implies* $\bigcap B \subseteq \bigcap A$;

(v) $\bigcap(A \cup B) = \bigcap A \cap \bigcap B$;

(vi) $(\bigcap A) \cup (\bigcap B) \subseteq \bigcap(A \cap B)$.

Remark. Most of the results for infinite unions and intersections are generalizations of analogous results for finite unions and intersections. There are notable exceptions, for example, 9(vi) and 10(vi); 10(i) is also surprising. $\square$

A.5　Algebra of Relations

Recall that a relation is a class every member of which is an ordered pair. It is usually the case for a relation R that one writes xRy instead of $(x, y) \in R$; for example, the relation, $<$ on **R**, where one writes $1 < x$, instead of $(0, 1) \in <$.

It is trivial $\emptyset$ is a relation and, to check whether two relations, R and S are equal, one has only to show equivalence of membership for ordered pairs.

Below is a list of some of the more useful properties of relations. A more complete list can be found in Monk (1969).

Definition. $R^{-1} = \{(y, x) : (x, y) \in R\}$. R^{-1} is called the **inverse** of R.

Remark. This definition together with the Boolean operations permit us to notice $a = \bigcap\bigcap(a, b)$ and $b = \bigcap\bigcap\bigcap\{(a, b)\}^{-1}$. As a consequence, there is a constructive method for obtaining the first and second coordinates of any ordered pair. $\square$

Theorem 11. *Let R and S be relations. Then*

(i) Dmn $(R \cup S) = $ Dmn $R \cup$ Dmn S *and* Rng $(R \cup S) = $ Rng $R \cup$ Rng S;

(ii) Dmn $(R \cap S) \subseteq $ Dmn $R \cap$ Dmn S *and similarly for* Rng ;

(iii) Dmn $R^{-1} = $ Rng R *and* Dmn $R = $ Rng R^{-1}.

Definition. Fld $R = $ Dmn $R \cup$ Rng R. Fld R is called the **field** of R. R is said to be **on** A if $A = $ Fld R.

Remark. It is straightforward that R and R^{-1} have the same field and trivial that R is on Fld R. $\square$

Definition. $A \times B = \{(a,b) : a \in A \wedge b \in B\}$. $A \times B$ is called the **cartesian product** of A and B. More generally, $A \times B \times C = (A \times B) \times C$.

Remark. A relation that can be viewed as as being a subset of $A \times B$ is called a binary relation. We stress 'viewed', since the ternary relation $(A \times B) \times C$ is also a subset of a cartesian product. $\square$

Theorem 12. *Let a and b be sets and A and B be classes. Then*

(i) *$a \times b$ is a set;*

(ii) *$A \times B$ is a relation;*

(iii) *$(A \times B)^{-1} = B \times A$;*

(iv) *if B is nonempty, then $\operatorname{Dmn} A \times B = A$; a similar statement holds for Rng.*

Theorem 13. *If r is a relation, then r^{-1}, $\operatorname{Dmn} r$, $\operatorname{Rng} r$, and $\operatorname{Fld} r$ are all sets.*

A.6   Functions

After ordered pairs, about the single most useful and prevalent entities in mathematics are functions. They are ubiquitous and their important properties and related definitions are listed below.

Definition. $F(A) = \{x : \forall y([A \text{ is a set } \wedge (A,y) \in F] \implies x \in y)\}$. Read '$F$ of A' for $F(A)$. $F(A)$ is the value of the function F with input A.

Theorem 14. *Let F be a function.*

(i) *If $x \in \operatorname{Dmn} F$, then $F(x)$ is the unique y such that $(x,y) \in F$, and hence $(x, F(x)) \in F$; in particular, $F(x)$ is a set if $x \in \operatorname{Dmn} F$.*

(ii) *If $A \notin \operatorname{Dmn} F$, then $F(A) = \mathcal{V}^*$.*

Remark. This theorem gives the main properties of $F(A)$; namely, if A is in the domain of F, then $F(A)$ is the second coordinate of the ordered pair in F having first coordinate, A. In any other case, $F(A)$ is $\mathcal{V}^*$, whence $F(A)$ is always defined. There is no possibility of confusion, since $\mathcal{V}^*$ cannot ever occur as a second coordinate of an ordered pair because it is a proper class. $\square$

Theorem 15. *Let F and G be functions. Then*

(i) *$F = G$ if and only if $\operatorname{Dmn} F = \operatorname{Dmn} G$ and $F(x) = G(x)$ for all $x \in \operatorname{Dmn} F$;*

(ii) *if $\operatorname{Dmn} F \cap \operatorname{Dmn} G = \emptyset$, then $F \cup G$ is a function.*

Definition. Let F and G be functions and A and B be classes.

(i) *$F \circ G = \{(x,z) : \exists y((x,y) \in G \wedge (y,z) \in F)\}$. $F \circ G$ is called the **composition** of F and G.*

(ii) F is **1–1** (**one-to-one**, **injection**) if and only if F and F^{-1} are both functions.

(iii) if Rng $F = A$, then F is said to be **onto** A, or a **surjection** on A.

(iv) If Rng $F \subseteq A$, then F is said to be **into** A.

(v) If $A = $ Dmn F, F is said to be **from** A.

(vi) If F is 1–1 and Dmn $F = A$ and Rng $F = B$, then F is said to be a **1–1 correspondence** (**bijection**) from A to B.

(vii) $^{A}B = \{f : f$ is a function from A into $B\}$.

(viii) A function, F, from A into A is called a **unary operation**.

(ix) A function, F, from $A \times A$ into A is called a **binary operation**.

Theorem 16. *Let F and G be functions. Then*

(i) Dmn $(F \circ G) = \{x : x \in$ Dmn $G \wedge \exists y \in$ Rng $G \cap$ Dmn $F\}$;

(ii) *if $x \in$ Dmn $(F \circ G)$, then $(F \circ G)(x) = F(G(x))$;*

(iii) *the following three conditions are equivalent:*

   (a) *F is 1–1;*
   (b) *for all $x, y \in$ Dmn F, if $F(x) = F(y)$, then $x = y$;*
   (c) *for all $x, y \in$ Dmn F, if $x \neq y$, then $F(x) \neq F(y)$.*

(iv) *if F is a 1–1 function from A onto b, then A is a set;*

(v) *^{a}b is a set.*

A.7   Equivalence Relations

One of the most useful kinds of relations are equivalence relations. These occur in various forms throughout all branches of mathematics. Most of the important facts concerning equivalence relations are presented below.

Definition. Let R be a relation.

(i) R is **transitive** if and only if

$$\forall x, y, z[(x, y) \in R \text{ and } (y, z) \in R \Longrightarrow (x, z) \in R].$$

(ii) R is **symmetric** if and only if $\forall x, y[(x, y) \in R \Longrightarrow (y, x) \in R]$.

(iii) R is an **equivalence relation** if and only if R is transitive and symmetric.

(iv) R is **reflexive on** A if and only if $\forall x[x \in A \Longrightarrow (x, x) \in R]$.

Remark. In the above definitions, R must be a relation. The simplest example of an equivalence relation is the equality relation.

The definition given requires only that to be an equivalence relation, R must be transitive and symmetric. These two conditions imply that R will be reflexive on its field. Thus, the issue of whether R is reflexive never arises.

Often one starts with a set, or class, A, and constructs a relation, R, with the intention that R will be an equivalence relation on A. To verify this, one must show not only that R is transitive and symmetric, but also must show R is a relation on A. One way of accomplishing this is showing

$$\forall x[x \in A \Longrightarrow (x,x) \in R],$$

in other words, R is reflexive on A. $\square$

Definition. Let R be an equivalence relation.

(i) $x/R = \{y : (x,y) \in R\}$. x/R is called the **equivalence class** of x under R.

(ii) $\pi_R = \{(x, x/R) : x \in \text{Fld } R\}$.

Theorem 17. *Let R be an equivalence relation.*

(i) $\text{Dmn } R = \text{Rng } R = \text{Fld } R$.

(ii) *R is reflexive on* $\text{Fld } R$.

(iii) *If $x \in \text{Fld } R$, then $x \in x/R$.*

(iv) *For any $x, y \in \text{Fld } R$, xRy if and only if $x/R = y/R$.*

(v) *For any $x, y \in \text{Fld } R$, if $x/R \cap y/R \neq \emptyset$, then $x/R = y/R$.*

(vi) *π_R is a function.*

(vii) *If R is a set, then π_R maps $\text{Fld } R$ onto $\{x/R : x \in \text{Fld } R\}$, and $\pi_R(x) = x/R$ for each $x \in \text{Fld } R$.*

Remark. This theorem contains the main properties of equivalence classes.

Part (iv) asserts x is related to y under **R** exactly if x and y belong to the same equivalence class. Part (v) asserts two equivalence classes are either disjoint or equal. These two facts mean that the nature of an equivalence class will be independent of any representative. It is for this reason that we can use x/R as a name for the equivalence class of x and realize that it refers not only to the equivalence class of x but also to the equivalence class of y, where y is any member of $\text{Fld } R$ such that xRy.

Finally, these facts imply π_R is a function, a fact that is found useful in many areas of mathematics. $\square$

Definition. P is a **partition** of A, if P is a family of pairwise disjoint nonempty sets and $\bigcup P = A$.

Theorem 18. *Let A be any set. Let*

$$E(A) = \{R : R \ \ is \ an \ equivalence \ relation \ with \ field \ A\}$$

and

$$P(A) = \{P : P \ \ is \ a \ partition \ of \ A\}.$$

Then there is a natural 1–1 correspondence between $E(A)$ and $P(A)$.

Remark. The important feature of this theorem is in the proof that is accomplished by noticing the equivalence classes form a partition. Once this has been observed, the details follow rather easily. But most important, the theorem tells us that we can usefully think of equivalence relations in terms of the partition generated by the equivalence classes. □

A.8   Ordering

Orderings also occur in all branches of mathematics. Orderings provide structure to sets. Perhaps the simplest example of such is the positive integers under $<$. Order provides not only structure to sets, but also a framework that can be used to establish proofs. Again the simplest example is induction on the positive integers.

Induction uses order in two ways to structure a proof. First, the ordering supplies a place to start a proof. Second, the ordering is used to establish the truth of the required fact for x, a member of the underlying set, provided the fact is known for all members of the underlying set that are less than x. It is important to realize, although not used in this book, that forms of induction can be used on many different types of orderings.

Some of the definitions and theorems related to orderings are presented below.

Definition. Let R be a relation.

(i) R is **antisymmetric** if and only if xRy and yRx implies $x = y$.

(ii) R is a **partial ordering** if and only if R is reflexive on Fld R, antisymmetric and transitive.

(iii) A is **partially ordered** by R if and only if $(A \times A) \cap R$ is a partial ordering with Fld $R = A$.

Theorem 19. *If R is any partial ordering, then $(A \times A) \cap R$ is also a partial ordering.*

Definition. *F is an **isomorphism from R onto S**, if and only if R and S are relations, F is a function mapping Fld R onto Fld S, and xRy if and only if $F(x)SF(y)$ for all $x, y \in$ Fld R.*

Remark. The notion of isomorphism is intended to identify structures, relations, and so forth, which are the same except for their names. □

Theorem 20.

(i) $\{(x, y) : x \subseteq y\}$ *is a partial ordering;*

(ii) *if $R \in \mathcal{V}^*$ is any partial ordering, then there is a set A and an isomorphism F from R onto $(A \times A) \cap \{(x, y) : x \subseteq y\}$.*

Remark. This theorem asserts, first, that the subset relation is a partial order and, second, that every partial ordering can be realized using the subset relation. □

Definition. Let R be a partial ordering with field A, and suppose that $X \subseteq A$ and $a \in A$.

(i) a is an R-**upper bound** of X if and only if xRa for each $x \in X$.

(ii) a is an R-**lower bound** for X if and only if aRx for each $x \in X$.

(iii) a is an R-**greatest element** of X if and only if a is an R-upper bound for X and $a \in X$.

(iv) a is an R-**least element** of X if and only if a is an R-lower bound for X and $a \in X$.

(v) a is an R-**least upper bound** (R-l.u.b.) for X if and only if a is an R-upper bound for X and a is a lower bound for the class of all R-upper bounds for X.

(vi) a is an R-**greatest lower bound** (R-g.l.b.) for X if and only if a is an R-lower bound for X and a is an upper bound for the class of all R-lower bounds for X.

(vii) a is an R-**minimal element** for X if and only if $a \in X$ and for all $x \in X$, xRa implies $x = a$.

(viii) a is an R-**maximal element** of X if and only if $a \in X$ and for all $x \in X$, aRx implies $x = a$.

Remark. All of these definitions occur in the context of real analysis in the discussions of suprema and infima. □

Theorem 21. *Let R be a partial ordering with field A, and suppose that every subclass $B \subseteq A$ has an R-l.u.b. If F maps A into A, and for all $x, y \in A$, xRy implies $F(x)RF(y)$, then $F(x) = x$, for some $x \in A$.*

Remark. This theorem asserts that if one has a partial ordering on A, then any order preserving map from A into A must have a fixed point, that is, a point x such that $F(x) = x$.

Fixed points can be very useful objects because they may have associated properties not available for arbitrary points. Such properties can then be used in proofs. □

Definition. Let R be a relation.

(i) R is a **simple ordering** or **linear ordering** if and only if R is a partial ordering and for all $x, y \in \text{Fld } R$, xRy or yRx.

(ii) A is **simply ordered** by R if and only if $(A \times A) \cap R$ is a simple ordering with field A.

(iii) A relation, R, is **well-founded** if and only if R is a relation and for every nonempty class, $A \subseteq \operatorname{Fld} R$, there is an $x \in A$ such that $A \cap \{y : yRx\} = \emptyset$.

(iv) $\leq$ is a **well-ordering** if and only if $\leq$ is a simple ordering and $<$ is well-founded where $< \, = \, \{(x,y) : (x,y) \in \leq \text{ and } x \neq y\}$.

Remark. The canonical well-founded relation is the membership relation $\in$. The axiom that forces this is the regularity axiom.

The symbols $\leq$ and $<$ in (iv) refer to general orders, not to the standard orders on **R**.

With respect to **R**, $\leq$ is a simple ordering, but it is not a well-ordering. If one considers $\leq$ on **N**, it is a well-ordering. $\square$

Theorem 22. *For any partial ordering, $\leq$, the following are equivalent:*

(i) $\leq$ *is a well-ordering;*

(ii) $\leq$ *is a simple ordering, and every nonempty class $A \subseteq \operatorname{Fld} \, \leq$ has a $\leq$-least element;*

(iii) *every nonempty class $A \subseteq \operatorname{Fld} \, \leq$ has a $\leq$-least element.*

Remark. Well-ordered sets are those on which there is a partial ordering satisfying every nonempty subset has a least element. In analysis, all subsets of the integers that are bounded below have this property, and so are well-ordered. If the reader studies the section on induction, she will realize that it is this fact that is at the heart of the method of proof by induction. $\square$

As stated, one use of order is in providing structure for proofs. The following theorems, which involve order, are all equivalent to the Axiom of Choice and are useful from time to time in constructing proofs.

Theorem 23 (Well-Ordering Principle). *Let A be a set. Then there is a well-ordering with field A.*

Theorem 24 (Zorn's Lemma). *Let R be a partial ordering of a set, A. If every simply ordered subset of A has an R-upper bound in A, then A has an R-maximal element.*

Theorem 25 (Maximality Principle). *Let $\subseteq$ be a partial ordering of a set, A. If every simply ordered subset of A has an $\subseteq$-upper bound in A, then A has an $\subseteq$-maximal element.*

A.9   Finite and Infinite Sets; Countable Sets

One of the main issues in set theory concerns the size of a set. For small sets, this issue can be settled by counting the elements. For large sets, counting, which amounts to well-ordering the elements of the set, fails to provide a unique answer to the question. The solution is equipotence.

Definition. Two sets X and Y are **equipotent** if there exists a 1–1 correspondence between X and Y. If X and Y are equipotent, we write $X \cong Y$.

Remark. The intuition behind this solution to the problem can be seen by thinking of a large auditorium. If every seat is filled and there are people standing at the back, we know there are more people than seats. If every seat is filled and no one is standing, we know there are the same number of seats as people. If there are unfilled seats and no one is standing, we know there are more seats than people. In no case do we have to under take a counting process to arrive at these conclusions. Thus, the notion of 1–1 correspondence is the perfect tool for addressing these questions.

The approach we are taking to cardinal numbers is straightforward, but has inherent difficulties. For an alternate approach that avoids these difficulties, see Monk (1969). $\square$

Theorem 26. $\cong$ *is an equivalence relation on the class of all sets.*

Definition. The equivalence classes corresponding to $\cong$ are called **cardinal numbers**.

Theorem 27 (Trichotomy Principle). *Let a and b be sets. Then there is a 1–1 function from a into b or a 1–1 function from b into a.*

Remark. As a result of the Trichotomy Principle, we can select a set from each of the equivalence classes determined by $\cong$, and this class will have a natural linear ordering on it determined by $a \leq b$ if and only if there is a 1–1 function from a into b.

The Trichotomy Principle is equivalent to the Axiom of Choice, which makes clear why the Axiom of Choice is essential to the development of the theory of cardinal numbers. $\square$

Theorem 28 (Schröder-Bernstein). *If X and Y are two sets such that X is equipotent with a subset of Y, and Y is equipotent with a subset of X, then X and Y are equipotent.*

Definition. $\mathcal{P}^*A = \{B : B \subseteq A\}$. $\mathcal{P}^*A$ is called the **power class** of A.

Theorem 29. *Let A and B be classes and a be any set.*

(i) $\emptyset \in \mathcal{P}^*A$.

(ii) $\mathcal{P}^*\emptyset = \{\emptyset\}$.

(iii) $A \subseteq B$ *implies* $\mathcal{P}^*A \subseteq \mathcal{P}^*B$.

(iv) $\mathcal{P}^*(A \cap B) = \mathcal{P}^*A \cap \mathcal{P}^*B$.

(v) $\mathcal{P}^*A\mathcal{P}^*B \subseteq \mathcal{P}^*(AB)$.

(vii) $\mathcal{P}^*a$ *is a set.*

Remark. This theorem gives some of the useful facts about power sets, the most important one being the power set of a set is again a set. $\square$

Theorem 30 (Cantor).   *There does not exist a function mapping the set a onto $\mathcal{P}^*a$.*

Proof. Let a be a set and $\mathcal{P}^*a$ be its power set. Suppose, for the sake of argument, that F is a function mapping a onto $\mathcal{P}^*a$. Set $B = \{x : x \in a \wedge x \notin F(x)\}$. Since F is onto and $B \subseteq a$, there exists $y \in a$ such that $F(y) = B$. The reader can check

$$y \in B \Longleftrightarrow y \notin B,$$

which is a contradiction. $\square$

Corollary. *If X is any set, then X is never equipotent with its power set $\mathcal{P}^*X$.*

Remark. It is an important fact that a set always has smaller cardinality that its power set. $\square$

Definition. A set is said to be **infinite** if it is equipotent with a proper subset of itself. A set is called **finite** if it is not infinite.

Remark. For an alternative approach to the problem of finite versus infinite, see Monk (1969). $\square$

Theorem 31.

(i) *The set $\mathbf{N} = \{1, 2, \ldots, \}$ of natural numbers forms an infinite set.*

(ii) *For each $n \in \mathbf{N}$, the set $I_n = \{1, 2, \ldots, n\}$ is a finite set.*

(iii) *If A is finite, then there is an I_n equipotent with A, or $A = \emptyset$.*

(iv) *For any set a, $\mathcal{P}^*a$ is equipotent with ${}^aI_2$, the set of all functions from a into I_2.*

Proof. The mapping determined by $f(n) = n + 1$ takes $\mathbf{N}$ into a proper subset of itself, whence $\mathbf{N}$ is infinite.

We prove the second statement by induction. Consider I_1. The only subset of I_1 is $\emptyset$. Since any function, F, from I_1 into $\emptyset$ would have $F(1) \in \emptyset$, we would have a contradiction. Thus, I_1 is finite. Suppose I_{n+1} is finite. If I_{n+1} is equipotent with a proper subset of itself via F, then we claim that one can assume $F(i) \in I_n$ for all $i \in I_n$. But F restricted to I_n establishes that I_n is equipotent with a proper subset of itself. $\square$

Remark. There are alternative, and much more satisfactory, approaches leading to these facts, see Monk (1969). Unfortunately, they require considerably more machinery than is presently available.

The important fact here is the positive integers and 0 list the cardinalities of all the finite sets. The proof of this fact is beyond this book. $\square$

Definition. A set is said to be **countable (countably infinite)** if it is equipotent with the set $\mathbf{N}$. An infinite set that is not countable is said to be **uncountable**.

Theorem 32.

(i) $\mathbf{N} \times \mathbf{N}$ *is countable;*

(ii) *if X is countable and Y is an infinite subset of X, then Y is countable;*

(iii) *if X and Y are countable, so is $X \cup Y$;*

(iv) *if A is a countable family of countable sets, $\bigcup A$ is countable;*

(v) *if X and Y are countable, $X \times Y$ is countable.*

(vi) *If X is uncountable and $X \subseteq Y$, then Y is uncountable.*

Proof. We give an indication of how to prove (i). Consider the following table which presents a portion of $\mathbf{N} \times \mathbf{N}$.

$(1,1)$	$(1,2)$	$(1,3)$	$(1,4)$	$(1,5)$	$(1,6)$	$(1,7)$	$\ldots$
$(2,1)$	$(2,2)$	$(2,3)$	$(2,4)$	$(2,5)$	$(2,6)$	$(2,7)$	$\ldots$
$(3,1)$	$(3,2)$	$(3,3)$	$(3,4)$	$(3,5)$	$(3,6)$	$(3,7)$	$\ldots$
$(4,1)$	$(4,2)$	$(4,3)$	$(4,4)$	$(4,5)$	$(4,6)$	$(4,7)$	$\ldots$
$(5,1)$	$(5,2)$	$(5,3)$	$(5,4)$	$(5,5)$	$(5,6)$	$(5,7)$	$\ldots$
$(6,1)$	$(6,2)$	$(6,3)$	$(6,4)$	$(6,5)$	$(6,6)$	$(6,7)$	$\ldots$
$(7,1)$	$(7,2)$	$(7,3)$	$(7,4)$	$(7,5)$	$(7,6)$	$(7,7)$	$\ldots$
$\vdots$	$\vdots$	$\vdots$	$\vdots$	$\vdots$	$\vdots$	$\vdots$	$\ddots$

Table 1

We count the elements of $\mathbf{N} \times \mathbf{N}$ by indicating how they can be listed in order. The ordering is as follows: $(1,1)$, $(2,1)$, $(1,2)$, $(3,1)$, $(2,2)$, $(1,3)$, $(4,1)$, $(3,2)$, $(2,3)$, $(1,4)$, $(5,1)$, …. This ordering amounts to listing the elements in consecutive diagonals. Each diagonal starts with $(n,1)$ and ends with $(1,n)$. It is left to the reader to give a precise description of the function from $\mathbf{N}$ onto $\mathbf{N} \times \mathbf{N}$, which is determined by this process. $\square$

Remark. The indicated argument presented above can be recast to prove (ii)–(v). $\square$

Theorem 33. *The set $\mathbf{Q}$ of rational numbers is countable.*

Remark. The reader can easily construct a mapping from $\mathbf{N} \times \mathbf{N}$ onto the positive rationals by observing that each positive rational can be thought of as an ordered pair in $\mathbf{N} \times \mathbf{N}$. The proof will follow.

A number is **algebraic** if it is the root of a polynomial of finite degree having integer coefficients. Clearly, every rational is algebraic. Using the results in Theorem 32, it can also be shown the algebraic numbers are countable, as well. $\square$

Theorem 34. *The set $\mathbf{R}$ of real numbers is uncountable.*

Proof. For the sake of argument, we suppose the reals in $(0,1)$ are countable. If this is the case, we can list them as a sequence, $\{a_n\}$. Each of these reals has a decimal expansion, $a_n = 0.a_{n,1}a_{n,2}a_{n,3}\ldots$, each digit of which is one of $0, 1, 2, \ldots, 9$. On this basis we list the sequences of digits in the decimal expansions of the reals in $(0,1)$ as in the following table.

$$
\begin{array}{llllllll}
a_{1,1} & a_{1,2} & a_{1,3} & a_{1,4} & a_{1,5} & a_{1,6} & a_{1,7} & \cdots \\
a_{2,1} & a_{2,2} & a_{2,3} & a_{2,4} & a_{2,5} & a_{2,6} & a_{2,7} & \cdots \\
a_{3,1} & a_{3,2} & a_{3,3} & a_{3,4} & a_{3,5} & a_{3,6} & a_{3,7} & \cdots \\
a_{4,1} & a_{4,2} & a_{4,3} & a_{4,4} & a_{4,5} & a_{4,6} & a_{4,7} & \cdots \\
a_{5,1} & a_{5,2} & a_{5,3} & a_{5,4} & a_{5,5} & a_{5,6} & a_{5,7} & \cdots \\
a_{6,1} & a_{6,2} & a_{6,3} & a_{6,4} & a_{6,5} & a_{6,6} & a_{6,7} & \cdots \\
a_{7,1} & a_{7,2} & a_{7,3} & a_{7,4} & a_{7,5} & a_{7,6} & a_{7,7} & \cdots \\
\vdots & \vdots & \vdots & \vdots & \vdots & \vdots & \vdots & \ddots
\end{array}
$$

Table 2

We define the sequence, $\{b_n\}$, by

$$
b_n = \begin{cases} a_{n,n} + 1, & \text{if } a_{n,n} < 8 \\ 3, & \text{otherwise.} \end{cases}
$$

The real number, $b = 0.b_1 b_2 b_3 \ldots$, is in $(0,1)$, but cannot be among the numbers listed in the countable collection, whence the assumption that the reals in $(0,1)$ are countable must be false. $\square$

Theorem 35. $\mathbf{R}$ *and* $\mathcal{P}^*\mathbf{N}$ *are equipotent.*

Proof. A real number a can be uniquely identified by

$$
a = \sup\{x : x \in \mathbf{Q} \wedge x \leq a\}.
$$

It follows that $\mathbf{R}$ is equipotent with a subset of the power set of $\mathbf{N}$. Finding a way to identify all subsets of $\mathcal{P}^*\mathbf{N}$ with a subset of $\mathbf{R}$ is left to the reader. $\square$

A.10 Direct Products and the Axiom of Choice

Product spaces appear regularly in many branches of mathematics. While finite products can be treated as generalizations of the cartesian product, infinite dimensional spaces (not treated in this text) require more machinery.

Definition. Let A be a function with Dmn $A = I$.

$$
\prod A = \{f : f \text{ is a function, Dmn } f = I, \text{ and } f(i) \in A_i \text{ for each } i \in I\}.
$$

$\prod A$ is called the **direct product** of the family A.

Remark. We emphasize the use of A_i, instead of $A(i)$ to denote the value of the function A at i. This approach is particularly fruitful if I is an ordered set. $\square$

Theorem 36.

(i) $\prod \emptyset = \{\emptyset\}$.

(ii) *If A is a function with Dmn $A = I$ and $A_i \neq \emptyset$ for each $i \in I$, then $\prod A$ is not empty.*

(iii) *If A is a set, then so is $\prod A$.*

Remark. The statement in (ii) is the Axiom of Choice for sets, which asserts that a product of nonempty sets is nonempty. This means that there will be a function $f \in \prod A$ such that $f(i) \in A_i$ for each $i \in I$. The function f is called a choice function because it chooses an element out of each member of the infinite family of sets, A_i, $i \in I$.

The Axiom of Choice is known to be independent of the other axioms. Since it is a very powerful axiom with some disconcerting consequences, for example, the existence of a nonmeasurable set, some prefer not to assume its truth. Unfortunately, this also has consequences; for example, one can no longer prove such desirable and plausible results as the Hahn-Banach theorem or that every vector space has a Hamel basis. The examples mentioned all relate to analysis, but analogous examples can be given in almost any other branch of mathematics.

The Axiom of Choice has many equivalent forms. The simplest is perhaps the Trichotomy Principle. This statement, which asserts that for any pair of sets, a and b either there is a 1–1 function from a into b, or vice versa, seems too plausible not to be true. For this reason, many would argue the Axiom of Choice is obviously true. For additional discussion of these ideas, see Monk (1969) or Rubin and Rubin (1963). □

Appendix B

Suggested Reading

Calculus

1. Ellis, R., and Gulick, D., *Calculus with Analytic Geometry*, 5th ed., Harcourt Brace Jovanovich, New York, 1994.

2. Larson, R. E., Hostetler, R. P., and Edwards, R. H., *Calculus of a Single Variable*, 5th ed., D. C. Heath, Lexington. MA, 1994.

3. Larson, R. E., Hostetler, R. P., and Edwards, R. H., *Multivariable Calculus*, 5th ed., D. C. Heath, Lexington, MA, 1994.

4. Leithold, L., *The Calculus of Single Variables with Analytic Geometry*, 5th ed., Harper & Row, New York, 1986.

Elementary Developments

1. Binmore, K. G., *Mathematical Analysis: A Straightforward Approach*, Cambridge University Press, Cambridge, 1977.

2. Clark, C., *Elementary Mathematical Analysis*, Wadsworth, Belmont, CA, 1982.

3. Gaughan, E. D., *Introduction to Analysis*, Brooks/Cole, San Francisco, 1975.

4. Gemignani, M., *Introduction to Real Analysis*, W. B. Saunders, Philadelphia, 1971.

5. Goffman, C., *Introduction to Real Analysis*, Harper & Row, New York, 1966.

6. Lay, S. R., *Analysis—An Introduction to Proof*, Prentice Hall, Englewood Cliffs, NJ, 1986.

7. Moss, R. M., and Roberts, G. T., *A Preliminary Course in Analysis*, Chapman & Hall, Englewood Cliffs, NJ, 1968.

8. Ramanujan, M. S., and Thomas, E. S., *Intermediate Analysis*, Collier Macmillan Limited, London, 1970.

9. Youse, B. K., *Introduction to Real Analysis*, Allyn & Bacon, Boston, 1972.

Intermediate Analysis

1. Anderson, J. A., *Real Analysis*, Logos Press, London, 1969.

2. Apostol, T. M., *Mathematical Analysis*, Addison-Wesley, Reading, MA, 1974.

3. Bartle, R. G., *The Elements of Real Analysis*, 2nd ed., John Wiley, New York, 1976.

4. Bartle, R. G., and Sherbert, D. R., *Introduction to Real Analysis*, John Wiley, New York, 1982.

5. Burril C. W., and Knusden, J. R., *Real Variables*, Holt, Rinehart and Winston, New York, 1969.

6. Fulks, W., *Advanced Calculus*, John Wiley, New York, 1978.

7. Goldberg, R. R., *Methods of Real Analysis*, John Wiley, New York, 1976.

8. Johnsonbaugh, R., and Pfaffenberger W. E., *Foundations of Mathematical Analysis*, Marcel Dekker, New York, 1981.

9. Lang S,, *Analysis I*, Addison-Wesley, Reading, MA, 1968.

10. Marsden, J. E., *Elementary Classical Analysis*, W. H. Freeman, San Francisco, 1974.

11. Olmstead, J. M. H., *Advanced Calculus*, Prentice Hall, Englewood Cliffs, NJ, 1961.

12. Protter, M. H., and Morrey, C. B., *A First Course in Real Analysis*, Springer-Verlag, New York, 1977.

13. Ross, K. A., *Elementary Analysis: The Theory of Calculus*, Springer-Verlag, New York, 1980.

14. Rudin, W., *Principles of Mathematical Analysis*, McGraw-Hill, New York, 1976.

15. Barner, M., and Flohr F., *Analysis I*, Walter de Gruyter, Berlin, 1982.

16. Rosentlicht, M., *Introduction to Analysis*, Scott, Foresman, Glendale, IL, 1968.

17. Sprecher, D. A., *Elements of Real Analysis*, Academic Press, New York, 1970.

18. White, A. J., *Real Analysis, An Introduction*, Addison-Wesley, Reading, MA, 1968.

Advanced Analysis

1. Aliprantis, C. D., and Burkinshaw, O., *Principles of Real Analysis,* North Holland, Amsterdam, 1981.

2. Fischer, E., *Intermediate Real Analysis*, Springer-Verlag, New York, 1983.

3. Folland, G. B., *Real Analysis*, Wiley Interscience, New York, 1984.

4. Lang, S., *Analysis II*, Addison-Wesley, Reading, MA, 1968.

5. Hewitt, E., and Stromberg, K., *Real and Abstract Analysis*, Springer-Verlag, New York, 1955.

6. Phillip, E. R., *An Introduction to Analysis and Integration Theory*, Intext Educational Publishers, Scranton, PA, 1971.

7. Royden, H. L., *Real Analysis*, 3rd ed.,, Macmillan, New York, 1987.

8. Simmons, G. G., *Introduction to Topology and Modern Analysis*, McGraw-Hill, New York, 1963.

9. Stromberg, K., *An Introduction to Classical Real Analysis*, PWS, Boston, MA, 1980; Wadsworth, Belmont, CA, 1987.

10. Torchinsky, A., *Real Variables*, Addison-Wesley, Reading, MA, 1988.

Set Theory

1. Halmos, P. R., *Naive Set Theory*, Van Nostrand, Princeton, NJ, 1960.

2. Monk, J. D., *Introduction to Set Theory*, McGraw-Hill, New York, 1969.

3. Rubin, H., and Rubin, J., *Equivalents of the Axiom of Choice*, North-Holland, Amsterdam, 1963.

4. Suppes, P., *Axiomatic Set Theory*, Van Nostrand, Princeton, NJ, 1960.

History

1. Bell, E. T., *The Development of Mathematics*, McGraw-Hill, New York, 1945.

2. Boyer, C. B., *A History of Mathematics*, John Wiley, New York, 1968.

3. Eves, H., *An Introduction to the History of Mathematics*, W. B. Saunders, Philadelphia, 1983.

4. Kline, M., *Mathematical Thought from Ancient to Modern Times*, Oxford University Press, Oxford, 1972.

5. Priestley, W. M., *Calculus: A Historic Approach*, Springer-Verlag, New York, 1979.

Number Theory

1. Cohen, L., and Ehrlich, L., *The Structure of the Real Number System*, Van Nostrand, Princeton, NJ, 1963.

2. Landau, E., *Foundations of Analysis*, Chelsea House, New York, 1951.

3. Niven, I., *Irrational Numbers*, Carus Mathematical Monographs, No. 11, Mathematical Association of America, 1956.

4. Thurston, H. A., *The Number System*, Blackie, London, 1956.

Counterexamples

1. Gelbaum, B. R., and Olmstead, J. M. H., *Counterexamples in Analysis*, Holden Day, San Francisco, 1964.

2. Steen, L. A., and Seebach, J. A., *Counterexamples in Topology*, Springer-Verlag, New York, 1978.

Papers

1. Hewitt, E., "The Role of Compactness in Analysis," *American Math. Monthly* 67 (1960), pp. 499–516.

2. Saari, D. G., and Urenko, J. B., "Newton's Method, Circle Maps, and Chaotic Motion," *American Math. Monthly* 9, No. 1 (1984), pp. 3–17.

3. Stone, M. H., "A Generalized Weierstrass Approximation Theorem," Studies in Mathematics, Vol 1, Mathematical Association of America, 1962.

4. Tong, J., "Kummer's Test Gives Characterizations for Convergence of All Positive Series," *American Math. Monthly* (1994) pp. 450–452.

Others

1. Baker, A., *Transcendental Number Theory*, Cambridge University Press, Cambridge, 1972.

2. Bromwich, T. J. I'a., *Introduction to the Theory of Infinite Series*, 2nd ed., Macmillan, New York, 1965.

3. Conte, S. D., and de Boor, C., *Elementary Numerical Analysis: An Algorithmic Approach*, McGraw-Hill, New York, 1972.

4. Corwin, L. J., and Szczarba, R. H., *Multivariable Calculus*, Marcel Dekker, New York, 1982.

5. Dieudonne, J., *Foundations of Modern Analysis*, Academic Press, New York, 1961.

6. Fleming, W., *Functions of Several Variables*, Springer-Verlag, New York, 1977.

7. Hardy, G. H., *A Course in Pure Mathematics*, 10th ed., Cambridge University Press, Cambridge, 1952.

8. Hardy, G. H., and Wright, E. M., *An Introduction to the Theory of Numbers*, 3rd ed., Clarendon Press, Oxford, 1954.

9. Hirschmann, I., *Infinite Series,* Holt, Rinehart, and Winston, New York, 1960.

10. Kelley, J. L., *General Topology*, Van Nostrand, Princeton, NJ., 1955.

11. Knopp, K., *Theory and Applications of Infinite Series*, 2nd ed., Hafner, New York, 1951.

12. Wilder, R. L., *Introduction to Foundations of Mathematics*, 2nd ed., John Wiley, New York, 1965.

Index